AF293990

Extended Irreversible Thermodynamics

D. Jou   J. Casas-Vázquez   G. Lebon

Extended Irreversible Thermodynamics

Third, Revised and Enlarged Edition
With 30 Figures and 15 Tables

Springer

Professor Dr. David Jou
Professor Dr. José Casas-Vázquez
Departament de Física
Grup de Física Estadística
Universitat Autònoma de Barcelona
Edifici C
08193 Bellaterra, Catalonia, Spain

Professor Dr. Georgy Lebon
Departement de Physique
Université de Liège
Sart Tilman B5
4000 Liège, Belgium

Library of Congress Cataloging-in-Publication Data. Jou, D. (David). Extended irreversible thermodynamics/ D. Jou, J. Casas-Vázquez, G. Lebon. – 3rd ed. p. cm. Includes bibliographical references and index. ISBN 978-3-642-62505-3 1. Irreversible processes. I. Casas-Vazquez, J. (José), 1938– II. Lebon, G. (Georgy) III. Title. QC318.I7 J67 2000 536'.7–dc21 00-053784

ISBN 978-3-642-62505-3   ISBN 978-3-642-56565-6 (eBook)
DOI 10.1007/978-3-642-56565-6

http://www.springer.de

© Springer-Verlag Berlin Heidelberg 1993, 1996, 2001
Originally published by Springer-Verlag Berlin Heidelberg New York in 2001
Softcover reprint of the hardcover 3rd edition 2001

Typesetting: Camera-ready copy from the authors
Cover design: Erich Kirchner, Heidelberg

Printed on acid-free paper   SPIN 10720979   55/3141/ba 5 4 3 2 1 0

Preface to the Third Edition

In writing this third edition, we proceeded having in mind two main objectives: pedagogy and physics. One of our essential concerns is indeed to open the subject to a large audience of scientists with a diversity of interests. This is the reason we have emphasized the pedagogical aspects, starting with the simplest situations before treating more complex problems. Our aim is to show that thermodynamics is not restricted to the study of purely thermal phenomena but that it largely covers many different areas, going from continuum mechanics to statistical physics, from nuclear collisions to cosmology, by passing, for instance, through chemistry, rheology and biology.

Our second objective is to stress the physical aspects in their broadest sense, i.e. trying to clarify the foundations and relating the theoretical results with experimental data and applications, rather than introducing formal and lengthy developments. Important sections of the book are devoted to the discussion of fundamental notions, such as the selection of basic variables, the definition of entropy, temperature, pressure, and chemical potential outside equilibrium and the statement of the second law. We feel that these crucial questions deserve special attention when they cross the borders of classical theories. We have included new original results and opened the door to new extensions and applications.

For the student or the researcher it may be stimulating to go beyond the classical theories and to discover a domain of new ideas, new applications, and new problems. This is what extended irreversible thermodynamics (EIT) offers nowadays: it does not pretend to solve all the problems raised in continuum physics and statistical mechanics, but it can be viewed as an emergent new global framework of non-equilibrium thermodynamics and a rapidly advancing frontier with new applications being treated and fundamental questions being asked and tentatively clarified.

This edition has been extensively remodelled compared to the first two. We have gone from 12 to 18 chapters as a result of splitting some of the former chapters into shorter ones and focusing better on the foundations and on the introduction to the basic ideas of the latest developments. As for the two preceding editions, we have clearly separated the structure of the book into three parts, namely the general theory (Part I), microscopic foundations (Part II), and selected applications (Part III). Amongst the most visible changes, we have split the general presentation of EIT into two chapters: Chap. 2 devoted to establishing the transport equations and dynamical aspects, and Chap. 3 in which we introduce the non-equilibrium equations of state emphasizing the meaning of entropy and temperature in non-equilibrium states. Comparison with the rational version of extended thermodynamics has been made more explicit. Chapter 4 shows an overview of Hamiltonian formulations, which have developed in recent years into a promising domain.

The microscopic foundations are discussed in Chaps. 5–9, with two new chapters concerning information theory and computer simulations. Both topics were already present in the former editions as subsections of other chapters, but recent developments have justified a renewed and updated presentation.

Applications cover half of the book: nine chapters from the total eigthteen. They show that the exploration of new theoretical grounds is both an intellectual challenge and a source of new practical possibilities. Since contemporary technology strives towards higher speed, power, and miniaturization, the transport equations must incorporate memory, non-local, and non-linear effects. Compared to the generalised transport equations incorporating the aforementioned effects, EIT plays a role similar to that played by classical irreversible thermodynamics with respect to the classical transport equa-tions. Diffusion and electrical phenomena, treated in a single chapter in the previous editions, are the subject of two different, updated chapters; the relativistic formulation has also been split into two different chapters, one devoted to the foundations and the other to cosmological applications. Finally, let us mention that a wide overview of the literature on EIT as well as the solutions to the proposed problems can be found on the www.uab.es/dep-fisica/eit website.

This book, in its successive editions, has widely benefited from the fruitful suggestions and comments by colleagues, including M. Anile, J. Camacho, M. Criado-Sancho, L.S. García-Colín, M. Grmela, P.T. Landsberg, R. Luzzi, W. Muschik, D. Pavón, M. Torrisi, A. Valenti and many others. We acknowledge the financial support of the DGICyT of the Spanish Ministry of Education and Culture under Grant Nos. PB90-0676 and PB94-0718, of the DGR of the Generalitat of Catalonia, under Grant Nos. 1997SGR 00378 and 2000SGR 00095, of the Belgian Ministry of Scientific Policy under PAI Grant Nos. 21, 29 and IV 6, and a grant from the European Union in the framework of the Program of Human Capital and Mobility (European Thermodynamic Network ERB-CHR XCT 920 007).

Bellaterra, Barcelona
Liège   *David Jou*
March 2001   *José Casas-Vázquez*
                                                                              Georgy Lebon

Preface to the First Edition

Classical irreversible thermodynamics, as developed by Onsager, Prigogine and many other authors, is based on the local-equilibrium hypothesis. Out of equilibrium, any system is assumed to depend locally on the same set of variables as when it is in equilibrium. This leads to a formal thermodynamic structure identical to that of equilibrium: intensive parameters such as temperature, pressure and chemical potentials are well-defined quantities keeping their usual meaning, thermodynamic potentials are derived as Legendre transformations and all equilibrium thermodynamic relations retain their validity. The theory based on this hypothesis has turned out to be very useful and has collected a number of successes in many practical situations.

However, the recent decade has witnessed a surge of interest in going beyond the classical formulation. There are several reasons for this. One of them is the development of experimental methods able to deal with the response of systems to high-frequency and short-wavelength perturbations, such as ultrasound propagation and light and neutron scattering. The observed results have led to generalisations of the classical hydrodynamical theories, by including memory functions or generalised transport coefficients depending on the frequency and the wavevector. This field has generated impressive progress in non-equilibrium statistical mechanics, but for the moment it has not brought about a parallel development in non-equilibrium thermodynamics. An extension of thermodynamics compatible with generalised hydrodynamics therefore appears to be a natural subject of research.

An additional reason has fostered an interest in generalising the classical transport equations, like Fourier's law for heat conduction, Fick's law for diffusion, and Newton's law for viscous flow. It is well known that after introducing these relations in the balance equations, one is led to parabolic partial-differential equations which imply that perturbations propagate with infinite speed. This behaviour is incompatible with experimental evidence and it is also disturbing from a theoretical point of view, because collective molecular effects should be expected to propagate at finite velocity, not only in a relativistic framework, but even from a non-relativistic point of view. This unpleasant property can be avoided by taking into account the finite nonvanishing relaxation time of the respective fluxes, e.g. heat flux, diffusion flux, momentum flux, sometimes generically called dissipative fluxes. The subsequent equations are however not compatible with the classical non-equilibrium thermodynamics, since they lead in some circumstances to negative entropy production. Thus, a thermodynamic theory compatible with these phenomena is highly desirable, because it may provide new insights into the meaning and definition of fundamental thermodynamic quantities, as entropy and temperature, and may clarify the limits of validity of the local-equilibrium hypothesis and of the usual formulations of the second law out of equilibrium.

The former problems are not merely academic. It has been observed in several systems that the dissipative fluxes are characterized by long relaxation times. Typical examples are polymeric fluids, heat and electric conductors at low temperature, superconductors, and so on. An accurate understanding of these systems may thus be important not only from a theoretical point of view, but also for practical purposes. In real situations, these systems are out of equilibrium. Accordingly, there is an urgent need for a non-equilibrium thermodynamic theory able, on the one hand, to cope with the effects of long relaxation times and, on the other, to complement other formalisms based on the use of internal variables.

There are other reasons for the present study. One should be aware that classical irreversible thermodynamics is not the only non-equilibrium thermodynamic theory: other theories, in particular the so-called rational thermodynamics, have achieved some valuable results. To reconcile the classical and the rational points of view, it would be of interest to have a theory able to provide a sufficiently wide ground for discussion, thus making their common points evident and their main differences understandable. Extended irreversible thermodynamics is a promising candidate.

Extended irreversible thermodynamics received a strong impetus in the past decade. Besides the classical thermodynamic variables, this theory introduces as new independent variables the dissipative fluxes and aims to obtain for them evolution equations compatible with the second law of thermodynamics. The central quantity is a generalised non-equilibrium entropy, depending on both the conserved variables and the fluxes, which sheds new light on the content of the second law. This generalised theory is corroborated from a microscopic point of view by the kinetic theory, non-equilibrium information theory and other formulations of non-equilibrium statistical mechanics.

The purpose of this book is to provide an introduction to the foundations of extended irreversible thermodynamics, to discuss the main results and to present some of its applications. After more than twenty years of research and several hundreds of papers published by many groups in several countries we feel such a book is sorely needed. Guided by the aim to be as illustrative and pedagogical as possible, a relatively simple formulation of the theory is presented, but this is nevertheless more than sufficient for the description of several phenomena not accessible to the classical theory. The various topics treated in this book range from thermal waves and phonon hydrodynamics to material and electrical transport, from ultrasound propagation and generalised hydrodynamics to rheology, from kinetic theory to cosmology. Of course, other formulations of extended thermodynamics and other kinds of applications are possible. They are likely to arise in the near future. We hope that this book will be useful in providing a general view of present achievements and in stimulating future research.

We are very pleased to acknowledge many stimulating discussions with our colleagues Carlos Pérez-García, Josep-Enric Llebot, Diego Pavón, José-Miguel Rubí and Joseph Lambermont for more than fifteen years of joint research, and also with many other colleagues from the different groups which have devoted their attention to extended irreversible thermodynamics. We also acknowledge the financial support of the Comisión Asesora para la Investigación Científica y Técnica of the Spanish Government during the years 1979-1986 under grants 3913/79 and 2389/83, and of the Dirección General de Investigación Científica y Técnica of the Spanish Ministry of Education, under grants PB86-0287, PB89-0290, and PB90-0676. The collaboration between our groups in Bellaterra and Liège has been made economically possible because of the NATO grant 0355/83.

March 1993

David Jou
José Casas-Vázquez
Georgy Lebon

Contents

Part III. Selected Applications

Part I

General Theory

Chapter 1

Classical and Rational Formulations of Non-equilibrium Thermodynamics

The nineteenth century was the golden age for classical thermodynamics: as this theory concerns essentially systems at equilibrium, we shall refer to it as equilibrium thermodynamics. It was developed by a pleiad of exceptionally brilliant scientists as Carnot, Mayer, Joule, Helmholtz, Clausius, Lord Kelvin, Maxwell, Boltzmann, Gibbs, Planck, Duhem, etc.

Equilibrium thermodynamics is based on two major statements:

1. The energy of the universe is constant (first law).

2. The entropy of the universe never decreases (second law).

Only in the first half of the twentieth century it was felt necessary to go beyond the equilibrium approach. Lars Onsager published two seminal and pioneering papers on non-equilibrium thermodynamics in the Physical Review in 1931 [1.1]. The merit of Onsager was recognized by the Nobel Foundation, which awarded him the Nobel prize for chemistry in 1968. Other fundamental contributions to non-equilibrium thermodynamics are due to Eckart, Meixner and Prigogine (Nobel prize for chemistry in 1977). The formalism proposed by Onsager, Eckart, Meixner, and Prigogine is usually referred to as classical irreversible thermodynamics (CIT). This theory has seen an extraordinary expansion since the 1940s and has been widely applied in physics, biophysics, chemistry and engineering science. Excellent papers and monographs on the subject are those of Eckart [1.2], Meixner and Reik [1.3], Prigogine [1.4], and De Groot and Mazur [1.5]. Other interesting contributions are to be found in references [1.6–12].

In this chapter, a broad outline of classical non-relativistic irreversible thermodynamics is presented. Although CIT has been able to describe a large number of non-equilibrium processes in complete agreement with experiment, it suffers from some limitations: several phenomena do not comply with the framework of CIT. Efforts have been made to enlarge the range of application of CIT and these emerge in the formulation of a new theory, called by its authors rational thermodynamics. This formalism, mainly developed by Coleman [1.13], Truesdell [1.14] and Noll [1.15], has significantly contributed to the advancement of the knowledge of continuum thermodynamics, and it is our opinion that this subject deserves a place in this review section.

1.1 The General Balance Laws of Continuum Physics

This section is preparatory to all the other sections and will concern the establishment of the general balance laws of mass, momentum, and energy. Here, for the sake of simplicity, relativistic effects are not considered. We consider a portion of matter of mass M occupying a volume V limited by a boundary Σ. We suppose that the continuum hypothesis holds and that for any extensive quantity B one has

$$\lim_{\Delta M \to 0} \frac{\Delta B}{\Delta M} = b, \quad \lim_{\Delta M \to 0} \frac{\Delta B}{\Delta V} = \rho b \quad \text{or} \quad B = \int_V \rho b \, dV,$$

where ρ is the total mass density and b the specific value of B referred to per unit mass.

The most general form of the balance equation for any quantity b is given by

$$\int_{V(t)} \frac{\partial}{\partial t}(\rho b) \, dV = -\int_{\Sigma(t)} \boldsymbol{J}^b \cdot \boldsymbol{n} \, d\Sigma + \int_{V(t)} \sigma^b \, dV, \tag{1.1}$$

where $\partial/\partial t$ is the local or Eulerian time derivative, a middot means the scalar product, $\boldsymbol{J}^b$ represents the quantity B flowing per unit area and unit time through the boundary Σ, $\boldsymbol{n}$ is the positive unit normal to Σ, oriented by convention in the outward direction of the volume V, and σ^b is the rate of production (> 0) or destruction (< 0) per unit volume of the quantity B. The flux $\boldsymbol{J}^b$ is a vector (or tensor) if b is a scalar (or vector), and σ^b has the same tensorial rank as b. Assuming that (1.1) is valid for any volume V and that the integrands are continuous functions of position, one obtains the following local form of the balance equation:

$$\frac{\partial}{\partial t}(\rho b) = -\nabla \cdot \boldsymbol{J}^b + \sigma^b, \tag{1.2}$$

after use is made of Gauss' theorem,

$$\int \boldsymbol{J}^b \cdot \boldsymbol{n} \, d\Sigma = \int \nabla \cdot \boldsymbol{J}^b \, dV,$$

where ∇ denotes the nabla operator. When σ^b is zero, the quantity b is said to be conserved.

If the integrands in (1.1) are not continuous, there exist discontinuities inside the volume taking for example the form of a surface. By imposing balance laws on a volume surrounding the discontinuity surface and letting the volume go to zero, one obtains the so-called jump conditions. Under rather general conditions [1.16], it is possible to show that relation (1.2) leads to the jump condition

$$-[\rho v_n b] + [\boldsymbol{n} \cdot \boldsymbol{J}^b] = \sigma_\Sigma^b. \tag{1.3}$$

Brackets denote the jump of the corresponding quantity across the discontinuity; v_n is the normal speed of displacement of the surface with respect to the material and σ_Σ^b the rate of production per unit area. On a material discontinuity surface one has $v_n = 0$. Obviously, the production term σ_Σ^b vanishes for a conserved quantity.

An elegant and quick derivation of the balance equations of mass, momentum and energy can be achieved by starting from the first law of thermodynamics and imposing the principle of Galilean invariance.

1.1.1 The One-component System

The first law of thermodynamics, which expresses the balance of total energy, takes for an electrically neutral system the form

$$\frac{dE}{dt} = \frac{dW}{dt} + \frac{dQ}{dt} , \tag{1.4}$$

with $E = U + K$; E is the total energy, equal to the sum of the internal energy U and the kinetic energy K, W is the work performed by the body forces per unit mass $\boldsymbol{F}$ and the contact forces $\boldsymbol{T}$, and Q is the heat exchanged with the outside world through the boundary. More explicitly, U, K, dQ/dt, and dW/dt are given by

$$U = \int_{V(t)} \rho u \, dV , \qquad K = \tfrac{1}{2} \int_{V(t)} \rho \boldsymbol{v} \cdot \boldsymbol{v} \, dV , \tag{1.5a}$$

$$\frac{dQ}{dt} = -\int_{\Sigma(t)} \boldsymbol{q} \cdot \boldsymbol{n} \, d\Sigma , \qquad \frac{dW}{dt} = \int_{\Sigma(t)} \boldsymbol{T} \cdot \boldsymbol{v} \, d\Sigma + \int_{V(t)} \rho \boldsymbol{F} \cdot \boldsymbol{v} \, dV , \tag{1.5b}$$

where $\boldsymbol{T}$ is related to the pressure tensor $\mathbf{P}$ by means of Cauchy's relation $\boldsymbol{T} = -\mathbf{P} \cdot \boldsymbol{n}$. The notation in (1.5a, b) is classical: u is the specific internal energy, $\boldsymbol{v}$ the velocity field, and q the heat flux vector. We now make use of the Reynolds transport theorem, which, for an arbitrary quantity b, can be written as

$$\frac{\mathrm{d}}{\mathrm{d}t}\int_{V(t)} b\,\mathrm{d}V = \int_{V(t)} \frac{\partial b}{\partial t}\,\mathrm{d}V + \int_{\Sigma(t)} b\mathbf{v}\cdot\mathbf{n}\,\mathrm{d}\Sigma\ .$$

After substitutions of expressions (1.5) in the first law (1.4), one is led to the local form of the energy balance

$$\rho(\dot{u}+\mathbf{v}\cdot\dot{\mathbf{v}})+\left(u+\tfrac{1}{2}\mathbf{v}\cdot\mathbf{v}\right)(\dot{\rho}+\rho\nabla\cdot\mathbf{v})=-\nabla\cdot\mathbf{q}-\mathbf{P}^{T}:\nabla\mathbf{v}-\mathbf{v}\cdot(\nabla\cdot\mathbf{P})+\rho\mathbf{F}\cdot\mathbf{v}, \quad (1.6)$$

where superscript T means transposition, an upper dot stands for the material or Lagrangian time derivative (i.e., $\mathrm{d}/\mathrm{d}t = \partial/\partial t + \mathbf{v}\cdot\nabla$) and a colon the double scalar product

$$\mathbf{A}:\mathbf{B}=\sum_{i,j} A_{ij}B_{ji}\ .$$

A short summary of basic concepts in tensorial calculus may be found in Appendix A.

According to the Galileo principle, (1.6) must be invariant with respect to the transformation $\mathbf{v}\to\mathbf{v}+\mathbf{v}_0$, where $\mathbf{v}_0$ is a constant and uniform velocity. After substitution in (1.6) of $\mathbf{v}$ by $\mathbf{v}+\mathbf{v}_0$ and subtraction of (1.6), one obtains

$$\tfrac{1}{2}\mathbf{v}_0\cdot\mathbf{v}_0(\dot{\rho}+\rho\nabla\cdot\mathbf{v})+\mathbf{v}_0\cdot\left[(\dot{\rho}+\rho\nabla\cdot\mathbf{v})\mathbf{v}+\rho\dot{\mathbf{v}}+\nabla\cdot\mathbf{P}-\rho\mathbf{F}\right]=0. \quad (1.7)$$

This relation could be invariant with respect to $\mathbf{v}_0$ on condition that the following equations are satisfied:

$$\dot{\rho}=-\rho\nabla\cdot\mathbf{v}, \quad (1.8)$$

$$\rho\dot{\mathbf{v}}=-\nabla\cdot\mathbf{P}+\rho\mathbf{F}\ . \quad (1.9)$$

With these results in mind, (1.6) reduces to

$$\rho\dot{u}=-\nabla\cdot\mathbf{q}-\mathbf{P}^{T}:\nabla\mathbf{v}\ . \quad (1.10)$$

It is also convenient to express the balance of mass (1.8) in terms of the specific volume $v\ (=1/\rho)$; this leads to

$$\rho\dot{v}=\nabla\cdot\mathbf{v}\ . \quad (1.11)$$

Relations (1.8–10) are the laws of balance of mass, momentum, and internal energy respectively, written in the Lagrange representation; the balance of mass is also known as the continuity equation.

In terms of the local time derivative (Euler representation), equations (1.8–10) take the form

$$\frac{\partial \rho}{\partial t} = -\nabla \cdot (\rho v), \tag{1.12}$$

$$\frac{\partial (\rho v)}{\partial t} = -\nabla \cdot (\mathbf{P} + \rho v v) + \rho F, \tag{1.13}$$

$$\frac{\partial (\rho u)}{\partial t} = -\nabla \cdot (q + \rho u v) - \mathbf{P}^T : \nabla v, \tag{1.14}$$

where vv is a dyadic product. Expressions (1.12–14) are useful because they allow one to identify the various fluxes and supply terms corresponding respectively to the mass, momentum, and energy, as shown in Table 1.1.

Table 1.1. Fluxes and supply terms (sources)

Quantity, b	Flux, J^b	Source, σ^b
mass	ρv	0
momentum	$\mathbf{P} + \rho v v$	ρF
internal energy	$q + \rho u v$	$-\mathbf{P}^T : \nabla v$

Clearly the mass is conserved. The quantities $\mathbf{P}$ and $\rho v v$ are the conductive and convective transport of momentum respectively; if no external (or body) force is acting on the system, momentum is conserved. Similarly, q describes the transport of internal energy due to conduction, while $\rho u v$ is the contribution arising from convection; the term $-\mathbf{P}^T : \nabla v$ represents the internal energy supply.

Addition of (1.13), after scalar multiplication by v, and (1.14) yields the balance equation of total energy:

$$\frac{\partial}{\partial t}\left[\rho\left(u + \tfrac{1}{2} v \cdot v\right)\right] = -\nabla \cdot \left[q + \rho\left(u + \tfrac{1}{2} v \cdot v\right)v + \mathbf{P} \cdot v\right] + \rho F \cdot v, \tag{1.15}$$

from which it follows that in the absence of external force the total energy is conserved.

For later use, we split the velocity gradient into a symmetric and an antisymmetric part

$$\nabla v = \mathbf{V} + \mathbf{W}, \tag{1.16}$$

with

$$\mathbf{V} = \tfrac{1}{2}\left[\nabla v + (\nabla v)^T\right], \qquad \mathbf{W} = \tfrac{1}{2}\left[\nabla v - (\nabla v)^T\right],$$

or, in Cartesian coordinates,

$$V_{ij} = \frac{1}{2}\left(\frac{\partial v_j}{\partial x_i} + \frac{\partial v_i}{\partial x_j}\right), \qquad W_{ij} = \frac{1}{2}\left(\frac{\partial v_j}{\partial x_i} - \frac{\partial v_i}{\partial x_j}\right).$$

The rate of deformation tensor $\mathbf{V}$ may be further decomposed as

$$\mathbf{V} = \tfrac{1}{3}(\nabla.v)\mathbf{U} + \overset{0}{\mathbf{V}}, \tag{1.17}$$

where $\mathbf{U}$ is the identity tensor and $\overset{0}{\mathbf{V}}$ the deviatoric traceless tensor.

Without loss of generality, the pressure tensor can be decomposed into an equilibrium part p and a viscous part $\mathbf{P}^v$:

$$\mathbf{P} = p\mathbf{U} + \mathbf{P}^v. \tag{1.18}$$

Further, the viscous pressure tensor $\mathbf{P}^v$ can be split into a scalar bulk viscous pressure p^v and a traceless deviator $\overset{0}{\mathbf{P}^v}$ according to

$$\mathbf{P}^v = p^v\mathbf{U} + \overset{0}{\mathbf{P}^v}, \tag{1.19}$$

with

$$p^v = \tfrac{1}{3}\mathrm{Tr}\,\mathbf{P}^v.$$

In the absence of intrinsic rotational motions and external couples, conservation of the angular momentum implies the symmetry of the pressure tensor [1.5, 9]: $\mathbf{P} = \mathbf{P}^T$. Along this book, tensor $\mathbf{P}$ is assumed to be symmetric. For a more general situation where $\mathbf{P}$ has an antisymmetric part, see Problem 1.7.

1.1.2 The Multicomponent Mixture

Let us consider a system containing N different constituents labelled $k = 1, 2, \ldots, N$, with mass density ρ_k, among which take place n chemical reactions. It is assumed that every point in space is occupied simultaneously by particles of all the constituents. The balance equations of mass, momentum, and energy of the individual constituents are derived in many books and papers (e.g. [1.5, 10, 14]) and we just list the results:

$$\frac{\partial}{\partial t}\rho_k = -\nabla\cdot(\rho_k v_k) + \rho\sum_{l=1}^{n} v_{kl}\dot{\xi}_l \qquad (k=1,2,...,N), \qquad (1.20)$$

$$\frac{\partial}{\partial t}(\rho_k v_k) = -\nabla\cdot(\mathbf{P}_k + \rho_k v_k v_k) + \rho_k F_k + \Gamma_k \qquad (k=1,2,...,N), \qquad (1.21)$$

$$\frac{\partial}{\partial t}\left[\rho_k\left(u_k + \tfrac{1}{2}v_k\cdot v_k\right)\right] = -\nabla\cdot\left[q_k + \rho_k\left(u_k + \tfrac{1}{2}v_k\cdot v_k\right)v_k + \mathbf{P}_k\cdot v_k\right]$$

$$+\rho_k F_k\cdot v_k + e_k \qquad (k=1,2,...,N). \qquad (1.22)$$

In (1.20), the term $v_{kl}\dot{\xi}_l$ represents the rate of production of constituent k in the lth chemical reaction, v_{kl} is the stoichiometric coefficient of constituent k in the chemical reaction l times the ratio of the molecular mass of k and a constant mass, say the total mass of the reactants (v_{kl} is positive for products of the reaction and negative for reactants), and $\dot{\xi}_l$ is the rate of advancement of reaction l. In (1.21) and (1.22), F_k is the specific body force acting on constituent k, and the quantities $\mathbf{P}_k$, q_k, u_k denote respectively the partial pressure tensor, heat flux, and internal energy corresponding to component k. The production terms Γ_k in the momentum law (1.21) and e_k in (1.22) play similar roles as the mass production due to chemical reactions in (1.20): the productions Γ_k and e_k of momentum and energy contain contributions due to the interaction forces and the exchange of momentum and energy between the various components, respectively.

Let us define the total density ρ and the centre-of-mass velocity v respectively by

$$\rho = \sum_{k=1}^{N}\rho_k \qquad (1.23)$$

and

$$\rho v = \sum_{k=1}^{N}\rho_k v_k. \qquad (1.24)$$

By mass fraction of the constituent k is meant the ratio

$$c_k = \frac{\rho_k}{\rho}, \qquad (1.25)$$

with, obviously, $\sum_{k=1}^{N} c_k = 1$.

It is customary to introduce the following diffusion velocity w_k and diffusion flux J_k:

$$w_k = v_k - v,$$

$$J_k = \rho_k (v_k - v), \tag{1.26}$$

with the property

$$\sum_{k=1}^{N} J_k = 0. \tag{1.27}$$

Bearing in mind these definitions, the balance law for the mass fraction takes the form

$$\rho \dot{c}_k = -\nabla \cdot J_k + \rho \sum_{l=1}^{n} v_{kl} \dot{\xi}_l. \tag{1.28}$$

The balance equations (1.12), (1.13), and (1.15) for the total mass, momentum, and energy must be satisfied, and this requirement is reflected in the following constraint equations:

$$\sum_{k=1}^{N} v_{kl} = 0 \qquad (l = 1, 2, \dots, n), \tag{1.29a}$$

$$\sum_{k=1}^{N} (\mathbf{P}_k + \rho_k w_k w_k) = \mathbf{P}, \qquad \sum_{k=1}^{N} \rho_k F_k = \rho F, \qquad \sum_{k=1}^{N} \Gamma_k = 0, \tag{1.29b}$$

$$\sum_{k=1}^{N} \rho_k \left(u_k + \tfrac{1}{2} w_k \cdot w_k \right) = \rho u, \qquad \sum_{k=1}^{N} \left[q_k + \rho w_k \left(u_k + \tfrac{1}{2} w_k \cdot w_k \right) + \mathbf{P}_k \cdot w_k \right] = q, \tag{1.29c}$$

$$\sum_{k=1}^{N} e_k = 0. \tag{1.29d}$$

Equation (1.29a) expresses the conservation of mass in each chemical reaction. Equations (1.29b) and (1.29c) indicate that the sum of the partial pressure tensors, internal energies, and heat fluxes represent only partial contributions to the total pressure tensor, internal energy, and heat flux respectively.

1.1.3 Charged Systems

In this subsection, we shall reformulate the balance equations for a mixture of N charged components subject to an electromagnetic field. Denoting by z_k the charge per unit mass of constituent k, the total charge per unit mass is given by

$$\rho z = \sum_{k=1}^{N} \rho_k z_k.$$
(1.30)

The current density is defined by

$$I = \sum_{k=1}^{N} \rho_k z_k v_k,$$
(1.31)

which can be cast into

$$I = \sum_{k=1}^{N} \rho_k z_k (v_k - v) + \sum_{k=1}^{N} \rho_k z_k v.$$
(1.32)

Defining the conduction current by

$$i = \sum_{k=1}^{N} \rho_k z_k (v_k - v) = \sum_{k=1}^{N} z_k J_k$$
(1.33)

and using expression (1.30), we find that (1.31) becomes

$$I = i + \rho z v,$$
(1.34)

where $\rho z v$ is the convection current.

We now assume, for simplicity, that the components are chemically inert and that polarization effects are negligible. The charge conservation law is directly derived from the mass conservation law (1.28). After multiplying (1.28) by z_k and adding up all the N constituents one obtains

$$\rho \dot{z} = -\nabla \cdot i,$$
(1.35)

or, equivalently,

$$\frac{\partial}{\partial t}(\rho z) = -\nabla \cdot I.$$
(1.36)

The laws of momentum and energy are obtained by using the same procedure as in Subsect. 1.1.1. The only difference with the above developments is expression dW/dt, which now contains an additional term arising from the presence of electrical forces and is given by

$$\frac{dW_{el}}{dt} = \int_{V(t)} \sum_{k=1}^{N} \rho_k F_k \cdot v_k dV, \qquad (1.37)$$

where W_{el} is the work performed by electromagnetic forces and F_k stands for the Lorentz force acting per unit mass of constituent k,

$$F_k = z_k(E + v_k \times B), \qquad (1.38)$$

E being the electrical field and B the magnetic induction. It is a simple exercise to show that (1.37) can be rewritten as

$$\frac{dW_{el}}{dt} = \int_{V(t)} \left[\sum_{k=1}^{N} \rho_k z_k (E + v_k \times B) \cdot v + i \cdot E + i \cdot (v \times B) \right] dV. \qquad (1.39)$$

Denoting by U_t the sum of the total internal energy and the diffusion energy,

$$U_t = U + \tfrac{1}{2} \int_{V(t)} \sum_{k=1}^{N} \rho_k w_k \cdot w_k \, dV, \qquad (1.40)$$

one can write the balance of total energy (1.4) as

$$\frac{dU_t}{dt} + \frac{dK}{dt} = \frac{dQ}{dt} + \frac{dW_{mec}}{dt} + \frac{dW_{el}}{dt}, \qquad (1.41)$$

where K is the kinetic energy of the centre of mass,

$$K = \tfrac{1}{2} \int_{V(t)} \rho v \cdot v dV, \qquad (1.42)$$

and dW_{mec}/dt is the power developed by the mechanical forces,

$$\frac{dW_{mec}}{dt} = -\int_{V(t)} [\nabla \cdot (v \cdot P) - \rho F \cdot v] dV,$$

where (1.13) and (1.14) have been used.

It is readily checked that (1.41) is locally given by

$$\rho \dot{u}_t + \rho v \cdot \dot{v} = -\nabla \cdot q + \left[\rho F - \nabla \cdot \mathbf{P} + \sum_{k=1}^{N} \rho_k z_k (E + v_k \times B) \right] \cdot v$$

$$+ i \cdot (E + v \times B) - \mathbf{P} : \nabla v. \tag{1.43}$$

Let ε be the Galilean invariant quantity

$$\varepsilon = E + v \times B, \tag{1.44}$$

representing the electric field measured in a moving reference frame with velocity v. In terms of ε, (1.43) becomes

$$(-\rho \dot{v} + \rho F - \nabla \cdot \mathbf{P} + \rho z \varepsilon + i \times B) \cdot v = \rho \dot{u}_t + \nabla \cdot q + \mathbf{P} : \nabla v - \varepsilon \cdot i. \tag{1.45}$$

The invariance with respect to the Galilean transformation results in the following balance equations for momentum and internal energy:

$$\rho \dot{v} = -\nabla \cdot \mathbf{P} + \rho F + \rho z \varepsilon + i \times B, \tag{1.46}$$

$$\rho \dot{u}_t = -\nabla \cdot q - \mathbf{P} : \nabla v + \varepsilon \cdot i. \tag{1.47}$$

In addition to the mechanical forces, the momentum balance (1.46) contains extra contributions coming from the Lorentz force $\rho z \varepsilon$ and the Laplace force $i \times B$ acting on current i. In the energy law, the supplementary contribution $\varepsilon \cdot i$ is identified as the rate of dissipated energy.

1.2 The Law of Balance of Entropy

In analogy with equilibrium thermodynamics, it is assumed that outside equilibrium there exists an extensive quantity S, called entropy, which is a sole function of the state variables. In general, the rate of change of S can be written as the sum of two terms:

$$\frac{\mathrm{d}S}{\mathrm{d}t} = \frac{\mathrm{d}^e S}{\mathrm{d}t} + \frac{\mathrm{d}^i S}{\mathrm{d}t}, \tag{1.48}$$

where $\mathrm{d}^e S/\mathrm{d}t$ is the rate of entropy exchanged with the surroundings, which may be invariably zero, positive, or negative, and $\mathrm{d}^i S/\mathrm{d}t$ derives from processes occurring inside the system. According to the second law of thermodynamics, d^iS/dt is a non-negative quantity:

$$\frac{d^iS}{dt} \geq 0. \tag{1.49}$$

d^iS/dt is zero at equilibrium or for reversible transformations and positive for irreversible processes.

Without loss of generality, we can define a local specific entropy, a local entropy flux $\boldsymbol{J}^s$, and a local rate of production σ^s, respectively:

$$S = \int_{V(t)} \rho s\, dV,$$

$$\frac{d^eS}{dt} = -\int_{\Sigma(t)} \boldsymbol{J}^s \cdot \boldsymbol{n}\, d\Sigma, \tag{1.50}$$

$$\frac{d^iS}{dt} = \int_{V(t)} \sigma^s\, dV.$$

After replacing (1.50) in (1.48) and making use of the Gauss and Reynolds theorems, one obtains the local Lagrangian form of the entropy balance:

$$\rho \dot{s} = -\nabla \cdot \boldsymbol{J}^s + \sigma^s, \tag{1.51}$$

and, in Eulerian form, one has

$$\frac{\partial}{\partial t}(\rho s) = -\nabla \cdot (\boldsymbol{J}^s + \rho s \boldsymbol{v}) + \sigma^s, \tag{1.52}$$

with

$$\sigma^s \geq 0 \tag{1.53}$$

in either case. Inequality (1.53) goes beyond the usual formulation of the second law in equilibrium thermodynamics where only the global increase of entropy between two equilibrium states in an isolated system is considered. Here it is assumed that the statement (1.53) holds at any position in space and any instant of time, for whatever the evolution of the system.

One of the main objectives of non-equilibrium thermodynamics is to express σ^s as a function of the quantities characterizing the irreversible processes. This is important because it displays the sources of irreversibility occurring in a process. Moreover, it will be shown later that the very form of the local rate of entropy production may serve as a guide to determine the constitutive relations describing the dynamical response of the system to external or internal solicitations.

1.3 Classical Irreversible Thermodynamics

This section is devoted to the derivation and discussion of the main results of classical irreversible thermodynamics (CIT). Here we only examine the macroscopic aspects, but it must be realized that these are deeply rooted in the microscopic point of view. The reader interested in a detailed analysis is referred to the authoritative treatises [1.3–12]. The range of application of CIT comprises those systems satisfying the hypothesis of local equilibrium, which is analysed in full in the next subsection.

1.3.1 The Local-equilibrium Hypothesis

The fundamental hypothesis underlying CIT is that of local equilibrium. It postulates that the local and instantaneous relations between the thermal and mechanical properties of a physical system are the same as for a uniform system at equilibrium. It is assumed that the system under study can be mentally split into a series of cells sufficiently large to allow them to be treated as macroscopic thermodynamic subsystems, but sufficiently small that equilibrium is very close to being realized in each cell.

The local-equilibrium hypothesis implies that

1. All the variables defined in equilibrium thermodynamics remain significant. Variables such as the temperature and the entropy are rigorously and unambiguously defined just as they are in equilibrium. In each cell, these quantities remain uniform but they take different values from cell to cell; they are also allowed to change in the course of time in such a way that they depend continuously on the space and time coordinates (r,t).

2. The relationships in equilibrium thermodynamics between state variables remain valid outside equilibrium provided that they are stated locally at each instant of time. Thus, the entropy outside equilibrium will depend on the same state variables as at equilibrium.

For an N-component fluid, the specific entropy s will be a function of the specific internal energy u, the specific volume v, and the mass fractions c_k of the different components, i.e. $s(r,t) = s[u(r,t),v(r,t),c_k(r,t)]$ and, in differential form,

$$ds = \left(\frac{\partial s}{\partial u}\right)_{v,c_k} du + \left(\frac{\partial s}{\partial v}\right)_{u,c_k} dv + \sum_{k=1}^{N} \left(\frac{\partial s}{\partial c_k}\right)_{u,v,c_{k'}} dc_k \quad (\text{for } k' \neq k). \quad (1.54)$$

Defining, as in equilibrium thermodynamics, the absolute temperature T, the pressure p, and the chemical potential μ_k by

$$T^{-1} = \left(\frac{\partial s}{\partial u}\right)_{v,c_k} , \quad T^{-1}p = \left(\frac{\partial s}{\partial v}\right)_{u,c_k} , \quad T^{-1}\mu_k = -\left(\frac{\partial s}{\partial c_k}\right)_{u,v,c_{k'}} , \quad (1.55)$$

respectively, one obtains from (1.54) the local form of the Gibbs equation, namely

$$T ds = du + p dv - \sum_{k=1}^{N} \mu_k dc_k . \qquad (1.56)$$

This equation is fundamental for finding out the rate of entropy production, as shown in the next subsection. Note that (1.56) is written in the centre-of-mass reference frame, since equilibrium thermodynamics cannot cope with convective phenomena. A simple procedure allowing one to determine the form of the Gibbs equation for a wide variety of systems (one-component fluid, mixtures, electromagnetic systems) was proposed by Lambermont and Lebon [1.18].

3. A third consequence of the local-equilibrium hypothesis is that it permits one, from the convexity property of entropy, to derive locally the thermodynamic conditions of stability, such as the positiveness of the specific heat and the isothermal compressibility.

A precise limitation of the domain of validity of CIT cannot be obtained from the macroscopic formalism itself: it requires either a wider macroscopic or a microscopic theory like the kinetic theory of gases. Starting from the Chapman–Enskog development, Prigogine [1.19] has shown that the hypothesis of local equilibrium is satisfactory provided that the distribution function is limited to the first-order term. Explicit conditions under which the local-equilibrium hypothesis holds are established in Chap. 5.

1.3.2 Entropy Production and Entropy Flux

Our objective is to explicitly calculate the entropy flux and entropy production in a system in which different irreversible processes are under way. Consider a mixture of N charged components among which n chemical reactions may take place. In terms of time derivatives, the Gibbs equation (1.56) can be written as

$$T\dot{s} = \dot{u} + p\dot{v} - \sum_{k=1}^{N} \mu_k \dot{c}_k . \qquad (1.57)$$

By multiplying (1.57) by ρ and replacing $\dot{u}$, $\dot{v}$, and $\dot{c}_k$ by their values determined from the energy balance equation (1.47), the total mass conservation equation (1.11), and the mass fraction balance equation (1.28) respectively, we obtain the following expression for the rate of change of the entropy:

$$\rho\dot{s} = -\frac{1}{T}\nabla\cdot\boldsymbol{q} - \frac{1}{T}\mathbf{P}^v:\nabla\boldsymbol{v} + \frac{1}{T}\sum_{k=1}^{N}\mu_k\nabla\cdot\boldsymbol{J}_k + \frac{\rho}{T}\sum_{l=1}^{n}\mathcal{A}_l\dot{\xi}_l + \frac{1}{T}\boldsymbol{\varepsilon}\cdot\boldsymbol{i}. \qquad (1.58)$$

In establishing (1.58), it was assumed that the external forces applied to each species $k = 1, 2,..., N$ are all identical and that the time derivative of the diffusion velocities w_k may be neglected, which is a hypothesis frequently used in CIT [1.5]. Otherwise, (1.58) would contain a supplementary contribution of the form $-\Sigma_k(\boldsymbol{J}_k\cdot\dot{\boldsymbol{w}}_k)$ [1.5]. The quantity $\mathcal{A}_l$ is the affinity of the lth chemical reaction, defined by

$$\mathcal{A}_l = -\sum_{k=1}^{N}v_{kl}\mu_k \qquad (l=1,2,...,n). \qquad (1.59)$$

By using the decomposition (1.19) of the pressure tensor, (1.58) may be also rearranged as follows:

$$\rho\dot{s} = -\nabla\cdot\left[\frac{1}{T}\left(\boldsymbol{q} - \sum_{k=1}^{N}\mu_k\boldsymbol{J}_k\right)\right] + \boldsymbol{q}\cdot\nabla\left(\frac{1}{T}\right) - \sum_{k=1}^{N}\boldsymbol{J}_k\cdot\nabla\left(\frac{\mu_k}{T}\right)$$

$$-\frac{1}{T}p^v\nabla\cdot\boldsymbol{v} - \frac{1}{T}\overset{0}{\mathbf{P}}{}^v:\overset{0}{\mathbf{V}} + \frac{\rho}{T}\sum_{l=1}^{n}\mathcal{A}_l\dot{\xi}_l + \frac{1}{T}\boldsymbol{\varepsilon}\cdot\boldsymbol{i}.$$

A comparison of this expression and the general balance equation of entropy (1.52) reveals that $\boldsymbol{J}^s$ and σ^s are respectively given by

$$\boldsymbol{J}^s = \frac{1}{T}\left(\boldsymbol{q} - \sum_{k=1}^{N}\mu_k\boldsymbol{J}_k\right) \qquad (1.60)$$

and

$$\sigma^s = \boldsymbol{q}\cdot\nabla\left(\frac{1}{T}\right) - \sum_{k=1}^{N}\boldsymbol{J}_k\cdot\nabla\left(\frac{\mu_k}{T}\right) - \frac{1}{T}p^v\nabla\cdot\boldsymbol{v} - \frac{1}{T}\overset{0}{\mathbf{P}}{}^v:\overset{0}{\mathbf{V}} + \frac{\rho}{T}\sum_{l=1}^{n}\mathcal{A}_l\dot{\xi}_l + \frac{1}{T}\boldsymbol{\varepsilon}\cdot\boldsymbol{i}. \quad (1.61)$$

Expression (1.60) shows that the entropy flux splits into two parts: the first is connected with heat conduction and the second arises from the diffusion of matter. From (1.61) it is concluded that six different effects contribute to the rate of entropy production: the first is related to heat conduction, the second to matter flow, the third and fourth to mechanical dissipation, the fifth is due to chemical reactions, and the sixth to electric currents. Relation (1.61) for σ^s is a sum of products of two factors called respectively thermodynamic flux J_α and thermodynamic force X_α (explicit expressions of these quantities are given in Table 1.2). In terms of them the rate of entropy production presents the bilinear structure

$$\sigma^s = \sum_\alpha J_\alpha X_\alpha . \tag{1.62}$$

The fluxes J_α and forces X_α in (1.62) are not necessarily scalar quantities: they represent vectorial and tensorial quantities as well. Each individual flux and force has the property of vanishing at equilibrium. It must be stressed that the decomposition into thermodynamic fluxes and forces is arbitrary to a certain extent: one could, for instance, include the factor $1/T$ in the flux instead of in the force. Likewise, one could permute the definitions of fluxes and forces. However, these various choices are not crucial and have no direct consequences for the interpretation of the final results.

Table 1.2. Independent fluxes and forces

Flux: J_α	q	J_k	p^v	$\overset{0}{\mathbf{p}}{}^v$	$\rho\dot{\xi}_l$	i
Force: X_α	∇T^{-1}	$-\nabla(T^{-1}\mu_k)$	$-T^{-1}\nabla\cdot v$	$-T^{-1}\overset{0}{\mathbf{V}}$	$T^{-1}\mathcal{A}_l$	$T^{-1}\varepsilon$

Nevertheless, as shown by Meixner [1.20], it is essential to select independent fluxes and independent forces as well. This can be achieved in particular by choosing the quantities of Table 1.2. This splitting is quite natural, since it meets the requirements of cause and effect: the cause is provided by the driving thermodynamic force, which elicits the effect manifested through the conjugated flux.

1.3.3 Linear Constitutive Equations

The fluxes are unknown quantities, in contrast to the forces, which are known functions of the state variables or (and) their gradients. It has been found experimentally that fluxes and forces are interwoven. In general, a given flux does not only depend on its own conjugated force but may depend on the whole set of forces acting on the system. Furthermore, the flux may depend on all the thermodynamic state variables T, p and c_k as well:

$$J_\alpha = J_\alpha(X_1, X_2, ..., X_\alpha, ...; T, p, c_k). \tag{1.63}$$

A relation like (1.63) between fluxes and forces is called a phenomenological or constitutive equation: it expresses specific properties of the material involved in an irreversible process. After expansion of (1.63) around the equilibrium values $J_{\alpha,eq} = 0$ and $X_{\alpha,eq} = 0$, one has

$$J_\alpha = \sum_\beta \left(\frac{\partial J_\alpha}{\partial X_\beta} \right)_{eq} X_\beta + O\left(X_\beta X_\gamma \right) + \dots \qquad (1.64)$$

Neglecting the second-order and subsequent terms and setting

$$L_{\alpha\beta} = \left(\frac{\partial J_\alpha}{\partial X_\beta} \right)_{eq},$$

we find that (1.64) reduces to

$$J_\alpha = \sum_\beta L_{\alpha\beta} X_\beta. \qquad (1.65)$$

The $L_{\alpha\beta}$ quantities are called phenomenological coefficients and depend generally on T, p, and c_k. The constitutive equations (1.65), together with the balance equations of mass, momentum, and energy, constitute a closed set of equations which can be solved when initial and boundary conditions are specified. Experimental evidence and theoretical considerations in statistical mechanics have confirmed that a wide class of phenomena can be described by means of linear flux–force relations. This is true in particular for transport processes where the macroscopic gradients vary on a much larger scale than the mean free path.

It must be realized that the symmetry properties of the material have an influence on the form of the constitutive equations. For instance, in isotropic systems, some couplings between fluxes and forces are forbidden. As a consequence of the representation theorem of isotropic tensors [1.21], it can be shown that fluxes and forces of different tensorial rank do not couple so far as linear relations are involved. For example, a temperature gradient cannot give rise to a viscous pressure in a linear description. The independence of processes of different tensorial rank is often referred to as the 'Curie principle' in the CIT literature. As Truesdell acidly observes [1.14], it is redundant to invoke the name of Curie and the term 'principle' to establish a result which comes directly from tensor algebra.

For isotropic systems, the most general linear constitutive relations between the fluxes and forces of Table 1.2 are

$$q = L_{qq} \nabla\left(\frac{1}{T} \right) - \sum_{k=1}^{N} L_{qk} \nabla\left(\frac{\mu_k}{T} \right) + L_{qe} \frac{\varepsilon}{T}, \qquad (1.66a)$$

$$J_k = L_{kq} \nabla\left(\frac{1}{T} \right) - \sum_{j=1}^{N} L_{kj} \nabla\left(\frac{\mu_j}{T} \right) + L_{ke} \frac{\varepsilon}{T} \quad (k = 1, 2, \dots, N), \qquad (1.66b)$$

$$i = L_{eq}\nabla\left(\frac{1}{T}\right) - \sum_{k=1}^{N} L_{ek}\nabla\left(\frac{\mu_k}{T}\right) + L_{ee}\frac{\varepsilon}{T},$$

(1.66c)

$$\rho\dot{\xi}_j = \frac{l_{jv}}{T}\nabla\cdot v + \sum_{l=1}^{n} l_{jl}\frac{\mathcal{A}_l}{T} \quad (j=1,2,...,n),$$

(1.66d)

$$p^v = -\frac{l_{vv}}{T}\nabla\cdot v + \sum_{l=1}^{n} l_{vl}\frac{\mathcal{A}_l}{T},$$

(1.66e)

$$\overset{0}{\mathbf{P}}{}^v = -\frac{L}{T}\overset{0}{\mathbf{V}}.$$

(1.66f)

In these relations, all the phenomenological coefficients are scalar quantities. The phenomenological coefficients L_{qq}, L_{kj}, L_{ee}, l_{vv}, and L are related to the usual transport coefficients of thermal conductivity λ, diffusion D_{kj}, electrical resistivity r_e, bulk viscosity ζ, and shear viscosity η by

$$L_{qq} = \lambda T^2, \quad D_{kj} = \frac{1}{T}\sum_{r=1}^{N} L_{kr}\left(\frac{\partial\mu_r}{\partial c_j}\right)_{T,p,c_{j'}}, \quad L_{ee} = \frac{T}{r_e}, \quad l_{vv} = \zeta T, \quad L = 2\eta T. \quad (1.67)$$

By using the identifications (1.67) and omitting in (1.66a,e,f) the coupling coefficients, one recovers the Fourier and Newton–Stokes laws:

$$q = -\lambda\nabla T \qquad \text{(Fourier's law)},$$

(1.68)

$$p^v = -\zeta\nabla\cdot v \qquad \text{(Stokes' law)},$$

(1.69)

$$\overset{0}{\mathbf{P}}{}^v = -2\eta\overset{0}{\mathbf{V}} \qquad \text{(Newton's law)}.$$

(1.70)

The Ohm and Fick laws are obtained by introducing supplementary constraints; for instance, the classical expression for Fick's law,

$$J_k = -\sum_{j=1}^{N} D_{kj}\nabla c_j,$$

(1.71)

demands that one works at constant temperature and pressure, and Ohm's law

$$i = \frac{1}{r_e}E,$$

is derived from (1.66c), provided that the magnetic induction and all couplings are ignored. It is worthwhile pointing out that (1.66d) predicts a linear relation between the rate of advancement of a chemical reaction and the affinities. Such a linear law is unrealistic, since it is only correct in a very narrow domain around the equilibrium. It is indeed known from chemical kinetics that in a multicomponent incompressible system ($\nabla \cdot v = 0$) in which just one chemical reaction takes place, one has

$$\rho \dot{\xi} = l\left[1 - \exp\left(-\frac{\mathcal{A}}{RT}\right)\right],\tag{1.72}$$

where the coefficient l is a function of the temperature and mass fractions. The derivation of expression (1.72) is outlined in Problem 1.2. It is only in the limiting case $\mathcal{A} << RT$ that (1.72) reduces to

$$\rho \dot{\xi} = \frac{l}{RT}\mathcal{A},$$

from which it may be concluded that the linear relation (1.66d) between $\dot{\xi}$ and $\mathcal{A}$ is only satisfied in the close vicinity of equilibrium.

1.3.4 Constraints on the Phenomenological Coefficients

The linear flux–force relations are the simplest constitutive equations which guarantee the semi-positiveness of the rate of entropy production. Indeed, by substitution of (1.60) into (1.61), one gets

$$\sigma^s = \sum_{\alpha,\beta} L_{\alpha\beta} X_\alpha X_\beta \geq 0.\tag{1.73}$$

By writing (1.73) in the form

$$\sigma^s = \sum_\alpha L_{\alpha\alpha} X_\alpha X_\alpha + \sum_{\alpha,\beta\neq\alpha} \tfrac{1}{2}\left(L_{\alpha\beta} + L_{\beta\alpha}\right) X_\alpha X_\beta,$$

it can be seen that the necessary and sufficient conditions for $\sigma^s \geq 0$ to be held are that the determinant of the symmetric part of $L_{\alpha\beta}$ and all its principal minors are positive; in particular, necessary but not sufficient conditions are

$$L_{\alpha\alpha} \geq 0,$$

$$L_{\alpha\alpha} L_{\beta\beta} \geq \tfrac{1}{4}\left(L_{\alpha\beta} + L_{\beta\alpha}\right)^2.\tag{1.74}$$

As a consequence of (1.74), it is seen that the heat conductivity λ, the bulk and shear viscosity coefficients ζ and η, and the electrical resistivity r_e are all semi-positive definite quantities.

1.3.5 The Onsager–Casimir Reciprocal Relations

Another important kind of constraint on the coefficients $L_{\alpha\beta}$ concerns their symmetry property and was established by Onsager [1.1]. Under the three conditions that (a) the fluxes are identified as time rates of state variables a_α, (b) the forces X_α are identified as the derivatives of the entropy with respect to the state variables a_α, and (c) there exists between these so-defined fluxes and forces linear constitutive relations of the form

$$\dot{a}_\alpha = \sum_\beta L_{\alpha\beta} X_\beta , \tag{1.75}$$

in which the $L_{\alpha\beta}$ obey the reciprocal relations

$$L_{\alpha\beta} = L_{\beta\alpha}. \tag{1.76}$$

This result is a consequence of the time-reversal invariance of the microscopic dynamics demanding that the particles retrace their former path when the velocities are reversed, as it is shown in Sect. 6.6. The Onsager reciprocal relations (1.76) are very useful in studying coupled phenomena, such as thermodiffusion, thermoelectricity, and thermoelectromagnetic effects. The Onsager original derivation was only valid for state variables that are even under microscopic time-reversal. An extension to variables with odd parities was carried out by Casimir [1.22], who demonstrated that in full generality

$$L_{\alpha\beta} = \varepsilon_\alpha \varepsilon_\beta L_{\beta\alpha}, \tag{1.77}$$

where ε_α, ε_β are equal to $+1$ or -1 whether the state variable is even or odd under time-reversal.

If an external magnetic induction $\boldsymbol{B}$ is acting, one must not only reverse the velocities but also the magnetic field if it is desired that the particles retrace their former path: this is a consequence of the expression of the Lorentz force. The same reasoning can be applied for processes taking place in non-inertial frames rotating with an angular velocity $\boldsymbol{\omega}$. It follows from the form of the Coriolis force that, in this case, the velocity of particles $\boldsymbol{v}$ and $\boldsymbol{\omega}$ must be reversed. The reciprocal relations (1.77) have now to be replaced by the following expression:

$$L_{\alpha\beta}(\boldsymbol{B},\boldsymbol{\omega}) = \varepsilon_\alpha \varepsilon_\beta L_{\beta\alpha}(-\boldsymbol{B},-\boldsymbol{\omega}). \tag{1.78}$$

It must be mentioned that the validity of the Onsager–Casimir relations has been challenged by some people working in continuum thermodynamics [1.14]. In this respect, we wish to make the following comments.

Although various proofs of the Onsager reciprocal relations have been proposed so far, all of them are based on microscopic theories: statistical mechanics of fluctuations or kinetic theory. Nevertheless, the Onsager relations are generally accepted to be correct at the macroscopic level, even when the thermodynamic fluxes cannot be expressed in the form of time derivatives of state variables. Typical quantities that do not meet this condition are the heat flux vector and the viscous pressure tensor.

A crucial point in the derivation of Onsager's relations is that the regressions of fluctuations are assumed to follow the same linear dynamical laws as the macroscopic equations. This assertion is questionable because, quoting Truesdell [1.14], '... not even the form of the constitutive equation is derived from the molecular theory, rather the molecular theory, so-called, is forced into agreement with preconceived phenomenological ideas'.

These criticisms have been the motivation behind submitting the Onsager–Casimir relations to severe experimental scrutiny. Careful experimental tests have been performed, especially in thermodiffusion and thermoelectricity. They do confirm the symmetry property of the coefficients $L_{\alpha\beta}$ within reasonable limits of experimental errors [1.23]. In spite of these encouraging observations, it is our opinion that, unless a complete macroscopic proof of the Onsager relations is proposed, one should regard them as postulates at the macroscopic level.

Finally we summarize that the main points underlying CIT are:
- the local-equilibrium hypothesis that allows one to write the Gibbs equation locally for any time;
- the existence of a non-negative rate of entropy production;
- the existence of linear constitutive laws;
- the Onsager–Casimir reciprocal relations.

1.3.6 Limitations

The classical description has been undoubtedly useful, and has led to an impressive production of scientific work. Nevertheless, it presents some drawbacks like the mentioned below.

1. It is based on the local-equilibrium hypothesis, which may be too restrictive for a wide class of phenomena. It is conceivable, indeed, that other variables, not found in equilibrium, may influence the thermodynamic equations in non-equilibrium situations. To illustrate this observation, we mention an old example quoted by O. Reynolds in 1885 [1.24]. He pointed out that when a leather bag is filled with marbles, topped up with

water, and then twisted, the marble density decreases when the rate of shear is increased, at constant temperature and pressure. This means that, in contradiction with the local-equilibrium assumption, the density does not only depend on temperature and pressure but also on the shearing rate.

2. Statistical and kinetic analyses show that the local-equilibrium hypothesis is only consistent with linear and instantaneous relations between fluxes and forces. In many problems, the assumption of linear and stationary constitutive relations are too stringent. This is particularly true in chemistry, as mentioned above, and in rheology as well.

3. The linear steady constitutive equations are not satisfactory at high frequencies and short wavelengths, as manifested in experiments on sound absorption and dispersion in dilute gases. The dispersion relation obtained from CIT is in agreement with experimental observations at low frequencies only.

4. The classical Fourier law of heat conduction leads, when introduced into the energy conservation law, to a partial differential parabolic equation for the temperature. This implies that disturbances propagate with boundless speed. This unpleasant physical property is also observed with other quantities, such as concentration and viscous signals.

5. Recently very intensive work has been performed in the so-called generalised hydrodynamics. According to this formalism, the transport coefficients in the Stokes–Navier–Fourier constitutive equations are frequency and wavelength dependent, as confirmed by neutron scattering techniques. Such a result is at variance with the local-equilibrium assumption assessing that the transport coefficients are frequency and wavelength independent.

1.4 Rational Thermodynamics

This formalism was essentially developed by Coleman [1.13], Truesdell [1.14], and Noll [1.15] and follows a line of thought drastically different from CIT. Its main objective is to provide a method for deriving constitutive equations. The basic hypotheses underlying rational thermodynamics can be summarized as follows.

1. Absolute temperature and entropy are considered primitive concepts. They are introduced a priori in order to ensure the coherence of the theory and do not have a precise physical interpretation.

2. It is assumed that materials have a memory, i.e. the behaviour of a system at a given instant of time is determined not only by the values of the characteristic parameters at the present time, but also by their past history. The local-equilibrium hypothesis is no longer assumed since a knowledge of the values of the parameters at the present time is not enough to specify unambiguously the behaviour of the system.

3. The general expressions previously formulated for the balance of mass, momentum, and energy are however retained. Nevertheless, there are two essential nuances. The first is the introduction of a specific rate of energy supply r in the balance of internal energy, which in local form is written as

$$\rho \dot{u} = -\nabla \cdot \boldsymbol{q} - \mathbf{P} : \nabla \boldsymbol{v} + \rho r; \tag{1.79}$$

r is generally referred to as the power supplied or lost by radiation. The second crucial point is that the body forces $\boldsymbol{F}$ and the radiation term r are not given a priori as a function of position and time but are computed from the laws of momentum and internal energy respectively.

4. Another capital point is the mathematical formulation of the second law of thermodynamics, which serves essentially as a restriction on the form of the constitutive equations. The starting relation is the Clausius–Planck inequality, which states that between two equilibrium states A and B one has

$$\Delta S \geq \int_A^B \frac{\mathrm{d}Q}{T}. \tag{1.80}$$

In rational thermodynamics, inequality (1.80) is written as

$$\frac{\mathrm{d}}{\mathrm{d}t} \int_{V(t)} \rho s \mathrm{d}V \geq -\int_{\Sigma(t)} \frac{1}{T} \boldsymbol{q} \cdot \boldsymbol{n} \mathrm{d}\Sigma + \int_{V(t)} \rho \frac{r}{T} \mathrm{d}V, \tag{1.81}$$

or, in local form, as

$$\rho \dot{s} + \nabla \cdot \frac{\boldsymbol{q}}{T} - \rho \frac{r}{T} \geq 0. \tag{1.82}$$

By introducing the Helmholtz free energy $f (= u - Ts)$ and eliminating r between the energy balance equation (1.79) and the inequality (1.82) leads to

$$-\rho \left(\dot{f} + s \dot{T} \right) - \mathbf{P} : \mathbf{V} - \frac{1}{T} \boldsymbol{q} \cdot \nabla T \geq 0. \tag{1.83}$$

This inequality, established here for a one-component uncharged system, is either known as the *Clausius–Duhem* or the *fundamental inequality*.

An important problem is certainly the selection of the constitutive independent variables. This choice is subordinated to the type of material one deals with. In hydrodynamics, it is customary to take as variables the density, velocity, and temperature fields. It is also known that the balance laws and Clausius–Duhem inequality introduce complementary variables, such as the internal energy, the heat flux, the pressure tensor, and the entropy. The latter are expressed in terms of the former by means of constitutive equations. By an *admissible process* is meant a solution of the balance laws when the constitutive relations are taken into account and the Clausius–Duhem inequality holds.

1.4.1 The Basic Axioms of Rational Thermodynamics

Before deriving the constitutive equations, let us briefly examine the main principles they must meet. A word of caution is required about the use and abuse of the term 'principle' in rational thermodynamics. In most cases, this term is employed to designate merely convenient assumptions.

The principle of equipresence. This principle asserts that if a variable is present in one constitutive equation, it will be a priori present in all the constitutive equations. However, the condition for the final presence or absence of a dependent variable in a constitutive relation derives from the Clausius–Duhem inequality.

The principle of memory or heredity. According to this principle the present effects are dictated not only by present causes but by past causes as well. Consequently, the set of independent variables is no longer formed by the variables at the present time but by their whole history. If Φ is an arbitrary variable, e.g. v, $\mathbf{v}$ and T, we shall denote its history up to time t by

$$\Phi^t = \Phi(t - t') \qquad (0 < t' < \infty).$$

The principles of equipresence and memory when applied to hydrodynamics assert that

$$\left.\begin{array}{c} u\,(\text{or } f) \\[4pt] s \\[4pt] q \\[4pt] \mathbf{P} \end{array}\right\} \quad \text{at}\,(\mathbf{r}, t) \text{ are functionals of} \quad \left\{\begin{array}{c} v^t \\[4pt] \mathbf{v}^t \\[4pt] T^t \end{array}\right.$$

Of course, the choice of the dependent and independent variables is not unique. One could for instance permute the roles of u and T, but, since the usual attitude in rational thermodynamics is to select T as the independent quantity, here we shall follow this point of view.

The principle of local action. This principle establishes that the behaviour of a material point should only be influenced by its immediate neighbourhood. Otherwise stated, the values of the constitutive equation at a given point are insensitive to what happens at distant points; accordingly, in a first-order theory, second-order and higher-order space derivatives should be omitted.

The principle of material frame-indifference. As a preliminary to the formulation of this principle, it is useful to introduce the notion of objectivity. Consider two reference frames (or observers) moving with respect to each other arbitrarily. Let r be the position vector of a material point at time t in one of the reference frames (say an inertial one) and r^* its position vector at the same time in the other frame (say a non-inertial one). We impose that the relation between r and r^* be such that the distance between two arbitrary points in the body and the angle between two directions are preserved. The most general transformation law that satisfies these requirements is the one given by the Euclidean transformation

$$r^* = \mathbf{Q}(t) \cdot r + c(t), \qquad (1.84)$$

where $\mathbf{Q}(t)$ is a real, proper orthogonal, time-dependent tensor

$$\mathbf{Q} \cdot \mathbf{Q}^T = \mathbf{Q}^T \cdot \mathbf{Q} = \mathbf{U}, \quad \det \mathbf{Q} = 1, \qquad (1.85)$$

and $c(t)$ is the distance between the origins of the two frames. In this book we have not considered the more general transformation $\det \mathbf{Q} = \pm 1$, which includes symmetry under reflection and a possible translation in time $t^* = t - \tau$, with τ being a real constant. The general transformation $\det \mathbf{Q} = \pm 1$ has been introduced and discussed in length in specific treatises on rational thermodynamics [1.13–16] to which the reader is referred for further details. When the Euclidean group (1.84) acts on a tensor of rank n ($n = 0, 1, 2, ...$), the latter is said to be objective, if it transforms according to

$$A^*_{ij...k} = Q_{i\alpha} Q_{j\beta} \cdots Q_{k\gamma} A_{\alpha\beta...\gamma}. \qquad (1.86)$$

For tensors of rank zero (scalar), rank one (vector), and rank two, one has

$$a^* = a \qquad \text{(objective scalar)}, \qquad (1.87a)$$

$$a^* = \mathbf{Q} \cdot a \qquad \text{(objective vector)}, \qquad (1.87b)$$

$$\mathbf{A}^* = \mathbf{Q} \cdot \mathbf{A} \cdot \mathbf{Q}^T \qquad \text{(objective tensor)}, \qquad (1.87c)$$

respectively. According to (1.87a), a scalar is objective if it keeps the same value in all moving reference frames. The velocity vector is not objective because it transforms like

$$v^* = \mathbf{Q} \cdot v + \dot{\mathbf{Q}} \cdot r + \dot{c}, \qquad (1.88)$$

which is not of the form (1.87b). Similarly, the acceleration $\dot{v}$ is not objective either. It can also be shown that the symmetric and antisymmetric parts of the velocity gradient transform like

$$\mathbf{V}^* = \mathbf{Q} \cdot \mathbf{V} \cdot \mathbf{Q}^T \qquad \mathbf{W}^* = \mathbf{Q} \cdot \mathbf{W} \cdot \mathbf{Q}^T - \dot{\mathbf{Q}} \cdot \mathbf{Q}^T. \qquad (1.89)$$

Thus $\mathbf{V}$ is objective and $\mathbf{W}$ is not. The term responsible for the non-objective character of $\mathbf{W}$ is the angular velocity $\mathbf{\Omega} = \dot{\mathbf{Q}} \cdot \mathbf{Q}^T$ of the moving frame with respect to the inertial one.

It can also be seen that the material time derivatives of objective vectors and tensors are not objective because they transform like

$$\dot{a}^* = \mathbf{Q} \cdot \dot{a} + \dot{\mathbf{Q}} \cdot a,$$

$$\dot{\mathbf{A}}^* = \mathbf{Q} \cdot \dot{\mathbf{A}} \cdot \mathbf{Q}^T + \dot{\mathbf{Q}} \cdot \mathbf{A} \cdot \mathbf{Q}^T + \mathbf{Q} \cdot \mathbf{A} \cdot \dot{\mathbf{Q}}^T.$$

The failure of the objectivity of the material time derivatives of vectors and tensors arises obviously from the time-dependence of the orthogonal tensor $\mathbf{Q}(t)$. This has motivated the search for objective time derivatives. The answer is not unique and several objective time derivatives satisfying (1.87b) and (1.87c) have been proposed. Among the most frequently used, let us mention the Jaumann or co-rotational derivative:

$$D_J a = \dot{a} + \mathbf{W} \cdot a,$$

$$D_J \mathbf{A} = \dot{\mathbf{A}} + \mathbf{W} \cdot \mathbf{A} - \mathbf{A} \cdot \mathbf{W}; \qquad (1.90)$$

the covariant or lower convected derivative:

$$D_{\downarrow} a = \dot{a} + (\nabla v) \cdot a,$$

$$D_{\downarrow} \mathbf{A} = \dot{\mathbf{A}} + (\nabla v) \cdot \mathbf{A} + \mathbf{A} \cdot (\nabla v)^T; \qquad (1.91)$$

and the contravariant or upper convected derivative:

$$D^{\uparrow} a = \dot{a} - (\nabla v)^T \cdot a,$$

$$D^{\uparrow} \mathbf{A} = \dot{\mathbf{A}} - (\nabla v)^T \cdot \mathbf{A} - \mathbf{A} \cdot (\nabla v). \qquad (1.92)$$

The Jaumann derivative has a simple physical interpretation, since it is the derivative measured by an observer whose frame of reference is carried by the medium and rotates with it; upper and lower convected derivatives correspond to non-orthogonal time-dependent reference frames deforming with the medium and moving with it. From a pure continuum mechanics standpoint, there is no preference for any of the above objective time derivatives. However, it may happen that the formulation of some particular class of constitutive equations becomes more simple and elegant when one particular objective time rate is selected.

We are now in a position to formulate the principle of material frame indifference. Generally stated, this principle demands that the constitutive equations be independent of the observer. This statement implies two requirements. First, the constitutive equations should be objective, i.e. form-invariant under arbitrary time-dependent rotations and translations of the reference frames as expressed by the Euclidean transformations (1.84). This statement amounts to the requirement that the form of the constitutive relations is left unaffected by the superposition of any arbitrary rigid body motion. Second, the constitutive equations should be independent of the frame, in particular its angular velocity. To give an example, Newton's equation of motion is form invariant, but at the same time it depends on the frame through the inertial forces; therefore, it fulfills the first requirement but not the second one.

Of course, before examining the effect of a change of frame on a constitutive equation, it is necessary to specify how the basic variables such as temperature, energy, entropy, heat flux, pressure tensor, etc. behave under such a transformation. Since these quantities have an intrinsic meaning, they are expected to be objective and at the same time frame independent. It is thus taken for granted that

$u, s, f,...$ are objective scalars,

q is an objective vector,

$\mathbf{P}$ is an objective tensor of order two.

1.4.2 Constitutive Equations

We proceed further with the establishment of the constitutive equations. For simplicity, we consider a very particular thermomechanical material, namely the Stokesian fluid. It is characterized by the absence of memory and described by the following set of constitutive equations:

$$\phi = \phi(v, \mathbf{v}, T, \nabla \mathbf{v}, \nabla T), \tag{1.93}$$

where ϕ stands for any constitutive dependent variable. The absence of memory allows us to express the dependence of ϕ by means of ordinary functions instead of functionals. In explicit form, the constitutive relations (1.93) will be given by

$$f = f(v, T, \mathbf{V}, \nabla T), \tag{1.94a}$$

$$s = s(v, T, \mathbf{V}, \nabla T), \tag{1.94b}$$

$$q = q(v,T,\mathbf{V},\nabla T), \tag{1.94c}$$

$$\mathbf{P} = \mathbf{P}(v,T,\mathbf{V},\nabla T). \tag{1.94d}$$

By formulating (1.94a–d), the axiom of equipresence was used. In (1.94), the non-objective velocity v has been eliminated and the velocity gradient ∇v replaced by its objective symmetric part $\mathbf{V}$ to ensure material frame-indifference.

A further constraint is imposed by the second law of thermodynamics. This is achieved by substituting the constitutive laws (1.94) in the Clausius–Duhem inequality (1.83). Using the chain differentiation rule for calculating $\dot{f}$, inequality (1.83) reads

$$-\rho\left(\frac{\partial f}{\partial T}+s\right)\dot{T} - \rho\frac{\partial f}{\partial \mathbf{V}}:\dot{\mathbf{V}} - \rho\frac{\partial f}{\partial(\nabla T)}\cdot(\dot{\nabla T}) - \frac{1}{T}q\cdot\nabla T - \left(\frac{\partial f}{\partial v}\mathbf{U}+\mathbf{P}\right):\mathbf{V}\geq 0, \tag{1.95}$$

where use has been made of the mass conservation law

$$\rho\dot{v} = \nabla\cdot v = \mathbf{V}:\mathbf{U}.$$

It is worth noticing that inequality (1.95) is linear in $\dot{T}$, $\dot{\mathbf{V}}$, and $(\dot{\nabla T})$. Now it is assumed [1.13] that there always exist body forces and energy supplies that ensure that the balance equations of momentum and internal energy are identically satisfied. Therefore the balance laws do not impose constraints on the set $\dot{T}$, $\dot{\mathbf{V}}$, $(\dot{\nabla T})$, and we assign these time derivatives arbitrary and independent values. It then appears that unless the coefficients of these terms vanish, (1.95) could be violated. This gives the following results:

$$\frac{\partial f}{\partial T}+s=0 \quad\text{(a)}, \qquad \frac{\partial f}{\partial \mathbf{V}}=0 \quad\text{(b)}, \qquad \frac{\partial f}{\partial(\nabla T)}=0 \quad\text{(c)}. \tag{1.96}$$

Equation (1.96a) is classical while (1.96b) and (1.96c) express the idea that f is independent of $\mathbf{V}$ and ∇T. As a consequence, the constitutive equations for f and s simply read

$$f = f(v,T),$$
$$\tag{1.97}$$
$$s = s(v,T).$$

The above derivation rests on the controversial argument [1.25] that the body forces F and the radiation sources r can be assigned arbitrarily in order that the balance laws are identically satisfied. Although this procedure is at variance with the usual way of thinking, where F and r are regarded as assigned a priori, it may also be asked what happens when F and (or) r are zero. To circumvent these difficulties, an alternative method was proposed by Liu [1.26], who considers the balance laws as constraints for the Clausius–Duhem inequality and accounts for them by means of the well-known Lagrange multipliers. The delicate points in Liu's method are the derivation and the physical interpretation of the Lagrange multipliers, but the results (1.96) remain unchanged.

Defining the equilibrium pressure by

$$p = -\frac{\partial f}{\partial v} \tag{1.98}$$

and decomposing $\mathbf{P}$ according to (1.18), (1.95) reduces to

$$-\frac{1}{T}\boldsymbol{q} \cdot \nabla T - \mathbf{P}^v : \mathbf{V} \geq 0. \tag{1.99}$$

This expression is nothing but the rate of energy dissipation $(= T\sigma^s)$ calculated earlier in the framework of CIT [see (1.61), where diffusive, chemical, and electrical effects were ignored].

Using the representation theorems of tensors, one obtains within the linear approximation the following constitutive equations for $\boldsymbol{q}$ and $\mathbf{P}^v$:

$$\boldsymbol{q} = -\lambda(v,T)\nabla T, \tag{1.100a}$$

$$\mathbf{P}^v = -\eta_1(v,T)(\nabla \cdot v)\mathbf{U} - 2\eta(v,T)\mathbf{V}. \tag{1.100b}$$

The coefficients λ, η_1, and η may depend on v, T, and the first invariant of $\mathbf{V}$. After splitting $\mathbf{V}$ into its bulk and its deviatoric part, one obtains

$$\mathbf{P}^v = -\zeta(\nabla \cdot v)\mathbf{U} - 2\eta\overset{\circ}{\mathbf{V}}, \tag{1.101}$$

where ζ stands for $(\eta_1 + 2\eta/3)$. We recognize (1.100a) as the classical constitutive Fourier's equation and (1.100b) as the classical constitutive Newton–Stokes' equation, with λ the heat conductivity, ζ the bulk viscosity, and η the shear viscosity. Substitution of (1.100a) and (1.101) into (1.99) gives

$$\frac{\lambda}{T}(\nabla T) \cdot (\nabla T) + \zeta(\nabla \cdot v)^2 + 2\eta\overset{\circ}{\mathbf{V}} : \overset{\circ}{\mathbf{V}} \geq 0, \tag{1.102}$$

from which it follows that $\lambda > 0$, $\zeta > 0$, $\eta > 0$.

Likewise, it is an easy matter to recover the Gibbs equation. From (1.97), one has

$$df = \frac{\partial f}{\partial T}dT + \frac{\partial f}{\partial v}dv,$$

and, in view of (1.96a) and (1.98),

$$df = -s\,dT - p\,dv. \qquad\qquad (1.103)$$

Unlike CIT, where the Gibbs equation is postulated at the outset, it can be said that in rational thermodynamics the Gibbs relation is derived.

The steps leading to the establishment of the constitutive relations and the constraints on the signs of the transport coefficients are elegant and employ a minimum of hypotheses. At no time does one call upon symmetry relations of the Onsager type. Besides, the theory is not limited to linear constitutive equations. Attempts to develop the bases of rational thermodynamics further have been made recently [1.27].

1.4.3 Critical Remarks

Rational thermodynamics has not been free of criticisms, such as:

1. Temperature and entropy remain undefined objects. For example, it is not possible to check whether or not the temperature measured by a thermocouple corresponds to the temperature T used in rational thermodynamics. Concerning the entropy, no prescription is given for determining its actual functional dependence, either by experiment or by calculation from a physical model. Furthermore, it has been demonstrated by Day [1.28], who examined the problem of the temperature distribution in a rigid heat conductor with memory, that the value of the entropy is not unique.

2. The fundamental inequality (1.82) used in rational thermodynamics is not, strictly speaking, the Clausius inequality [1.29]. Indeed, the latter is given by (1.80) and connects two equilibrium states. In rational thermodynamics, Clausius' expression is generalised to arbitrary non-equilibrium states. When dealing with the Clausius–Duhem inequality in rational thermodynamics, it must be understood that the existence of a specific entropy that satisfies the fundamental inequality (1.83) has been postulated. The latter also implies that the entropy flux is given by the heat flux divided by the temperature, a result only valid in the vicinity of equilibrium as shown in the kinetic theory of gases.

3. Likewise, rational thermodynamics predicts unphysical properties in some classes of rheological materials. If it is admitted that the Rivlin–Ericksen model (see Chap. 15) provides a good description of rheological bodies, then the signs of some material coefficients, as given by the rational approach, are found to be in contradiction with experimental data.

4. Although the principle of material frame-indifference has revealed itself as a useful tool in establishing constitutive equations in continuum mechanics, it has recently been stressed that the two requirements of the principle, namely form invariance and frame independence, are not satisfied in several disciplines, such as classical mechanics, kinetic theory of gases, turbulence, rheology, and molecular hydrodynamics. In kinetic theory, it has been shown that the Burnett constitutive relations are frame-dependent [1.30, 31];

the origin of the frame-dependence lies in the Coriolis force of the rotating frame. It was also noticed that when the objective time derivatives (1.90–92) are used in stress–rate constitutive equations, one obtains results that are contradicted by Grad's kinetic model [1.32]. A similar problem arises in turbulence theory [1.33]; there is an ample experimental confirmation that turbulence, in a non-inertial frame, is quite different from turbulence in an inertial frame, owing to the dependence of the turbulent viscosity on the angular velocity of the reference frame. Recently, the validity of frame-indifference in viscoelastic materials has been discussed by Bird and de Gennes [1.34]. It was concluded that inertial forces can contribute to the material functions of viscoelastic media and that frame indifference is useful only whenever inertial effects are negligible. Another example of violation of the principle of material frame-indifference is provided by the phenomenological coefficients $L_{\alpha\beta}$ of CIT. When measured in a rotating frame, the $L_{\alpha\beta}$ are known to depend on the angular velocity, as pointed out earlier. Recently Hoover *et al.* [1.35] performed a molecular dynamics simulation for a fluid modelled by two-dimensional rotating disks: they found an angular component for the heat flux, in contradiction with the material frame-indifference. This result is confirmed in Sect. 10.5, where it is shown that heat conduction in fast rotating rigid cylinders depends on the angular velocity of the body. All these observations have cast serious doubts about the general validity of the principle of material frame-indifference [1.36–39].

5. From a practical point of view, the constitutive equations, when they are written in their general form involving functionals dependent on the whole history of the variables, are not easily tractable and generally require the knowledge of too vast an amount of information.

Problems

Solutions of the problems proposed in this book may be found, as Solutions Manual for Extended Irreversible Thermodynamics, on the www.uab.es/dep-fisica/eit website

1.1 *Onsager's reciprocity relations in a triangular chemical reaction scheme.* The cycle of chemical reactions A $\rightleftarrows$ B $\rightleftarrows$ C $\rightleftarrows$ A was analysed by Onsager in his papers of 1931 as an illustration of the reciprocity relations. Let k_i ($i = 1,2,3$) be the kinetic constants, J_i the respective fluxes of the reactions, which, according to the mass action law, are given by $J_1 = k_1 c_A - k_{-1} c_B$, $J_2 = k_2 c_B - k_{-2} c_C$, $J_3 = k_3 c_C - k_{-3} c_A$, and $\mathcal{A}_i$ the respective affinities, i.e. $\mathcal{A}_1 = \mu_A - \mu_B$, $\mathcal{A}_2 = \mu_B - \mu_C$, $\mathcal{A}_3 = \mu_C - \mu_A$. Since the process is cyclic, only two reactions are independent and $\mathcal{A}_1 + \mathcal{A}_2 + \mathcal{A}_3 = 0$. Show that when the relations between fluxes and forces are expressed in the form

$$J_1 - J_3 = L_{11}\mathcal{A}_1 + L_{12}\mathcal{A}_2 ,$$

$$J_2 - J_3 = L_{21}\mathcal{A}_1 + L_{22}\mathcal{A}_2 ,$$

the Onsager reciprocal relation $L_{12} = L_{21}$ is automatically satisfied near equilibrium if the principle of detailed balance is valid, i.e. if one assumes that in equilibrium $J_1 = J_2 = J_3 = 0$. Note that this assumption does not follow directly from the constitutive equations, since the non-zero values $J_1 = J_2 = J_3 \neq 0$ are compatible with $\mathcal{A}_1 = \mathcal{A}_2 = \mathcal{A}_3 = 0$.

1.2 *The mass action law.* Consider the chemical reaction

$$v_X X + v_Y\, Y \;\rightleftarrows\; v_Z\, Z + v_W\, W ,$$

where v_i are the stoichiometric coefficients. According to the mass action law, the reaction rate will be given by

$$J = \frac{1}{v_Z}\frac{dZ}{dt} = k_+ [X]^{v_X}[Y]^{v_Y} - k_-[Z]^{v_Z}[W]^{v_W} ,$$

with k_+ and k_- being the forward and backward kinetic constants, respectively. (a) Show that for ideal gases, whose chemical potential is of the form $\mu_i = RT \ln[i] + \zeta_i(T,p)$, where $[i]$ is the molar concentration and $\zeta_i(T,p)$ is an arbitrary function of T and p, the constitutive relation between J and the affinity $\mathcal{A}$ is

$$J = k_-[Z]^{v_Z}[W]^{v_W}\big[\exp(\mathcal{A}/RT) - 1\big].$$

(b) Show that when $\mathcal{A}/RT \ll 1$, this relation reduces to $J = L(\mathcal{A}/T)$ with $L = (k_-/R)$ $[Z_{eq}]^{v_Z}[W_{eq}]^{v_W} = (k_+/R)[X_{eq}]^{v_X}[Y_{eq}]^{v_Y}$, where subscript eq refers to equilibrium concentrations.

1.3 Show that in the phenomenological equations (1.66a–f), the following Onsager reciprocal relations are verified:

$$L_{qk} = L_{kq} , \quad L_{kj} = L_{jk}, \quad L_{ek} = L_{ke}, \quad l_{jl} = l_{lj}, \quad l_{jv} = -l_{vj}.$$

1.4 *The Einstein relation.* A dilute suspension of small particles in a viscous fluid at homogeneous temperature T is under the action of the gravitational field. The friction coefficient of the particles with respect to the fluid is α ($\alpha = 6\pi\eta r$ for spherical particles of radius r in a solvent with viscosity η). Owing to gravity, the particles have a sedimentation velocity $v_{sed} = m'g/\alpha$, with m' the mass of one particle minus the mass of the fluid displaced by one particle (Archimedes' principle); the corresponding sedimentation flux is $J_{sed} = nv_{sed}$, with n being the number of particles per unit volume. Against the sedimentation flux a diffusion flux $J_{dif} = -D\nabla n$ acts, D being the diffusion coefficient. (a) Find the vertical distribution of the concentration $n(z)$ of particles in equilibrium when the upward diffusion flux cancels exactly the downward sedimentation flux. (b) Compare this expression with Boltzmann's general expression

$$n(z) = n(0) \exp\left[-(m'gz/k_BT)\right],$$

where k_B is the Boltzmann constant, and demonstrate Einstein's relation

$$D = k_BT/\alpha.$$

1.5 Calculate the entropy production per unit volume in an anisotropic rigid heat conductor subjected to a temperature gradient. Formulate for this problem the corresponding phenomenological laws and the Onsager reciprocal relations. Assume that the thermal conductivity has an antisymmetric part. What would be its consequences on the temperature distribution in the body?

1.6 (a) Determine the entropy production per unit volume in a two-component diffusing mixture at rest; the system is chemically inert and the viscous effects are assumed to be negligible. (b) Show that the relevant phenomenological equations are

$$q_1 = -\lambda\nabla T - \frac{\mu_{11}}{c_2}D_F\nabla c_1 \qquad \text{(Dufour law)},$$

$$J_1 = -D_S\nabla T - D\nabla c_1 \qquad \text{(Soret law)},$$

where $D_F = D_S$; q_1 is the reduced heat flux vector, $q_1 = q - \sum_k h_k J_k$, J_1 is the flux of matter of component 1, subscripts 1, 2 refer to components 1, 2 respectively, μ_{11} stands for $\partial\mu_1/\partial c_1$, and D, D_F and D_S are respectively given by

$$D = L_{11}\mu_{11}/c_2, \quad D_F = L_{q1}, \quad D_S = L_{1q}.$$

1.7 *Micropolar fluids.* In some fluids (composed of elongated particles or rough spheres) the pressure tensor is non-symmetric. Its antisymmetric part is related to the rate of variation of an intrinsic angular momentum, and thus, it contributes to the balance equation of angular momentum [see, for instance, R. F. Snider and K. S. Lewchuk, J. Chem. Phys. **46** (1967) 3163, or J. M. Rubí and J. Casas-Vázquez, J. Non-Equilib. Thermodyn. **5** (1980) 155]. The antisymmetric part of the tensor is usually related to an axial vector $\boldsymbol{P}^{va}$, whose components are defined as $P_1^{va} = P_{23}^{va}$, $P_2^{va} = P_{31}^{va}$, $P_3^{va} = P_{12}^{va}$. The equation of balance for the internal angular momentum is

$$\rho j \dot{\omega} + \nabla \cdot \mathbf{Q} = -2 \boldsymbol{P}^{va},$$

with j the microinertia per unit mass of the fluid, $\boldsymbol{\omega}$ the angular velocity and $\mathbf{Q}$ the flux of the intrinsic angular momentum, which is usually neglected. (a) Show that the entropy production is given by

$$\sigma^s = -T^{-1} p^v \nabla \cdot \boldsymbol{v} - T^{-1} \overset{0}{\mathbf{P}}{}^{vs} : (\overset{0}{\nabla}\boldsymbol{v})^s - T^{-1} \boldsymbol{P}^{va} \cdot (\nabla \times \boldsymbol{v} - 2\boldsymbol{\omega}) + \boldsymbol{q} \cdot \nabla T^{-1},$$

where $\mathbf{P}^{vs}$ is the symmetric part of $\mathbf{P}^v$. [*Hint*: Note that $\partial s / \partial \boldsymbol{\omega} = -\rho j T^{-1} \boldsymbol{\omega}$]. (b) Show that the corresponding constitutive equation for $\boldsymbol{P}^{va}$ is

$$\boldsymbol{P}^{va} = -\eta_r (\nabla \times \boldsymbol{v} - 2\boldsymbol{\omega}),$$

where η_r is the so-called rotational viscosity; explain why there is no coupling between $\boldsymbol{P}^{va}$ and $\boldsymbol{q}$.

1.8 Prove that the symmetric part $\mathbf{V}$ of the velocity gradient tensor is objective; prove that the following quantities are not objective: $\nabla \boldsymbol{v}$ (the velocity gradient tensor), $\mathbf{W}$ (the skew-symmetric velocity gradient tensor), $\dot{\boldsymbol{q}}$ (the material time derivative of the heat flux vector).

1.9 Consider an incompressible fluid characterized by a specific entropy s, a specific Helmholtz free energy f, a heat flux $\boldsymbol{q}$, and a pressure tensor $\mathbf{P}$ depending on the temperature T, the temperature gradient ∇T, the symmetric velocity gradient tensor $\mathbf{V}$, and their first-order material time derivatives. Determine the corresponding constitutive equations in the framework of rational thermodynamics.

1.10 Derive, from rational thermodynamics, the linear constitutive equations of a two-component diffusing fluid mixture, when a non-uniform temperature acts upon the system.

1.11 *The efficiency of energy conversion.* Consider two coupled chemical reactions, with rates J_1 and J_2 and affinities $\mathcal{A}_1$ and $\mathcal{A}_2$ respectively. The entropy production is given by $T\sigma^s = J_1\mathcal{A}_1 + J_2\mathcal{A}_2 > 0$. Assume $J_1\mathcal{A}_1 < 0$ and $J_2\mathcal{A}_2 > 0$, which means that reaction 2 liberates an amount of free energy, which is used in reaction 1. This situation is common in biology: the free energy liberated by ATP is used to pump ions against their chemical potential gradient, or the free-energy liberated by oxidation-reduction reactions in the respiratory process is utilized to produce ATP by phosphorylation of ADP. [See D. Jou and J. E. Llebot, *Introduction to Thermodynamics of Biological Processes*, Prentice Hall, Englewood Cliffs, 1990]. Assume linear constitutive laws of the form

$$J_1 = L_{11}\mathcal{A}_1 + L_{12}\mathcal{A}_2,$$

$$J_2 = L_{21}\mathcal{A}_1 + L_{22}\mathcal{A}_2.$$

The degree of coupling of the process is defined as $q = L_{12}(L_{11}L_{22})^{-1/2}$ and the efficiency of the energy conversion is given by

$$\eta = -(J_1\mathcal{A}_1)/(J_2\mathcal{A}_2) \, .$$

(a) Show that $-1 \leq q \leq 1$. (b) Show that the maximum possible value of the efficiency of the energy conversion is

$$\eta_{max} = \frac{q^2}{\left[1 + (1 - q^2)^{1/2}\right]^2}.$$

1.12 It is known that by writing a linear relation of the form $J_\alpha = \Sigma_\beta L_{\alpha\beta} X_\beta$ between n independent fluxes J_α ($\alpha = 1,..., n$) and n independent thermodynamic forces X_β, the matrix $L_{\alpha\beta}$ of phenomenological coefficients is symmetric, according to Onsager. (a) Show that the matrix $L_{\alpha\beta}$ is still symmetric when not all the fluxes are independent but one of them is a linear combination of the other ones, i.e. $J_n = \Sigma_\alpha a_\alpha J_\alpha$ ($\alpha = 1,..., n - 1$). (b) Does this conclusion remain true when instead of a linear relation between the fluxes there exists a linear relation between the forces $X_n = \Sigma_\beta b_\beta X_\beta$? Check this result on the cyclic reaction of Problem 1.1.

1.13 *Non-linear constitutive relations.* Some authors have postulated an extension of the Onsager relations to non-linear situations in the form

$$(\partial J_i/\partial X_j) = (\partial J_j/\partial X_i)$$

with J_i and X_j being the thermodynamic fluxes and forces. (a) Show that for linear flux-force relations, this result is equivalent to Onsager's reciprocal relations. (b) Consider now the following expansion:

$$J_1 = L_{11}X_1 + L_{12}X_2 + L_{111}X_1^2 + L_{112}X_1X_2 + L_{122}X_2^2,$$

$$J_2 = L_{21}X_1 + L_{22}X_2 + L_{211}X_1^2 + L_{212}X_1X_2 + L_{222}X_2^2 \, .$$

What are the relations between the phenomenological coefficients L according to relation (1)? (c) Consider the sequence of reactions A $\rightleftarrows$ B $\rightleftarrows$ C. Using the mass action law, express J_1 and J_2 in terms of $\mathcal{A}_1 = \mu_A - \mu_B$ and $\mathcal{A}_2 = \mu_B - \mu_C$. Expand these relations up to second order and show that the reciprocal relations obtained in (b) are not satisfied. This result is important, since it shows that there are not reciprocal properties for non-linear expansions.

1.14 *Minimum entropy production.* A system is described by the two linear phenomenological laws $J_1 = L_{11}X_1 + L_{12}X_2$ and $J_2 = L_{21}X_1 + L_{22}X_2$. Assume that the thermodynamic force X_2 is kept fixed at a non-vanishing value. (a) Show that when the L_{ij} are constant and satisfy Onsager's reciprocal relations the entropy production $\sigma^s = \Sigma_{\alpha\beta} L_{\alpha\beta}X_\alpha X_\beta$ is minimum in the steady state, i.e. for a value of X_1 such that $J_1 = 0$. (b) The entropy production in a rigid heat conductor is found to be given by $\sigma^s = L(\nabla T^{-1})\cdot(\nabla T^{-1})$ with $L = \lambda T^2$. Keeping the temperature T fixed at the boundaries of the sample, show that, for constant L, the total entropy production, i.e. $P = \int \sigma^s dV$, is a minimum in the steady state $\nabla \cdot \boldsymbol{q} = 0$.

1.15 *Cycles with finite time.* In a Carnot engine, the working fluid is kept in contact with heat reservoirs at respective temperatures T_1 and T_2 during the isothermal parts of the cycle. Assume that the temperatures of the reservoirs are different from the temperature of the fluid, i.e. $T_1 > T_1'$ and $T_2' > T_2$, in such a way that the heat exchanged per unit time during these processes is given by $dQ_1/dt = \alpha(T_1 - T_1')$, $dQ_2/dt = \alpha(T_2' - T_2)$, with α being a constant which depends on the thermal conductivity of the wall separating the thermal reservoirs from the working fluid. (a) Determine the power developed by this Carnot engine, assuming that the total duration of each cycle is proportional to the sum of the duration of the isothermal branches. (b) For given values of T_1 and T_2, find the values of T_1' and T_2' which maximise the power. (c) Show that the efficiency at maximum power is

$$\eta_{maximum\,power} = 1 - \left(\frac{T_2}{T_1}\right)^{1/2}.$$

Compare this expression with the efficiency of a reversible Carnot cycle $\eta = 1 - (T_2/T_1)$. (Note that for a reversible Carnot engine the power is zero, because a cycle lasts an infinite time.) [See F. L. Curzon and B. Ahlborn, Am. J. Phys. **43** (1975) 22].

Chapter 2

Extended Irreversible Thermodynamics:
Evolution Equations

Our general purpose is to propose a theory which goes beyond the classical formulation of irreversible thermodynamics (CIT). This is achieved by enlarging the space of basic independent variables through the introduction of non-equilibrium variables, such as the dissipative fluxes appearing in the balance equations of mass, momentum and energy. The next step is to find evolution equations for these extra variables. Whereas the evolution equations for the classical variables are given by the usual balance laws, no general criteria exist concerning the evolution equations of the dissipative fluxes, with the exception of the restrictions imposed on them by the second law of thermodynamics.

The independent character of the fluxes is made evident in high-frequency phenomena. In general, they are *fast* variables that decay to their local-equilibrium values after a short relaxation time. Whereas many authors have studied the elimination of such *fast* variables in order to obtain a description of the system in terms of slow variables, our objective is the opposite one. We want to describe phenomena at frequencies comparable to the inverse of the relaxation times of the fluxes. Therefore, at such time scales, it is natural to include the fast variables among the set of basic independent variables.

A simple way to obtain the evolution equations for the fluxes from a macroscopic basis is to generalise the classical theories presented in the previous chapter. In that spirit, we assume the existence of a generalised entropy which depends on the dissipative fluxes and on the classical variables as well. A physical interpretation of the different contributions to the generalised entropy is proposed. Once this expression is known, it is an easy matter to derive generalised equations of state, which are of interest in the description of non-equilibrium steady states.

For pedagogical reasons, we shall first study the simple problem of heat transport in a rigid isotropic body when only the heat flux is introduced as an extra variable. Afterwards, we shall consider the more general case of a one-component isotropic fluid, where the heat flux, the bulk viscous pressure, and the viscous pressure tensor are taken as supplementary independent variables, on the same footing as the classical ones. After having obtained the evolution equations for the flux variables by methods which generalise those of CIT, we will show how to obtain them in the framework of rational extended thermodynamics. The respective merits of both approaches are also discussed.

2.1 Heat Conduction

Let us first outline the motivations for introducing fluxes as independent variables for the simple problem of heat conduction in undeformable bodies. Afterwards, we will take advantage of this simple situation to introduce the main tenets of the theory.

2.1.1 Motivation

The best known model for heat conduction in undeformable solids is Fourier's law, which relates linearly the temperature gradient ∇T to the heat flux $\boldsymbol{q}$ according to

$$\boldsymbol{q} = -\lambda \nabla T, \tag{2.1}$$

where λ is the heat conductivity, depending generally on the temperature. By substituting (2.1) in the energy balance equation (written in absence of source terms and in a system at rest)

$$\rho \frac{\partial u}{\partial t} = -\nabla \cdot \boldsymbol{q}, \tag{2.2}$$

where the specific internal energy u is related to the temperature by $du = c_v\, dT$, with c_v being the heat capacity per unit mass at constant volume, one obtains a parabolic differential equation for the temperature given by

$$\rho c_v \frac{\partial T}{\partial t} = \nabla \cdot (\lambda \nabla T). \tag{2.3}$$

This equation shows excellent agreement with experiments for most practical problems, but it suffers from some main deficiencies over short times or at high frequencies. In particular, Onsager [1.1] noted that Fourier's model contradicts the principle of microscopic reversibility, but this contradiction '... *is removed when we recognize that [Fourier's law] is only an approximate description of the process of conduction, neglecting the time needed for acceleration of the heat flow'*. In other words, Fourier's law has the unphysical property that it lacks inertial effects: if a sudden temperature perturbation is applied at one point in the solid, it will be felt instantaneously and everywhere at distant points. Moreover, Fourier's model is not adequate for describing heat transport at very high frequencies and short wavelengths. Such situations are met when the phenomena are very fast or very steep (as ultrasound propagation, light scattering in gases, neutron scattering in liquids, heat propagation at low temperatures, shock waves, etc.) or when the relaxation times of the fluxes are very long (as in polymer solutions, suspensions, superfluids or superconductors). Historically, Maxwell [2.1] was the first to introduce inertia in transport equations.

To eliminate these anomalies, Cattaneo [2.2] proposed in 1948 a damped version of Fourier's law by introducing a heat-flux relaxation term, namely,

$$\tau \frac{\partial \boldsymbol{q}}{\partial t} = -(\boldsymbol{q} + \lambda \nabla T).$$ (2.4)

When the relaxation time τ of the heat flux is negligible or when the time variation of the heat flux is slow, this equation reduces to Fourier's law.

Introduction of (2.4) into (2.2) results in a hyperbolic equation of the telegrapher type,

$$\tau \frac{\partial^2 T}{\partial t^2} + \frac{\partial T}{\partial t} - \chi \nabla^2 T = 0.$$ (2.5)

By establishing (2.5) it is assumed that τ and λ are constant and positive, while the quantity $\chi = \lambda/\rho c_v$ designates the heat diffusivity. The dynamical properties of this equation have been thoroughly analysed. However, the thermodynamic consequences are less known and are therefore worth examining.

Indeed, from the expression for the classical entropy production, namely, $\sigma^s = \boldsymbol{q} \cdot \nabla T^{-1}$, one obtains

$$\sigma^s = \frac{\lambda}{T^2} (\nabla T)^2 + \frac{\tau}{T^2} \frac{\partial \boldsymbol{q}}{\partial t} \cdot \nabla T,$$ (2.6)

which is no longer definite positive, because of the presence of the second term. To illustrate the problems raised by the local-equilibrium entropy, let us examine the time evolution of entropy in an isolated rigid body with an initial sinusoidal temperature profile by using (2.5); one obtains non-monotonic behaviour, as exhibited by the dashed curve shown in Fig. 2.1. The details of the calculations are found in the Appendix at the end of this chapter, where, in addition, an analysis of a discrete system composed of two rigid solids in contact at different temperatures is carried out. It is seen that instead of being monotonically increasing, the classical entropy behaves in an oscillatory way. Strictly speaking this result is not incompatible with the Clausius formulation of the second law, which states that the entropy of the final equilibrium state must be higher than the entropy of the initial equilibrium state. However, the non-monotonic behaviour of the entropy is in contradiction with the local-equilibrium formulation of the second law, which requires that the entropy production must be positive everywhere at any time in the evolution.

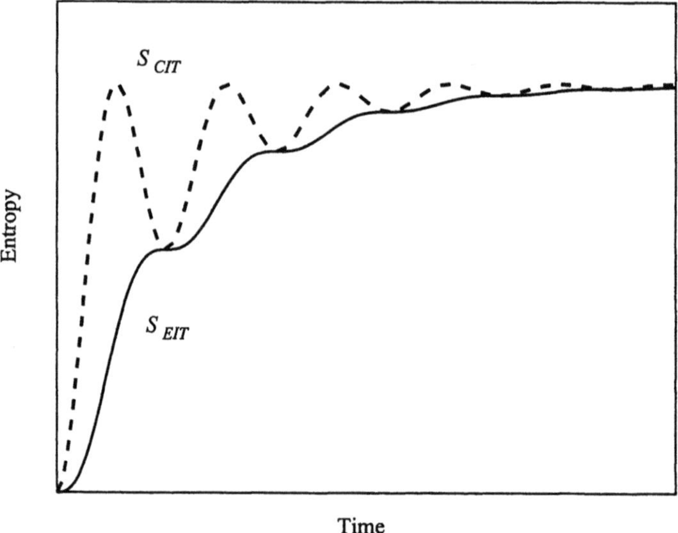

Fig. 2.1. The evolution of the classical entropy S_{CIT} during the equilibration of an isolated system described by the Maxwell–Cattaneo equation (2.4) is given by the dashed curve. The evolution of the extended entropy S_{EIT}, obtained from (2.32), is represented by the solid curve, which, in contrast with that of S_{CIT}, increases monotonically.

It is thus found that the Maxwell–Cattaneo law (2.4), which has been shown to be very applicable in the description of the results of many experiments on heat waves, is not compatible with the local-equilibrium hypothesis, which ought therefore to be reconsidered. This is precisely the aim of the Sect. 2.1.2.

2.1.2 The Generalised Gibbs Equation

As in CIT, the entropy and the Gibbs equation play a central role in extended irreversible thermodynamics (EIT). Here, it is assumed that the entropy will not only depend on the classical variable, namely the specific internal energy u, but in addition on the heat flux q, whose initial value must be specified in order to solve equation (2.4):

$$s = s(u, q).$$

The generalised entropy $s(u, q)$ will be assigned the following properties:
- it is an additive quantity;
- it is a convex function of the whole set of variables, which means that it is a function which lies everywhere below its family of tangent lines; and
- its rate of production is locally positive.

The hypothesis of a generalised macroscopic entropy depending on the fluxes was advanced by Machlup and Onsager [2.3] in an indirect way. During the 1960s a more direct formulation applied to fluids was developed by Nettleton [2.4] and Müller [2.5]. In the 1970s, new reformulations were proposed independently by several authors and have inspired much research [2.6–11], which has been reviewed in [2.12].

The differential form of the generalised entropy is written as follows:

$$\mathrm{d}s = \left(\frac{\partial s}{\partial u} \right) \mathrm{d}u + \left(\frac{\partial s}{\partial q} \right) \cdot \mathrm{d}q. \tag{2.7}$$

In analogy with the classical theory, we define the non-equilibrium temperature θ by

$$\theta^{-1}(u,q) = \left(\frac{\partial s}{\partial u} \right)_q. \tag{2.8}$$

The quantity θ^{-1} can be expanded around the inverse of the local-equilibrium temperature T and written as

$$\theta^{-1}(u,q) = T^{-1}(u) + a(u)q^2, \tag{2.9}$$

when terms of order higher than q^2 are neglected; the coefficient $a(u)$ depends generally on u, and q^2 stands for $q \cdot q$. The remaining partial derivative in (2.7) will be denoted as

$$\left(\frac{\partial s}{\partial q} \right) = -T^{-1}v\alpha_1(u,q), \tag{2.10}$$

wherein the minus sign and the factor $T^{-1}v$ are introduced for convenience. Still neglecting terms of order higher than q^2, α_1 takes the form

$$\alpha_1 = \alpha_{10}(u)q. \tag{2.11}$$

Substituting (2.8–10) into (2.7) results in the final expression of the generalised Gibbs equation when third-order terms in the heat flux are omitted:

$$ds = \theta^{-1}du - T^{-1}v\alpha_{10}q \cdot dq. \tag{2.12}$$

From this expression and the internal energy balance law (2.2), one obtains for the time derivative $\dot{s}$ of the entropy,

$$\rho\dot{s} = -\theta^{-1}\nabla \cdot q - T^{-1}\alpha_{10}q \cdot \dot{q}$$

or, equivalently,

$$\rho\dot{s} = -\nabla \cdot (\theta^{-1}q) + q \cdot (\nabla\theta^{-1} - T^{-1}\alpha_{10}q \cdot \dot{q}). \tag{2.13}$$

This equation can be cast in the general form of a balance equation,

$$\rho\dot{s} = -\nabla \cdot J^s + \sigma^s, \tag{2.14}$$

in order to identify the expressions for the entropy flux J^s and the (positive) entropy production σ^s. By comparison of (2.13) and (2.14), it can be inferred that

$$J^s = \theta^{-1}q \tag{2.15}$$

and

$$\sigma^s = q \cdot (\nabla\theta^{-1} - T^{-1}\alpha_{10}\dot{q}). \tag{2.16}$$

Relation (2.16) has the structure of a bilinear form in a flux q and a force X

$$\sigma^s = q \cdot X, \tag{2.17}$$

the force X may be identified with the quantity within parentheses in (2.16). To obtain an evolution equation for q compatible with the positiveness of σ^s, let us assume that the force X is linear in q, so that

$$X = \mu_1(u)q, \tag{2.18}$$

wherein μ_1 may depend on u but not on q. Replacing X by its value drawn from (2.16), relation (2.18) can be rewritten as

$$\nabla \theta^{-1} - T^{-1}\alpha_{10}\dot{q} = \mu_1 q, \tag{2.19}$$

by omitting, as previously, third-order contributions in q. Introducing (2.18) into (2.17) results in

$$\sigma^s = \mu_1 q \cdot q \geq 0, \tag{2.20}$$

and the requirement that σ^s is positive leads to the restriction $\mu_1 > 0$.

The relation (2.18) can be generalised to the non-linear domain, but this raises some questions. To be more explicit, let us consider as a simple example the non-linear law

$$q = \frac{X_1}{\mu_1\sqrt{1 + aX_1^2}}; \tag{2.21}$$

the corresponding entropy production is given by

$$\sigma^s = q \cdot X_1 = \frac{X_1^2}{\mu_1\sqrt{1 + aX_1^2}}. \tag{2.22}$$

The quantity σ^s is clearly positive for any value of X_1 provided that μ_1 and a are positive. However, if one expands the square root in series of X_1^2 one has

$$\sigma^s = q \cdot X_1 = \frac{X_1^2}{\mu_1}\left(1 - \frac{1}{2}aX_1^2 + \frac{1}{4}a^2X_1^4 + ...\right), \tag{2.23}$$

from which it can be seen that by truncating the expansion after a finite number of terms it may happen that σ^s could become negative. Therefore, if one deals with a series expansion, the condition $\sigma^s \geq 0$ only imposes restrictions on the sign of the quadratic term, but not on the fourth and higher-order terms in X_1.

In expression (2.19) there appear two non-defined coefficients α_{10} and μ_1 which must be identified on physical grounds. Assume first a situation characterized by a stationary flux so that (2.19) simplifies to

$$q = -\frac{1}{\mu_1 \theta^2}\nabla\theta \qquad\qquad (2.24a)$$

or, working at the lowest approximation, by identifying θ with T,

$$q = -\frac{1}{\mu_1 T^2}\nabla T. \qquad\qquad (2.24b)$$

A comparison with Fourier's law $q = -\lambda\nabla T$ yields

$$\mu_1 = \frac{1}{\lambda T^2}. \qquad\qquad (2.25)$$

In the next step, we compare (2.19) with Cattaneo's equation (2.4). Writing (2.19) in the form

$$\lambda T\alpha_{10}\dot{q} - q = -\lambda\nabla\theta, \qquad\qquad (2.26)$$

it is directly inferred that

$$\alpha_{10} = \frac{\tau}{\lambda T}. \qquad\qquad (2.27)$$

In virtue of this result, the generalised Gibbs equation (2.12) will be written as

$$ds = \theta^{-1}du - \frac{\tau}{\rho\lambda T^2}q\cdot dq, \qquad\qquad (2.28)$$

wherein the coefficient of the new term in dq has been completely identified in terms of physical quantities. When the contribution of the quadratic terms in q is negligible in the expression of θ^{-1}, the latter reduces to T^{-1} and the Gibbs equation (2.28) is given by

$$ds = T^{-1}du - \frac{\tau}{\rho\lambda T^2}q\cdot dq. \qquad\qquad (2.29)$$

In this case, the integrability condition, i.e. the equality of the second-order cross derivatives of s, namely $\partial^2 s/\partial u\partial q$ and $\partial^2 s/\partial q\partial u$, infers that $\tau/\rho\lambda T^2$ is a constant. Expression (2.29) was widely used in the first developments of EIT. In the local-equilibrium approximation, (2.28) reads simply as $ds = T^{-1}du$ as it should, and at this level of approximation θ is identical to T. The integrability condition of (2.28) yields in general

$$\frac{\partial \theta^{-1}}{\partial q} = -\frac{\partial}{\partial u}\left(\frac{\tau}{\rho \lambda T^2}\right)q, \tag{2.30}$$

and, after integration,

$$\theta^{-1}(u,q) = T^{-1}(u) - \frac{1}{2}\frac{\partial}{\partial u}\left(\frac{\tau}{\rho \lambda T^2}\right)q \cdot q. \tag{2.31}$$

After substitution of this result in (2.28) and integration, one obtains an explicit expression for entropy outside (local) equilibrium up to second-order terms in q

$$\rho s(u,q) = \rho s_{eq}(u) - \frac{1}{2}\frac{\tau}{\lambda T^2}q \cdot q. \tag{2.32}$$

In virtue of (2.25), the corresponding entropy production (2.20) is

$$\sigma^s = \frac{1}{\lambda T^2}q \cdot q, \tag{2.33}$$

and its positiveness implies that $\lambda > 0$. This expression of σ^s is to be compared with σ^s obtained from CIT when Cattaneo's law is used instead of Fourier's, namely

$$\sigma^s_{CIT} = \frac{1}{\lambda T^2}(q \cdot q - \tau q \cdot \dot{q}). \tag{2.34}$$

Clearly, the positiveness of the entropy production is no longer guaranteed due to the presence of the second term on the right-hand side of (2.34).

Going back to Fig. 2.1 it is seen that the extended entropy (2.32) has a monotonic increase, in contrast with the classical local-equilibrium entropy s_{eq}. It can thus be concluded that the Cattaneo law (2.4) is not compatible with the local-equilibrium version of the second law, but is consistent with EIT.

2.2 One-Component Viscous Fluid

We now proceed to the more general case of a compressible one-component fluid. According to EIT, the space of the thermodynamic state variables is the union of the classical one (internal energy u, specific volume v) and the space of flux variables q, p^v, and $\overset{0}{\mathbf{P}}{}^{v}$.

The generalised Gibbs equation takes the form

$$ds = \left(\frac{\partial s}{\partial u}\right)_{v,q,p^v,\overset{0}{\mathbf{P}}{}^v} du + \left(\frac{\partial s}{\partial v}\right)_{u,q,p^v,\overset{0}{\mathbf{P}}{}^v} dv + \left(\frac{\partial s}{\partial q}\right)_{u,v,p^v,\overset{0}{\mathbf{P}}{}^v} \cdot dq$$

$$+ \left(\frac{\partial s}{\partial p^v}\right)_{u,v,q,\overset{0}{\mathbf{P}}{}^v} dp^v + \left(\frac{\partial s}{\partial \overset{0}{\mathbf{P}}{}^v}\right)_{u,v,q,p^v} : d\overset{0}{\mathbf{P}}{}^{v}. \tag{2.35}$$

In analogy with the classical theory of irreversible processes, we define a non-equilibrium absolute temperature θ and a non-equilibrium thermodynamic pressure π by respectively

$$\theta^{-1}(u,v,q,p^v,\overset{0}{\mathbf{P}}{}^v) = (\partial s / \partial u)_{v,q,p^v,\overset{0}{\mathbf{P}}{}^v}\,,$$

$$\tag{2.36}$$

$$\theta^{-1}\pi(u,v,q,p^v,\overset{0}{\mathbf{P}}{}^v) = (\partial s / \partial v)_{u,q,p^v,\overset{0}{\mathbf{P}}{}^v}\,.$$

The remaining partial derivatives in (2.35) are assumed to be linear in the fluxes and are denoted as

$$(\partial s / \partial q)_{u,v,p^v,\overset{0}{\mathbf{P}}{}^v} = -T^{-1}v\alpha_{10}(u,v)q,$$

$$(\partial s / \partial p^v)_{u,v,q,\overset{0}{\mathbf{P}}{}^v} = -T^{-1}v\alpha_{00}(u,v)p^v, \tag{2.37}$$

$$(\partial s / \partial \overset{0}{\mathbf{P}}{}^v)_{u,v,q,p^v} = -T^{-1}v\alpha_{21}(u,v)\overset{0}{\mathbf{P}}{}^{v},$$

wherein the coefficients α_{10}, α_{00} and α_{21} are unknown scalar functions of u and v. Introducing (2.36–37) into (2.35) yields the generalised Gibbs equation

$$ds = \theta^{-1}du + \theta^{-1}\pi dv - T^{-1}v\alpha_{00}p^v dp^v - T^{-1}v\alpha_{10}q \cdot dq - T^{-1}v\alpha_{21}\overset{0}{\mathbf{P}}{}^v : d\overset{0}{\mathbf{P}}{}^{v}. \tag{2.38}$$

The evolution of entropy is governed by the balance law (2.14), and our objective is to determine the corresponding expressions for the entropy flux J^s and entropy production

σ^s. To this end, let us substitute into (2.38) the expressions of $\dot{u}$ and $\dot{v}$ as derived from the balance laws of energy and mass (1.10–11). After direct calculations it is found that

$$\rho\dot{s} = -\theta^{-1}\nabla\cdot\boldsymbol{q} - T^{-1}p^v\nabla\cdot\boldsymbol{v} - T^{-1}\overset{0}{\mathbf{P}}{}^v : \overset{0}{\mathbf{V}} - T^{-1}\alpha_{00}p^v\dot{p}^v$$

$$-T^{-1}\alpha_{10}\boldsymbol{q}\cdot\dot{\boldsymbol{q}} - T^{-1}\alpha_{21}\overset{0}{\mathbf{P}}{}^v : (\overset{0}{\mathbf{P}}{}^v)^{\cdot}. \tag{2.39}$$

To derive this result it was assumed that the total pressure tensor $\mathbf{P}$ was decomposed as follows

$$\mathbf{P} = \pi\mathbf{U} + p^v\mathbf{U} + \overset{0}{\mathbf{P}}{}^v, \tag{2.40}$$

where π in the first term on the right-hand side has replaced p as found usually in classical hydrodynamics and in Chap. 1. It should be realized that at equilibrium π reduces to the classical hydrostatic equilibrium pressure p and that when second-order contributions in the fluxes are negligible π is still identical to p. The motivation for substituting p with π in (2.40) will be discussed in detail in Chap. 3; for the moment, we may accept it is as analogous to changing T with θ in the generalised Gibbs equation (2.38).

2.3 The Generalised Entropy Flux and Entropy Production

Before we determine the expression for entropy production, defined by

$$\sigma^s = \rho\dot{s} + \nabla\cdot\boldsymbol{J}^s \geq 0, \tag{2.41}$$

we need a relation for the entropy flux $\boldsymbol{J}^s$. For isotropic systems, the most general vector depending on the variables u, v, $\boldsymbol{q}$, $\overset{0}{\mathbf{P}}{}^v$, and p^v is, up to second order terms in the fluxes,

$$\boldsymbol{J}^s = \beta\boldsymbol{q} + \beta'p^v\boldsymbol{q} + \beta''\overset{0}{\mathbf{P}}{}^v\cdot\boldsymbol{q}, \tag{2.42}$$

where the coefficients β' and β'' are generally functions of u and v; to recover the results of the heat conduction problem, the coefficient β must be made θ^{-1}. Accordingly

$$\boldsymbol{J}^s = \theta^{-1}\boldsymbol{q} + \beta'p^v\boldsymbol{q} + \beta''\overset{0}{\mathbf{P}}{}^v\cdot\boldsymbol{q}. \tag{2.43}$$

The entropy production is easily derived from (2.41) by replacing $\rho\dot{s}$ and $\boldsymbol{J}^s$ respectively by their expressions in (2.39) and (2.42). The final result is

$$\sigma^s = \boldsymbol{q} \cdot (\nabla\theta^{-1} + \beta''\nabla \cdot \overset{0}{\dot{\mathbf{P}}}{}^{\nu} + \beta'\nabla p^{\nu} - T^{-1}\alpha_{10}\dot{\boldsymbol{q}})$$

$$+ p^{\nu}[-T^{-1}\nabla \cdot \boldsymbol{v} - T^{-1}\alpha_{00}\dot{p}^{\nu} + \nabla \cdot (\beta'\boldsymbol{q})]$$

$$+ \overset{0}{\mathbf{P}}{}^{\nu} : \{-T^{-1}\overset{0}{\mathbf{V}} - T^{-1}\alpha_{21}(\overset{0}{\mathbf{P}}{}^{\nu})^{\cdot} + [\nabla(\beta''\boldsymbol{q})^s]\}. \tag{2.44}$$

One observes that (2.44) has the structure of a bilinear form,

$$\sigma^s = \boldsymbol{q} \cdot \boldsymbol{X}_1 + p^{\nu}X_0 + \overset{0}{\mathbf{P}}{}^{\nu} : \overset{0}{\mathbf{X}}_2, \tag{2.45}$$

consisting of a sum of products of the fluxes $\boldsymbol{q}$, p^{ν}, and $\overset{0}{\mathbf{P}}{}^{\nu}$ and their conjugate generalised forces $\boldsymbol{X}_1, X_0$, and $\overset{0}{\mathbf{X}}_2$. The latter follows from direct comparison of (2.45) with (2.44). They are similar to the expressions obtained in CIT, except for the substitution of T by θ and the fact that they contain additional terms depending on the time and space derivatives of the fluxes.

Upon defining the proper form of the forces $\boldsymbol{X}_1$, $\overset{0}{\mathbf{X}}_2$, and X_0, it can be noted that there exists a class of transformations of the time derivatives of $\boldsymbol{q}$ and $\overset{0}{\mathbf{P}}{}^{\nu}$ which leave the entropy production invariant. An example of such a transformation is provided by

$$\mathbf{D}_a\boldsymbol{q} = \dot{\boldsymbol{q}} + a\mathbf{W} \cdot \boldsymbol{q},$$

$$\tag{2.46}$$

$$\mathbf{D}_b\overset{0}{\mathbf{P}}{}^{\nu} = (\overset{0}{\mathbf{P}}{}^{\nu})^{\cdot} + b(\mathbf{W} \cdot \overset{0}{\mathbf{P}}{}^{\nu} - \overset{0}{\mathbf{P}}{}^{\nu} \cdot \mathbf{W}),$$

where a and b are constants and $\mathbf{W}$ is an antisymmetric tensor, for instance, the antisymmetric part of the velocity gradient. Indeed, it is easy to verify that

$$\boldsymbol{q} \cdot \mathbf{D}_a\boldsymbol{q} = \boldsymbol{q} \cdot \dot{\boldsymbol{q}}, \qquad \overset{0}{\mathbf{P}}{}^{\nu} : \mathbf{D}_b\overset{0}{\mathbf{P}}{}^{\nu} = \overset{0}{\mathbf{P}}{}^{\nu} : (\overset{0}{\mathbf{P}}{}^{\nu})^{\cdot}. \tag{2.47}$$

This means that the expression of the entropy production remains unchanged when general derivatives of the form (2.46) are used instead of the material time derivatives of the fluxes.

Obviously, thermodynamics cannot give any information about the coefficients a and b in (2.46), since they do not appear explicitly either in the entropy production or in the Gibbs equation. However, they can be determined by other means: (a) From general invariance requirements, such as the frame-indifference principle, which leads to $a = b = 1$, and in this case the derivatives defined by (2.46) coincide with the corotational time derivative. (b) From a microscopic description, e.g. the kinetic theory of gases; a comparison with kinetic theory yields $a = b = -1$ (for a detailed discussion on the compatibility between kinetic theory and frame indifference see [1.32]). (c) From experiments on rotating systems, such as those discussed in Chap. 10; such experiments turn out to be very difficult in view of the smallness of the terms involved.

In order to obtain evolution equations for the fluxes compatible with the positiveness of σ^s, we express the forces X_1, X_0, and $\overset{0}{X}_2$ as functions of the fluxes. Limiting as above the development to the second order in the fluxes, one has

$$X_1 = \mu_1 \boldsymbol{q} + \mu_{10} p^v \boldsymbol{q} + \mu_{12} \overset{0}{\mathbf{P}}{}^v \cdot \boldsymbol{q} , \tag{2.48a}$$

$$X_0 = \mu_0 p^v + \mu_{00} p^{v2} + \mu_{01} \boldsymbol{q} \cdot \boldsymbol{q} + \mu_{02} \overset{0}{\mathbf{P}}{}^v : \overset{0}{\mathbf{P}}{}^v , \tag{2.48b}$$

$$\mathbf{X}_2 = \mu_2 \mathbf{P}^v + \mu_{20} p^v \mathbf{P}^v + \mu_{22} \overset{0}{\mathbf{P}}{}^v \cdot \overset{0}{\mathbf{P}}{}^v + \mu_{21} \boldsymbol{q} \boldsymbol{q} , \tag{2.48c}$$

where all the coefficients μ may depend on u and v.

Later on, we will examine some particular situations where the non-linear terms are relevant. But from now on, we shall limit our analysis to linear flux–force relationships in order to avoid cumbersome and unduly lengthy mathematical expressions; at this approximation, one has simply $X_1 = \mu_1 \boldsymbol{q}$, $X_0 = \mu_0 p^v$, and $\overset{0}{X}_2 = \mu_2 \overset{0}{\mathbf{P}}{}^v$. These are the simplest flux–force relations ensuring the positiveness of σ^s. Indeed when these expressions are introduced into (2.45), we are led to

$$\sigma^s = \mu_1 \boldsymbol{q} \cdot \boldsymbol{q} + \mu_0 p^v p^v + \mu_2 \overset{0}{\mathbf{P}}{}^v : \overset{0}{\mathbf{P}}{}^v , \tag{2.49}$$

while the requirement that σ^s must be positive leads to the restrictions

$$\mu_1 \geq 0, \qquad \mu_0 \geq 0, \qquad \mu_2 \geq 0. \tag{2.50}$$

As in the problem of heat conduction, inclusion of non-linear terms raises some important conceptual questions concerning the interpretation of the second law. Consider, for instance, a force given by the form $X_0 = \mu_0 p^v + \mu_{00}(p^v)^2 + ...$, such that the entropy production becomes $\sigma^s = \mu_0 (p^v)^2 + \mu_{00}(p^v)^3 + ...$ It is clear that the positiveness of σ^s implies that the coefficient of $(p^v)^2$ must be positive, i.e. $\mu_0 > 0$. However, the restrictions on the coefficients of $(p^v)^3$ and higher-order terms depend on the interpretation given to the status of the constitutive equations. If one considers a given material which satisfies exactly $X_0 = \mu_0(p^v)^2 + \mu_{00}(p^v)^3$, then the positiveness of σ^s for every value of p^v would require that $\mu_{00} = 0$. By assuming that $X_0 = \mu_0 p^v + \mu_{00}(p^v)^2 + \ldots$ is only a second-order approximation with respect to an unknown exact constitutive relation, then one cannot conclude anything about the sign of μ_{00}. Nevertheless, the requirement that σ^s must be positive provides a useful limitation on the domain of validity of the constitutive equations and on their possible forms.

2.4 Linearized Evolution Equations of the Fluxes

Identifying the forces as the conjugate terms of the fluxes in (2.44) and substituting these expressions into (2.48), one obtains in the linear approximation (products such as $q \cdot \nabla u$, $q.\nabla v$, $(\nabla u) \cdot \overset{0}{\mathbf{P}}{}^v$, $(\nabla v) \cdot \overset{0}{\mathbf{P}}{}^v$ are omitted, and θ is identified with T) the following set of evolution equations:

$$\nabla T^{-1} - T^{-1}\alpha_{10}\dot{q} = \mu_1 q - \beta''\nabla \cdot \overset{0}{\mathbf{P}}{}^v - \beta'\nabla p^v, \tag{2.51}$$

$$-T^{-1}\nabla \cdot v - T^{-1}\alpha_{00}\dot{p}^v = \mu_0 p^v - \beta'\nabla \cdot q, \tag{2.52}$$

$$-T^{-1}\overset{0}{\mathbf{V}} - T^{-1}\alpha_{21}(\overset{0}{\mathbf{P}}{}^v)^{\cdot} = \mu_2 \overset{0}{\mathbf{P}}{}^v - \beta''(\overset{0}{\nabla q})^s. \tag{2.53}$$

The main features issued from the above thermodynamic formalism are:
- The positiveness of the coefficients μ_1, μ_0, and μ_2.
- The equality of the cross terms relating q with $\nabla \cdot \overset{0}{\mathbf{P}}{}^v$ and $\overset{0}{\mathbf{P}}{}^v$ with $(\overset{0}{\nabla q})^s$ on the one side, and q with ∇p^v and p^v with $\nabla \cdot q$ on the other. The equality of these coefficients, confirmed by the kinetic theory, belongs to a class of higher-order Onsager relations. However, in contrast with the usual Onsager relations, they have been obtained here from purely thermodynamic arguments.

- The coefficients β' and β'' appearing in the second-order terms of the entropy flux are the same as the coefficients of the cross terms in the evolution equations (2.19–21). This result is also confirmed by kinetic theory, as shown in Chap. 5.

As in the heat conduction problem, in (2.51–53) we have introduced several coefficients, which must receive a physical identification. Consider first a stationary and homogeneous situation, what means that the time and space derivatives of the fluxes are zero. Equations (2.51–53) then reduce to

$$\nabla T^{-1} = \mu_1 q, \qquad -T^{-1}\nabla \cdot v = \mu_0 p^v, \qquad -T^{-1}\overset{0}{\mathbf{V}} = \mu_2 \overset{0}{\mathbf{P}}{}^v. \tag{2.54}$$

Comparison with the Fourier and Newton–Stokes laws,

$$q = -\lambda \nabla T, \quad p^v = -\zeta \nabla \cdot v, \quad \overset{0}{\mathbf{P}}{}^v = -2\eta \overset{0}{\mathbf{V}}, \tag{2.55}$$

leads to the identifications

$$\mu_1 = (\lambda T^2)^{-1}, \quad \mu_0 = (\zeta T)^{-1}, \quad \mu_2 = (2\eta T)^{-1}, \tag{2.56}$$

with λ, ζ, and η being the thermal conductivity, bulk viscosity, and shear viscosity, respectively.

Consider now a non-stationary (but homogeneous) flow, so that (2.51–53) reduce to

$$\nabla T^{-1} - T^{-1}\alpha_{10}\dot{q} = (\lambda T^2)^{-1}q, \tag{2.57}$$

$$-T^{-1}\nabla \cdot v - T^{-1}\alpha_{00}\dot{p}^v = (\zeta T)^{-1}p^v, \tag{2.58}$$

$$-T^{-1}\overset{0}{\mathbf{V}} - T^{-1}\alpha_{21}(\overset{0}{\mathbf{P}}{}^v)^{\cdot} = (2\eta T)^{-1}\overset{0}{\mathbf{P}}{}^v. \tag{2.59}$$

These equations can be identified with the so-called Maxwell–Cattaneo laws [2.11–12]

$$\tau_1\dot{q} + q = -\lambda\nabla T, \tag{2.60}$$

$$\tau_0\dot{p}^v + p^v = -\zeta\nabla \cdot v, \tag{2.61}$$

$$\tau_2(\overset{0}{\mathbf{P}}{}^v)^{\cdot} + \overset{0}{\mathbf{P}}{}^v = -2\eta\overset{0}{\mathbf{V}}, \tag{2.62}$$

where τ_1, τ_0, and τ_2 are the relaxation times of the respective fluxes. We are then led to the identifications

$$\alpha_{10} = \tau_1(\lambda T)^{-1}, \quad \alpha_{00} = \tau_0\zeta^{-1}, \quad \alpha_{21} = \tau_2(2\eta)^{-1}. \tag{2.63}$$

In terms of λ, ζ, η, and the relaxation times τ_1, τ_0, and τ_2, the linearized evolution equations (2.51–53) take the following form:

$$\tau_1\dot{q} = -(q + \lambda\nabla T) + \beta''\lambda T^2\nabla \cdot \overset{0}{\mathbf{P}}{}^v + \beta'\lambda T^2\nabla p^v, \tag{2.64}$$

$$\tau_0 \dot{p}^v = -(p^v + \zeta \nabla \cdot v) + \beta' \zeta T \nabla \cdot q, \qquad (2.65)$$

$$\tau_2 (\overset{0}{\mathbf{P}}{}^v)^\cdot = -(\overset{0}{\mathbf{P}}{}^v + 2\eta \overset{0}{\mathbf{V}}) + 2\beta'' \eta T (\overset{0}{\nabla q})^s. \qquad (2.66)$$

In Table 2.1 the values of some of the coefficients appearing in (2.64–66) are reported.

Table 2.1. Values of τ_1, τ_2, λ, and η for some liquids at 20°C and 1 atm, according to Nettleton [2.4b] (first three columns) and international tables

Liquid	$10^{13}\tau_1$ (s)	$10\,\lambda$ (J/m s °C)	$10^{12}\tau_2$ (s)	$10^4\eta$ (N s/m^2)
Carbon tetrachloride	2.15	1.03	2.46	9.69
Chloroform	1.54	1.16	2.08	5.80
Carbon disulphide	1.43	1.61	1.38	3.63
Benzene	1.22	1.48	1.67	6.52
Toluene	1.63	1.35	1.60	5.90
Acetone	1.36	1.61	2.19	3.20

When relaxation times tend to infinity but their ratio to the respective transport coefficients (λ, ζ, η, etc.) remains finite (or in the high-frequency regime $\tau\omega \gg 1$ see Sect. 13.3), the Maxwell–Cattaneo equations become reversible (or time-reversal invariant) because the term involving the time derivative of the flux is more important than the flux itself. In these particular circumstances, there is no dissipation associated with the fluxes (e.g. electric current in superconductors). Nevertheless, for the sake of simplicity, we have kept dissipative fluxes as a generic term throughout this book.

2.5 Rational Extended Thermodynamics

EIT can be seen not only as an extension of CIT, but it may also be formulated along the line of thought of rational thermodynamics (RT); in this case, it is convenient to speak about rational extended thermodynamics (RET). The formalism of RT, the main objective of which is to provide a method for deriving constitutive equations, was essentially developed by Coleman, Truesdell and Noll during the 1960s [1.8] and offers an approach whose rationale is drastically different from the CIT approach. It is interesting to consider EIT from both perspectives of classical and rational irreversible thermodynamics because it provides a common ground for comparison in spite of the divergences of their original formulations.

Among the basic hypotheses underlying RT we can outline the following features: (a) absolute temperature and entropy are considered as primitive concepts, whose validity is not restricted to near-equilibrium situations; (b) it is assumed that systems have memory, i.e. their behaviour at a given instant of time is determined not only by the values of the variables at the present time, but also by their past history; (c) the second law of thermodynamics, which serves fundamentally as a restriction on the form of the constitutive equations, is expressed in mathematical terms by means of the so-called Clausius–Duhem inequality.

Although we borrow in this section some methods and concepts from RT, we depart from it in many other aspects. Essentially, the choice of the independent variables is different: in RT the variables are the histories of the classical state variables u, v, ..., while in RET the space of independent variables is enlarged in such a way that the history is no longer needed; accordingly, the response of the material system is described by evolution differential equations for the additional independent variables rather than in terms of the constitutive functionals as used in RT. Here, RT is used as a working method rather than as a theory.

2.5.1 Heat Conduction

To illustrate the RET approach, let us go back to the problem of heat conduction in a rigid isotropic body at rest. The relevant variables are the internal energy u and the heat flux q. The time evolution of u is governed by the balance law of energy (2.1) while the evolution equation of q will be cast in the general form

$$\rho \dot{q} = -\nabla \cdot \mathbf{Q} + \sigma^q. \tag{2.67}$$

The first term on the right-hand side of (2.67) expresses the exchange with the outside environment, with $\mathbf{Q}$ denoting the flux of the heat flux, while σ^q is a source term. The quantity $\mathbf{Q}$ is a tensor of second rank and σ^q is a vector. At this stage of the analysis, these quantities are unknowns and must be formulated by means of constitutive equations:

$$\mathbf{Q} = \mathbf{Q}(u,q), \tag{2.68}$$

$$\sigma^q = \sigma^q(u,q). \tag{2.69}$$

After substitution of (2.68–69) into (2.67), we are faced with a set of four scalar equations for the four unknowns u (scalar) and q (vector).

The evolution equations (2.1, 2.67) and the constitutive equations (2.68–69) cannot take an arbitrary form. They have to comply with the laws of thermodynamics, and in particular the second law. To satisfy the latter condition, it is assumed that there exists a regular and continuous function of the whole set of variables, called the non-equilibrium entropy s, which obeys a balance equation given by

$$\rho \dot{s} + \nabla \cdot \boldsymbol{J}^s = \sigma^s \geq 0, \tag{2.70}$$

where $\boldsymbol{J}^s$ is the entropy flux; the positive entropy production σ^s can be calculated by performing the operations indicated on the left-hand side of (2.70). As in RT, the positiveness of σ^s is used to place restrictions on the field equations.

At this point, let us emphasize some of the main differences between RET and RT. First, whereas in the latter theory the quantity q is given by a constitutive relation, in RET it is counted among the set of independent variables. Second, the balance law of energy is not regarded as a mere definition of the energy supply; in RET this quantity is given a priori. Third, the second law is not in the form of the Clausius–Duhem inequality as it is not imposed that the entropy flux is a priori to be given by the ratio of the heat flux and the temperature, but may contain extra terms. Fourth, the entropy is assumed to depend on the heat flux.

To take into account the restrictions placed by the second law (2.70) on the constitutive equations, we follow the method of Lagrange multipliers proposed by Liu and widely used by Müller and Ruggeri [2.5] in their formulation of RET. It must be observed that the inequality (2.70) does not hold for all the set of variables u and q but only for the solutions of the energy balance (2.1) and the evolution equation (2.67) for q. This means that we can consider the evolution equations as constraints for the entropy inequality to hold. To take these constraints into account is a difficult task and will lead to rather intricate calculations. An elegant way to circumvent this difficulty was proposed by Liu; he was able to show that the entropy inequality becomes completely arbitrary when inequality (2.70) is complemented with a linear combination of the balance laws. The factors multiplying the balance equations are called Lagrange multipliers by analogy with the extremization problem to constraints, although the present situation is not strictly a problem of extremals.

To be explicit, the entropy inequality will be formulated in such a way that the constraints imposed by the balance equations (2.1) and (2.67) are explicitly introduced via the Lagrange multipliers $\Lambda_0(u,q)$ (a scalar) and $\Lambda_1(u,q)$ (a vector), so that the inequality (2.70) will take the form

$$\rho \dot{s} + \nabla \cdot \boldsymbol{J}^s - \Lambda_0(\rho \dot{u} + \nabla \cdot \boldsymbol{q}) - \Lambda_1 \cdot (\rho \dot{q} + \nabla \cdot \boldsymbol{Q} - \sigma^q) \geq 0. \tag{2.71}$$

By differentiating s and J^s with respect to u and q, and rearranging the various terms one obtains the following from (2.71):

$$\left(\frac{\partial s}{\partial u} - \Lambda_0\right)\rho\dot{u} + \left(\frac{\partial s}{\partial q} - \Lambda_1\right)\cdot\rho\dot{q} + \frac{\partial J^s}{\partial u}\cdot\nabla u + \frac{\partial J^s}{\partial q}:\nabla q$$

$$-\Lambda_0\nabla\cdot q - \Lambda_1\cdot(\nabla\cdot\mathbf{Q}) + \Lambda_1\cdot\sigma^q \geq 0. \tag{2.72}$$

As it is wished to work at the same order of approximation as in Sect. 2.1, it is assumed that $\mathbf{Q}$ and σ^q will be of the form

$$\mathbf{Q} = \left[a_1(u) + a_2(u)q^2\right]\mathbf{U}, \tag{2.73}$$

$$\sigma^q = -b_1(u)q - 2b_2(u)q\cdot\nabla q, \tag{2.74}$$

where $a_1(u)$, $a_2(u)$, $b_1(u)$, and $b_2(u)$ are unknown fuctions of u. In the expression of $\mathbf{Q}$, we have not included the dyadic product qq because such a term will give rise to additional contributions in the expression of the entropy flux and the Cattaneo law [2.13]. Concerning σ^q, we have neglected terms in q^2q as they will contribute to an order of approximation higher than these introduced in Sect. 2.1.

Since inequality (2.72) is linear in $\dot{u}$ and $\dot{q}$, and since $\dot{u}$ and $\dot{q}$ are arbitrary and independent, the positiveness of this inequality requires that the first two parentheses on the left-hand side vanish, i.e.

$$\frac{\partial s}{\partial u} = \Lambda_0(\equiv \theta^{-1}), \tag{2.75}$$

$$\frac{\partial s}{\partial q} = \Lambda_1, \tag{2.76}$$

where we have identified Λ_0 with θ^{-1} as it has the dimension of the inverse of temperature, while at equilibrium, $\partial s/\partial u$ is the inverse of the equilibrium temperature.

After substitution of (2.73–74) in the remaining terms of (2.72) one is led to

$$\left(\frac{\partial J^s}{\partial u} - (a_1' + a_2'q^2)\Lambda_1\right)\cdot\nabla u + \left(\frac{\partial J^s}{\partial q} - \theta^{-1}U - 2(a_2 + b_2)\Lambda_1 q\right):\nabla q - b_1\Lambda_1\cdot q \geq 0, \tag{2.77}$$

where a prime designates differentiation with respect to u. Since (2.77) is linear in the arbitrary gradients ∇u and ∇q, it is inferred that the quantities within parentheses must vanish, i.e.

$$\frac{\partial J^s}{\partial u} = (a'_1 + a'_2 q^2)\Lambda_1, \tag{2.78}$$

$$\frac{\partial J^s}{\partial q} = \theta^{-1}U + 2(a_2 + b_2)\Lambda_1 q. \tag{2.79}$$

Note that the last term on the left-hand side of inequality (2.77) is all that remains, so we can write

$$\sigma^s = -b_1\Lambda_1 \cdot q \geq 0. \tag{2.80}$$

This is as far as we may go without further hypotheses on the form of J^s and Λ_1. To be explicit, let us write for these two quantities the following constitutive equations

$$J^s = \left[\phi_1(u) + \phi_2(u)q^2\right]q, \tag{2.81}$$

$$\Lambda_1 = \left[f_1(u) + f_2(u)q^2\right]q, \tag{2.82}$$

where, as above, the coefficients ϕ_1, ϕ_2, f_1, and f_2 are unknown functions of u. Substituting (2.81–82) into (2.79) results in

$$\left(\theta^{-1} - \phi_1 - \phi_2 q^2\right)U + \left[2(a_2 + b_2)(f_1 + f_2 q^2) - 2\phi_2\right]qq = 0, \tag{2.83}$$

from which it follows that

$$\theta^{-1} = \phi_1 + \phi_2 q^2, \tag{2.84}$$

$$\phi_2 = (a_2 + b_2)(f_1 + f_2 q^2). \tag{2.85}$$

A simple way to meet the requirement that ϕ_2 is independent of q, but non-zero, is to make

$$f_2 = 0, \tag{2.86}$$

so that (2.85) reduces to

$$\phi_2 = (a_2 + b_2)f_1. \tag{2.87}$$

As a consequence of (2.84), expression (2.81) of the entropy flux can be written as

$$J^s = \theta^{-1} q, \tag{2.88}$$

in agreement with (2.15).

What remains is to introduce the relation (2.82) into (2.78), and by comparing with the derivative obtained from (2.81) it is found that

$$(\phi_1' - f_1 a_1') + (\phi_2' - f_1 a_2')q^2 = 0,$$

which gives

$$\phi_1' = f_1 a_1', \qquad \phi_2' = f_1 a_2'. \tag{2.89}$$

Let us now formulate the Gibbs equation, which in general may be written as

$$ds = \frac{\partial s}{\partial u} du + \frac{\partial s}{\partial q} \cdot dq. \tag{2.90}$$

In view of (2.75), (2.76), (2.82), and (2.86), it is found that

$$ds = \theta^{-1} du + f_1 q \cdot dq. \tag{2.91}$$

The equality of second-order mixed derivatives yields

$$\frac{\partial \theta^{-1}}{\partial q} = f_1' q, \tag{2.92}$$

and, after integration,

$$\theta^{-1} = \tfrac{1}{2} f_1' q^2 + T^{-1}(u), \tag{2.93}$$

where $T(u)$ is defined as the local-equilibrium temperature as it corresponds to $q = 0$. Comparing (2.84) and (2.93) results in

$$\phi_1 = T^{-1}, \qquad \phi_2 = \tfrac{1}{2} f_1'. \tag{2.94}$$

It is interesting to notice that J^s reduces to its classical expression $T^{-1} q$ when f_1 is a constant.

Let us finally go back to the evolution equation (2.67) for q. With the constitutive equations (2.73–74), it reads as

$$\rho \dot{q} = -\left(a_1' + a_2' q^2\right)\nabla u - 2(a_2 + b_2)q \cdot \nabla q - b_1 q. \tag{2.95}$$

On the other hand, in virtue of (2.84) the gradient of θ^{-1} is given by

$$\nabla \theta^{-1} = \left(\phi_1' + \phi_2' q^2\right)\nabla u + 2\phi_2 q \cdot \nabla q. \tag{2.96}$$

Replacing ϕ_2, ϕ_1', and ϕ_2' by their expressions (2.87) and (2.89) respectively, one can write $\nabla \theta^{-1}$ as follows

$$\nabla \theta^{-1} = \left(a_1' + a_2' q^2\right)f_1 \nabla u + 2(a_2 + b_2)f_1 q \cdot \nabla q, \tag{2.97}$$

so that (2.95) takes the more familiar form

$$\frac{\rho}{b_1} \dot{q} = -\frac{1}{f_1 b_1} \nabla \theta^{-1} - q. \tag{2.98}$$

This expression is clearly identical to the generalised Cattaneo equation on condition that the following identifications are made

$$\frac{\rho}{b_1} = \tau, \qquad -\frac{1}{f_1 b_1} = \lambda, \tag{2.99}$$

with τ being the relaxation time and λ the heat conductivity. The results (2.99) are important as they allow the Lagrange multiplier Λ_1 to be identified by eliminating f_1 between (2.82) and (2.99), namely

$$\Lambda_1 = -\frac{\tau}{\rho \lambda T^2} q. \tag{2.100}$$

Finally, by substituting into inequality (2.80) the quantity $f_1 b_1$ given by (2.99), it is found that the entropy production takes the form

$$\sigma^s = \frac{1}{\lambda T^2} q \cdot q \geq 0, \tag{2.101}$$

and thus $\lambda \geq 0$.

Summarizing, we have obtained the following expressions for the Gibbs equation, the entropy flux, and the evolution equation respectively:

$$ds = \theta^{-1}du - \frac{\tau}{\rho\lambda T^2}\boldsymbol{q}\cdot d\boldsymbol{q},$$
(2.102)

$$\boldsymbol{J}^s = \theta^{-1}\boldsymbol{q},$$
(2.103)

$$\tau\dot{\boldsymbol{q}} = -\lambda\nabla\theta - \boldsymbol{q}.$$
(2.104)

These results comply with those derived by following the procedure of Sect. 2.1. Clearly, by using more intricate expressions for $\mathbf{Q}$, σ^q and Λ_1, one would obtain more complicated expressions for the Gibbs equation, the entropy flow and the generalised Cattaneo equation [2.13].

2.5.2 Viscous Fluids

The above procedure is easily generalised to other systems, such as viscous heat-conducting fluids in motion. The space of the variables, denoted $\mathcal{V}$, is formed by the union of the space of the classical variables C (the density $\rho = v^{-1}$, the specific internal energy u, the velocity v) and the space of the fluxes $\mathcal{F}$ (here the heat flux $\boldsymbol{q}$, and the viscous pressure $\mathbf{P}^v$). In total, the space $\mathcal{V}$ contains 14 independent variables (ρ, u, and p^v, plus three components of v and three components of q and five components of the symmetric traceless tensor $\overset{0}{\mathbf{P}}{}^v$). The evolution of the classical variables is governed by the balance equations of mass, momentum and energy, while the evolution of the flux variables obeys equations of the form

$$\rho\dot{\boldsymbol{q}} = -\nabla\cdot\mathbf{J}^q + \sigma^q,$$

$$\rho\dot{p}^v = -\nabla\cdot j^v + \sigma^v,$$
(2.105)

$$\rho(\overset{0}{\mathbf{P}}{}^v)^{\cdot} = -\nabla\cdot\overset{0}{\mathbf{J}}{}^v + \overset{0}{\sigma}{}^v,$$

$\mathbf{J}^q$ is a tensor of rank two representing the flux of the heat flux, and σ^q is a vector corresponding to the supply of heat flux; j^v is a vector denoting the flux of the scalar viscous pressure, and σ^v is the corresponding scalar source term; $\overset{0}{\mathbf{J}}{}^v$ is a third-rank tensor designating the flux of the traceless viscous pressure tensor, and $\overset{0}{\sigma}{}^v$ is its source term. Of course, at this stage of the analysis these quantities are not determined and must be specified by means of constitutive relations, which in view of the principle of equipresence will be written as

$$\mathbf{J}^q = \mathbf{J}^q(\mathcal{V}), \qquad j^v = j^v(\mathcal{V}), \qquad \overset{0}{j}{}^v = \overset{0}{j}{}^v(\mathcal{V}),$$

$$\sigma^q = \sigma^q(\mathcal{V}), \qquad \sigma^v = \sigma^v(\mathcal{V}), \qquad \overset{0}{\sigma}{}^v = \overset{0}{\sigma}{}^v(\mathcal{V}). \tag{2.106}$$

The evolution equations (2.105) and the constitutive relations (2.106) are not arbitrary. They have to comply with the following three constraints:

- Euclidean invariance (criterion of objectivity).
- Positiveness of the rate of entropy production.
- Convexity of entropy.

As a consequence of objectivity, the material time rates must be replaced by objective ones. Therefore, in the first and third of equations (2.105), the material time derivatives should be replaced by $\mathrm{D}q$ and $\mathrm{D}\overset{0}{\mathbf{P}}{}^v$, respectively, where D denotes an objective time derivative, say Jaumann's derivative.

As above, we will introduce Lagrange multipliers so that entropy inequality will take the form

$$\rho\dot{s} + \nabla \cdot \boldsymbol{J}^s + \Lambda_0(\rho\dot{v} + \nabla \cdot v) + \Lambda_1\left(\rho\dot{u} + \nabla \cdot \boldsymbol{q} + \mathbf{P}^T : \nabla v\right)$$

$$+ \boldsymbol{\Lambda}_2 \cdot (\rho\dot{v} + \nabla \cdot \mathbf{P} + \rho F) + \boldsymbol{\Lambda}_3 \cdot (\rho\dot{q} + \nabla \cdot \mathbf{J}^q + \boldsymbol{\sigma}^q)$$

$$+ \Lambda_4(\rho\dot{p}^v + \nabla \cdot j^v + \sigma^v) + \boldsymbol{\Lambda}_5 : [\rho(\overset{0}{\mathbf{P}}{}^v)^{\cdot} + \nabla \cdot \overset{0}{\boldsymbol{J}}{}^v + \overset{0}{\sigma}{}^v] \geq 0 \tag{2.107}$$

wherein the Lagrange multipliers Λ_i ($i = 0, 1, ..., 5$) are unknown functions of the set of variables $\mathcal{V}$. The procedure is the same as in Sect. 2.5.1 but will not be pursued further because the calculations are long and cumbersome outside the linear approximation.

It is interesting to summarize the main differences between the standard and rational presentations of EIT in Sects. 2.2 and 2.4: (a) in RET, the restrictions on the balance equations are explicitly taken into account in the formulation of the second law by means of Lagrange multipliers; (b) the Gibbs equation is not postulated a priori, as in the previous formulation, but is derived from the restrictions placed by the second law. However, in the linear approximation, the results of the two approaches will be identical both for the evolution equations of the fluxes and for the Gibbs equation. In the non-linear range, the complete equivalence is not yet established. In our opinion, the existence of both formalisms is useful and stimulating, as each point of view has its own advantages. For instance, the use of Lagrange multipliers in RET is elegant and appealing, but the standard description of EIT provides for them a physical identification in a more direct way and gives a more straightforward description of fluctuations, as will be shown in Chap. 6.

2.6 Some Comments and Perspectives

To shed further light on the scope and perspectives of EIT, let us add some general comments.

1. In EIT, the state variables are the classical hydrodynamic fields supplemented by the fluxes appearing on the right-hand side of the governing equations of classical hydrodynamics, i.e. the extra stress (or pressure) tensor (the total stress tensor minus its hydrostatic part) and the extra energy flux (the total energy flux minus the fluxes due to advection of energy and due to the mechanical work). EIT can be viewed as an extension of classical thermohydrodynamics wherein inertial effects are included. We recall that in classical mechanics inertia is introduced by considering the right-hand side of the evolution equation $\dot{r} = v$ (r is the position vector of a particle and v its velocity) as an independent state variable, whose behaviour is governed by an additional equation (Newton's equation of motion).

2. It could be asked why it is not preferable to select internal variables rather than dissipative fluxes as additional variables. Indeed, thermodynamics with internal variables has been successfully applied to a wide variety of problems [2.16]. Conceptually, however, there are several differences between the two approaches. First, internal variables are measurable but not controllable, whereas the fluxes are controllable and measurable. Quoting Kestin [2.16], by controllable variables we mean 'that they are coupled to no external force variable which might provide the means of control. And, not being coupled to a force variable, they cannot part in the mechanical work.' Second, internal variables are related to the microstructure of the system. Third, they are usually related to local relaxation, whereas the fluxes are more closely connected to non-local effects. Actually, the two points of view are not so opposite as it might appear at first sight. Indeed, in polymer solutions, for instance, the viscous pressure tensor is directly related to the macromolecular conformation tensor introduced in thermodynamics with internal variables. According to the specific problem to be investigated, one of the two theories may be preferable. For instance, for systems in steady states, the controllable character of the viscous pressure or the heat flux presents an advantage, as it allows the system to be mantained in the required state; this is no longer true when one is working with the conformation tensor. Finally, the use of the viscous pressure tensor as an independent variable allows us to present under a unified formalism the viscoelastic properties of such different systems as ideal and real gases and of polymer solutions (see Chaps. 12 and 15, respectively). This would not be possible in the framework of

thermodynamics with internal variables, because monatomic ideal gases without internal degrees of freedom are not characterized by particular internal variables.

3. There are also several reasons for choosing the fluxes rather than the gradients of the classical variables as independent quantities. (a) The fluxes are associated with well-defined microscopic operators, and as such allow for a more direct comparison with non-equilibrium statistical mechanics. (b) The use of the fluxes is more convenient than use of the gradients for fast processes, whereas for slow or steady-state phenomena the use of both sets of variables is equivalent because under these conditions the former ones are directly related to the latter. (c) By expressing the entropy in terms of the fluxes, the classical theory of fluctuations can be easily generalised to evaluate the coefficients of the non-classical part of the entropy. This would not be possible by taking the gradients as variables. (d) Finally, the selection of the gradients as extra variables leads to divergence problems in the expansion of constitutive equations, as well known from the kinetic theory.

4. Every dissipative flux has been considered in this chapter to be a quantity characterized by a single evolution equation. It is easily conceivable that fluxes may be split into several independent contributions, each with its own evolution equation. This is not exceptional, and typical examples are treated in Chap. 5 for non-ideal gases and in Chap. 15 for polymers. EIT is shown there to be able to cope in a quite natural way with these situations.

5. The space of the extra variables is not generally restricted to the ordinary dissipative fluxes, such as the heat flux, the viscous pressure, or the flux of matter. To cope with the complexity of some fast non-equilibrium phenomena, it is necessary to introduce more variables. It is shown in Chap. 5 how EIT is able to account for such more general descriptions, when higher-order fluxes, such as the fluxes of the fluxes, are considered as supplementary variables.

6. An important issue is the measurability of the dissipative fluxes. The heat flux may be simply evaluated by measuring the amount of energy transported per unit area and time through the boundaries of the system. The viscous pressure can be measured from the tangential shear force exerted per unit area. In practice, it may be difficult to evaluate these quantities at each instant of time and at every point in space. Nevertheless, for several problems of practical interest, such as wave propagation, the fluxes are eliminated from the final equations, although the corresponding dispersion relations contain explicitly the whole set of parameters appearing in the evolution equations of the fluxes. Thus the predictions of EIT concerning these parameters may be checked without direct measurement of the fluxes.

7. EIT provides a connection between thermodynamics and dynamics. In EIT, the fluxes are no longer considered as mere control parameters but as independent vari-

ables. The fact that EIT makes a connection between dynamics and thermodynamics should be underlined. Moreover, EIT does not only provide a natural framework for a wide class of dynamical models, it also generates original results, as shown in Chaps. 10 and 13. Most of the dynamical models dealt with in this book did not receive thermodynamical foundations before EIT was proposed; one reason is that non-equilibrium thermodynamics was simply ignored and investigations were exclusively performed on dynamical aspects.

8. EIT enlarges the range of applicability of non-equilibrium thermodynamics to a vast domain of phenomena where memory, non-local, and non-linear effects are relevant. A non-exhaustive list of applications in various fields of physics is given in Table 2.2. Many of them are finding increasing use in technology, which, in turn, enlarges the experimental possibilities for the observation of non-classical effects in a wider range of non-equilibrium situations.

Table 2.2. Some examples of application of EIT

High-frequency phenomena	*Short-wavelength phenomena*
Ultrasounds in gases	Light scattering in gases
Light scattering in gases	Neutron scattering in liquids
Neutron scattering in liquids	Ballistic phonon propagation
Second sound in solids	Phonon hydrodynamics
Heating of solids by laser pulses	Submicronic electronic devices
Nuclear collisions	Shock waves
Long relaxation times	*Long correlation lengths*
Polyatomic molecules	Rarefied gases
Suspensions, polymer solutions	Transport in harmonic chains
Diffusion in polymers	Cosmological decoupling eras
Propagation of fast crystallization fronts	Transport near critical points
Superfluids, superconductors	

9. It should be emphasised that EIT has fostered the use of generalised causal transport equations in several domains where, up to now, only non-causal transport equations were used. In connection with this, it is interesting to recall that, historically, one of the motivations for formulating EIT was to circumvent the problem met in classical theories: the prediction that the application of a perturbation should be felt instantaneously everywhere inside the whole system. Thus, the number of motivations is much

increased, many of them arising from experimental observations and practical applications rather than from purely theoretical arguments.

2.7 Entropy Evolution in an Isolated System: An Illustrative Example

In this section we compare the behaviour of classical entropy and extended entropy shown in Fig. 2.1 during thermal equilibration, for a discrete system. Consider heat transfer between two rigid bodies at different temperatures and isolated from the outside world [2.15]. Initially, the bodies are separated by an adiabatic wall. If the adiabatic constraint is removed, heat will flow from one body to the other without either work being performed or mass carried. Let us start with the classical local-equilibrium formulation. Each of the two subsystems, at temperatures T_1 and $T_2\,(< T_1)$ respectively, is in internal equilibrium, i.e. the internal temperature changes are negligible compared to the temperature difference $T_1 - T_2$. In virtue of the extensivity property, the total entropy of the whole system S is the sum of the entropies of each subsystem, S_1 and S_2, which are functions of the internal energies U_1 and U_2 of subsystems 1 and 2 respectively, even during the heat transfer process, namely $S(U_1, U_2) = S_1(U_1) + S_2(U_2)$. When the adiabatic wall separating the two subsystems is replaced by a diathermal one, the time-rate variation of total entropy is

$$\frac{\mathrm{d}S}{\mathrm{d}t} = \frac{\mathrm{d}S_1}{\mathrm{d}t} + \frac{\mathrm{d}S_2}{\mathrm{d}t} = T_1^{-1}\frac{\mathrm{d}U_1}{\mathrm{d}t} + T_2^{-1}\frac{\mathrm{d}U_2}{\mathrm{d}t}. \tag{2.108}$$

Since the global system is isolated, $\mathrm{d}U_1 + \mathrm{d}U_2 = 0$, and using the first law of thermodynamics we can write

$$\frac{\mathrm{d}U_1}{\mathrm{d}t} = -\frac{\mathrm{d}U_2}{\mathrm{d}t} = -\dot{Q}, \tag{2.109}$$

where $\dot{Q}$ is the amount of heat exchanged between subsystems 1 and 2 per unit time, i.e. the heat flux integrated over the surface separating the two bodies. Recall that, by convention, energy input is positive and energy release negative.

For an isolated system, (2.108) represents the rate of entropy produced inside the system; in virtue of (2.109), it can be written as

$$\frac{\mathrm{d}S}{\mathrm{d}t} = -\left(T_1^{-1} - T_2^{-1}\right)\dot{Q}. \tag{2.110}$$

According to the second law of thermodynamics, this quantity must be non-negative. This implies that heat only flows from the region of highest temperature to the region of lowest temperature, which is the original Clausius formulation of the second law.

The simplest hypothesis ensuring the positiveness of (2.110) is to assume that the heat flux $\dot{Q}$ is proportional to the driving force $(T_1^{-1} - T_2^{-1})$, so that

$$\dot{Q} = -K\left(T_1^{-1} - T_2^{-1}\right), \tag{2.111}$$

with K being a positive coefficient. If the temperature difference is small, one may linearize this relation and write it in the usual form

$$\dot{Q} = K'(T_1 - T_2), \tag{2.112}$$

with $K' = KT^{-2}$ assumed to be a constant and T an intermediate temperature between T_1 and T_2.

In the course of time, the evolution equation for the temperature of each subsystem is easily found by recalling that $dU_1 = C_1 dT_1$ and $dU_2 = C_2 dT_2$, with C_1 and C_2 being the heat capacities of the respective subsystems. Combining this result with (2.109), it is found that the temperature difference $\varepsilon = T_1 - T_2$ varies as

$$\frac{d\varepsilon}{dt} = -\dot{Q} C_{eff}^{-1}, \tag{2.113}$$

with $C_{eff}^{-1} = C_1^{-1} + C_2^{-1}$. When (2.112) is introduced into (2.113) one finds that

$$\frac{d\varepsilon}{dt} = -K''\varepsilon, \tag{2.114}$$

with $K'' = K'C_{eff}^{-1}$. Thus ε decays exponentially as $\varepsilon = \varepsilon_0 \exp(-K''t)$, and after an infinite lapse of time the temperature inside the system becomes uniform. In view of (2.110) and (2.113) the rate of evolution of entropy may be written in terms of ε as

$$\frac{dS}{dt} = -T^{-2}\frac{K'}{K''}\varepsilon\frac{d\varepsilon}{dt}. \tag{2.115}$$

It is thus seen that the entropy is a monotonically increasing function of time.

Consider now the more general situation in which heat transfer is described by

$$\tau\frac{d\dot{Q}}{dt} + \dot{Q} = -K\left(T_1^{-1} - T_2^{-1}\right), \tag{2.116}$$

where τ is the relaxation time of the heat flux $\dot{Q}$. This expression is analogous to the Maxwell–Cattaneo equation (2.4). After combining (2.113) and (2.116), one finds for the evolution of ε that

$$\tau\frac{d^2\varepsilon}{dt^2} + \frac{d\varepsilon}{dt} + K''\varepsilon = 0. \tag{2.117}$$

This equation is similar to the equation of motion of a damped pendulum. The decay of ε will not in general be exponential but may exhibit oscillatory behaviour when $4\tau K'' > 1$. In continuous systems, the exponential decay observed in the discrete model would correspond to a diffusive perturbation, whereas an oscillation in ε would correspond to the propagation of a heat wave. According to (2.34), in the case of oscillatory decay of ε, the classical entropy S behaves as a non-monotonic function of time, as indicated by the dashed curve of Fig. 2.1.

In extended irreversible thermodynamics (EIT), the heat flux $\dot{Q}$ is viewed as an independent variable of the entropy. The proposed form of the generalised entropy in EIT is therefore $S_{EIT}^*(U_1, U_2, \dot{Q})$, where the asterisk refers to the extended entropy. The rate of variation of S_{EIT} is, up to the second order in Q,

$$\frac{dS_{EIT}}{dt} = T_1^{-1}\frac{dU_1}{dt} + T_2^{-1}\frac{dU_2}{dt} - a\dot{Q}\frac{d\dot{Q}}{dt}. \tag{2.118}$$

Using the conservation of energy (2.109) one may write (2.118) as

$$\frac{dS^*}{dt} = -\left(T_1^{-1} - T_2^{-1}\right)\dot{Q} - a\dot{Q}\frac{d\dot{Q}}{dt}. \tag{2.119}$$

For small temperature differences, dS_{EIT}/dt simplifies to

$$\frac{dS_{EIT}}{dt} = \left(T^{-2}\varepsilon - a\frac{d\dot{Q}}{dt}\right)\dot{Q}. \tag{2.120}$$

The simplest way to guarantee the positiveness of the entropy production dS_{EIT}/dt is to assume the linear relation

$$\dot{Q} = K\left(T^{-2}\varepsilon - a\frac{d\dot{Q}}{dt}\right), \tag{2.121}$$

where K is a positive constant and, in analogy with the Maxwell–Cattaneo equation

(2.116), aK may be interpreted as a relaxation time, namely $aK = \tau$. In view of this identification, the entropy variation can be written as

$$\frac{\mathrm{d}S_{EIT}}{\mathrm{d}t} = -\left(T_1^{-1} - T_2^{-1}\right)\dot{Q} - \frac{\tau}{K}\dot{Q}\frac{\mathrm{d}\dot{Q}}{\mathrm{d}t}, \tag{2.122}$$

or, in an integrated form,

$$S_{EIT}(U_1, U_2, \dot{Q}) = S_1(U_1) + S_2(U_2) - \frac{\tau}{2K}\dot{Q}^2. \tag{2.123}$$

Observe that in the limiting case $\tau = 0$ the above expression reduces to the local-equilibrium entropy. The last term on the right-hand side of (2.123) may be viewed as expressing the interaction between subsystems 1 and 2.

According to (2.120) and (2.121), the time variation of S_{EIT} may simply be written as

$$\frac{\mathrm{d}S_{EIT}}{\mathrm{d}t} = K^{-1}\dot{Q}^2 = K^{-1}C_{eff}^2\left(\frac{\mathrm{d}\varepsilon}{\mathrm{d}t}\right)^2. \tag{2.124}$$

This expression is either positive or zero, but never negative, since it was found that $K > 0$, and therefore $S_{EIT}[U_1(t), U_2(t), \dot{Q}(t)]$, whose evolution is plotted in the continuous curve of Fig. 2.1, increases monotonically in the course of time and is thus compatible with evolution equations of the Maxwell–Cattaneo type. The corresponding analysis for a continuous system may be found in [2.17].

We will sketch, finally, the analysis for a continuous system. We consider an isolated undeformable wire of length L and assume an initial temperature profile of the form

$$T(x,0) = T_0 + \delta T_0 \cos\frac{2\pi nx}{L}, \qquad (n = 0, 1, \ldots) \tag{2.125}$$

where T_0 is a uniform temperature reference. We want to study the decay of temperature towards equilibrium. Assume that the perturbation δT has the form $\delta T = \delta T_0 f(t) \times \cos kx$, with $f(t)$ being a function of time to be determined and k the wave number $k = 2\pi n/L$; introducing this expression in (2.5), one obtains

$$f(t) = A\exp(a_+t) + B\exp(a_-t) \tag{2.126}$$

where A and B are constants fixed by the initial conditions of T and $\partial T/\partial t$, and

$$a_{\pm} = \frac{1}{2\tau}\left[-1 \pm (1 - 4\chi k^2 \tau)^{1/2}\right].$$

(2.127)

For $4\chi k^2 \tau > 1$ the decay is oscillatory and δT is given by

$$\delta T = \delta T_0 (A \sin \omega t + \cos \omega t)\exp\left(-\frac{t}{2\tau}\right)\cos kx,$$

(2.128)

with

$$\omega = \frac{1}{2\tau}\left(4\chi \tau k^2 - 1\right)^{1/2}.$$

(2.129)

According to classical irreversible thermodynamics, the entropy of the wire at temperature T is given by

$$S(t) = \rho c_v \int_0^L \ln \frac{T(x,t)}{T_0}\, dx,$$

(2.130)

where it has been assumed that the system is comprised between $x = 0$ and $x = L$ and ρc_v is constant. Combination of the evolution (2.126–127) with the classical expression (2.128) for the entropy yields non-monotonic behaviour. In contrast, the evolution of the extended entropy increases as

$$\frac{dS_{EIT}(t)}{dt} = \int_0^L \frac{1}{\lambda T^2(x,t)} q(x,t) \cdot q(x,t)\, dx.$$

(2.131)

This expression is always positive and, therefore, the entropy of EIT increases in a monotonic way, with some stationary points, corresponding to the minima of the local-equilibrium entropy, where $q(x, t) = 0$. The change of classical entropy and of extended entropy in the course of time has been calculated explicitly by Criado-Sancho and Llebot [2.17] and its characteristic behaviour is plotted in Fig. 2.1.

Problems

2.1 Assume that the entropy S is a function of a variable α and its time derivative $\eta = d\alpha/dt$, and that α satisfies the differential equation

$$M\frac{\mathrm{d}^2\alpha}{\mathrm{d}t^2}+\frac{\mathrm{d}\alpha}{\mathrm{d}t}=L\frac{\partial S}{\partial\alpha}.$$

(a) Show that the positiveness of the entropy production demands that $L>0$ but does not imply any restriction on the sign of M. (*Hint*: Write $\mathrm{d}\alpha/\mathrm{d}t$ and $\mathrm{d}\eta/\mathrm{d}t$ in terms of $\partial S/\partial\alpha$ and $\partial S/\partial\eta$.) (b) Assume that $\partial S/\partial\eta=a\eta$, with a being a constant. Show that the stability condition $\delta^2 S<0$ implies that $M>0$ and $a=-(M/L)$.

2.2 The positiveness of the entropy production requires the positiveness of the second-order terms in the expression of the entropy production, but does not imply any criterion about the sign of the fourth- and higher-order terms in the entropy production. (a) As an illustration, study the sign of the entropy production and its second- and fourth-order terms in ∇T for a hypothetical system where $\sigma^s=\boldsymbol{q}\cdot\nabla T^{-1}$ and

$$\boldsymbol{q}=-\lambda\exp\!\left[-a(\nabla T)^2\right]\nabla T,$$

with a being a constant parameter. (b) The fact that no criterion may be obtained about the sign of the higher-order terms in the entropy production does not imply that the positiveness of the entropy production must be abandoned. As an illustration, study the sign of the entropy production and its second-order terms in ∇T for a constitutive law of the form

$$\boldsymbol{q}=-\lambda\sin\!\left[a(\nabla T)^2\right]\nabla T,$$

with a being a (positive) constant. Despite the fact that the second-order terms are positive, is a constitutive equation of this form admissible?

2.3 Show that

$$q(t)=\int_{-\infty}^{t}G(t-t')\nabla T(t')\mathrm{d}t',$$

with $G(t-t')$ being a memory function given by $G(t-t')=-(\lambda/\tau)\exp\!\left[-(t-t')/\tau\right]$, is a solution of the Maxwell–Cattaneo equation.

2.4 Show that by combining the generalised Fourier law

$$\boldsymbol{q}=-\lambda\nabla T+\lambda_1\frac{\partial}{\partial t}(\nabla T)$$

with the energy balance law of a rigid heat conductor one obtains a parabolic evolution equation for the temperature. Compare with the Maxwell–Cattaneo equation.

2.5 In order to reconcile the results obtained from the material frame-indifference principle (see Sect. 1.4) and from the kinetic theory, Lebon and Boukary [1.32] proposed replacing the Jaumann derivative D_J by the following time derivative

$$\mathcal{D}q = D_J q - 2(\Omega + \mathbf{W}) \cdot q,$$

where Ω is the angular velocity tensor and $\mathbf{W}$ is the antisymmetric part of the velocity gradient ∇v. The two quantities are related to the angular velocity vector $\boldsymbol{\omega}$ and the vorticity $\nabla \times v$ by $\omega_i = \frac{1}{2}\varepsilon_{ijk}\Omega_{jk}$, $(\nabla \times v)_i = \varepsilon_{ijk}W_{jk}$, respectively. In both expressions ε_{ijk} is the permutation tensor. Prove that the time derivative $\mathcal{D}$ is objective.

2.6 (a) Show that in the presence of several fluxes J_i the generalised Gibbs equation is

$$ds = ds_{eq} - v \sum_{ikl} J_i (L^{-1})_{ik} \tau_{kl} dJ_l,$$

when the quantities J_i obey generalised constitutive laws

$$\sum_j \tau_{ij} \frac{dJ_j}{dt} + J_i = \sum_j L_{ij} X_j,$$

where X_j are the usual thermodynamic forces of the classical theory, L_{ij} the elements of the matrix of phenomenological coefficients $\mathbf{L}$, and τ_{ij} the elements of the matrix of relaxation times $\boldsymbol{\tau}$. (b) Assume linear transformations of the fluxes $J_i^* = \Sigma_j A_{ij} J_j$ and of the forces $X_i^* = \Sigma_j [(A^T)^{-1}]_{ij} X_j$. Find the new tensors $\mathbf{L}^*$ and $\boldsymbol{\tau}^*$.

2.7 Internal-variable theories [2.16] include additional variables which allow for a more detailed description than the local-equilibrium approach. Suppose a vectorial internal variable g, related to the heat flux, and an entropy s depending on the internal energy u and g. The corresponding Gibbs equation is

$$ds = \theta^{-1}du - v\alpha' g \cdot dg$$

Show that if the heat flux is required to obey a relaxational equation ot the Maxwell–Cattaneo type, then g can be identified with q and the Gibbs equation coincides with (2.29) (D. Jou *et al.*, Rep. Prog. Phys. **62** (1999) 1035).

Chapter 3

Extended Irreversible Thermodynamics: Non-equilibrium Equations of State

In Chap. 2, we postulated the existence of a generalised entropy which is compatible with some classes of evolution equations for the fluxes. Otherwise stated, our formalism aims to describe the class of processes which are compatible with the existence of a non-equilibrium entropy whose rate of production is non-negative. Once the expression of the entropy is known, there is no difficulty in deriving the corresponding equations of state, which are directly obtained as the first derivatives of the entropy with respect to the basic variables. A natural question concerns the physical meaning of these equations of state, which, of course, depend on the fluxes and therefore differ from their analogous local-equilibrium expressions. In classical thermodynamics, it is known that the derivative of the entropy with respect to the internal energy (by keeping fixed the volume and the composition of the system) is the reciprocal of the absolute temperature; the derivatives with respect to the volume and to the number of moles yield the equilibrium pressure and (with a minus sign) the chemical potentials respectively (divided by the absolute temperature). It may then be asked whether the derivatives of the generalised entropy introduced in extended irreversible thermodynamics (EIT) still allow an absolute temperature to be defined, as well as a non-equilibrium pressure and a non-equilibrium chemical potential. Another important problem is to determine whether the non-equilibrium temperature and pressure are measurable by a thermometer and a manometer. These are subtle and unsolved problems which have however received partial answers during the last years. The objective of the present chapter is to better apprehend the physical meaning of the generalised entropy and to pay detailed attention to the nature of the corresponding equations of state.

3.1 Physical Interpretation of the Non-equilibrium Entropy

Consider a volume $\mathcal{V}$ of a fluid which is sufficiently small so that within it the spatial variations of pressure and temperature are negligible; if the fluid element is subject to a heat flux q and a viscous pressure $\overset{0}{\mathbf{P}}{}^{v}$ (viscous bulk effects are ignored here for the sake of simplicity), it can then be asked which entropy may be ascribed to it. To answer this question, the volume element, at $t = 0$, is suddenly isolated, i.e. bounded by adiabatic and rigid walls, and allowed to decay to equilibrium. The decay of q and $\overset{0}{\mathbf{P}}{}^{v}$ to their final vanishing equilibrium values is accompanied by a production of entropy, so that the final equilibrium entropy value is given by

$$S_{eq,f} = S_{neq,i} + \mathcal{V}\int_{0}^{\infty} \sigma^{s} \, dt. \tag{3.1}$$

Indices i and f refer to the initial non-equilibrium state and the final equilibrium state respectively, S is the entropy of the small system of volume $\mathcal{V}$ and σ^{s} is the rate of entropy production per unit volume. A sketch of the situation when only heat flux is present is given in Fig. 3.1.

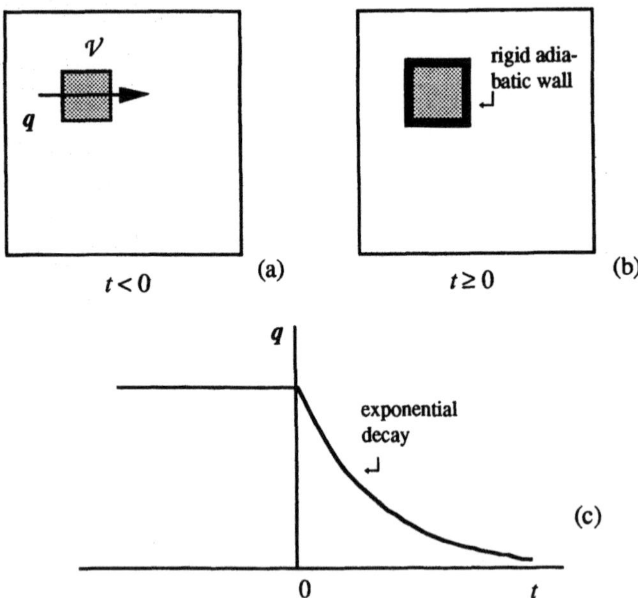

Fig. 3.1. (a) System under a heat flux; (b) at time $t = 0$ the system is completely isolated; (c) time evolution of heat flux.

In Fig. 3.1, the non-equilibrium state (a) and the local-equilibrium state (b) corresponding to the same values of U and $\mathcal{V}$ are represented, respectively, before and after insulation. The second term in (3.1) corresponds to the non-compensated heat introduced by Clausius and will be commented on below.

The explicit form of the non-equilibrium contribution in the case of the Maxwell–Cattaneo equations may be obtained in a rather straightforward way. According to (2.49) and the identifications $\mu_1 = (\lambda T^2)^{-1}$ and $\mu_2 = (2\eta T)^{-1}$, σ^s is given by

$$\sigma^s = (\lambda T^2)^{-1} \boldsymbol{q} \cdot \boldsymbol{q} + (2\eta T)^{-1} \overset{0}{\mathbf{P}}{}^{v} : \overset{0}{\mathbf{P}}{}^{v}. \tag{3.2}$$

If the decay of $\boldsymbol{q}$ and $\overset{0}{\mathbf{P}}{}^{v}$ is governed by the Maxwell–Cattaneo equations (2.60–62), we have

$$\boldsymbol{q}(t) = \boldsymbol{q}(0)\exp(-t/\tau_1), \qquad \overset{0}{\mathbf{P}}{}^{v}(t) = \overset{0}{\mathbf{P}}{}^{v}(0)\exp(-t/\tau_2). \tag{3.3}$$

By inserting these expressions into (3.1) and integrating with respect to the time, one obtains for the non-equilibrium entropy S

$$S = S_{eq} - (\tau_1 \mathcal{V}/2\lambda T^2)\boldsymbol{q} \cdot \boldsymbol{q} - (\tau_2 \mathcal{V}/4\eta T)\overset{0}{\mathbf{P}}{}^{v} : \overset{0}{\mathbf{P}}{}^{v} \tag{3.4}$$

which is the integrated form of the entropy (2.38) in the absence of bulk viscous pressure. Note that S and S_{eq} correspond to entropies of the fluid element of volume $\mathcal{V}$ and that s and s_{eq} in (2.38) are given per unit mass, and consequently the ratio S/s and $\mathcal{V}/v$, with v being the specific volume, is the mass of the fluid element. The above derivation is suggestive as it assigns a meaning to the non-classical terms in the equation for the entropy, by relating them to a physical operational definition.

However, it must be realized that the above procedure exhibits three limitations: it is based on the hypotheses that $\boldsymbol{q}$ and $\overset{0}{\mathbf{P}}{}^{v}$ are the relevant additional variables, it is assumed that $\boldsymbol{q}$ and $\overset{0}{\mathbf{P}}{}^{v}$ decay exponentially, and it is restricted to a given non-equilibrium process (namely, relaxation of the fluxes after sudden isolation of the elementary volume). It would be of interest to get rid of these restrictions.

In this respect, the idea of the uncompensated heat may be helpful to clarify in more general terms the nature of the non-equilibrium contributions [3.1]. In 1865, Clausius wrote his famous inequality for a cyclic process:

$$\oint \frac{\mathrm{d}Q}{\theta} \leq 0, \tag{3.5}$$

where θ is the absolute temperature of the heat reservoir which is in thermal contact with the system during the infinitesimal process of exchange of a quantity of heat $\mathrm{d}Q$

(we write θ rather than T to stress that it is not necessary that the heat reservoir is itself in equilibrium). Inequality (3.5) may still be expressed as

$$-N \equiv \oint \frac{dQ}{\theta} \le 0 , \tag{3.6}$$

where N is the (positive) uncompensated heat. By expressing N in the form of a contour integral over an irreversible cycle $N = \oint dN$, Clausius inequality may be rewritten as

$$\oint \left(\frac{dQ}{\theta} + dN \right) = 0 . \tag{3.7}$$

The assumption of a cyclic integral implies the assumption of a good choice of variables, i.e. a complete set of relevant macroscopic variables. The vanishing of integral (3.7) for arbitrary cycles implies the existence of the exact differential of a given quantity which is called the generalised entropy in our formalism (or the calortropy in Eu's work [3.1]), such that

$$dS = \frac{dQ}{\theta} + dN . \tag{3.8}$$

Note that neither dQ/θ nor dN are exact differentials, only their sum is. Furthermore, dN is always positive, and it vanishes for reversible processes. In the latter case, $dN = 0$, $dQ = dQ_{rev}$, $\theta = T$ and $S = S_{eq}$, and one recovers Clausius definition of entropy

$$dS_{eq} = \frac{dQ_{rev}}{T} . \tag{3.9}$$

The quantity S defined in (3.8) generalises Clausius definition of entropy, and it is important to note that $dS \ne dS_{eq}$.

The above argument is more general than the one given before, as it does not require a priori specification of either the nature of the non-equilibrium variables, or the dynamics of the variables, or the nature of the process being involved. It can be objected that a theory based on the choice of fluxes $\boldsymbol{q}$ and $\overset{0}{\mathbf{P}}^{\nu}$ as variables is too restrictive. Indeed, a more refined formalism would demand inclusion of more non-equilibrium higher-order fluxes. The choice of $\boldsymbol{q}$ and $\overset{0}{\mathbf{P}}^{\nu}$ is useful when the evolution equations for $\boldsymbol{q}$ and $\overset{0}{\mathbf{P}}^{\nu}$ are those given by (2.60) and (2.62). The choice of the relevant variables must always be motivated either by experimental considerations or by microscopic theories when these are available. Note that the family of states whose adiabatic projection (i.e. the set of states which when isolated decay to a given equilibrium state) is a given equilibrium state may be considered as a fibre in the general thermodynamic space, which by this procedure is given the structure of a fibred space [3.2].

3.2 Non-equilibrium Equations of State: Temperature

Having identified in Sect. 2.3 the parameters α_{10}, α_{00}, and α_{21} in physical terms, we are now in a position to evaluate explicitly the contributions of the fluxes to the equations of state. Up to the second order terms in the fluxes, the Gibbs equation (2.38) may be written as

$$ds = \theta^{-1}du + \theta^{-1}\pi dv - \frac{v\tau_1}{\lambda T^2}\boldsymbol{q}\cdot d\boldsymbol{q} - \frac{v\tau_0}{\zeta T}p^v dp^v - \frac{v\tau_2}{2\eta T}\overset{0}{\mathbf{P}}{}^v : d\overset{0}{\mathbf{P}}{}^v. \qquad (3.11)$$

From the integrability condition of (3.11), i.e. equality of the second crossed derivatives as for instance $\partial^2 s/\partial u \partial q = \partial^2 s/\partial q \partial u$, it follows that

$$\frac{\partial \theta^{-1}}{\partial q} = -\left(\frac{\partial(v\tau_1/\lambda T^2)}{\partial u}\right)\boldsymbol{q}, \qquad \frac{\partial \theta^{-1}}{\partial \overset{0}{\mathbf{P}}{}^v} = -\left(\frac{\partial(v\tau_2/2\eta T)}{\partial u}\right)\overset{0}{\mathbf{P}}{}^v,$$

$$\frac{\partial \theta^{-1}}{\partial p^v} = -\left(\frac{\partial(v\tau_0/\zeta T)}{\partial u}\right)p^v, \qquad (3.12a)$$

$$\frac{\partial(\theta^{-1}\pi)}{\partial q} = -\left(\frac{\partial(v\tau_1/\lambda T^2)}{\partial v}\right)\boldsymbol{q}, \qquad \frac{\partial(\theta^{-1}\pi)}{\partial \overset{0}{\mathbf{P}}{}^v} = -\left(\frac{\partial(v\tau_2/2\eta T)}{\partial v}\right)\overset{0}{\mathbf{P}}{}^v,$$

$$\frac{\partial(\theta^{-1}\pi)}{\partial p^v} = -\left(\frac{\partial(v\tau_0/\zeta T)}{\partial v}\right)p^v. \qquad (3.12b)$$

As a consequence of (3.12a,b), and keeping in mind that for vanishing values of the fluxes one must recover the local-equilibrium values of T and p, one obtains

$$\theta^{-1} = T^{-1} - \frac{1}{2}\left[\frac{\partial(v\tau_1/\lambda T^2)}{\partial u}\boldsymbol{q}\cdot\boldsymbol{q} + \frac{\partial(v\tau_0/\zeta T)}{\partial u}(p^v)^2 + \frac{\partial(v\tau_2/2\eta T)}{\partial u}\overset{0}{\mathbf{P}}{}^v : \overset{0}{\mathbf{P}}{}^v\right]$$

$$\qquad (3.13)$$

$$\theta^{-1}\pi = T^{-1}p - \frac{1}{2}\left[\frac{\partial(v\tau_1/\lambda T^2)}{\partial v}\boldsymbol{q}\cdot\boldsymbol{q} + \frac{\partial(v\tau_0/\zeta T)}{\partial v}(p^v)^2 + \frac{\partial(v\tau_2/2\eta T)}{\partial v}\overset{0}{\mathbf{P}}{}^v : \overset{0}{\mathbf{P}}{}^v\right].$$

These expressions can be viewed as non-equilibrium equations of state for the temperature and pressure. In this section, we focus our analysis on temperature, which allows us a rather extensive discussion of the conceptual problems found in this context. In the Sect. 3.3, we will deal with the non-equilibrium pressure.

3.2.1 Zeroth Law, Second Law and Temperature

As it is well known, the concept of absolute temperature is closely connected to two principles of thermodynamics. The zeroth principle expresses the transitivity of thermal equilibrium: it states that if a system C is in thermal equilibrium with two systems A and B, then A and B will be in mutual thermal equilibrium if they are put in direct thermal contact. This principle allows an infinite number of empirical temperatures to be defined, which characterize the different classes of equivalence established in the space of states by the condition of mutual thermal equilibrium. To define an absolute temperature scale independent of the thermometric working substance, the second law is needed (for instance, in the form of Carnot's theorem, which states that the efficiency of Carnot's reversible heat engines is independent of the working substance). This result was used by W. Thomson (Lord Kelvin) in 1848 to introduce for the first time the absolute temperature. Here we will clarify the concept of non-equilibrium temperature in the light of these two principles of thermodynamics.

First of all, recall the definition of the temperature θ which, according to (3.11), is

$$\theta^{-1} = \left(\frac{\partial s}{\partial u} \right)_{v,q,p^v,\overset{0}{\mathbf{P}}{}^v} .\tag{3.14}$$

If the entropy is known in terms of the various variables, (3.14) gives directly the equation of state of θ. Conversely, when all the equations of state are known, they may be integrated to obtain the expression of the entropy. From a geometrical point of view, the difference between θ and T may be easily understood as exhibited in Fig. 3.2.

The actual non-equilibrium state is A, the equilibrium state reached by an adiabatic projection is B. The slope of the non-equilibrium entropy at A (corresponding to $1/\theta$) is different from the slope of the equilibrium entropy at B (corresponding to $1/T$). Thus, when one refers to the local-equilibrium temperature T, one is taking as reference state the accompanying local-equilibrium state B rather than the actual non-equilibrium state A. Note that $\theta^{-1} > T^{-1}$. This is a rather general feature because $s(u,q) \leq s_{eq}(u)$; it is also worth noticing that $s(u,q) \to s_{eq}(u)$ when $u \to \infty$ at constant q (this results from the fact that the value of $\tau v \, q^2/\lambda T^2$ decreases with increasing u at constant q). Therefore, the curve $s(u,q)$ will be steeper than $s_{eq}(u)$ yielding generally $\theta \leq T$. Since a non-equilibrium steady state is characterized by less entropy than the corresponding equilibrium state, it is therefore more ordered than the equilibrium state (recall that entropy may be related to molecular disorder). One could qualitatively split the internal energy per unit volume as $\rho u = \rho u_{ord} + \rho u_{dis}$ where u_{ord} and u_{dis} can be interpreted as the 'ordered' and 'disordered' parts of the internal energy per unit mass, respectively. These energies can be expressed as $\rho u_{dis} = \frac{3}{2} n k_B \theta$ and $\rho u_{ord} = \frac{3}{2} n k_B (T - \theta)$, with n the number density and k_B the Boltzmann constant, so that in total one finds the classical expression for the internal energy of monatomic gases per unit volume, given by $\rho u = \frac{3}{2} n k_B T$ (see Sect. 3.3 for additional comments on this interpretation). In equilibrium, for which $T = \theta$, one has $u_{ord} = 0$ and $u_{dis} = u$.

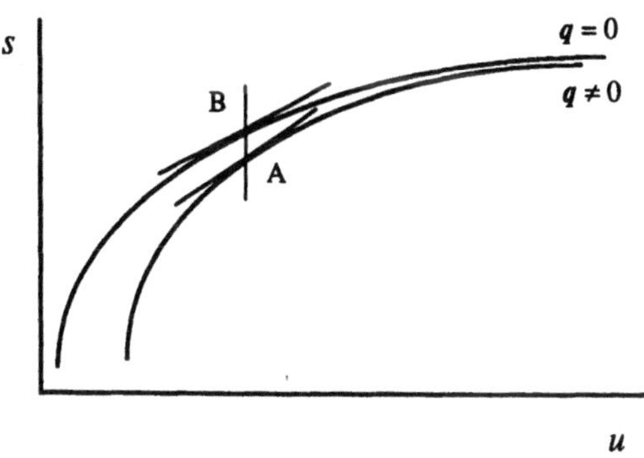

Fig. 3.2. The local-equilibrium entropy $s_{eq}(u)$ (*upper curve*) and the generalised non-equilibrium entropy $s(u, q)$ (*lower curve*) are shown. The two temperatures θ and T mentioned in the text are given by the inverse of the slopes of the curves at points A and B respectively.

To discuss the connection of temperature with the zeroth law, consider the experimental setup depicted in Fig. 3.3 [3.3]. A reference system Σ_r and the system under consideration Σ_s are thermally connected through a highly conducting rod of length L. We denote by q_s and q_{rs} the heat fluxes flowing through Σ_s and the rod respectively. Then we introduce the following statement: Σ_r and Σ_s are in mutual thermal equilibrium when $q_{rs} = 0$. This does not necessarily mean that $q_s = 0$.

Each of the two systems may be internally out of equilibrium notwithstanding the condition $q_{rs} = 0$. Naturally, in complete equilibrium $q_s = q_{rs} = 0$. Note that heat is the only quantity allowed to be exchanged between both systems along the rod. Otherwise, the coupling between several thermodynamic forces (for instance, temperature and concentration gradients) could give a vanishing net heat flow between systems at different temperatures.

Of course, these considerations do not directly help the decision of whether θ or T is truly the (non-equilibrium) temperature. To answer this question it is necessary to know whether the relation between heat flux and temperature gradient is $q = -\lambda \nabla \theta$

rather than $q = -\lambda \nabla T$. The answer is provided by the second law; indeed, as shown in Sect. 2.1, the first alternative is the only one following from the positive definite character of the entropy production. This means that the heat exchange between Σ_r and Σ_s is directly governed by the non-equilibrium temperature θ. If Σ_r and Σ_s are both at equilibrium, θ coincides with the local-equilibrium temperature T. However, if Σ_r and Σ_s are out of equilibrium, with $q_{rs} = 0$, then one will observe that $\theta_r = \theta_s$ rather than $T_r = T_s$.

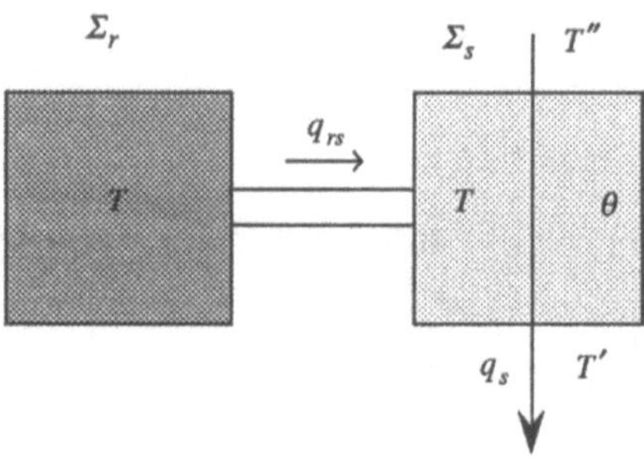

Fig. 3.3. Gedanken experiment to illustrate the difference between the local-equilibrium temperature and the non-equilibrium temperature.

To be more specific, assume that the system Σ_r on the left in Fig. 3.3 is at equilibrium at temperature T. The system on the right is in a non-equilibrium steady state under a heat flux q_s generated by a temperature difference. Assume that both ends of the conducting rod are at the same local-equilibrium absolute temperature T. According to the classical theory, no heat flow will be observed from one system to another. In contrast, EIT predicts a flow q_{rs} proportional to the gradient of the non-equilibrium temperature θ, namely $(\theta_r - \theta_s)/L$. Since the system Σ_r is in equilibrium, one has $\theta_r = T_r$, but $\theta_s = T_s[1 - \gamma q_s^2]$, with $\gamma = \frac{2}{5}(m/n^2 k_B^3 T^3)$ for ideal monatomic gases, so that $\theta_s < T_s$. Consequently, a heat flux q_{rs} will flow from the left to the right. This example corroborates the statement that a thermometer will measure the generalised temperature θ rather than the local-equilibrium T: heat transfer between the two systems will take place until they have reached the same generalised temperature, rather than the same local-equilibrium temperature. A confirmation is found in the kinetic theory; indeed, it may be shown [3.3b] that when both systems Σ_s and Σ_r of Fig. 3.3 are composed of an ideal gas with the same mean kinetic energy at the position of the thermal contact, energy flows from the equilibrium gas to the gas submitted to the vertical heat flux. Another confirmation is provided by computer simulations, as discussed in Sect. 9.4.

3.2.2 Evaluation of θ in Some Special Situations

To be specific, we focus on situations where the heat flux $\boldsymbol{q}$ is the only non-equilibrium flux, and we try to give a numerical estimation of the differences between θ and T. The first of the expressions (3.12) reduces to

$$\theta^{-1}(u,\boldsymbol{q}) = T^{-1}(u) - \frac{1}{2}\frac{\partial(\tau_1 v/\lambda T^2)}{\partial u}\boldsymbol{q}\cdot\boldsymbol{q}, \tag{3.15}$$

wherein $T(u)$ only depends on the internal energy, not on the heat flux.

For monatomic ideal gases obeying the Boltzmann equation, a well-known result of kinetic theory is that $\tau_1/\lambda T^2 = \frac{2}{5}(m/k_B^2 T^3 n)$, with m being the mass of a molecule, k_B the Boltzmann constant and n the number density, so that the inverse of the temperature is given by

$$\theta^{-1}(u,\boldsymbol{q}) = T^{-1}(u) + \frac{2}{5}\frac{\rho}{p^3 T}\boldsymbol{q}\cdot\boldsymbol{q}. \tag{3.16}$$

This expression provides an explicit estimation of the difference between T and θ.

For metallic rigid conductors, (3.15) is given by [3.3]

$$\theta^{-1}(u,\boldsymbol{q}) = T^{-1}(u) + \frac{9}{\pi^4}\frac{m^2\varepsilon_F}{nk_B^4 T^5}\boldsymbol{q}\cdot\boldsymbol{q}, \tag{3.17}$$

with m being the electron mass, n the electron number density, and ε_F the Fermi energy of the metal. Since the coefficients of the term in $\boldsymbol{q}\cdot\boldsymbol{q}$ are very small, it turns out that in many practical situations the corrective term may be neglected. Indeed, for CO_2 at 300 K and 0.1 atm, and for a heat flux of the order of 10^9 W/m^2, the difference $T - \theta$ is 9.6×10^{-2} K. However, in other circumstances the difference is not minute: in nuclear collisions (as studied in Sect. 17.4) $T - \theta$ is of the order of 7% of T, and in radiation near the surface of stars, the difference is close to 3% [3.4].

3.2.3 Alternative Definitions of Generalised Temperature

The problems of the definition and measurement of a non-equilibrium temperature have raised some questions [3.5]. A first one is related to the nature of the variables to be kept constant during the differentiation of the entropy in (3.14). With v and $\boldsymbol{q}$ fixed, one recovers (3.12a). If instead one keeps constant v and the quantity $(\tau_1/\rho\lambda T^2)^{1/2}\boldsymbol{q}$ [3.6], the derivatives of the generalised entropy and of the local-equilibrium entropy coincide. Indeed, since

$$s(u,v,\boldsymbol{q}) = s_{eq}(u,v) - \tau v (2\lambda T^2)^{-1}\boldsymbol{q}\cdot\boldsymbol{q}, \tag{3.18}$$

then of course

$$\left(\frac{\partial s}{\partial u}\right)_{v,\sqrt{\tau/(\lambda T^2)}\,\boldsymbol{q}} = \left(\frac{\partial s_{eq}}{\partial u}\right)_v. \tag{3.19}$$

Still another possibility [3.7] is to maintain the temperature gradient fixed and, since (3.18) implies that for steady situations

$$s = s_{eq} - \frac{\tau v \lambda}{2T^2}\nabla T \cdot \nabla T, \tag{3.20}$$

one is led to [3.3]

$$\left(\frac{\partial s}{\partial u}\right)_{v,\nabla T} = T^{-1} + (1+2b)(\tau\lambda/2\rho c_v T^3)\nabla T \cdot \nabla T, \tag{3.21}$$

where b is the exponent which characterizes the dependence of τ with T according to $\tau \approx T^{-b}$ (i.e. $b = 0$ for Maxwell molecules and $b = 1/2$ for hard spheres). It can be argued that not all these definitions may be valid: recall, indeed, that in equilibrium thermodynamics $(\partial s/\partial u)_v = T^{-1}$, but if one keeps pressure constant instead of volume during the differentiation, $(\partial s/\partial u)_p$ is different from T^{-1}; clearly $(\partial s/\partial u)_p = T^{-1}[1-(pv\alpha/c_p)]^{-1}$ with α being the coefficient of thermal expansion and c_p the specific heat at constant pressure. At the present time, it is not evident which among the above restrictions is the most suitable to define the non-equilibrium temperature.

The presence of a generalised temperature is not exclusive to EIT. In his entropy-free formulation of non-equilibrium thermodynamics, Meixner [3.8] postulated the existence of a dynamical temperature depending on the interactions of the system with the outside. Müller [3.9] used a 'coldness' function assumed to depend on the empirical temperature and its time derivative: in a steady state, the 'coldness' reduces to the local-equilibrium temperature, in contrast with the generalised temperature appearing in EIT. More recently, Muschik [3.10] introduced the notion of contact temperatures and explored the conceptual difficulties of their measurement; this contact temperature is not identical to the local-equilibrium temperature. Finally, Keizer [3.11] proposed a non-equilibrium temperature defined as the derivative with respect to the internal energy of a generalised entropy based on statistical considerations of molecular fluctuations; this temperature depends not only on the classical variables but also on the second moments of fluctuations. It is worth stressing that the EIT generalised temperature may also be expressed in terms of the second moments of the energy fluctuations, in analogy with Keizer's approach [2.12].

3.2.4 Experimental Hints for the Non-equilibrium Temperature

There is some experimental confirmation of the difference between the non-equilibrium temperature and the local-equilibrium temperature. There are strong indications from experiments on modulation optical spectrometry in laser-induced plasma in semiconductors submitted to an external electric field, that the temperature appearing in the non-equilibrium distribution function depends not only on the local-equilibrium variables but also on the fluxes [3.12]. In these experiments, one measures the intensity of radiative recombination $I(\omega)$, as a function of the frequency ω. It is found that

$$I(\omega) = g(\omega)\exp\left(-\frac{\hbar\omega - E_G}{k_B\theta}\right), \tag{3.22}$$

wherein E_G is the energy gap of the semiconductor and $g(\omega)$ the absorption coefficient, which depends only weakly on ω. Thus, a plot of $\ln I(\omega)$ as a function of ω allows θ to be determined. When the system is perturbed around an equilibrium state, it is seen that θ coincides with T. However, when the system is perturbed around a non-equilibrium steady state in the presence of an electric field E, it turns out that θ is indeed different from T and depends on the electric flux. For more details, the reader is referred to [3.12].

3.3 Non-equilibrium Equations of State: Thermodynamic Pressure

In (2.40) we wrote for the pressure tensor the expression $\mathbf{P} = \pi\mathbf{U} + p^v\mathbf{U} + \overset{0}{\mathbf{P}}{}^v$, where π on the right-hand side is replacing the equilibrium pressure p, which is usually found in the expressions for $\mathbf{P}$ in classical irreversible thermodynamics (CIT) [see, for instance, (1.18)]. In Sect. 2.3 we have provisionally accepted (2.40) as a consistent analogy with the replacement of temperature T by temperature θ and of pressure p by π in the generalised Gibbs equation (2.38). Here, we analyse this hypothesis in more depth.

The definition of a non-equilibrium pressure requires subtler considerations than that of the non-equilibrium temperature. It must be recalled from equilibrium thermodynamics that the entropy is a characteristic function (i.e. all conceivable thermodynamic information about the system is ascertainable from it) on condition that it is expressed in terms of extensive variables. Therefore, rather than the fluxes themselves, one should use vq, vp^v and $v\overset{0}{\mathbf{P}}{}^v$ as independent variables. Indeed, these variables are extensive in the following sense: if we have two systems of volumes V_1 and V_2 crossed by the same heat flux q, the variable Vq is additive, i.e. $V_{tot}q = V_1q + V_2q$, although q itself is not additive. The consequence of this choice will be reflected in the definition of the non-equilibrium pressure, which is obtained as the derivative of the entropy with respect to the volume at constant vq rather than at constant q. Since the temperature is related to the differential of the entropy at constant v, the fact that vq or q are kept constant does not modify the results, because v is a constant.

In view of the above considerations, the most suitable definition for the non-equilibrium pressure is

$$\pi\theta^{-1} = \left(\frac{\partial s}{\partial v}\right)_{u,vq}.$$

(3.23)

However, now a conceptual problem arises, because in an ideal gas the equilibrium pressure is defined as one third of the trace of the pressure tensor. Since for an ideal gas $p = \frac{2}{3}\rho u$ is independent of the fluxes, it seems that we are faced with a contradiction. It is however easy to convince oneself that there is no problem. Consider, for instance, an ideal gas submitted to a heat flux and that its contribution to $\mathbf{P}^v$ has the form $\mathbf{P}^v = \alpha qq$, where α is a phenomenological coefficient function of u and v. Expression (2.40) then takes the form

$$\mathbf{P} = \pi\mathbf{U} + \alpha qq.$$

(3.24)

Since the pressure tensor must satisfy $\mathrm{Tr}\,\mathbf{P} = 3p$, it is required that the coefficient α in (3.24) to be given by the condition $\mathrm{Tr}\,\mathbf{P} = 3\pi + \alpha q \cdot q = 3p$. This result makes clear that although the derivative of the entropy with respect to the volume depends on q, it is not in contradiction with the property that the trace of the tensor is not dependent on the flux. We add that the expression (3.24) is supported by other approaches, such as microscopic analyses of electromagnetic radiation [3.13], the Hamiltonian methods introduced in Chap. 4 and the information methods presented in Chap. 7.

It is helpful to write $\mathbf{P}$ explicitly in a simple situation, namely, when the heat flux is in the y direction, so

$$\mathbf{P} = \begin{pmatrix} \pi & 0 & 0 \\ 0 & \pi & 0 \\ 0 & 0 & \pi \end{pmatrix} + \begin{pmatrix} 0 & 0 & 0 \\ 0 & \alpha q_y^2 & 0 \\ 0 & 0 & 0 \end{pmatrix}.$$

(3.25)

This expression implies that the work of compression or expansion will depend on the relative direction between the axis of compression and the heat flux. This has been found in some computer simulations (see Sect. 9.4), where the thermodynamic pressure is seen to be equal to the minimum eigenvalue of the pressure tensor. This is indeed the situation found in (3.25), because $\pi < p$ and therefore $\alpha q^2 > 0$. A Gedankenexperiment analogous to that in Fig. 3.3 could be devised for the non-equilibrium pressure, by replacing the rigid conducting rod connecting the systems in Fig. 3.3 by a mobile piston.

Note that for an ideal gas under a heat flux, the definition (3.23) taken with vq as a constant yields $\pi/\theta = p/T$. Indeed,

$$\frac{\pi}{\theta} = \left(\frac{\partial s}{\partial v}\right)_{u,vq} = \left(\frac{\partial s_{eq}}{\partial v}\right)_{u} - \frac{1}{2}\frac{\partial}{\partial v}\left(\frac{\tau_1}{v\lambda T^2}\right)vq \cdot vq. \tag{3.26}$$

But, according to kinetic theory (see (5.44)) $\lambda = \frac{5}{2}(nk_BT^2/m)\tau_1$, and therefore $\tau_1/v\lambda$ is independent of v, as $vn = 1/m$. Therefore, the second term on the right-hand side in (3.26) vanishes and it is found that

$$\frac{\pi}{\theta} = \left(\frac{\partial s_{eq}}{\partial v}\right)_{u} = \frac{p}{T} = nk_B. \tag{3.27}$$

Note that the result would have been different if the derivative was performed at constant q. In this case one would have

$$\left(\frac{\partial s}{\partial v}\right)_{u,q} = \left(\frac{\partial s_{eq}}{\partial v}\right)_{u} - \frac{1}{2}\frac{\partial}{\partial v}\left(\frac{\tau_1 v}{\lambda T^2}\right)q \cdot q = \frac{p}{T} - \frac{\tau_1}{\lambda T^2}q \cdot q. \tag{3.28}$$

Let us go back to expression (3.25) for the pressure tensor when q is directed along the y axis; it is easy to see that $P_{xx} = P_{zz} = \pi < p$, and $P_{yy} = 3p - 2\pi > p$. Since for an ideal gas the components of the pressure tensor are related to the second moments of the velocity, we have

$$\left\langle \tfrac{1}{2}mv_xv_x\right\rangle = \left\langle \tfrac{1}{2}mv_zv_z\right\rangle = \tfrac{1}{2}k_B\theta < \tfrac{1}{2}k_BT,$$

$$\left\langle \tfrac{1}{2}mv_yv_y\right\rangle = \tfrac{1}{2}k_B(3T - 2\theta) > \tfrac{1}{2}k_BT. \tag{3.29}$$

Three points are worth noticing: (a) the average molecular kinetic energy $\left\langle \tfrac{1}{2}mv^2\right\rangle$ is given by $\tfrac{3}{2}k_BT$, in agreement with the definition of T in kinetic theory; (b) out of equilibrium, the equipartition theorem, implying that the average kinetic energy along the three axes is the same, is no longer valid: the average energy is lower in the directions orthogonal to the heat flux. This consequence could in principle be checked, for instance, by analysing the Doppler broadening of emission lines in excited rarefied gases along the direction of the heat flux and perpendicularly to it [3.14]; and (c) a fraction $\frac{3}{2}k_B\theta$ is equally distributed in the three degrees of freedom along the three axes, whereas the remaining fraction $\frac{3}{2}k_B(T-\theta)$ is ordered along the axis corresponding to the heat flux. This observation supports the remark in Sect. 3.2.1 about the splitting of the internal energy in an 'ordered' and a 'disordered' part.

These considerations remind us that thermometry is not a trivial subject. Since the mean kinetic molecular energy along the direction x is not the same along the direction y, a thermometer will indicate different temperatures according to its relative position with respect to the direction of q. This is not totally surprising because in non-equilibrium there is no longer equipartition, and therefore, thermometers which are sensitive to different degrees of freedom will indicate different values for the temperature. For instance, if we have a mixture of matter and radiation at different temperatures, a completely reflecting thermometer will be sensitive only to matter, and it will indicate the temperature of matter, whereas a thermometer with perfectly black walls will read a value which is the average temperature of the mixture.

Another situation worthy of attention is found in fluids at uniform temperature subjected to a steady shear flow in the x direction. Since the non-equilibrium contributions to the pressure are of second order in the shear rate $\dot{\gamma}$, we must also go to this same order in the expression of the viscous pressure tensor, which will be given by $\mathbf{P}^v = -2\eta\mathbf{V} + \mu_{22}\mathbf{V}\cdot\mathbf{V}$. According to (2.40), the total pressure tensor will be written

$$\mathbf{P} = \begin{pmatrix} \pi & 0 & 0 \\ 0 & \pi & 0 \\ 0 & 0 & \pi \end{pmatrix} + \begin{pmatrix} \mu_{22}\eta^2\dot{\gamma}^2 & -\eta\dot{\gamma} & 0 \\ -\eta\dot{\gamma} & \mu_{22}\eta^2\dot{\gamma}^2 & 0 \\ 0 & 0 & 0 \end{pmatrix}. \tag{3.30}$$

For purely Newtonian fluids, $\mu_{22} = 0$, and for non-Newtonian fluids, the terms in μ_{22} are related to normal pressure effects (see Sect. 15.1). Ideal gases are Newtonian at low shear rates but exhibit normal pressure effects at higher values of $\dot{\gamma}$. By writing $\pi = p - a\dot{\gamma}^2$, the condition $\mathrm{Tr}\,\mathbf{P} = 3p$ imposes that

$$-3a\dot{\gamma}^2 + 2\mu_{22}\eta^2\dot{\gamma}^2 = 0, \tag{3.31}$$

which provides a relation between η and μ_{22}. According to (3.4), the entropy has the form

$$s(u,v,\mathbf{P}^v) = s_{eq}(u,v) - \frac{\tau_2 v}{2\eta T}(P_{12}^v)^2. \tag{3.32}$$

Taking into account that according to kinetic theory [see (5.44)], $\eta = p\tau_2$ and $pv =$ constant at constant internal energy, one has

$$\frac{\pi}{\theta} = \left(\frac{\partial s}{\partial v}\right)_{u,vP^v} = \frac{p}{T} = nk_B, \tag{3.33}$$

which is analogous to the result (3.27). From $\theta < T$, it follows that $\pi < p$; therefore $a > 0$ in (3.31) and henceforth $\mu_{22} > 0$ in virtue of (3.31). It turns out that the diagonal term of tensor $\mathbf{P}$ with the lowest value is precisely π. We comment further on this point in Sect. 9.4.

Similar considerations apply to the equations of state for the chemical potential and have important consequences in several phenomena. Note that the following occur: a shift of the critical point and the coexistence lines in the phase diagram of polymer solutions under a shear rate, and the displacement of chemical equilibrium under a flow or a heat flux. Some examples are analysed in Chap. 16.

3.4 Convexity Requirements and Stability

Now we turn our attention to the requirements stemming from the convexity of entropy, namely, the requirement that the second differential $\delta^2 s$ of the entropy must be negative definite. After integrating the generalised Gibbs equation (2.11) and expanding the entropy around the local equilibrium value up to the second order in the fluxes, one obtains

$$s(u,v,\boldsymbol{q},p^v,\overset{0}{\mathbf{P}}{}^v) = s_{eq}(u,v) - \frac{v\tau_1}{2\lambda T^2}\boldsymbol{q}\cdot\boldsymbol{q} - \frac{v\tau_0}{2\zeta T}p^v p^v - \frac{v\tau_2}{4\eta T}\overset{0}{\mathbf{P}}{}^v : \overset{0}{\mathbf{P}}{}^v. \tag{3.34}$$

When the dissipative fluxes are zero, one recovers the classical conditions for stability,

$$c_v = T(\partial s/\partial T)_v > 0, \qquad k_T = -(1/v)(\partial v/\partial p)_T > 0, \tag{3.35a}$$

with c_v being the specific heat at constant volume and k_T the isothermal compressibility. When the non-classical terms containing the fluxes are taken into account, the convexity of the entropy (3.34) implies that

$$\frac{v\tau_1}{\lambda T^2} > 0, \qquad \frac{v\tau_0}{\zeta T} > 0, \qquad \frac{v\tau_2}{2\eta T} > 0. \tag{3.35b}$$

Because the local-equilibrium temperature T is positive, and since the second law requires that the dissipative coefficients λ, ζ, and η are all positive, it turns out from (3.35b) that the relaxation times τ_1, τ_0, and τ_2 must be positive, otherwise causality would be violated.

The property of convexity of entropy is equivalent to the requirement that the field equations form a symmetric hyperbolic system. Such a property ensures that the problem is well posed and that the characteristic speeds are real and finite. The relation between symmetric hyperbolicity and the existence of a convex function whose production has a definite sign, that is, the existence of an entropy function, has been examined from a general mathematical point of view by several authors [3.15–16]. Such a result is important as it indicates that the condition of hyperbolicity implies the existence of an entropy function. Here we have adopted the reciprocal point of view: starting from the hypothesis of the existence of a generalised entropy, we have arrived at hyperbolic field equations.

Let us now show using simple examples, such as heat conduction in a rigid solid and an ideal gas, and Couette flow in a viscous fluid, that convexity is satisfied only for values of the fluxes smaller than some critical values. Up to now, we have developed entropy around the local-equilibrium value. Here, we discuss the convexity requirements when the reference state is a non-equilibrium steady state, characterized by the values u (or T), $\mathbf{q}$ and $\mathbf{P}^v$ of the state variables.

3.4.1 Heat Conduction in a Rigid Solid

Let us first consider heat conduction in a rigid solid for which $s = s(u, \mathbf{q})$. The second differential of the generalised entropy (2.32) is then given by

$$\delta^2 s = -\left(\frac{1}{c_v T^2} + \frac{1}{2}q^2\frac{\partial^2\alpha}{\partial u^2}\right)(\delta u)^2 - \alpha\delta\mathbf{q}\cdot\delta\mathbf{q} - 2q\frac{\partial\alpha}{\partial u}\delta u\cdot\delta\mathbf{q}, \qquad (3.36)$$

where use is made of (2.31) with $c_v = \partial u/\partial T$ and $\alpha = \tau_1 v/\lambda T^2$. If v, τ_1, c_v and λ are constant, one has $\partial\alpha/\partial u = -2\alpha(c_v T)^{-1}$ and $\partial^2\alpha/\partial u^2 = 6\alpha(c_v T)^{-2}$. The expression for $\delta^2 s$ is definite negative if the matrix Δ of its coefficients, given by

$$\Delta = -\begin{pmatrix} \dfrac{1}{c_v T^2} + \dfrac{1}{2}\dfrac{\partial^2\alpha}{\partial u^2}q^2 & \dfrac{\partial\alpha}{\partial u}q \\[2ex] \dfrac{\partial\alpha}{\partial u}q & \alpha \end{pmatrix}, \qquad (3.37)$$

is negative. This requires that the diagonal components and the determinant of Δ are negative. The first condition is valid for all values of the heat flux. The second condition is expressed by

$$\frac{\alpha}{c_v T^2} + \left[\frac{\alpha}{2} \frac{\partial^2 \alpha}{\partial u^2} - \left(\frac{\partial \alpha}{\partial u} \right)^2 \right] q^2 \geq 0 ; \tag{3.38}$$

This inequality is satisfied for values of q such that (see Problem 3.5)

$$q \leq \left(\frac{c_v}{\alpha} \right)^{1/2} = \rho c_v T \left(\frac{\lambda}{\rho c_v \tau_1} \right)^{1/2} , \tag{3.39}$$

where $(\lambda/\rho c_v \tau_1)^{1/2} = U$ is the maximum speed of the thermal waves, as shown in Sect. 10.1. Thus, thermodynamic stability is only guaranteed when the heat flux is lower than the critical value (3.39).

Furthermore, when this condition is satisfied, one avoids an unsatisfactory property of the telegrapher equation, namely, that an initial absolute temperature profile which is positive everywhere may take negative values during some time intervals [3.17]; this unpleasant result is not found with parabolic diffusion equations, where the positive character of the temperature field is always respected. In the context of hyperbolic heat transport, one should impose as initial conditions not only a positive temperature profile but also that the heat flux is bounded everywhere by the critical value $\rho u U$.

3.4.2 Heat Conduction in Ideal Gases

A more general situation taking into account that the generalised entropy $s(u,v,q)$ depends also on v yields the following matrix for the coefficients of the second-order derivatives of s:

$$\delta^2 s = \begin{pmatrix} \dfrac{\partial^2 s_{eq}}{\partial u^2} - \dfrac{1}{2} \dfrac{\partial^2 \alpha}{\partial u^2} q^2 & \dfrac{\partial^2 s_{eq}}{\partial u \partial v} - \dfrac{1}{2} \dfrac{\partial^2 \alpha}{\partial u \partial v} q^2 & -\dfrac{\partial \alpha_1}{\partial u} q \\[2ex] \dfrac{\partial^2 s_{eq}}{\partial u \partial v} - \dfrac{1}{2} \dfrac{\partial^2 \alpha}{\partial u \partial v} q^2 & \dfrac{\partial^2 s_{eq}}{\partial v^2} - \dfrac{1}{2} \dfrac{\partial^2 \alpha}{\partial v^2} q^2 & -\dfrac{\partial \alpha_1}{\partial v} q \\[2ex] -\dfrac{\partial \alpha}{\partial u} q & -\dfrac{\partial \alpha}{\partial v} q & -\alpha_1 \end{pmatrix} , \tag{3.40}$$

where, as in (3.36), α stands for $\alpha = \tau_1 v/\lambda T^2$. To obtain a more explicit expression, consider an ideal monatomic gas for which, according to the kinetic theory of gases,

$\lambda = \frac{5}{2}(nk_BT^2/m)\tau_1$ and therefore $\alpha = \frac{2}{5}(mv/pk_BT^2)$ [see (5.44)]. Furthermore, the caloric and thermal equations of state are $u = \frac{3}{2}k_BT/m$ and $p = nk_BT$, where n is the number density of particles and m the mass of the particles. Simple calculations lead to

$$
\delta^2 s = \begin{pmatrix}
-\dfrac{2m}{3k_BT^2} - \dfrac{16m^2}{15n^2k_B^4T^5}q^2 & \dfrac{4m^2}{5nk_B^3T^4}q^2 & \dfrac{4m}{5n^2k_B^3T^4}q \\[4mm]
\dfrac{4m^2}{5nk_B^3T^4}q^2 & -mn^2k_B - \dfrac{2m^2}{5k_B^2T^3}q^2 & -\dfrac{4m}{5nk_B^2T^3}q \\[4mm]
\dfrac{4m}{5n^2k_B^3T^4}q & -\dfrac{4m}{5nk_B^2T^3}q & -\dfrac{2}{5n^2k_B^2T^3}
\end{pmatrix}. \tag{3.41}
$$

Necessary conditions for a matrix to be negative definite are that the diagonal terms are negative and the complete and partial determinants along the diagonal have alternate signs. The first condition is always fulfilled by (3.41), but not the latter. Such conditions state that the minor M formed by the first two columns and the first two rows is positive. This quantity is found to be

$$
M = \tfrac{2}{3}(m^2n^2/T^2) + \tfrac{4}{3}(m^3/k_B^3T^5)q^2 - \tfrac{16}{75}(m^4/n^2k_B^6T^8)q^4. \tag{3.42}
$$

It is positive for q not exceeding $q_{c1} = 2.67nk_BT(k_BT/m)^{1/2} = 1.03\rho u v_{rms}$, with $\rho u = \frac{3}{2}nk_BT$ and v_{rms} being the root mean-square velocity given by $v_{rms} = (3k_BT/m)^{1/2}$. This corresponds to a critical temperature gradient $\partial \ln T/\partial x = (1.85/\ell)$, with $\ell = \tau v_{rms}$ being the mean free path. The determinant of the matrix (3.41) is

$$
\det \Delta = -\tfrac{4}{15}(m^2/k_B^2T^5) + \tfrac{8}{15}(m^3/n^2k_B^5T^8)q^2. \tag{3.43}
$$

It remains negative for heat fluxes smaller than $q_{c2} = 0.27\rho u v_{rms}$. This second value puts a more severe limitation on the maximum allowable value of the temperature gradient, ensuring the validity of EIT. The limiting value is now $\partial \ln T/\partial x = (0.49/\ell)$, which corresponds in dilute gases at room temperature to a temperature gradient of the order of 10^5 K/cm. Accordingly, $\det\Delta$ will remain negative for perturbations of wavelength not smaller than $\ell/0.49 = 2.04\ell$. The above result is interesting as it gives a hint about the domain of applicability of EIT, which appears to be well suited for wavelengths slightly greater than the mean free path.

3.4.3 Shear Viscous Pressure in Viscous Fluids

Further information on the stability restrictions on the fluxes is provided by the study of an ordinary viscous fluid when heat effects are negligible. The entropy (3.4) is now given by

$$s(u,v,\overset{0}{\mathbf{P}}{}^{v}) = s_{eq}(u,v) - (\tau_2 v/4\eta T)\overset{0}{\mathbf{P}}{}^{v} : \overset{0}{\mathbf{P}}{}^{v}. \qquad (3.44)$$

Comparison with the kinetic theory of gases shows that $\eta = p\tau$ and as a consequence $\tau_2 v/4\eta T = (mv^2/4k_B T^2)$. By restricting the analysis to a plane Couette flow, for which $\overset{0}{\mathbf{P}}{}^{v} : \overset{0}{\mathbf{P}}{}^{v} = 2\eta^2\dot\gamma^2$, $\dot\gamma$ being as before the shear rate, a calculation parallel to the previous one yields

$$\Delta = \begin{pmatrix} -\dfrac{2m}{3k_B T^2} - \dfrac{4m^3 v^2}{3k_B^3 T^4}\eta^2\dot\gamma^2 & \dfrac{4m^2 v}{3k_B^2 T^3}\eta^2\dot\gamma^2 & \dfrac{4m^2 v^2}{3k_B^2 T^3}\eta\dot\gamma \\[3ex] \dfrac{4m^2 v}{3k_B^2 T^3}\eta^2\dot\gamma^2 & -mn^2 k_B - \dfrac{m}{k_B T^2}\eta^2\dot\gamma^2 & -\dfrac{2mv}{k_B T^2}\eta\dot\gamma \\[3ex] \dfrac{4m^2 v^2}{3k_B^2 T^3}\eta\dot\gamma & -\dfrac{2mv}{k_B T^2}\eta\dot\gamma & -\dfrac{mv^2}{k_B T^2} \end{pmatrix}. \qquad (3.45)$$

As before, the diagonal elements are always negative, whereas the first minor formed by the first two rows and the first two columns,

$$M = \tfrac{2}{3}(m^2 n^2/T^2) + 2(m^2/k_B^2 T^4)(\eta\dot\gamma)^2 - \tfrac{4}{9}(m^2/n^2 k_B^4 T^6)(\eta\dot\gamma)^4, \qquad (3.46)$$

remains positive up to a critical value $(\eta\dot\gamma)^2 = 4.81(nk_B T)^2$. In view of the kinetic result $\eta = p\tau$, one has $(\dot\gamma\,\tau)_c = 2.19$. On the other hand, the determinant of (3.45) results in

$$\det\Delta = -\tfrac{1}{3}(mk_B/T^4) - \tfrac{7}{9}(m/n^2 k_B^3 T^6)(\eta\dot\gamma) - \tfrac{10}{9}(m/n^4 k_B^5 T^8)(\eta\dot\gamma)^2 \qquad (3.47)$$

which is negative for $\dot\gamma\,\tau$ lower than $(\dot\gamma\,\tau)_c = 0.51$. It is thus concluded that (3.47) is negative for values of $\dot\gamma^{-1} > 1.96\tau$. This gives the domain of validity of the relaxational equation for the viscous pressure.

The above results are comforting. Indeed, it would not be realistic to conceive a thermodynamic description for length-scales shorter than the mean free path, or for time-scales shorter than the mean collision time; at these length-scales and time-scales the behaviour of the particles is individual and ballistic rather than collective, because of the lack of a sufficient number of collisions. Therefore, the result that the generalised entropy remains convex for lengths and time intervals at least of the order of the mean free path and the collision time, respectively, makes plausible the hypotheses underlying EIT. However, it must be emphasized that this self-consistency does not imply at all that the simple model based on the Maxwell–Cattaneo equations has a general validity, it merely exhibits its self-consistency as a thermodynamic model.

Problems

3.1 In a monatomic ideal gas, the thermal conductivity is given by $\lambda = \frac{5}{2}(nk_B^2 T/m)\tau_1$. (a) Determine θ^{-1} in the presence of a heat flux q by using (3.14). (b) Calculate the difference between θ and the local-equilibrium temperature T for Ar (atomic weight 40) at 0.1 atm and 300 K. (c) According to (3.23), obtain the generalised pressure π under a heat flux, and evaluate the difference between it and the local-equilibrium pressure p for Ar at 0.1 atm and 300 K.

3.2 (a) For an ideal monatomic gas, the internal energy is given by $\rho u = (3/2)nk_B T$, whether in equilibrium or in non-equilibrium. Write ρu in terms of the generalised absolute temperature θ and the heat flux q. (b) Show that, up to the second-order terms in q, the caloric equation of state can be given in the form

$$\rho u(\theta, q) = \rho u_{eq}(\theta) + a(\theta) q \cdot q,$$

with $a(\theta) = (1/\theta)(\tau/\lambda) - \frac{1}{2} d(\tau v/\lambda)/d\theta$, and where $u_{eq}(\theta)$ is the value of u for $q = 0$.

3.3 The thermal conductivity of a dielectric solid is $\lambda = \frac{1}{3}\rho c_v c_0^2 \tau_1$, with c_0 being the phonon speed, τ_1 the relaxation time due to resistive phonon collisions, c_v the heat capacity per unit mass, and ρ the mass density. In the Debye model, c_v is proportional to T^3 at low temperature. Determine the difference $\theta - T$ when the solid is subjected to a heat flux q.

3.4 Show that inequality (3.38) is indeed satisfied for values of the heat flux lower than the critical value (3.39) for $\alpha = \tau_1 v / \lambda T^2$, with τ_1, v and λ being constant.

3.5 Consider a monatomic gas with internal energy U. Assume that the molecular motions may be organized in such a way that all the molecules have the same velocity in the same direction, so that the total molecular energy E is equal to U. Evaluate the energy flux when all the molecules are oriented in the same direction and compare its value with the value of the critical heat flux derived from (3.43).

3.6 For a monatomic gas at temperature T subject to a heat flux q, the specific internal energy per unit mass is $u = \frac{3}{2}(k_B T/m)$, and its specific entropy is (2.32). Determine the internal energy u' of the gas in such a way that its equilibrium entropy $s_{eq}(u')$ is equal to the entropy $s(u, q)$. What would be the corresponding temperature T' of the gas? Compare T' with T and the generalised absolute temperature θ.

Chapter 4

Hamiltonian Formulations

Hamiltonian formulations have been identified at various levels of description: the microscopic one (classical mechanics of particles), the kinetic theory (based on a distribution function rather than on precise values for the mechanical variables of particles), and macroscopic approaches (as, for instance, hydrodynamics or equilibrium thermodynamics) [4.1–5]. Therefore, it is natural to ask whether such a structure is preserved in mesoscopic intermediate descriptions and more particularly in extended irreversible thermodynamics (EIT). This compatibility condition among different levels of description is especially useful in the non-linear domain. There are several reasons that militate in favour of a Hamiltonian description: it is attractive because of its conciseness and its physical content. Indeed, the whole set of balance equations is now expressed in terms of one or several generating functionals which may be generally identified with a well-defined physical quantity such as the energy, the entropy, or the Gibbs free energy. Moreover, there exist many elegant results and powerful methods of solution typically developed for general Hamiltonian systems which can be of direct use in analysing the solutions of the basic equations of EIT. Finally, Hamiltonian techniques have a wide range of validity – they are not restricted to the linear regime – and provide supplementary restrictions which complement those of the second law as well as a link between thermodynamics and dynamics at several levels of description.

Another tool providing a complementary point of view with respect to thermodynamics is the search for variational principles; this allows a whole set of evolution equations to be condensed into one single expression and to exploit in an optimal way the invariance properties of the Lagrangian. The requirement that the evolution equations may derive from a variational principle places on them restrictions which may be interesting from the physical point of view. These topics are of high interest but will not be covered in the present monograph; the reader is referred to the appropriate literature [4.6].

In view of the non-universality of selecting variables out of equilibrium, the Hamiltonian approach leaves the state variables undetermined (denoted by the symbol x) and concentrates first on the structure of the governing equations. This structure is expected to be universal, since it is what guarantees that solutions of the governing equations agree with the universal experience that externally unforced systems reach states at which their behaviour is found to be well described by equilibrium thermo-dynamics. In Sect. 4.1 we present the general structure, and in Sect. 4.2 we choose x to be the set of classical hydrodynamic fields or the fields used in EIT.

4.1 GENERIC Formulation

Early Hamiltonian versions of non-equilibrium thermodynamics were proposed by Grmela [4.1–2] and were based on one single generator; the time evolution of the basic variables was expressed in terms of a non-dissipative Poisson bracket plus a dissipative bracket. Some papers have been devoted to show the compatibility of EIT with this early Hamiltonian formulation [4.4]. Though the latter has been very successful (see [4.5] for a wide review of applications and analysis of the foundations), it has been recently superseded in some aspects by the so-called GENERIC (General Equation for the Non-Equilibrium Reversible-Irreversible Coupling) formulation by Grmela and Öttinger [4.7], which is especially elegant and useful in Rheology. Here, we will present this approach and discuss its compatibility with EIT.

In GENERIC it is assumed that the time evolution equations can be given the form

$$\frac{\mathrm{d}x}{\mathrm{d}t} = \mathbf{L} \cdot \frac{\delta E}{\delta x} + \mathbf{M} \cdot \frac{\delta S}{\delta x}, \tag{4.1}$$

where x represents a set of independent variables required for the complete description of the non-equilibrium system (for instance, hydrodynamical fields and additional structural variables), E and S are the total energy and entropy of the system expressed in terms of x, and $\mathbf{L}$ and $\mathbf{M}$ are linear functional operators, expressing reversible and irre-versible kinematics, respectively. The dot indicates the multiplication of a vector by a matrix and $\delta/\delta x$ stands for the Volterra functional derivative which reduces to the usual partial derivative in the absence of non-local effects. The first term in the right-hand side of (4.1) is purely Hamiltonian and represents the reversible contribution to the time evolution equations of x, whereas the second term corresponds to the irreversible contributions.

Equation (4.1) is supplemented by the following degeneracy requirements

$$\mathbf{L} \cdot \frac{\delta S}{\delta x} = 0, \quad \mathbf{M} \cdot \frac{\delta E}{\delta x} = 0. \tag{4.2}$$

The first condition expresses that $\delta S/\delta x$ is in the null space of $\mathbf{L}$ and that the functional form of the entropy cannot contribute to the reversible part of the dynamics. The second one expresses that the total energy is not affected by the irreversible contribution to the dynamics. Note that these degeneracy requirements express fundamental and general physical aspects.

Furthermore, one associates to the matrices $\mathbf{L}$ and $\mathbf{M}$ the following brackets:

$$\{A,B\} = \left\langle \frac{\delta A}{\delta x}, \mathbf{L} \cdot \frac{\delta B}{\delta x} \right\rangle,$$

$$\tag{4.3}$$

$$[A,B] = \left\langle \frac{\delta A}{\delta x}, \mathbf{M} \cdot \frac{\delta B}{\delta x} \right\rangle,$$

where $\langle\,,\,\rangle$ denotes a scalar product. Braces $\{\,,\,\}$ stand for the Poisson bracket of classical mechanics [4.8], whereas the brackets $[\,,\,]$ are intended to describe the dissipative behaviour [4.1, 4.5].

In terms of these brackets and the chain rule, (4.1) yields the following form for the evolution equation of an arbitrary function $A(x)$:

$$\frac{\mathrm{d}A}{\mathrm{d}t} = \{A,E\} + [A,S]. \tag{4.4}$$

Further conditions about $\mathbf{L}$ and $\mathbf{M}$ are obtained from the following properties of the brackets:

$$\{A,B\} = -\{B,A\}\,, \qquad \text{(antisymmetry)} \tag{4.5a}$$

$$\{A,\{B,C\}\} + \{B,\{C,A\}\} + \{C,\{A,B\}\} = 0\,, \qquad \text{(Jacobi identity)} \tag{4.5b}$$

which are the well-known properties of Poisson brackets in classical mechanics. Note that the Jacobi identity guarantees that the bracket $\{A,B\}$ is preserved in the reversible time evolution. Indeed, by replacing A by $\{A,B\}$ in (4.4) one has

$$\frac{d\{A,B\}}{dt} = \{\{A,B\},E\}.$$

(4.6)

But we also have, of course,

$$\frac{d\{A,B\}}{dt} = \left\{\frac{dA}{dt},B\right\} + \left\{A,\frac{dB}{dt}\right\} = \{\{A,E\},B\} + \{A,\{B,E\}\}.$$

(4.7)

The equality of the right-hand side of (4.6) and (4.7) is the Jacobi identity (4.5b). In some problems, it may be a difficult task to prove the Jacobi identity which in certain circumstances may require to be numerically checked.

From condition (4.5a) it follows that $\mathbf{L}$ is antisymmetric:

$$\mathbf{L}(x) = -\mathbf{L}^T(x).$$

(4.8)

The properties of $\mathbf{M}$ are derived from the symmetry condition

$$[A,B] = [B,A]$$

(4.9)

and the non-negative condition

$$[A,A] \geq 0.$$

(4.10)

This implies that $\mathbf{M}$ is symmetric (provided that all the variables x have the same time-reversal parity) and semipositive definite:

$$\mathbf{M}(x) = \mathbf{M}^T(x),$$

(4.11a)

$$y \cdot \mathbf{M}(x) \cdot y \geq 0 \qquad \text{for all } y.$$

(4.11b)

This non-negativeness, together with the degeneracy requirement (4.2a), guarantees that $dS/dt \geq 0$, while the symmetry property (4.11a) is directly related to Onsager's reciprocal relations. The symmetry of $\mathbf{M}$ is only valid for variables with the same parity. Otherwise, $\mathbf{M}$ should contain a non-symmetrical part in agreement with Onsager–Casimir theory.

GENERIC may be viewed as an extension of Ginzburg–Landau equation [4.9] describing evolution towards equilibrium and given by

$$\frac{dx}{dt} = -\mathbf{M} \cdot \frac{\delta \phi}{\delta x},$$

(4.12)

where $\mathbf{M}$ is a positive-definite linear operator and ϕ a given potential. The interest of GENERIC lies in its compact, abstract and general framework. Details of the evolution equation (4.1) depend of course on the specific choice made for the two generators E and S and the two matrices $\mathbf{L}$ and $\mathbf{M}$. Restrictions on $\mathbf{L}$ and $\mathbf{M}$ are imposed by (4.5b), (4.8) and (4.11a,b).

The main innovation of GENERIC with respect to earlier bracket formalisms is the use of two different generators, E and S, instead of a single one [4.7]: this gives more flexibility to the theory and emphasizes the central role played by the thermodynamic potentials E and S. The matrix $\mathbf{L}$ is determined by the behaviour of the variables x under space transformations, whereas the dynamical material information enters in the friction matrix $\mathbf{M}$, which is related to transport coefficients.

From the properties $\{A, A\} = 0$ and $[A, A] \geq 0$ and the degeneracy conditions (4.2), one sees that

$$\frac{dE}{dt} = 0 \ , \qquad \frac{dS}{dt} \geq 0. \tag{4.13}$$

Now, given the variables E and N (total number of particles), let us determine how their equilibrium values are expressed in terms of x. Since $\mathbf{L}$ and $\mathbf{M}$ are degenerate, we will have $\mathbf{L} \cdot \partial N/\partial x = \mathbf{M} \cdot \partial N/\partial x = 0$ besides the constraints (4.2). The properties (4.13) suggest that the equilibrium solution x_{eq} [which is the time-independent solution of (4.1) when t approaches infinity] is the solution that maximizes S under the constraints $E =$ constant and $N =$ constant. To show this, we introduce the potential

$$\Phi(x,T,\mu) = -S(x) + \frac{1}{T} E(x) - \frac{\mu}{T} N(x), \tag{4.14}$$

where T^{-1} and $-\mu T^{-1}$ play the role of Lagrange multipliers, with T and μ the temperature and the chemical potential. The state x_{eq} that maximizes S under the above constraints is a solution of

$$\frac{\delta \Phi}{\delta x} = 0. \tag{4.15}$$

Indeed, the degeneracy conditions allow one to express (4.1) in the form

$$\frac{dx}{dt} = T\mathbf{L} \cdot \frac{\partial \Phi}{\partial x} - \mathbf{M} \cdot \frac{\partial \Phi}{\partial x}, \tag{4.16}$$

from which follows clearly that the solution x_{eq} of (4.15) is the time-independent solution of (4.16), and therefore of the GENERIC equation (4.1). In view of Euler's relation, it can also be stated that the potential (4.14) evaluated at equilibrium is given by

$$\Phi\big(x_{eq}(T,\mu),T,\mu\big) = -\frac{pV}{T},\qquad(4.17)$$

with V being the total volume and p the pressure.

4.2 Reversible and Irreversible Kinematics

In this section we shall focus on a particular realization of the GENERIC structure, i.e. a specification of x, $E(x)$, $S(x)$, $\mathbf{L}(x)$ and $\mathbf{M}(x)$ for which (4.1), if appropriately linearized, becomes equivalent to the governing equations of EIT. In this section we shall also discuss in more detail the physical meaning of the operators $\mathbf{L}(x)$ and $\mathbf{M}(x)$ and some mathematical concepts.

4.2.1 State Variables x and Reversible Kinematics $\mathbf{L}(x)$

Since we want to compare GENERIC with EIT, we will select the same variables as in EIT. To be explicit, let us first consider classical hydrodynamics and specify how the state variables x are chosen and how $\mathbf{L}(x)$ is determined. In this situation, the reversible motion of a fluid is seen as a continuous group of transformations $\mathbb{R}^3 \to \mathbb{R}^3$. Mass density and entropy fields are allowed to be simply advected by this motion. It is useful to choose the entropy field rather than the energy field as a state variable since it is then easier to construct the $\mathbf{L}(x)$ that satisfies the degeneracy requirement $\mathbf{L}(x)\cdot\partial S/\partial x = 0$.

In order to find the expression for $\mathbf{L}$, let us introduce the following Poisson bracket [4.8]:

$$\{A,B\} = \int dV\, u_\gamma\Big(\partial_\alpha(A_{u_\gamma})b_{u_\alpha} - \partial_\alpha(b_{u_\gamma})A_{u_\alpha}\Big).\qquad(4.18)$$

In (4.18) A and B $(= \int b\,dV)$ denote functionals of the momentum $u(r)$ $[= \rho v(r)]$, r is the position vector, $\partial_\alpha := \partial/\partial r_\alpha$ and $A_{u_\gamma} := \delta A/\delta u_\gamma(r)$, with $\delta/\delta u_\gamma(r)$ the Volterra functional derivative and the summation convention over repeated indices is used. It can be verified that conditions (4.5) and (4.6) are satisfied so that (4.18) is a Poisson bracket. According to (4.3), the Poisson operator $\mathbf{L}$ is obtained from (4.18) by rewriting $\{A, B\}$ as

$$\{A,B\} = \int dV\, A_{u_\alpha} L_{\alpha\beta}(u)b_{u_\beta}.\qquad(4.19)$$

4.2.1.1 Euler Hydrodynamics

If in addition to the momentum field u we introduce the mass density field $\rho(r)$ and the entropy field $s(r)$ as variables, the Poisson bracket (4.18) generalises to

$$\{A,B\} = \int dV \Big[u_\gamma \Big(\partial_\alpha(A_{u_\gamma}) b_{u_\alpha} - \partial_\alpha(b_{u_\gamma}) A_{u_\alpha} \Big)$$

$$+ \rho \Big(\partial_\alpha(A_\rho) b_{u_\alpha} - \partial_\alpha(b_\rho) A_{u_\alpha} \Big)$$

$$+ s \Big(\partial_\alpha(A_s) b_{u_\alpha} - \partial_\alpha(b_s) A_{u_\alpha} \Big) \Big]. \tag{4.20}$$

The reversible (Euler) hydrodynamic equations are easily obtained from (4.20) [i.e. from (4.1) with the last term on the right-hand side missing]. Indeed, let us start from (4.4), which reduces to

$$\frac{dA}{dt} = \{A, E\} \tag{4.21}$$

and is required to hold for all functionals A. We note that the left-hand side of (4.21) can be written in the form

$$\frac{dA}{dt} = \int dV \left(A_\rho \frac{\partial \rho}{\partial t} + A_{u_\alpha} \frac{\partial u_\alpha}{\partial t} + A_s \frac{\partial s}{\partial t} \right). \tag{4.22}$$

By replacing in (4.20) the quantity A by E, the right-hand side of (4.21) can be rewritten (by using integrations by parts and assuming that the integrals over the boundary that arise in the calculations are equal to zero) into the form

$$\{A,E\} = \int dV \Big\{ A_\rho \Big[-\partial_\alpha(\rho e_{u_\alpha}) \Big] + A_{u_\alpha} \Big[-\partial_\gamma(u_\alpha e_{u_\gamma})$$

$$- \rho \partial_\alpha e_\rho - s \partial_\alpha e_s - u_\gamma \partial_\alpha e_{u_\gamma} \Big]$$

$$+ A_s \Big[-\partial_\alpha(s e_{u_\alpha}) \Big] \Big\}. \tag{4.23}$$

The boundary conditions satisfying the above requirement are realized either in an infinite domain when the fields decrease sufficiently rapidly to zero as $|r| \to \infty$ or in a finite domain with periodic boundary conditions. In both cases one restricts the analysis to externally unforced systems (recall that boundary conditions represent an external influence).

If (4.21) is required to hold for all A, then we must have

$$\frac{\partial \rho}{\partial t} = -\partial_\gamma \left(\rho e_{u_\gamma} \right),$$

$$\frac{\partial u_\alpha}{\partial t} = -\partial_\gamma \left(u_\alpha e_{u_\gamma} \right) - \rho \partial_\alpha e_\rho - s \partial_\alpha e_s - u_\gamma \partial_\alpha e_{u_\gamma}, \tag{4.24}$$

$$\frac{\partial s}{\partial t} = -\partial_\gamma \left(s e_{u_\gamma} \right).$$

We see easily that $\mathbf{L} \cdot \delta S / \delta x = 0$ for $S = \int d\mathbf{r}\, s(\mathbf{r})$ and $\mathbf{L} \cdot \delta N / \delta x = 0$.

The function E has the physical meaning of the total energy. By choosing

$$E = \int dV\, e(\rho, s, \mathbf{u}; \mathbf{r}) = \int dV \left(\tfrac{1}{2} (u^2 / \rho) + \varepsilon(\rho, s; \mathbf{r}) \right) \tag{4.25}$$

where ε is the internal energy per unit volume, then

$$e_{u_\gamma} = \rho^{-1} u_\gamma \tag{4.26}$$

is the velocity and

$$-\rho \partial_\alpha e_\rho - s \partial_\alpha e_s - u_\gamma \partial_\alpha e_{u_\gamma} = -\partial_\alpha p,$$

where

$$p(\mathbf{r}) = -e(\mathbf{r}) + \rho(\mathbf{r}) e_{\rho(\mathbf{r})} + s(\mathbf{r}) e_{s(\mathbf{r})} + u_\gamma(\mathbf{r}) e_{u_\gamma(\mathbf{r})}, \tag{4.27}$$

is the local pressure. We see now that (4.24) take the familiar form of Euler equations

$$\frac{\partial \rho}{\partial t} = -\partial_\gamma (u_\gamma),$$

$$\frac{\partial u_\alpha}{\partial t} = -\partial_\gamma \left(\frac{u_\alpha u_\gamma}{\rho} \right) - \partial_\alpha p, \tag{4.28}$$

$$\frac{\partial s}{\partial t} = -\partial_\gamma \left(\frac{s u_\gamma}{\rho} \right).$$

It is left as an exercice (see Problem 4.2) to verify that the Poisson operator of the evolution equations for ρ, $\boldsymbol{u}$ and ε is given by

$$
\mathbf{L}(r) = - \begin{pmatrix} 0 & \nabla\rho & 0 \\ \rho\nabla & (\nabla\boldsymbol{u}+\boldsymbol{u}\nabla)^T & \varepsilon\nabla+\nabla p \\ 0 & \nabla\varepsilon+p\nabla & 0 \end{pmatrix},
\tag{4.29}
$$

while

$$
\frac{\delta e}{\delta x} = \begin{pmatrix} \dfrac{\partial}{\partial\rho} \\[4pt] \dfrac{\partial}{\partial\boldsymbol{u}} \\[4pt] \dfrac{\partial}{\partial\varepsilon} \end{pmatrix} e(\rho,\boldsymbol{u},\varepsilon) = \begin{pmatrix} -(1/2)v^2 \\ v \\ 1 \end{pmatrix}.
\tag{4.30}
$$

We note that the usual local-equilibrium relation (4.27) has appeared as a result. We can thus expect that in the context of an extended state space we shall arrive in the same way at an extension of the usual local-equilibrium relation.

4.2.1.2 Extended Irreversible Thermodynamics

We propose now an EIT-type extension by following the GENERIC approach. First we suggest a symmetric tensor field $\mathbf{p}(r)$ be introduced as a new state variable (its relation to the pressure tensor will become clear later on). To take this new contribution into account, the Poisson bracket (4.20) must be complemented by the additional terms

$$
\int dV \left\{ p_{\gamma\beta} \left[\partial_\alpha (A_{p_{\gamma\beta}}) b_{u_\alpha} - \partial_\alpha (b_{p_{\gamma\beta}}) A_{u_\alpha} \right] \right.
$$

$$
- p_{\alpha\beta} \left[A_{p_{\gamma\beta}} \partial_\alpha (b_{u_\gamma}) - b_{p_{\gamma\beta}} \partial_\alpha (A_{u_\gamma}) \right]
$$

$$
\left. - p_{\alpha\beta} \left[A_{p_{\gamma\alpha}} \partial_\beta (b_{u_\gamma}) - b_{p_{\gamma\alpha}} \partial_\beta (A_{u_\gamma}) \right] \right\}.
\tag{4.31}
$$

With the Poisson brackets (4.20) and (4.31), and using the same type of calculations to arrive at (4.24) from (4.20), the Euler hydrodynamic equations (4.20) will be replaced by

$$
\frac{\partial\rho}{\partial t} = -\partial_\gamma \left(\rho e_{u_\gamma} \right),
\tag{4.32a}
$$

$$\frac{\partial u_\alpha}{\partial t} = -\partial_\gamma \left(u_\alpha e_{u_\gamma} \right) - \partial_\alpha p - \partial_\gamma P_{\alpha\gamma}, \tag{4.32b}$$

$$\frac{\partial p_{\alpha\beta}}{\partial t} = -\partial_\gamma \left(p_{\alpha\beta} e_{u_\gamma} \right) - p_{\gamma\beta}\partial_\gamma \left(e_{u_\alpha} \right) - p_{\alpha\gamma}\partial_\gamma \left(e_{u_\beta} \right), \tag{4.32c}$$

$$\frac{\partial s}{\partial t} = -\partial_\gamma \left(s e_{u_\gamma} \right), \tag{4.32d}$$

where

$$p = -e + \rho e_\rho + s e_s + u_\gamma e_{u_\gamma} + p_{\alpha\beta} e_{p_{\alpha\beta}}. \tag{4.33}$$

The usual pressure tensor $P_{\alpha\gamma}$ is related to $p_{\alpha\beta}$ by

$$P_{\alpha\beta} = 2 p_{\alpha\gamma} e_{p_{\gamma\beta}}. \tag{4.34}$$

Expression (4.33) is the extension of the local-equilibrium relation (4.27), while (4.34) represents the relation between the pressure tensor **P** and the tensor **p** introduced as an additional state variable. We shall show that (4.32a–d) together with the appropriate irreversible contribution do reduce, after linearization, to the corresponding equations of EIT.

The final step would be to introduce the energy flux as an additional state variable. We could follow the same procedure as previously used with the field $p(r)$, i.e. to introduce a vector field and allow it to be simply advected by the flow, but in this way we would only enter into more details of the 'mechanical' part of dynamics. This would result in a modified expression for the pressure tensor and a change in that for the hydrostatic pressure; however, and this is important, (4.32d), governing the time evolution of the entropy field, would remain unchanged. But, from a physical point of view, it is the 'thermal or entropic' part of dynamics that we want to singularize by taking the energy flux as an independent state variable. To achieve this objective, we shall follow the methodology of [4.10].

We consider first heat transport without material flow. In the context of classical thermodynamics, the heat flow is only of irreversible nature (i.e. it will give rise to a positive production of entropy). But here we will adopt a different point of view. We recall that, from the microscopic point of view, we are focusing our attention on the rapid oscillations of atoms that we can picture as a phonon gas. Following [4.10], we shall consider this motion in the same way as the material motion; but the mass density $\rho(r)$ is now replaced by the entropy density $s(r)$ and the momentum field $u(r)$ by another momentum field that we denote $w(r)$. The Poisson bracket expressing the kinematics of $s(r)$ and $w(r)$ will therefore be [see (4.18) and (4.20)]

$$\{A,B\} = \int dV \left\{ w_\gamma \left[\partial_\alpha (A_{w_\gamma}) b_{w_\alpha} - \partial_\alpha (b_{w_\gamma}) A_{w_\alpha} \right] \right.$$

$$\left. + s \left[\partial_\alpha (A_s) b_{w_\alpha} - \partial_\alpha (b_s) A_{w_\alpha} \right] \right\}. \tag{4.35}$$

If we now consider both the mass flow as well as the heat flow, we shall require that the mass will be advected with the mass momentum u and the entropy by the sum of the mass momentum u and the momentum w. We thus make the transformation $(u, w) \to (u', w')$, where $u' \equiv u + w$. Under this transformation, the Poisson bracket (4.20), (4.31) [without the last term on the right-hand side of (4.20)] and (4.35) transforms into another Poisson bracket [to simplify the notation in (4.36) we use u instead of u')

$$\{A,B\} = \int dV \left\{ u_\gamma \left[\partial_\alpha (A_{u_\gamma}) b_{u_\alpha} - \partial_\alpha (b_{u_\gamma}) A_{u_\alpha} \right] \right.$$

$$+ \rho \left[\partial_\alpha (A_\rho) b_{u_\alpha} - \partial_\alpha (b_\rho) A_{u_\alpha} \right] + p_{\gamma\beta} \left[\partial_\alpha (A_{p_{\gamma\beta}}) b_{u_\alpha} - \partial_\alpha (b_{p_{\gamma\beta}}) A_{u_\alpha} \right]$$

$$- p_{\alpha\beta} \left[A_{p_{\gamma\beta}} \partial_\alpha (b_{u_\gamma}) - b_{p_{\gamma\beta}} \partial_\alpha (A_{u_\gamma}) \right] - p_{\alpha\beta} \left[A_{p_{\gamma\alpha}} \partial_\beta (b_{u_\gamma}) - b_{p_{\alpha\gamma}} \partial_\beta (A_{u_\gamma}) \right]$$

$$+ s \left[\partial_\alpha (A_s) b_{u_\alpha} - \partial_\alpha (b_s) A_{u_\alpha} \right] + s \left[\partial_\alpha (A_s) B_{w_\alpha} - \partial_\alpha (B_s) A_{w_\alpha} \right]$$

$$+ w_\gamma \left[\partial_\alpha (A_{u_\gamma}) b_{w_\alpha} - \partial_\alpha (b_{u_\gamma}) A_{w_\alpha} \right] + w_\gamma \left[\partial_\alpha (A_{w_\gamma}) b_{u_\alpha} - \partial_\alpha (b_{w_\gamma}) A_{u_\alpha} \right]$$

$$\left. + w_\gamma \left[\partial_\alpha (A_{w_\gamma}) b_{w_\alpha} - \partial_\alpha (b_{w_\gamma}) A_{w_\alpha} \right] \right\}. \tag{4.36}$$

By repeating the same type of calculations that led from (4.20) to (4.24), the corresponding reversible time evolution equations are shown to be

$$\frac{\partial \rho}{\partial t} = -\partial_\gamma \left(\rho e_{u_\gamma} \right), \tag{4.37a}$$

$$\frac{\partial u_\alpha}{\partial t} = -\partial_\gamma \left(u_\alpha e_{u_\gamma} \right) - \partial_\alpha p - \partial_\gamma P_{\alpha\gamma}, \tag{4.37b}$$

$$\frac{\partial P_{\alpha\beta}}{\partial t} = -\partial_\gamma\left(P_{\alpha\beta}e_{u_\gamma}\right) - P_{\gamma\beta}\partial_\gamma\left(e_{u_\alpha}\right) - P_{\gamma\alpha}\partial_\gamma\left(e_{u_\beta}\right), \tag{4.37c}$$

$$\frac{\partial s}{\partial t} = -\partial_\gamma\left(se_{u_\gamma}\right) - \partial_\gamma\left(se_{w_\gamma}\right), \tag{4.37d}$$

$$\frac{\partial w_\alpha}{\partial t} = -\partial_\gamma\left(w_\alpha e_{u_\gamma}\right) - \partial_\gamma\left(w_\alpha e_{w_\gamma}\right) - s\partial_\alpha(e_s) - w_\gamma\partial_\alpha\left(e_{u_\gamma}\right) - w_\gamma\partial_\alpha\left(e_{w_\gamma}\right), \tag{4.37e}$$

where

$$e = \varepsilon + \frac{u_\gamma^2}{\rho} + \frac{w_\gamma^2}{\rho}, \tag{4.38}$$

$$p = -e + \rho e_\rho + s e_s + u_\gamma e_{u_\gamma} + w_\gamma e_{w_\gamma} + P_{\alpha\beta}e_{P_{\alpha\beta}}, \tag{4.39}$$

and

$$P_{\alpha\beta} = 2P_{\beta\gamma}e_{P_{\gamma\alpha}} + w_\alpha e_{w_\beta}. \tag{4.40}$$

If we express the time derivative of the total energy as follows

$$\frac{\partial e}{\partial t} = e_\rho\frac{\partial\rho}{\partial t} + e_{u_\alpha}\frac{\partial u_\alpha}{\partial t} + e_s\frac{\partial s}{\partial t} + e_{w_\gamma}\frac{\partial w_\gamma}{\partial t} + e_{P_{\alpha\beta}}\frac{\partial P_{\alpha\beta}}{\partial t},$$

we obtain from (4.37–39) the energy balance law

$$\frac{\partial e}{\partial t} = -\partial_\gamma\left(ee_{u_\gamma} + pe_{u_\gamma} + P_{\gamma\beta}e_{u_\beta} + q_\gamma\right), \tag{4.41}$$

where q_γ is defined as

$$q_\gamma = \left(se_s + w_\alpha e_{w_\alpha}\right)e_{w_\gamma}. \tag{4.42}$$

Expressions (4.40) and (4.42) give the relations between the state variables $\mathbf{p}$ and w and the EIT fields $(\mathbf{P}, q)$ ($\mathbf{P}$ is the reversible pressure tensor and q is the extra reversible energy flux) and we also see from (4.37d) that the entropy flux is se_w. It should be emphasized that these extra fluxes arise as a direct consequence of the Hamiltonian formalism described in this section. The price paid for writing the evolution equations in GENERIC form (4.1) is to substitute the usual variables of EIT by some combination of them; see in that respect expressions (4.40) and (4.42) of $P_{\alpha\beta}$ and q_γ. Moreover, the physical meaning of the new GENERIC variables $p_{\alpha\beta}$ and w_γ is not directly evident.

In addition, a word of caution about the terminology is in order. In GENERIC, the distribution between reversible and irreversible processes is not always very clear and is different from EIT. In the GENERIC literature, it is admitted that the reversible and irreversible contributions are those generated respectively by the operator $\mathbf{L}$ and the operator $\mathbf{M}$ in (4.1).

4.2.2 Irreversible Kinematics $\mathbf{M}(x)$

In the previous subsection we determined the first term on the right-hand side of (4.1) [see (4.37)]; what remains is to determine the second term involving the operator $\mathbf{M}(x)$. The procedure for obtaining $\mathbf{M}(x)$ is given in the following.

First, we recall that we have to satisfy the degeneracy requirements expressing the conservation of total energy and total number of particles, namely $\mathbf{M} \cdot \delta E/\delta x = 0$ and $\mathbf{M} \cdot \delta N/\delta x = 0$. From the GENERIC point of view, we expect that only the extra fields $\mathbf{p}$ and w will dissipate, which means that irreversible contributions will appear only in their evolution equations. The rest of the state variables will dissipate only indirectly through the dependence of their time evolution on $\mathbf{p}$ and w. Next, we shall assume that $\mathbf{p}$ and w are not far from their equilibrium values, which means that $\delta s/\delta p$ and $\delta s/\delta w$ are small. The simplest way to account for irreversible time evolution is thus to add $\Lambda_{\alpha\beta}\delta s/\delta w_\beta$ to the right-hand side of the equation governing the time evolution of w and $\lambda_{\alpha\beta\gamma\delta}\delta s/\delta p_{\gamma\delta} +\tilde{\lambda}(\delta s/\delta p_{\gamma\gamma})\delta_{\alpha\beta}$ to the right-hand side of the evolution equation of $\mathbf{p}$. We require that $\tilde{\lambda}$, λ and Λ are positive definite, in order to guarantee that $\mathbf{M}$ is non-negative. The resulting complete set of time evolution equations for $(\rho, u, s, \mathbf{p}, w)$ is therefore

$$\frac{\partial \rho}{\partial t} = -\partial_\gamma\left(\rho e_{u_\gamma}\right), \tag{4.43a}$$

$$\frac{\partial u_\alpha}{\partial t} = -\partial_\gamma\left(u_\alpha e_{u_\gamma}\right) - \partial_\alpha p - \partial_\gamma P_{\alpha\gamma}, \tag{4.43b}$$

$$\frac{\partial s}{\partial t} = -\partial_\gamma\left(s e_{u_\gamma}\right) - \partial_\gamma\left(s e_{w_\gamma}\right) + \sigma^s., \tag{4.43c}$$

$$\frac{\partial p_{\alpha\beta}}{\partial t} = -\partial_\gamma\left(p_{\alpha\beta} e_{u_\gamma}\right) - p_{\gamma\beta}\partial_\gamma\left(e_{u_\alpha}\right) - p_{\gamma\alpha}\partial_\gamma\left(e_{u_\beta}\right) - e_s\lambda_{\alpha\beta\gamma\delta}e_{p_{\gamma\delta}}$$

$$-e_s\lambda_{\alpha\beta\gamma\delta}e_{p_{\gamma\delta}} - e_s\tilde{\lambda}e_{p_{\gamma\gamma}}\delta_{\alpha\beta}, \tag{4.43d}$$

$$\frac{\partial w_\alpha}{\partial t} = -\partial_\gamma\left(w_\alpha e_{u_\gamma}\right) - \partial_\gamma\left(w_\alpha e_{w_\gamma}\right) - s\partial_\alpha(e_s)$$

$$-w_\gamma\partial_\alpha\left(e_{u_\gamma}\right) - w_\gamma\partial_\alpha\left(e_{w_\gamma}\right) - e_s\Lambda_{\alpha\beta}e_{w_\beta}. \tag{4.43e}$$

The quantities p and $P_{\alpha\beta}$ are given by (4.39) and (4.40), while the entropy production σ^s is given by

$$\sigma^s = e_{p_{\alpha\beta}}\lambda_{\alpha\beta\gamma\delta}e_{p_{\gamma\delta}} + e_{w_\alpha}\Lambda_{\alpha\beta}e_{w_\beta} + e_{p_{\gamma\gamma}}\tilde{\lambda}e_{p_{\gamma\gamma}}, \tag{4.44}$$

from which it follows that $\sigma^s \geq 0$. The time evolution of the energy remains unchanged and is still governed by (4.41).

The limitation to linear terms in $\delta e/\delta \mathbf{p}$ and $\delta e/\delta w$ can be removed by introducing the concept of a dissipation potential Ψ. We say that Ψ is a dissipation potential if Ψ is a function of $\delta e/\delta \mathbf{p}$ and $\delta e/\delta w$ satisfying the following properties: $\Psi(0, 0) = 0$, Ψ reaches its minimum at $(0, 0)$ and Ψ is a concave function in the vicinity of $(0, 0)$. By addition of the terms $-e_s(\delta\Psi/\delta p_{\alpha\beta})$ and $-e_s(\delta\Psi/\delta w_\alpha)$ to the right-hand side of (4.43d) and (4.43e), respectively (instead of the terms $-e_s\lambda_{\alpha\beta\gamma\delta}E_{p_{\gamma\delta}} - e_s\tilde{\lambda}E_{p_{\gamma\gamma}}\delta_{\alpha\beta}$ and $-e_s\Lambda_{\alpha\beta}E_{w_\beta}$), we still obtain equations possessing the GENERIC structure. We note that the choice made in (4.43) corresponds to the following dissipative potential

$$\Psi = \tfrac{1}{2}\int dV\, e_{p_{\alpha\beta}}\lambda_{\alpha\beta\gamma\delta}e_{p_{\gamma\delta}} + \tfrac{1}{2}\int dV\, e_{w_\alpha}\Lambda_{\alpha\beta}e_{w_\beta} + \tfrac{1}{2}\int dV\, e_{p_{\gamma\gamma}}\tilde{\lambda}e_{p_{\gamma\gamma}}.$$

Finally, we recapitulate the relations between the new state variables $(\mathbf{p}, w)$ and the pressure tensor $\mathbf{P}$, the heat flux q and the extra entropy flux $J^{(s)}$ [see (4.40), (4.42) and (4.43c)]:

$$P_{\alpha\beta} = 2p_{\gamma\beta}e_{p_{\gamma\alpha}} + w_\alpha e_{w_\beta}, \tag{4.45a}$$

$$q_\alpha = \left(se_s + w_\gamma e_{w_\gamma}\right)e_{w_\alpha}, \tag{4.45b}$$

$$J_\alpha^{(s)} = se_{w_\alpha}. \tag{4.45c}$$

We note that for a given energy e we can always pass from $(\mathbf{p}, w)$ to $(\mathbf{P}, q, J^{(s)})$. This also means, of course, that only two of the quantities $\mathbf{P}$, q and $J^{(s)}$ are independent. The existence of the inverse is however not guaranteed. It can be stated that the formulation based on $(\mathbf{P}, \boldsymbol{q}, \boldsymbol{J}^{(s)})$ is more 'macroscopic' than the one based on $(\mathbf{p}, \boldsymbol{w})$, as the macroscopic meaning of the latter variables is not clearly specified. It appears that $\mathbf{p}$ and $\boldsymbol{w}$ are more directly related to microscopic quantities. Indeed, in some situations $\mathbf{p}$ has been identified as the average conformation tensor in polymer solutions, which is a quantity closer to a microscopic interpretation than $\mathbf{P}$ itself. In a linear approximation, the use of $\mathbf{p}$ or $\mathbf{P}$ is immaterial, but in a non-linear approach, different microscopic configurations (implying different $\mathbf{p}$'s) could yield the same macroscopic $\mathbf{P}$. In fact, we expect that the relation (4.45) is one-to-one only exceptionally. One example of such a particular situation, corresponding to EIT, is discussed in the subsequent section.

4.3 Governing Equations of EIT

We return now to EIT. The next step, after choosing the state variables, is the formulation of equations governing their time evolution. We recall that the state variables chosen in EIT are $(\rho(\boldsymbol{r}), \boldsymbol{u}(\boldsymbol{r}), s(\boldsymbol{r}), \mathbf{P}(\boldsymbol{r}), \boldsymbol{q}(\boldsymbol{r}))$, where ρ, $\boldsymbol{u}$ and s are the standard hydrodynamic fields, $\mathbf{P}$ is the viscous pressure tensor, and $\boldsymbol{q}$ is the heat flux. Working in the context of GENERIC, the governing equations have to comply with the requirements of compatibility with the conservation laws $dE/dt = 0$ and $dN/dt = 0$, and the dissipation law $dS/dt > 0$. A possible choice for the energy e, the entropy s and the entropy production σ^s is the following

$$e(\rho, \boldsymbol{u}, s, \mathbf{P}, \boldsymbol{q}) = e_0(\rho, \boldsymbol{u}, s) + e_1(\mathbf{P}, \boldsymbol{q}), \tag{4.46}$$

$$s(\rho, \boldsymbol{u}, s, \mathbf{P}, \boldsymbol{q}) = s_0(\rho, \boldsymbol{u}, s) + s_1(\mathbf{P}, \boldsymbol{q}), \tag{4.47}$$

$$\sigma^s(\rho, \boldsymbol{u}, s, \mathbf{P}, \boldsymbol{q}) = \sigma_1^s(\mathbf{P}, \boldsymbol{q}), \tag{4.48}$$

where e_1, s_1 and σ_1^s are quadratic functions of $\mathbf{P}$, $\boldsymbol{q}$, with $\sigma^s \geq 0$.

In the previous section we formulated the time evolution equations that possess the GENERIC structure, but we did not restrict the choice of e, s, and σ^s. We now show that, by selecting the expressions (4.46)–(4.48), one recovers the governing equations of EIT.

Let the free energy Φ (4.14) be chosen in such a way that the equilibrium states [i.e. solutions to (4.15)] are $\left(\rho^{(eq)}, \boldsymbol{u}^{(eq)}, \mathbf{p}^{(eq)}, \boldsymbol{w}^{(eq)}\right)$, where

$$p_{\alpha\beta}^{(eq)} = p^{(0)}\delta_{\alpha\beta}, \qquad w_\alpha^{(eq)} = 0, \tag{4.49}$$

and $p^{(0)}$ is a constant. Let us introduce

$$P_{\alpha\beta} = p^{(eq)}_{\alpha\beta} + \pi_{\alpha\beta}, \tag{4.50}$$

$$w_\alpha = w^{(eq)}_\alpha + v_\alpha. \tag{4.51}$$

Assuming that Φ is quadratic in $\pi_{\alpha\beta}$ and $\mathbf{v}_\alpha$, one has

$$\Phi_{P_{\alpha\beta}} = a\pi_{\alpha\beta}, \qquad \Phi_{w_\alpha} = bv_\alpha, \tag{4.52}$$

where a and b are constants. If we now substitute (4.50–51) into (4.45), limit ourselves to terms linear in $\boldsymbol{\pi}$ and $\mathbf{v}$, and replace $e_{P_{\alpha\beta}}$ and e_{w_α} in (4.45) with $T\Phi_{P_{\alpha\beta}}$ and $T\Phi_{w_\alpha}$ (T is the equilibrium temperature), we obtain, after making use of (4.51) and (4.52),

$$P_{\alpha\beta} = 2Tp^{(0)}a\pi_{\alpha\beta}, \tag{4.53}$$

$$q_\alpha = Tbs\theta v_\alpha, \tag{4.54}$$

$$J^{(s)}_\alpha = Tbsv_\alpha, \tag{4.55}$$

where $e_s = \theta$ is the non-equilibrium temperature. We note that (4.53–54) represent a one-to-one relation between $(\mathbf{P}, \mathbf{q})$ and $(\boldsymbol{\pi}, \mathbf{v})$. Moreover, by comparing (4.54) and (4.55), we recognize the familiar relation $J^{(s)}_\alpha = q_\alpha\theta^{-1}$ between the entropy flux and the energy flux.

Now we turn our attention to the evolution equations (4.43). If we limit the developments to linear terms in $\mathbf{P}$ and $\mathbf{q}$, the first three equations are the standard hydrodynamic equations,

$$\frac{\partial\rho}{\partial t} = -\partial_\gamma\left(\rho e_{u_\gamma}\right), \tag{4.56a}$$

$$\frac{\partial u_\alpha}{\partial t} = -\partial_\gamma\left(u_\alpha e_{u_\gamma}\right) - \partial_\alpha p - \partial_\gamma P_{\alpha\gamma}, \tag{4.56b}$$

$$\frac{\partial s}{\partial t} = -\partial_\gamma\left(s e_{u_\gamma}\right) - \partial_\gamma\left(q_\gamma\theta^{-1}\right) + \sigma^s, \tag{4.56c}$$

with a modified local-equilibrium relation. Instead of the local-equilibrium relation (4.27) we now have the relation [see (4.39)]

$$p = -e + \rho e_\rho + s e_s + u_\gamma e_{u_\gamma} + \tfrac{1}{2} P_{\gamma\gamma}. \qquad (4.57)$$

The modification of the classical definition of the pressure is related to the non-equilibrium pressure π introduced in Sect. 3.3. It is interesting to notice that the Hamiltonian formulation gives a clear-cut expression for the modified pressure. This feature outlines the usefulness of the latter formulation in connection with the problem of extending thermodynamics far from the local-equilibrium approach.

The equations that govern the time evolution of **P** and q are easily obtained from (4.43d,e) and (4.49–52); in the linear approximation it is found that

$$\frac{\partial P_{\alpha\beta}}{\partial t} = -2T\left(p^{(0)}\right)^2 a\left(\delta_{\alpha\beta}\partial_\gamma(e_{u_\gamma}) + \partial_\beta(e_{u_\alpha}) + \partial_\alpha(e_{u_\beta})\right)$$

$$-a\theta\lambda_{\alpha\beta\gamma\delta}P_{\gamma\delta} - a\theta\tilde{\lambda}P_{\gamma\gamma}\delta_{\alpha\beta}, \qquad (4.58)$$

$$\frac{\partial q_\alpha}{\partial t} = -\theta T b s^2 \partial_\alpha\theta - \theta b \Lambda_{\alpha\beta}q_\beta. \qquad (4.59)$$

Expressions (4.58–59) are similar to the basic equations (2.82) and (2.84) established in the framework of EIT, under the conditions to define appropriately the heat conductivity, the viscosity and the relaxation times. It can therefore be stated that EIT shares the properties necessary to possess a Hamiltonian structure. Observe also the absence of terms in $\nabla \cdot \mathbf{P}$ and ∇q in (4.58) and (4.59); to obtain them in the Hamiltonian approach it would be necessary to admit that the potential Φ (or s) is not only a function of the basic set ρ, u, s, **p**, w, but also of their derivatives with respect to the position coordinates [4.12].

Problems

4.1 Show that (4.18) satisfies the properties (4.5) defining a Poisson bracket.

4.2 Determine the **L** and **M** operators for classical hydrodynamics, namely, for

$$\frac{\partial\rho}{\partial t} = -\partial_\gamma\left(\rho v_\gamma\right)$$

$$\frac{\partial u_\alpha}{\partial t} = -\partial_\gamma\left(\rho v_\alpha v_\gamma\right) - \partial_\alpha p + \eta\partial_\gamma\partial_\gamma v$$

$$\frac{\partial \varepsilon}{\partial t} = -\partial_\gamma(\varepsilon v_\gamma) - p\partial_\gamma v_\gamma + \lambda \partial_\gamma \partial_\gamma T + \eta(\partial_\gamma v\alpha)(\partial_\gamma v\alpha)$$

Check that the consistency conditions (4.2) are satisfied.

4.3 In the Gaussian approximation, the polymer entropy S_p in a dilute polymer solution is given by [see (15.35)]

$$S_p = \tfrac{1}{2}n_p k_B \int \left\{ \mathrm{Tr}[\mathbf{U} - \alpha \mathbf{C}(r)] + \ln[\det \alpha \mathbf{C}(r)] \right\} dr, \qquad (4.60)$$

where $\mathbf{C}$ is the configuration tensor defined by

$$\mathbf{C}(r) = (n_p)^{-1} \int \mathbf{RR}\psi(r, \mathbf{R}) d\mathbf{R}, \qquad (4.61)$$

with $\mathbf{R}$ being the end-to-end vector of macromolecules, ψ the configurational distribution function, n_p the number of polymer molecules per unit volume of the solution; the constant α is chosen such that $\mathbf{C}(r) = \mathbf{U}$ at equilibrium (i.e. $\alpha = H/k_B T$, with H being the elastic constant of the macromolecule). According to the consistency conditions of GENERIC, the contribution $\mathbf{P}_p$ of the polymer to the pressure tensor $\mathbf{P}$ should be

$$\mathbf{P} = T\left(2\mathbf{C} \cdot \frac{\delta S_p}{\delta \mathbf{C}} + S_p \mathbf{U} \right), \qquad (4.62)$$

where a divergence-free term may be added to $\mathbf{P}/T$. (a) Show that

$$\frac{\delta S}{\delta \mathbf{C}} = \frac{1}{2} n_p k_B \left[\mathbf{C}^{-1}(r) - \alpha \mathbf{U} \right]. \qquad (4.63)$$

(b) Apply this result to show that

$$\mathbf{P} = n_p k_B T(\mathbf{U} - \alpha \mathbf{C}). \qquad (4.64)$$

This is called the Giesekus form of the pressure tensor, which will be used in (15.81).

4.4 (a) Expand the entropy (4.60) of the above problem up to the second order in $\mathbf{C}$. (b) Show that application of (4.62) of the above problem yields for $\mathbf{P}_p$ the result

$$\mathbf{P} = n_p k_B T\alpha \mathbf{C} \cdot (\mathbf{U} - \alpha \mathbf{C}),$$

which is not supported by the microscopic theory and which only coincides with (4.64) of the previous problem for small values of $\mathbf{C}$.

Part II

Microscopic Foundations

Chapter 5

The Kinetic Theory of Gases

The aim of this chapter is to provide a microscopic interpretation of extended irreversible thermodynamics (EIT) by means of the kinetic theory of gases. The interface between the macroscopic description and the kinetic theory is shown to be much wider in EIT than in the classical irreversible thermodynamics (CIT). As a consequence, the comparison provides more information in the extended case than in the classical situation.

Our purpose is to justify the hypotheses and the main results of EIT. To be explicit, we have to substantiate (a) the choice of the dissipative fluxes as extra independent variables in the description of systems out of equilibrium, (b) the form of the generalised Gibbs equation and (c) the form of the entropy flux. Moreover, the evolution equations for the fluxes as well as the relations between the transport coefficients in these equations demand a sound justification.

The main part of this chapter addresses the study of the linear terms of the evolution equations. We have not systematically considered the non-linear terms, and will only occasionally insist on them. Our attitude is not dictated by convenience, but it has a more fundamental justification: as noted in Chap. 2, thermodynamics, in its present state of development, is not able to impose explicit restrictions on the non-linear terms.

This chapter concerns essentially ideal non-relativistic monatomic gases, whereas the corrections arising from the interaction potential in non-ideal gases are also examined up to the lowest order in the density. Finally, for pedagogical reasons, we postpone the study of the relativistic kinetic theory to Chap. 17.

5.1 The Basic Concepts of Kinetic Theory

We first consider ideal or highly diluted monatomic gases. The basis for the analysis is the distribution function $f(r,c,t)$, which accounts for the number of particles between r and $r + dr$ with velocity between c and $c + dc$ at time t. The evolution of $f(r,c,t)$ is described by the well-known Boltzmann equation, which takes into account the effects of binary collisions between particles, and neglects collisions involving more than two particles, a quite plausible hypothesis in dilute gases [5.1–3].

The Boltzmann equation has the form

$$\frac{\partial f}{\partial t} + c \cdot \frac{\partial f}{\partial r} + \frac{F}{m} \cdot \frac{\partial f}{\partial c} = \int d\tilde{c} \int d\Omega |c - \tilde{c}| \sigma(c - \tilde{c}, \theta)[f'\tilde{f}' - f\tilde{f}]. \qquad (5.1)$$

Here, f, $\tilde{f}$, f', and $\tilde{f}'$ stand for $f(r,c,t)$, $f(r,\tilde{c},t)$, $f(r,c',t)$, and $f(r,\tilde{c}',t)$ respectively; m is the mass of the particles and F the external force acting on the particles; $\sigma(c - \tilde{c}, \theta)$ is the differential cross-section of the collisions between the particles, one of them with initial velocity c and the other with initial velocity $\tilde{c}$, which give as final velocities after collision c' and $\tilde{c}'$; θ is the angle between c and c'; $d\Omega$ is the differential solid angle around θ.

We consider only one single species of molecules, without internal degrees of freedom. If different types of molecules are concerned, a separate distribution function for each type must be introduced, and the collision term will couple them.

The Boltzmann equation is a non-linear, integro-differential equation which is very difficult to solve. However, several general consequences can be drawn from it even without solving it explicitly. This arises as a consequence of an important and very useful symmetry property of the collision term. Let us write (5.1) as

$$\frac{\partial f}{\partial t} + c \cdot \frac{\partial f}{\partial r} + \frac{F}{m} \cdot \frac{\partial f}{\partial c} = J(f), \qquad (5.2)$$

with $J(f)$ the collision term and let $\psi(c)$ be an arbitrary function of the molecular velocity c. Then, the following relation is satisfied

$$\int \psi(c) J(f) dc = \tfrac{1}{4} \int [\psi(c) + \psi(\tilde{c}) - \psi(c') - \psi(\tilde{c}')] J(f) dc. \qquad (5.3)$$

This equality is a consequence of the next three relations. First, it is immediately seen that

$$\int \psi(c) J(f) dc = \int dc \int d\tilde{c} \int d\Omega |c - \tilde{c}| \sigma [f'\tilde{f}' - f\tilde{f}] \psi(c) = \int \psi(\tilde{c}) J(f) dc. \qquad (5.4)$$

The equality is obvious, for in (5.4) c and $\tilde{c}$ are dummy quantities, since they are integrated over all their possible values; hence an exchange between c and $\tilde{c}$ is irrelevant.

A second equality is obtained after replacing c and $\tilde{c}$ by c' and $\tilde{c}'$; it yields

$$\int \psi(c')J(f')\mathrm{d}c' = \int \mathrm{d}c' \int \mathrm{d}\tilde{c}' \int \mathrm{d}\Omega' \, |c' - \tilde{c}'| \sigma' [f\tilde{f} - f'\tilde{f}']\psi(c')$$

$$= -\int \mathrm{d}c \int \mathrm{d}\tilde{c} \int \mathrm{d}\Omega \, |c - \tilde{c}| \sigma [f'\tilde{f}' - f\tilde{f}]\psi(\tilde{c}) = -\int \psi(\tilde{c})J(f)\mathrm{d}c. \tag{5.5}$$

This equality results clearly from the properties $\mathrm{d}c\,\mathrm{d}\tilde{c} = \mathrm{d}c'\mathrm{d}\tilde{c}'$, $c - \tilde{c} = c' - \tilde{c}'$, and $\sigma = \sigma(c - \tilde{c}, \theta) = \sigma(c' - \tilde{c}', \theta') = \sigma'$. The first two of these equalities are valid for elastic binary collisions, while the third is satisfied when the intermolecular potential is invariant under spatial rotations and reflections, and under time reversal.

A third equality follows from the exchange of the dummy quantities c' and $\tilde{c}'$:

$$\int \psi(c')J(f')\mathrm{d}c' = \int \psi(\tilde{c}')J(\tilde{f}')\mathrm{d}\tilde{c}'. \tag{5.6}$$

Now, by combining (5.4), (5.5) and (5.6), the key relation (5.3) is obtained.

5.1.1 Balance Equations

The first important consequence of (5.3) is the possibility of deriving the hydrodynamic balance equations for mass, momentum, and energy from Boltzmann's equation. The mass density ρ, the mean velocity v, and the internal energy u per unit mass are defined in terms of the distribution function as follows:

$$\rho(r,t) = \int m f(r,c,t)\mathrm{d}c, \tag{5.7}$$

$$\rho(r,t)v(r,t) = \int mc f(r,c,t)\mathrm{d}c, \tag{5.8}$$

$$\rho(r,t)u(r,t) = \int \tfrac{1}{2}m(c-v).(c-v) f(r,c,t)\mathrm{d}c. \tag{5.9}$$

Evolution equations for these quantities are obtained from the evolution equation for f. In order to do this, one must consider the influence of the collision term. Note, however, that in view of (5.3) one may write

$$\int m J(f)\mathrm{d}c = \tfrac{1}{4}\int [m + m - m - m]J(f)\mathrm{d}c = 0, \tag{5.10a}$$

$$\int mc J(f)\mathrm{d}c = \tfrac{1}{4}\int [mc + m\tilde{c} - mc' - m\tilde{c}']J(f)\mathrm{d}c = 0, \tag{5.10b}$$

$$\int mc^2\, J(f)\,\mathrm{d}c = \tfrac{1}{4}\int [mc^2 + m\tilde{c}^2 - mc'^2 - m\tilde{c}'^2]J(f)\,\mathrm{d}c = 0. \qquad (5.10c)$$

The vanishing of these integrals follows from the property that mass, momentum and kinetic energy are collisional invariants, i.e. they do not change in elastic binary collisions.

From definitions (5.7–9) and relations (5.10a–c), the balance equations for mass, momentum, and energy can be derived from the Boltzmann equation. Multiplying each term of (5.1) by m, mc, and $(1/2)mc^2$ respectively and integrating over c it is found that

$$\frac{\partial \rho}{\partial t} + \nabla \cdot (\rho v) = 0, \qquad (5.11)$$

$$\frac{\partial (\rho v)}{\partial t} + \nabla \cdot (\rho v v + \mathbf{P}) = \rho \mathbf{F}, \qquad (5.12)$$

$$\frac{\partial}{\partial t}\left(\rho u + \tfrac{1}{2}\rho v^2\right) + \nabla \cdot \left[\rho(u + \tfrac{1}{2}v^2)v + \mathbf{P}\cdot v + q\right] = 0, \qquad (5.13)$$

with F being the external force per unit mass and $\mathbf{P}$ and q defined as

$$\mathbf{P} = \int m\mathbf{C}\mathbf{C}f\,\mathrm{d}c, \qquad (5.14)$$

$$q = \int \tfrac{1}{2}m C^2 \mathbf{C}f\,\mathrm{d}c. \qquad (5.15)$$

$C = c - v$ is the relative velocity of the molecules with respect to the mean motion of the gas.

Equations (5.11–13) turn out to be the well-known balance equations of hydrodynamics, provided that one identifies $\mathbf{P}$ and q defined by (5.14) and (5.15) as the pressure tensor and the heat flux vector respectively.

Since at equilibrium f is an isotropic function of C, the pressure tensor reduces to

$$\mathbf{P} = p\mathbf{U}, \qquad (5.16)$$

with $\mathbf{U}$ the identity tensor and p, the equilibrium pressure, given by

$$p = \tfrac{1}{3}\int m C^2 f\,\mathrm{d}c. \qquad (5.17)$$

It is found from the definition (5.9) of the internal energy that $p = (2/3)\rho u$. The macroscopic thermal equation of state for ideal gases leads then to the following definition of the absolute equilibrium temperature:

$$p = \tfrac{2}{3}\rho u = nk_B T, \tag{5.18}$$

with n the number of particles per unit volume and k_B the Boltzmann constant.

5.1.2 The H-Theorem and the Second Law

Another important result from Boltzmann's equation is the so-called H-theorem, which is a kinetic version of the second law. We define η as follows:

$$\rho(r,t)\eta(r,t) = \int f(r,c,t)\ln f(r,c,t)\,\mathrm{d}c. \tag{5.19}$$

The evolution equation for η may be derived by multiplying term by term the Boltzmann equation (5.1) by $\ln f$ and integrating over c. In this way, one obtains

$$\frac{\partial(\rho\eta)}{\partial t} + \nabla\cdot(\boldsymbol{J}^\eta + \rho\eta v) = \sigma^\eta, \tag{5.20}$$

with the flux $\boldsymbol{J}^\eta$ given by

$$\boldsymbol{J}^\eta = \int \boldsymbol{C} f \ln f\,\mathrm{d}c. \tag{5.21}$$

The production term σ^η, defined as

$$\sigma^\eta = \int J(f)\ln f\,\mathrm{d}c, \tag{5.22}$$

can be written, in virtue of (5.3), as

$$\sigma^\eta = \tfrac{1}{4}\int \mathrm{d}c\int \mathrm{d}\tilde{c}\int \mathrm{d}\Omega\,|c-\tilde{c}|\sigma[f'\tilde{f}'-f\tilde{f}\,][\ln f + \ln\tilde{f} - \ln f' - \ln\tilde{f}'\,]$$

$$= \tfrac{1}{4}\int \mathrm{d}c\int \mathrm{d}\tilde{c}\int \mathrm{d}\Omega\,|c-\tilde{c}|\sigma[f'\tilde{f}'-f\tilde{f}\,]\ln[f\tilde{f}/f'\tilde{f}'\,]. \tag{5.23}$$

By inspecting the product of the quantities $[f'\tilde{f}'-f\tilde{f}\,]$ and $\ln[f\tilde{f}/f'\tilde{f}'\,]$ one sees immediately that

$$\sigma^\eta \le 0. \tag{5.24}$$

The equality sign holds only for $f'\tilde{f}' = f\tilde{f}$, i.e. for $J(f) = 0$, which corresponds to equilibrium situations.

The equilibrium distribution function is obtained by realizing that, since $f_{eq} \tilde{f}_{eq} = f'_{eq} \tilde{f}'_{eq}$, it follows that $\ln f_{eq}$ is a collisional invariant and can therefore be expressed as a linear combination of m, mc, and $(1/2)mc^2$:

$$\ln f_{eq} = Am + \boldsymbol{B} \cdot m\boldsymbol{c} + C \tfrac{1}{2} mc^2. \tag{5.25}$$

The five constants A, $\boldsymbol{B}$, and C appearing in (5.25) can be determined in terms of ρ, $\boldsymbol{v}$, and T, taking into account the definitions (5.7–9) and (5.18). The result is the well-known Maxwell–Boltzmann distribution function

$$f_{eq} = n\left(\frac{m}{2\pi k_B T}\right)^{3/2} \exp\left[-\frac{mC^2}{2k_B T}\right]. \tag{5.26}$$

The H-theorem, stating the negative character of σ^η, allows us to express the entropy in terms of the distribution function f. Indeed, it suggests that the entropy s per unit mass may be defined as $\rho s = -A'\rho\eta + B'$, where A' and B' are two constants. Their values can be determined by comparing the equilibrium value of s with the value of η when the equilibrium distribution function (5.26) is substituted in the definition (5.19). It turns out that $A' = k_B$, the Boltzmann constant. The value of B' is not so important because generally only entropy changes are relevant. The above considerations suggest the following expressions for the entropy and the entropy flux respectively:

$$\rho s = -k_B \int f \ln f \, d\boldsymbol{c}, \tag{5.27}$$

$$\boldsymbol{J}^s = -k_B \int \boldsymbol{C} f \ln f \, d\boldsymbol{c}. \tag{5.28}$$

For completeness let us recall the expression of the equilibrium entropy for monatomic gases, the so-called Sackur–Tetrode formula:

$$\rho s = n k_B \left\{ \frac{5}{2} + \ln\left[\frac{1}{n}\left(\frac{2\pi m k_B T}{h^2} \right)^{3/2} \right] \right\}, \tag{5.29}$$

where h stands for Planck's constant.

5.2 Non–equilibrium Entropy and the Entropy Flux

Up to now, we have derived microscopic expressions for all the quantities of interest for our study. The microscopic definitions (5.27) and (5.28) give explicit expressions for the entropy and the entropy flux in non-equilibrium situations in terms of the non-equilibrium distribution function. The latter may be expanded according to

$$f = f_{eq}\left[1 + \phi^{(1)} + \phi^{(2)} + ...\right],\qquad(5.30)$$

where $\phi^{(1)}$, $\phi^{(2)}$,... are expressed in terms of a small parameter, for instance the ratio of the relaxation time to the macroscopic time, the ratio of the mean free path to a characteristic length of the macroscopic inhomogeneities, the higher-order moments of the velocity distribution function, etc. The function f_{eq} is the equilibrium distribution function either in global or local equilibrium (in the first case, ρ, v, and T would not depend on position or time, and in the second case they would depend on these variables).

The quantities $\rho = nm$, v, and T are determined from the first five moments of the distribution function given by equations (5.7–9). This imposes on $\phi^{(i)}$ the closure conditions

$$\int f_{eq}\,\phi^{(i)}\,dc = 0, \qquad\qquad \int f_{eq}\,\phi^{(i)}C\,dc = 0,$$

$$\int f_{eq}\,\phi^{(i)}C^2\,dc = 0 \qquad (i = 1,2,...),\qquad(5.31)$$

and when (5.30) is introduced into (5.27) and (5.28), one obtains up to second order

$$\rho s = \rho s_{eq} - \tfrac{1}{2}k_B\int f_{eq}\,\phi^{(1)2}\,dc\qquad(5.32)$$

and

$$J^s = \frac{1}{T}q - \tfrac{1}{2}k_B\int f_{eq}\,\phi^{(1)2}\,C\,dc.\qquad(5.33)$$

Owing to the restrictions (5.31), $\phi^{(2)}$ does not contribute to the entropy up to the second order of approximation. Furthermore, it follows from the third of conditions (5.31) that the bulk viscous pressure for an ideal monatomic gas vanishes identically. The first terms on the right-hand side of (5.32) and (5.33) are the classical ones. The second terms are related to the non-classical corrections on which we shall focus our attention in the next section.

It must be noted that, in contrast to the exact solutions of the Boltzmann equation, approximate solutions like (5.30) do not necessarily satisfy the H-theorem beyond the linear approximation. The same problem arises in macroscopic theories when constitutive equations containing non-linear truncated approximations are used. The requirement of a positive entropy production may provide a criterion on the range of validity of a given approximation, both in microscopic and in macroscopic theories. The positive entropy production requirement may also be useful to model the non-linear terms in such a way that entropy production is always positive.

A second point worth outlining is that the thermodynamic entropy is a function of macroscopic variables. Therefore, assuming the existence of a macroscopic entropy implies that one selects among the several possible microscopic distribution functions those depending parametrically on quantities which have a clear macroscopic meaning or, at least, which are macroscopically measurable and controllable. In principle, one can construct a wide range of microscopic distribution functions but their physical meaning is not necessarily clear. Here, we always refer to the so-called thermodynamic branch of approximate solutions of the Boltzmann equation [5.5], i.e. to those solutions satisfying the above-mentioned requirement.

5.3 Grad's Solution

Two important models, namely the Chapman–Enskog and Grad ones, have been proposed to solve the Boltzmann equation in non-equilibrium situations. In the Chapman–Enskog approach [5.1], f is expressed in terms of the first five moments n, v, and T and their gradients. Then, $\phi^{(1)}$ is proportional to ∇v and ∇T, while $\phi^{(2)}$ includes terms in $\nabla \nabla v$, $\nabla \nabla T$ and so on. In Grad's model [5.2–3], f is developed in terms of its moments with respect to the molecular velocity. Note that, in view of definitions (5.14–15), $\mathbf{P}$ and q are directly related to the moments of the velocity distribution function (the scalar viscous pressure p^v vanishes in an ideal gas). Therefore, the mean values of q and of $\overset{0}{\mathbf{P}}{}^{v}$ are considered in Grad's theory as independent variables, so that Grad's theory is closer to the macroscopic developments of EIT than Chapman–Enskog's. It can thus be asserted that both EIT and Grad's thirteen-moment method make use of the same independent variables.

In Grad's method, the non-equilibrium distribution function $f(r,c,t)$ is replaced by the infinite set of variables $\rho = mn(r,t)$, $v(r,t)$, $T(r,t)$, $a_n(r,t)$, where a_n stand for the successive higher-order moments of the distribution function. These moments are chosen in such a way that they are mutually orthogonal, and they are given by Hermite polynomials. In the thirteen-moment approximation, the development is limited to all the second-order moments and to some of the third-order moments, those related to the heat flux.

In the thirteen-moment approximation, the distribution function is written as

$$f = f_{eq}[1 + A \cdot C + \overset{0}{\mathbf{B}}{}^v : \overset{0}{C}C + (C \cdot C)(D \cdot C)]. \tag{5.34}$$

The coefficients $A(r,t)$, $\overset{0}{\mathbf{B}}{}^v(r,t)$ and $D(r,t)$ are determined by introducing (5.34) into (5.14), (5.15), and the third of conditions (5.31). In order to help the reader, the general form of integrals appearing in these calculations are given in Appendix B. Such equations allow us to identify A, $\overset{0}{\mathbf{B}}{}^v$, and D in terms of the heat flux q and the viscous pressure tensor $\overset{0}{\mathbf{P}}{}^v$:

$$A = -\frac{m}{pk_BT}q, \qquad \overset{0}{\mathbf{B}}{}^v = \frac{m}{2pk_BT}\overset{0}{\mathbf{P}}{}^v, \qquad D = \frac{m^2}{5pk_B^2T^2}q. \tag{5.35}$$

As a consequence, f may be written explicitly as

$$f = f_{eq}\left[1 + \frac{m}{2pk_BT}\overset{0}{C}C : \overset{0}{\mathbf{P}}{}^v + \frac{2m}{5pk_B^2T^2}\left(\frac{1}{2}mC^2 - \frac{5}{2}k_BT\right)C \cdot q\right]. \tag{5.36}$$

After substitution of (5.36) into (5.32–33), the expressions for the entropy and entropy flux turn out to be

$$\rho s = \rho s_{eq} - \frac{1}{4pT}\overset{0}{\mathbf{P}}{}^v : \overset{0}{\mathbf{P}}{}^v - \frac{m}{5pk_BT^2}q \cdot q, \tag{5.37}$$

$$J^s = \frac{1}{T}q - \frac{2}{5pT}\overset{0}{\mathbf{P}}{}^v \cdot q. \tag{5.38}$$

These results confirm the plausibility of the hypotheses of EIT stating that the entropy may depend on the dissipative fluxes and that the entropy flux contains extra contributions besides $T^{-1}q$.

It remains to verify that the EIT expressions of the transport coefficients appearing in the non-classical parts of s and J^s are confirmed by the kinetic theory. Moreover, EIT predicts the existence of some relations between these coefficients and those found in the evolution equations for the fluxes.

To check these results, one needs the evolution equations for the fluxes. These can be obtained by inserting (5.36) into Boltzmann's equation. In the thirteen-moment approximation one is led to [5.2]

$$(\overset{0}{\mathbf{P}}{}^{v})^{\cdot} = -\tfrac{4}{5}(\overset{0}{\nabla}q)^{s} - 2p\,\overset{0}{\mathbf{V}} - \rho\gamma\overset{0}{\mathbf{P}}{}^{v} - \overset{0}{\mathbf{P}}{}^{v}.(\nabla v) - (\nabla v).(\overset{0}{\mathbf{P}}{}^{v})^{T}$$

$$- \overset{0}{\mathbf{P}}{}^{v}(\nabla \cdot v) + \tfrac{2}{3}[\overset{0}{\mathbf{P}}{}^{v} : (\nabla v)]\mathbf{U} \tag{5.39}$$

and

$$\dot{q} = -(k_B T/m)\nabla \cdot \overset{0}{\mathbf{P}}{}^{v} - \tfrac{5}{2}(pk_B/m)\nabla T - \tfrac{2}{3}\rho\gamma q - \tfrac{7}{5}q\cdot(\nabla v) - \tfrac{2}{5}q\cdot(\nabla v)^{T}$$

$$- \tfrac{7}{5}q(\nabla \cdot v) - \tfrac{7}{2}(k_B/m)\overset{0}{\mathbf{P}}{}^{v}.\nabla T + \rho^{-1}\overset{0}{\mathbf{P}}{}^{v}.(\nabla.\overset{0}{\mathbf{P}}{}^{v}). \tag{5.40}$$

The coefficient γ is given in terms of the collision integrals by

$$\gamma = \frac{2\sqrt{2\pi}}{5}\int_{0}^{\infty} x^{6}e^{-x^{2}/2}\left[\int_{0}^{\infty}\frac{1}{m}\sigma\left(\theta, x\sqrt{\frac{2kT}{m}}\right)\sin^{2}\theta\cos^{2}\theta\,d\theta\right]dx \tag{5.41}$$

and is shown to be a positive quantity [5.2].

We focus our attention on the linear terms of (5.39–40) and therefore omit non-linear contributions, such as $\overset{0}{\mathbf{P}}{}^{v}.\nabla T$ in (5.40). It is true that for a fixed temperature gradient this term is linear in $\overset{0}{\mathbf{P}}{}^{v}$, but when we take perturbations around equilibrium, both $\overset{0}{\mathbf{P}}{}^{v}$ and ∇T can be considered as perturbations, and consequently such a term is non-linear. In the linear approximation, (5.39–40) reduce to

$$\frac{1}{\rho\gamma}(\overset{0}{\mathbf{P}}{}^{v})^{\cdot} = -\left[\overset{0}{\mathbf{P}}{}^{v} + \frac{2p}{\rho\gamma}\overset{0}{\mathbf{V}}\right] - \frac{4}{5\rho\gamma}(\overset{0}{\nabla}q)^{s} \tag{5.42}$$

and

$$\frac{3}{2\rho\gamma}\dot{q} = -\left[q + \frac{15pk_B}{4m\rho\gamma}\nabla T\right] - \frac{3k_B T}{2m\rho\gamma}\nabla \cdot \overset{0}{\mathbf{P}}{}^{v}. \tag{5.43}$$

These relations may be directly compared with the linear evolution equations (2.66) and (2.64), respectively, derived from the macroscopic theory. One is then led to the identifications

$$\tau_1 = \frac{3}{2\rho\gamma} \text{ (a)}, \qquad \lambda = \frac{5pk_B}{2m}\tau_1 \text{ (b)}, \qquad -\lambda T^2\beta = \frac{k_B T}{m}\tau_1 \text{ (c)}, \tag{5.44}$$

$$\tau_2 = \frac{1}{\rho\gamma} \ \text{(a)}, \quad \eta = p\tau_2 \ \text{(b)}, \quad -2\eta T\beta = \frac{4}{5}\tau_2 \ \text{(c)}. \tag{5.45}$$

Expressions of the form (2.64–66) (or (5.42–43)) could have been guessed by purely dimensional arguments, but thermodynamics brings supplementary information and imposes specific limitations. The first two are $\tau_1 > 0$ and $\tau_2 > 0$, as a consequence of $\gamma > 0$. Furthermore, a direct consequence of (5.44b) and (5.45b) is that $\lambda > 0$ and $\eta > 0$. The symmetry relations between the cross coefficients in (2.64–66) are confirmed by comparing the two independent expressions obtained for β in (5.44c) and (5.45c): they are seen to coincide and are given by

$$\beta = -\frac{2}{5pT}. \tag{5.46}$$

This expression for β is nothing but the factor $- 2/(5pT)$ of the non-classical term of the entropy flux (5.38) and is the same as that predicted by the macroscopic theory. Note also that the relaxation times of q and $\overset{0}{\mathbf{P}}{}^{v}$ are not coincident but $\tau_1 = \frac{3}{2}\tau_2$.

We now turn our attention to the macroscopic Gibbs equation (2.38). After integration this yields

$$\rho s = \rho s_{eq} - \frac{\tau_1}{2\lambda T^2}\boldsymbol{q}\cdot\boldsymbol{q} - \frac{\tau_2}{4\eta T}\overset{0}{\mathbf{P}}{}^{v} : \overset{0}{\mathbf{P}}{}^{v}. \tag{5.47}$$

Making use of the results given by (5.44c), (5.45c), and (5.46), we see that $\tau_1/(2\lambda T^2) = m/(5pk_BT^2)$ and $\tau_2/(4\eta T) = 1/(4pT)$, from which it follows that (5.37) is strictly identical with the macroscopic result (5.47). Consequently, we can conclude that, within the linear range, there is a complete agreement between Grad's theory and the predictions of EIT, not only in regard to the choice of the variables and the expressions of the entropy and the entropy flux, but also concerning the expressions of the linearized evolution equations for the thermodynamic fluxes.

Expression (5.47) allows us to calculate the order of magnitude of the non-equilibrium correction with respect to the local-equilibrium entropy. In the absence of viscous effects, one has

$$\rho s = \rho s_{eq} - \frac{\tau_1}{2\lambda T^2}\boldsymbol{q}\cdot\boldsymbol{q}. \tag{5.48}$$

To evaluate the range of temperature gradients for which the local-equilibrium hypothesis is acceptable, we write the non-equilibrium contribution to (5.48) as

$$\frac{\tau_1}{2\lambda T^2}\, \boldsymbol{q}\cdot\boldsymbol{q} = \frac{\tau_1\lambda}{2T^2}(\nabla T)^2 = \frac{5nk_B^2 T}{4m}\tau_1^2\left(\frac{\nabla T}{T}\right)^2. \tag{5.49}$$

By defining the mean free path by $\ell = (\tfrac{3}{2}k_B T/m)^{1/2}\tau_1$, one may write the non-equilibrium correction as

$$\Delta(\rho s)_{neq} = \frac{5nk_B}{6}\ell^2\left(\frac{\nabla T}{T}\right)^2. \tag{5.50}$$

Moreover, according to the Sackur–Tetrode formula (5.29), the local-equilibrium entropy is of the order of $\rho s_{eq} \approx nk_B$. Thus the relative value of the non-equilibrium contribution with respect to the local-equilibrium value is of the order of

$$\frac{\Delta(\rho s)_{neq}}{\rho s_{eq}} \approx \ell^2\left(\frac{\nabla T}{T}\right)^2. \tag{5.51}$$

For gases like O_2 and N_2 at standard temperature and pressure, the mean free path ℓ is of the order of 10^{-4} cm, so that the relative non-equilibrium corrections will be less than 0.01% for temperature gradients lower than 10^4 K/cm. The conclusion is that in these situations the local-equilibrium entropy is a reasonable approximation. However, this agreement, achieved for steady state situations, does not necessarily imply that the local-equilibrium entropy remains a satisfactory concept in fast non-steady processes, for which $\boldsymbol{q}$ cannot be approximated by $\boldsymbol{q} = -\lambda\nabla T$, or for specific materials, such as polymer solutions where viscous effects play a dominant role.

It must be noted that the moment expansion of Grad lacks a smallness parameter allowing to control its domain of validity. In Sect. 5.7 we will present a wider approach taking into account an infinite number of higher-order moments. It is also interesting to observe that, since all the moments are mutually orthogonal, the incorporation of higher-order moments does not alter the value of the first thirteen ones; therefore, the specification of the first thirteen moments does not mean a univocal microscopic solution of the Boltzmann equation. Furthermore, the connection between the temperature defined in kinetic theory and the temperature measured by a thermometer in a non-equilibrium state needs also some clarifications. In that respect, the development of a theory more general than the local-equilibrium approximation may be useful for exploring some open problems raised by the introduction of higher-order expansions.

5.4 The Relaxation-Time Approximation

Since the Boltzmann equation is very complicated, simpler kinetic equations have been proposed in the literature. A very simple model is the well-known relaxation-time approximation, which is used here to analyse the thermodynamics of steady states.

In the relaxation-time approximation, the evolution equation of the distribution function is modelled by

$$\frac{\partial f}{\partial t} + c \cdot \frac{\partial f}{\partial r} + \frac{F}{m} \cdot \frac{\partial f}{\partial c} = -\frac{1}{\tau}\left(f - f_{eq}\right). \tag{5.52}$$

For steady non-equilibrium situations in the absence of external forces, it is found that

$$f = (1 + \tau c \cdot \nabla)^{-1} f_{eq}, \tag{5.53}$$

a formal expression to which an operative meaning can be attached by expanding it in powers of $\tau c \cdot \nabla$. In the simple case of temperature and velocity gradients, the first-order and second-order corrections to the equilibrium distribution function are

$$\phi^{(1)} = -\frac{\tau}{k_B T^2}\left(\frac{1}{2}mC^2 - \frac{5}{2}k_B T\right) C \cdot \nabla T - \frac{\tau}{k_B T} m \overset{0}{CC} : (\nabla v) + \mathrm{NL} \tag{5.54}$$

and

$$\phi^{(2)} = -\frac{\tau^2}{k_B T^2}\left(\frac{1}{2}mC^2 - \frac{5}{2}k_B T\right) \overset{0}{CC} \cdot \nabla\nabla T - \frac{\tau^2}{k_B T} m C \overset{0}{CC} : (\nabla \overset{0}{\nabla} v) + \mathrm{NL}, \tag{5.55}$$

where NL stands for non-linear terms involving gradient products. The introduction of (5.54) into (5.32) gives the following expression of the entropy:

$$\rho s = \rho s_{eq} - \frac{\tau}{2T^2}\left[\frac{\tau}{k_B T^2}\left\langle\left(\tfrac{1}{2}mC^2 - \tfrac{5}{2}k_B T\right)^2 C_1^2\right\rangle\right](\nabla T)^2$$

$$-\frac{\tau}{2T}\left[\frac{\tau}{k_B T}\left\langle m^2 C_1^2 C_2^2\right\rangle\right](\overset{0}{\nabla} v)^s : (\overset{0}{\nabla} v)^s, \tag{5.56}$$

and from (5.33) one finds for the entropy flux

$$J^s = \frac{1}{T} q - \frac{\tau^2}{k_B T^3}\left\langle\left(\tfrac{1}{2}mC^2 - \tfrac{5}{2}k_B T\right) m C_1^2 C_2^2\right\rangle(\overset{0}{\nabla} v)^s \cdot \nabla T, \tag{5.57}$$

where C_1, C_2 are components of C, and the quantity $\langle a \rangle$ stands for

$$\langle a \rangle = \int a(c) f_{eq}(c)\, dc. \tag{5.58}$$

By inserting (5.54–55) into (5.30) we obtain a distribution function which can be substituted into (5.14–15) and taking into account that $\langle (\frac{1}{2}mC^2 - \frac{5}{2}k_B T) C \cdot C \rangle = 0$, the expressions for the heat flux and the pressure tensor adopt the form

$$q = -\frac{\tau}{k_B T^2}\left\langle \left(\tfrac{1}{2}mC^2 - \tfrac{5}{2}k_B T\right)^2 C_1^2 \right\rangle \nabla T - \frac{\tau^2}{2T}\left\langle \left(\tfrac{1}{2}mC^2 - \tfrac{5}{2}k_B T\right)mC_1^2 C_2^2 \right\rangle \nabla \cdot (\overset{0}{\nabla}v)^s,$$

$$\tag{5.59}$$

$$\overset{0}{\mathbf{P}}{}^v = -\frac{\tau}{k_B T}\left\langle m^2 C_1^2 C_2^2 \right\rangle (\overset{0}{\nabla}v)^s - \frac{\tau^2}{k_B T^2}\left\langle \left(\tfrac{1}{2}mC^2 - \tfrac{5}{2}k_B T\right)^2 mC_1^2 C_2^2 \right\rangle \nabla\nabla T.$$

A limitation of the relaxation-time approximation is that the relaxation times of both fluxes coincide. Compare (5.59) with (2.64–66) in a steady state, with q and $\overset{0}{\mathbf{P}}{}^v$ substituted by their first-order expressions $q^{(1)} = -\lambda \nabla T$ and $\overset{0}{\mathbf{P}}{}^{v\,(1)} = -2\eta\,\overset{0}{\mathbf{V}}$. In this way one obtains a confirmation of the relation between the cross terms predicted by the thermodynamic theory and the relation of these coefficients with those of the second-order term in the entropy flux (5.57). Furthermore, the expression (5.56) for the entropy may equivalently be written as

$$\rho s = \rho s_{eq} - \frac{\tau_1 \lambda}{2T^2}(\nabla T)\cdot(\nabla T) - \frac{\tau_2 \eta}{T}(\overset{0}{\nabla}v)^s : (\overset{0}{\nabla}v)^s. \tag{5.60}$$

This is nothing but expression (3.4) when the fluxes are replaced by their first-order steady-state approximations. It is also interesting to observe that the entropy (5.60) does not coincide with the local-equilibrium entropy even in the steady case.

The agreement between the kinetic theory of gases and the macroscopic predictions undoubtedly reinforces the consistency of EIT. Although our analysis is limited to the linear range, the agreement between EIT and kinetic theory is much wider than the agreement of the latter with the usual local-equilibrium thermodynamics. It should be kept in mind that the previous comparisons have been achieved within the Boltzmann theory, which concerns only two-body collisions. A comparison at a higher level, where higher-order collisions are included, raises some notable features [5.10]. Computer experiments and careful microscopic studies of the correlation functions have shown that many-body processes play an important role in determining the long-time behaviour of correlation functions. Thus, whereas Boltzmann's equation leads to an exponential decay of the fluctuations, hydrodynamic arguments predict for a d-dimensional system a potential decay of the form $t^{-d/2}$. An appropriate thermodynamic extension to these situations remains a challenging problem.

5.5 Dilute Non–ideal Gases

The treatment of non-ideal gases is more complex, owing to intermolecular interactions which contribute both to the viscous pressure tensor and to the heat flux vector [5.11–12]. For simplicity, we deal here only with the viscous effects. In kinetic theory, the pressure tensor is usually split into a kinetic and a potential part

$$\mathbf{P} = \mathbf{P}_c + \mathbf{P}_p, \tag{5.61}$$

given respectively by

$$\mathbf{P}_c = \int m\mathbf{CC} f_1 \mathrm{d}\mathbf{c} \tag{5.62}$$

and

$$\mathbf{P}_p = -\frac{n^2}{2} \int \phi'(R)\frac{1}{R}\mathbf{RR}\, g(R)\mathrm{d}\mathbf{R}. \tag{5.63}$$

The kinetic part $\mathbf{P}_c$ is expressed in terms of the one-particle distribution function $f_1(r,c)$ while the potential part $\mathbf{P}_p$ depends on the two-particle distribution function $f_2(r_1,c_1,r_2,c_2)$. In (5.63), $\mathbf{R} = r_1 - r_2$ is the relative position of molecule 1 with respect to molecule 2, $\phi(R)$ the interaction potential, with a prime indicating the spatial derivative with respect to R; and $g(R)$ is the pair correlation function defined in terms of f_2 as

$$n^2 g(R) = \int f_2(r_1,c_1,r_2,c_2)\mathrm{d}c_1\mathrm{d}c_2. \tag{5.64}$$

Defining the thermodynamic equilibrium pressure as one-third of the trace of $\mathbf{P}$ at equilibrium, it follows from (5.61–63) that

$$p = nk_BT - \frac{n^2}{6} \int \phi'(R)g_{eq}(R)R\,\mathrm{d}\mathbf{R}, \tag{5.65}$$

with $g_{eq}(R)$ being the equilibrium pair correlation function.

Since the relaxation times of the one-particle distribution f_1 and the pair-correlation function g do not necessarily coincide, one should not regard $\mathbf{P}^v$ as one single physical quantity, but rather as the sum of two independent variables, $\mathbf{P}^v_c$ and $\mathbf{P}^v_p$.

5.5.1 Entropy and Evolution Equations

The thermodynamic phenomenological description of the non-ideal gas may be summarized as follows. The entropy has the form

$$\rho s = \rho s_{eq} - \frac{\tau_c}{4\eta_c T}\overset{\circ}{\mathbf{P}}{}^v_c : \overset{\circ}{\mathbf{P}}{}^v_c - \frac{\tau_p}{4\eta_p T}\overset{\circ}{\mathbf{P}}{}^v_p : \overset{\circ}{\mathbf{P}}{}^v_p - \frac{\tau_0}{2\zeta T}p^v_p p^v_p, \tag{5.66}$$

since p^v_c vanishes identically, as remarked in Sect. 5.2. The relaxation times and viscosities are defined through the evolution equations for their respective fluxes:

$$\frac{\mathrm{d}}{\mathrm{d}t}\overset{\circ}{\mathbf{P}}{}^v_c = -\frac{1}{\tau_c}(\overset{\circ}{\mathbf{P}}{}^v_c + 2\eta_c \overset{\circ}{\mathbf{V}}), \tag{5.67a}$$

$$\frac{\mathrm{d}}{\mathrm{d}t}\overset{\circ}{\mathbf{P}}{}^v_p = -\frac{1}{\tau_p}(\overset{\circ}{\mathbf{P}}{}^v_p + 2\eta_p \overset{\circ}{\mathbf{V}}), \tag{5.67b}$$

$$\frac{\mathrm{d}}{\mathrm{d}t}p^v_p = -\frac{1}{\tau_0}(p^v_p + \zeta\nabla\cdot\boldsymbol{v}). \tag{5.67c}$$

This is the simplest generalisation of the scheme proposed for ideal monatomic gases. A decomposition of the viscous pressure tensor in a sum of partial viscous pressure tensors is given in Chap. 15, where polymers are considered.

Our purpose is to explore the consistency of the thermodynamic scheme (5.66–67a,b,c) from a microscopic point of view. Therefore we need an expression for the entropy in terms of f_1 and f_2. This is supplied by [5.11]

$$\rho s = -k_B \int f_1(1)\ln f_1(1)\,\mathrm{d}\Gamma_1 - \tfrac{1}{2}k_B \int f_2(1,2)\ln \frac{f_2(1,2)}{f_1(1)f_1(2)}\,\mathrm{d}\Gamma_{12}, \tag{5.68}$$

where $\mathrm{d}\Gamma_1 = \mathrm{d}\boldsymbol{r}_1\,\mathrm{d}\boldsymbol{c}_1$ and $\mathrm{d}\Gamma_{12} = \mathrm{d}\boldsymbol{r}_1\,\mathrm{d}\boldsymbol{c}_1\,\mathrm{d}\boldsymbol{r}_2\,\mathrm{d}\boldsymbol{c}_2$. An explicit form for the entropy in terms of f_1 and g can be obtained by assuming that $f_2(1,2) = f_1(1)f_1(2)g(1,2)$. This result is exact at equilibrium and valid up to the first order in the shear rate for a wide class of interaction potentials. Within this approximation and setting $f_1 = f$ one has

$$\rho s = -k_B \int f \ln f \, dc - \tfrac{1}{2} n^2 k_B \int g \ln g \, d\mathbf{R} \, . \tag{5.69}$$

In analogy with Boltzmann's equation, the evolution equations for f and g can be written as

$$\frac{\partial f}{\partial t} + c \cdot \frac{\partial f}{\partial r} = J_c(f),$$

$$\tag{5.70}$$

$$\frac{\partial g}{\partial t} + \mathbf{R} \cdot (\nabla v) \cdot \nabla_R g = J_p(g),$$

where ∇_R stands for the gradient with respect to the relative position $\mathbf{R}$ between two molecules. It is not necessary to know the specific form of operators J_c and J_p on the right-hand sides of Eqs. (5.70). The only result of interest is that at equilibrium $J_c(f_{eq}) = 0$ and $J_p(g_{eq}) = 0$, with f_{eq} and g_{eq} given by

$$f_{eq} \approx \exp\left[-\frac{mC^2}{2k_B T}\right], \qquad g_{eq} \approx \exp\left[-\frac{w(R)}{k_B T}\right]. \tag{5.71}$$

The last expression defines an effective potential $w(R)$ which coincides with the interaction potential $\phi(R)$ only up to a first-order approximation in the density.

Now, by analogy with (5.34), one expands the non-equilibrium distribution functions f and g in terms of the moments of C and $\mathbf{R}$, respectively:

$$f = f_{eq}[1 + \overset{0}{CC} : \overset{0}{\mathbf{A}}(r,t)],$$

$$\tag{5.72}$$

$$g = g_{eq}[1 + \overset{0}{\mathbf{R}\mathbf{R}} : \overset{0}{\mathbf{B}}(r,t) + R^2 b(r,t)].$$

Here $\overset{0}{\mathbf{A}}$ and $\overset{0}{\mathbf{B}}$ are traceless symmetric tensors and b is a scalar, which may be related to $\overset{0}{\mathbf{P}}{}^v_c$, $\overset{0}{\mathbf{P}}{}^v_p$, and p^v_p by introducing (5.72) into (5.62–63). For a more general development, see [5.13–14]. Note that because of the third of conditions (5.31) the expansion of f is limited to the traceless term $\overset{0}{CC}$. It is found that

$$\overset{0}{\mathbf{P}}{}^v_c = \frac{2m}{15}\langle C^4 \rangle \overset{0}{\mathbf{A}}, \tag{5.73a}$$

$$\overset{0}{\mathbf{P}}{}^v_p = -\frac{n^2}{15}\langle \phi'(R) R^3 \rangle \overset{0}{\mathbf{B}}, \tag{5.73b}$$

$$p_p^v = -\frac{n^2}{6}\langle\phi'(R)R^3\rangle b, \tag{5.73c}$$

with $\langle...\rangle$ being the corresponding equilibrium average.

Furthermore, one can derive an expression for the entropy by substituting (5.72) into (5.69). Making $f = f_{eq}(1 + \psi_c)$ and $g = g_{eq}(1 + \psi_p)$, one obtains up to second order

$$\rho s = \rho s_{eq} - \tfrac{1}{2}k_B\int f_{eq}\psi_c^2\,d\mathbf{c} - \tfrac{1}{4}k_B n^2\int g_{eq}\psi_p^2\,d\mathbf{R}. \tag{5.74}$$

Relations (5.73) allow to express the entropy in terms of the dissipative fluxes, namely:

$$\rho s = \rho s_{eq} - \frac{\alpha_{cc}}{2T}\overset{0}{\mathbf{P}}{}_c^v:\overset{0}{\mathbf{P}}{}_c^v - \frac{\alpha_{pp}}{2T}\overset{0}{\mathbf{P}}{}_p^v:\overset{0}{\mathbf{P}}{}_p^v - \frac{\alpha_0}{2T}(p_p^v)^2, \tag{5.75}$$

with

$$\alpha_{cc} = \frac{15k_B T}{2m^2}\frac{1}{\langle C^4\rangle}, \tag{5.76a}$$

$$\alpha_{pp} = \frac{15k_B T}{n^2}\frac{\langle R^4\rangle}{\langle\phi'(R)R^3\rangle^2}, \tag{5.76b}$$

$$\alpha_0 = \frac{18k_B T}{n^2}\frac{\langle R^4\rangle}{\langle\phi'(R)R^3\rangle^2}. \tag{5.76c}$$

Expression (5.75) confirms the result that the entropy depends on $\overset{0}{\mathbf{P}}{}_c^v$, $\overset{0}{\mathbf{P}}{}_p^v$, and p_p^v. The relation between the coefficients α_{cc}, α_{pp}, and α_0 and their microscopic analogues are derived from the evolution equations for the fluxes. Introducing (5.72) into (5.70), and multiplying the resulting equations term by term by $\mathbf{CC}$, $\mathbf{RR}$, and R^2 respectively, one is led, after integration with respect to $\mathbf{C}$ and $\mathbf{R}$, to

$$\frac{2}{15}\langle C^4\rangle\frac{\partial\overset{0}{\mathbf{A}}}{\partial t} + \frac{2m}{15k_B T}\langle C^4\rangle\overset{0}{\mathbf{V}} = -\frac{1}{\tau_c}\overset{0}{\mathbf{V}} + \mathrm{NL}, \tag{5.77a}$$

$$\frac{2}{15}\langle R^4\rangle\frac{\partial\overset{0}{\mathbf{B}}}{\partial t} - \frac{2}{15k_B T}\langle w'(R)R^3\rangle\overset{0}{\mathbf{V}} = -\frac{1}{\tau_p}\overset{0}{\mathbf{B}} + \mathrm{NL}, \tag{5.77b}$$

$$\langle R^4\rangle\frac{\partial b}{\partial t} - \frac{1}{3k_B T}\langle w'(R)R^3\rangle\nabla\cdot v = -\frac{1}{\tau_0}b + \mathrm{NL}. \tag{5.77c}$$

In (5.77) NL stands for non-linear terms, while the relaxation times τ_c, τ_p, and τ_0 are related to the collision operators J_c and J_p of (5.70) by

$$-\frac{1}{\tau_c} = \langle C_1 C_2 J_c(C_1 C_2)\rangle, \quad -\frac{1}{\tau_p} = \langle R_1 R_2 J_p(R_1 R_2)\rangle, \quad -\frac{1}{\tau_0} = \langle R^2 J_p(R^2)\rangle. \quad (5.78)$$

Here we do not require the explicit expressions for the relaxation times. In the more general case, the relaxation times form a fourth-rank tensor; however, for simplicity we consider only the particular case for which they reduce to scalar quantities. After expressing $\overset{0}{A}$, $\overset{0}{B}$ and b in terms of the fluxes according to (5.73), one obtains from (5.77) the evolution equations for the fluxes. The ratios $\frac{1}{2}\tau/\eta$ may be derived directly from the ratio of the coefficients of the terms $d\overset{0}{P}{}^v/dt$ and the corresponding terms in $\overset{0}{V}$, as immediately seen by inspection of (5.67).

Comparison of (5.77a) with the ratio

$$\frac{\tau_c}{2\eta_c} = \frac{15k_B T}{m^2}\frac{1}{\langle C^4\rangle}, \tag{5.79a}$$

obtained from (5.76a) confirms the thermodynamic result $\alpha_{cc} = \frac{1}{2}\tau_c/\eta_c$. From (5.76b) and (5.76c) one obtains

$$\frac{\tau_p}{2\eta_p} = \frac{15k_B T}{n^2}\frac{\langle R^4\rangle}{\langle \phi' R^3\rangle\langle w' R^3\rangle}, \tag{5.79b}$$

$$\frac{\tau_0}{\zeta} = \frac{18k_B T}{n^2}\frac{\langle R^4\rangle}{\langle \phi' R^3\rangle\langle w' R^3\rangle}. \tag{5.79c}$$

Thus, the analogous thermodynamic predictions $\alpha_{pp} = \frac{1}{2}\tau_p/\eta_p$ and $\alpha_0 = \tau_0/\zeta$ are also confirmed, but only at first order in the density. This is so because the generalised potential $w(R)$ defined in (5.71b) and the interaction potential $\phi(R)$ (and consequently w' and ϕ') are identical only at this order of approximation: in this case, the ratios (5.79) coincide with their respective counterparts in (5.76), but not at higher orders in the density.

The restriction of the latter identifications to first-order terms in the density is not an important drawback, since the products $\alpha_{pp}\overset{0}{P}{}^v_p : \overset{0}{P}{}^v_p$ and $\alpha_0(p^v)^2$ in the entropy (5.75) are of order n^2, so that the differences between α_{pp} and $\frac{1}{2}\tau_p/\eta_p$ and α_0 and τ_0/ζ are of order n^3 in the expression of entropy. Such terms cannot be included in the present study because the definition (5.69) of entropy is valid up to order n^2 only. To incorporate terms in n^3 demands a description in terms of f_3, the three-particle distribution function in

(5.68), which is beyond the scope of the present book. Up to the order of approximation tested here, it can thus be claimed that the macroscopic predictions of EIT are in agreement with the kinetic approach.

The results of EIT are also confirmed by the kinetic theory for gases consisting of molecules with internal degrees of freedom [5.15] and for dilute solutions of dimers and polymers [5.16] (see also Chap. 15). The interested reader is referred to the original papers for the analysis of these systems.

5.6 Non–linear Transport

Eu [5.5] has developed a modified version of the moment method with the purpose of deriving non-linear evolution equations. Instead of using, like Grad, an expansion of the form

$$f = f_0 \left[1 + \sum_i a^{(i)} H^{(i)}(C) \right], \tag{5.80}$$

where $H^{(i)}(C)$ are Hermite polynomials expressing the moments of the distribution function, Eu proposes a canonical form

$$f_E \approx \exp \left[-\tfrac{1}{2} \beta m C^2 - \sum_i Y^{(i)} H^{(i)}(C) \right], \tag{5.81}$$

in which the expansion coefficients $Y^{(i)}$ are depending on r and t. The above expression for the distribution function is similar to that used in the so-called maximum-entropy approaches, which will be discussed in Sect. 7.3, but in Eu's formalism the coefficients $Y^{(i)}$ are required to satisfy some supplementary conditions which guarantee the positiveness of the entropy production at each step of the approximation. He also introduces a new thermodynamic quantity, the so-called calortropy Ψ (see Sect. 3.1), defined as

$$\rho \Psi = -k_B \int f (\ln f_E - 1) \mathrm{d}C, \tag{5.82}$$

rather than the Boltzmann definition for the entropy, and which plays in Eu's approach a central role.

From his modified moment method, Eu is able to determine specific forms for the non-linear terms of the evolution equations, which are important in several fields, such as rheology and electronics. We present here a brief analysis which provides explicit information about non-linearities in transport equations. Starting from (5.81), it can be seen that the distribution function will take the form

$$f_E = f_{eq}\exp[-X_1\cdot\hat{q} - \overset{0}{X}_2 : \overset{\wedge 0}{P}{}^v\,], \tag{5.83}$$

with $\hat{q}$ and $\overset{\wedge 0}{P}{}^v$ the microscopic operators for the respective macroscopic heat flux and pressure tensor; X_1 and $\overset{0}{X}_2$ are macroscopic quantities, which in the simplest version of EIT, can be identified as

$$X_1 = -\frac{\tau_1}{2k_B\lambda T^2}q, \qquad \overset{0}{X}_2 = -\frac{\tau_2}{4k_B\eta T}\overset{0}{P}{}^v. \tag{5.84}$$

The entropy production is given by

$$\sigma^s = -k_B\int \ln f_E\, J(f_E)\mathrm{d}c, \tag{5.85}$$

with $J(f_E)$ the collision operator defined in (5.1). It is convenient, for further purposes, to introduce a dimensionless form of (5.85), namely

$$\bar{\sigma}^s = \frac{\tau_E}{k_B n}\sigma^s, \tag{5.86}$$

where $\tau_E = [(2k_BT/m)^{1/2}nd^2]^{-1}$ is of the order of the collision time. Here d is the molecular diameter.

From the property (5.3) of the collision operator $J(f_E)$ and the fact that $\ln f_E$ is composed of collisional invariants, one can write (5.86) as

$$\bar{\sigma}^s = \tfrac{1}{4}\int (x - x')\, J(f_E)\mathrm{d}c, \tag{5.87}$$

with

$$x = \overset{0}{X}_2 : [\overset{\wedge 0}{P}{}^v(c) + \overset{\wedge 0}{P}{}^v(\tilde{c})] + X_1\cdot[\hat{q}(c) + \hat{q}(\tilde{c})] \tag{5.88}$$

and x' the corresponding value of x after the collision. In view of the form of (5.83) for the distribution function and that of $J(f)$ in (5.1), (5.87) can be rewritten as

$$\bar{\sigma}^s = \tfrac{1}{4}\langle\!\langle (x - x')[e^{-x'} - e^{-x}]\rangle\!\rangle, \tag{5.89}$$

with

$$\langle\!\langle A(c,\tilde{c})\rangle\!\rangle = \int \mathrm{d}c\int \mathrm{d}\tilde{c}\int \mathrm{d}\Omega\,\sigma|c - \tilde{c}|f_{eq}(\tilde{c})\,A(c,\tilde{c}).$$

For small values of x and x', (5.89) reduces to

$$\bar{\sigma}_0^s = \tfrac{1}{4}\langle\langle\langle(x-x')^2\rangle\rangle\rangle. \tag{5.90}$$

Going back to the general expression (5.89), Eu uses a cumulant expansion and writes

$$\langle\langle\langle(x-x')(e^{-x'}-1)\rangle\rangle\rangle = \langle\langle\langle(x-x')^2\rangle\rangle\rangle^{1/2}\left\{\exp\left[\sum_l \frac{(-1)^l}{l!}\kappa_l^{(+)}\right]-1\right\}, \tag{5.91a}$$

$$\langle\langle\langle(x-x')(e^{-x}-1)\rangle\rangle\rangle = \langle\langle\langle(x-x')^2\rangle\rangle\rangle^{1/2}\left\{\exp\left[\sum_l \frac{(-1)^l}{l!}\kappa_l^{(-)}\right]-1\right\}, \tag{5.91b}$$

where the cumulants $\kappa_l^{(+)}$ and $\kappa_l^{(-)}$ are identified by developing both sides of (5.91a,b) and comparing the respective powers in x'. Up to the first order, one immediately checks that

$$\kappa_1^{(+)} = \frac{\langle\langle\langle(x-x')x'\rangle\rangle\rangle}{\langle\langle\langle(x-x')^2\rangle\rangle\rangle^{1/2}},$$

$$\kappa_1^{(-)} = \frac{\langle\langle\langle(x-x')x\rangle\rangle\rangle}{\langle\langle\langle(x-x')^2\rangle\rangle\rangle^{1/2}}. \tag{5.92}$$

From the symmetry relations of the collision operator under time reversal and reversed collisions, the following relation is satisfied:

$$\langle\langle\langle(x-x')x'\rangle\rangle\rangle = -\langle\langle\langle(x-x')x\rangle\rangle\rangle, \tag{5.93}$$

Since the prime stands for the value after the collision, one has $(x')' = x$ in the reversed collision. It follows from (5.92) that

$$\kappa_1^{(+)} = -\kappa_1^{(-)} = -\tfrac{1}{2}\langle\langle\langle(x-x')^2\rangle\rangle\rangle^{1/2}. \tag{5.94}$$

At this order of approximation, (5.89) becomes

$$\bar{\sigma}^s = \tfrac{1}{4}\langle\langle\langle(x-x')^2\rangle\rangle\rangle^{1/2}\left\{\exp\left[\tfrac{1}{2}\langle\langle\langle(x-x')^2\rangle\rangle\rangle^{1/2}\right]-\exp\left[-\tfrac{1}{2}\langle\langle\langle(x-x')^2\rangle\rangle\rangle^{1/2}\right]\right\} \tag{5.95}$$

or

$$\bar{\sigma}^s = \chi \sinh \chi, \tag{5.96}$$

with

$$\chi = \tfrac{1}{2}\langle\langle (x-x')^2\rangle\rangle^{1/2} = (\bar{\sigma}_0^s)^{1/2} \tag{5.97}$$

and where $\bar{\sigma}_0^s$ is the expression (2.50) of the entropy production derived in Chap. 2.

The important result is that (5.96) is positive definite; however, in contrast to $\bar{\sigma}_0^s$, it is not a simple quadratic form. The essential difference between the conventional and the modified moment methods is that in the conventional case the entropy production is not proved to be positive definite beyond the second order of approximation, while in the modified moment approach it is rigorously positive definite at any order of approximation.

Let us rewrite the entropy production as follows

$$\sigma^s = \sigma_0^s \frac{\sinh \chi}{\chi}. \tag{5.98}$$

For small values of the fluxes, χ is small and (5.98) reduces to σ_0^s. Expression (5.98) is useful for determining the expressions for the non-linear thermal conductivity λ and shear viscosity η. From the form of σ_0^s derived in Chap. 2, it is inferred that

$$\sigma^s = \left(\frac{1}{\lambda_0 T^2}\boldsymbol{q}\cdot\boldsymbol{q} + \frac{1}{2\eta_0 T}\overset{\circ}{\mathbf{P}}{}^v : \overset{\circ}{\mathbf{P}}{}^v\right)\frac{\sinh \chi}{\chi}; \tag{5.99}$$

this suggests that we should define a non-linear λ and a non-linear η through

$$\lambda(\boldsymbol{q},\overset{\circ}{\mathbf{P}}{}^v) = \lambda_0 \frac{\chi}{\sinh \chi}, \qquad \eta(\boldsymbol{q},\overset{\circ}{\mathbf{P}}{}^v) = \eta_0 \frac{\chi}{\sinh \chi}, \tag{5.100}$$

where λ_0 and η_0 are the limits of λ and η for small values of the fluxes, and χ stands for

$$\chi = \left[\frac{\tau_E}{nk_B}\left(\frac{1}{\lambda_0 T^2}\boldsymbol{q}\cdot\boldsymbol{q} + \frac{1}{2\eta_0 T}\overset{\circ}{\mathbf{P}}{}^v : \overset{\circ}{\mathbf{P}}{}^v\right)\right]^{1/2}, \tag{5.101}$$

in view of (5.86) and (5.97).

To be more explicit consider the generalised viscosity (5.100) in the absence of heat flux. It takes the form

$$\eta(\overset{\circ}{\mathbf{P}}{}^{\nu}) = \eta_0 \frac{[(\tau_E/2nk_BT\eta_0)\overset{\circ}{\mathbf{P}}{}^{\nu}:\overset{\circ}{\mathbf{P}}{}^{\nu}]^{1/2}}{\sinh[(\tau_E/2nk_BT\eta_0)\overset{\circ}{\mathbf{P}}{}^{\nu}:\overset{\circ}{\mathbf{P}}{}^{\nu}]^{1/2}}. \tag{5.102}$$

For a plane Couette flow parallel to the x_1 axis, $\overset{\circ}{\mathbf{P}}{}^{\nu}:\overset{\circ}{\mathbf{P}}{}^{\nu} = 2(P_{12}^{\nu})^2$, where $P_{12}^{\nu} = -\eta(\dot{\gamma})\dot{\gamma}$ with $\dot{\gamma}$ the shear rate; in this case, (5.102) reduces to

$$\eta(\dot{\gamma}) = \eta_0 \frac{\sinh^{-1}(\tau^*\dot{\gamma})}{\tau^*\dot{\gamma}} = \eta_0 \frac{\ln\left\{\tau^*\dot{\gamma} + \left[1 + (\tau^*\dot{\gamma})^2\right]^{1/2}\right\}}{\tau^*\dot{\gamma}}, \tag{5.103}$$

with $\tau^* = (\tau_E\eta_0/nk_BT)^{1/2}$. Expression (5.103) is the well-known Eyring formula, which describes shear thinning, i.e. the decrease of the apparent viscosity η with increasing shear rate.

With regard to heat transport, the first of relations (5.100) predicts a reduction of the heat flux for very large values of the temperature gradient. This phenomenon, which is imperceptible under usual circumstances, becomes crucial in laser-driven plasmas devised to achieve controlled thermonuclear fusion. This reduction plays a very important role in the transport of laser energy from the deposition region to the ablative surface [5.18].

The consequences of non-linear transport coefficients for the velocity and the temperature profiles in Lennard-Jones fluids have been analysed in detail by Eu and co-workers in a variety of situations [5.19]. Their results are in good agreement with molecular dynamics and experimental data. Non-linear effects are especially important at very low densities, because τ varies as n^{-1}, and at very high density, where the viscosity η_0 increases exponentially with density. Both situations have been examined by Eu, who has shown that EIT provides a unified way to treat problems ranging from rarefied Knudsen gases to dense fluids.

Other non-linear extensions have been proposed where the dependence of the entropy on the fluxes is not limited to second order. An exact solution of the Boltzmann equation has been derived for Maxwell molecules under a heat flux [5.20], which confirms the basic idea that the generalised entropy will depend on the fluxes, but shows a complicated dependence of the entropy on q.

Other non-linearities may arise from the non-dissipative terms occurring in the entropy flux; as a matter of fact, the β coefficient in the expression for the entropy flux may depend, in principle, on the scalar invariants of the fluxes. Thermodynamics does not provide special restrictions on these coefficients in the non-linear domain. However, valuable information may be obtained if one requires that the non-dissipative part of the evolution equations be Hamiltonian, or, in other words, susceptible of being written in terms of Poisson brackets [5.21]. This hypothesis is quite reasonable, because both particle mechanics at the microscopic level and the macroscopic behaviour of a perfect fluid are Hamiltonian. Moreover, since the equations of EIT can be written in the form of Poisson brackets (see Chap. 4), this opens the way to the use of the powerful techniques of symplectic geometry. The formulation in terms of Poisson brackets appears therefore to be an interesting complement to the thermodynamic results.

5.7 Beyond the Thirteen-Moment Approximation: Continued-Fraction Expansions of Transport Coefficients

In the previous sections, one has selected as the extra variables the fluxes of heat, momentum, and matter. However, there exists no restriction on extending the space of variables by including a whole hierarchy of new quantities obeying linear or even non-linear evolution equations. Now, the question arises whether it is sufficient to base a mesoscopic description exclusively on the conserved variables plus the usual dissipative fluxes (or a small number of additional fluxes) or, on the contrary, whether one should include an infinite number of variables. As recalled earlier, Grad's method does not introduce a smallness parameter; therefore, it is difficult to find any rigorous justification to truncate the expansion at the thirteen-moment level, as it is shown in Sect. 5.3.

These questions have raised doubts about the foundations of EIT. The need to take into account higher-order fluxes has become increasingly urgent as shown in recent analyses [5.22]. In this section we propose a way out, which consists in including a priori an infinite number of fluxes in the description; in a second step most of these extra variables are then eliminated by introducing an effective relaxation time which takes their contribution into account. Such a reduction in the number of variables is needed as we wish to compare with experimental results, which are usually described by a limited number of variables. As an illustration, we consider the effect of introducing higher-order moments in heat conduction. The results can be directly extended to viscous flows, electrical conduction or any other transport phenomenon.

So far, we have taken as non-equilibrium variable the heat flux q; we now wish to incorporate higher-order variables, each one being defined as the flux of the preceding one. To be explicit, we take as variables $J^{(1)}, J^{(2)}, \dots, J^{(n)}$, where $J^{(n)}$, a tensor of order n, is the flux of $J^{(n-1)}$. This is not an academic game: at least two arguments can be put forward in favour of this extension. First, it follows from the kinetic theory that the relaxation times of q and $\mathbf{P}^v$ are of the order of magnitude of the collision time. Since the relaxation times of the higher-order fluxes cannot be much shorter than the collision time, all of them are of the same order. Therefore, using only the first-order fluxes as independent variables is probably not satisfactory to describe high-frequency processes; indeed, when the frequency becomes high enough to be comparable to the inverse of the relaxation time of the first-order fluxes, the higher-order fluxes behave themselves like independent variables and must be incorporated in the formalism. The second point is that, up to now, the analysis has been mostly limited to an exponential relaxation for the dynamics of the fluxes. The inclusion of higher-order fluxes opens the way to more complicated dynamics.

In microscopic terms, $J^{(1)} = q$ is a part of the third-order moment of the velocity distribution function and $J^{(n)}$ is, correspondingly, a part of its $(n+2)$th-order moment [5.23]. Up to the nth-order moment, the Gibbs equation takes the form

$$\mathrm{d}s = T^{-1}\mathrm{d}u - \alpha_1 v J^{(1)} \cdot \mathrm{d}J^{(1)} - ... - \alpha_n v J^{(n)} \otimes \mathrm{d}J^{(n)}, \tag{5.104}$$

and the entropy flux can be written as

$$J^s = \beta_0 J^{(1)} + \beta_1 J^{(2)} \cdot J^{(1)} + ... + \beta_{n-1} J^{(n)} \otimes J^{(n-1)}. \tag{5.105}$$

The symbol $\otimes$ denotes the contraction of the corresponding tensors, and the coefficient β_0 may be identified by comparison with the classical theory as $\beta_0 = T^{-1}$. We have limited ourselves to the simplest form of the entropy and the entropy flux and have not taken into account coupled contributions of the form $J^{(3)} : J^{(1)} J^{(1)}$, and similarly, in (5.104–105). This approximation is sufficient for the present purpose.

The entropy production is easily derived from (5.104–105) and is given by

$$\sigma^s = -\left[-\nabla T^{-1} + \alpha_1 j^{(1)} - \beta_1 \nabla \cdot J^{(2)}\right] \cdot J^{(1)} ...$$

$$-\sum_{n=3}^{N} J^{(n-1)} \otimes \left[\alpha_{n-1} j^{(n-1)} - \beta_{n-1} \nabla \cdot J^{(n)} - \beta_{n-2} \nabla J^{(n-2)}\right]. \tag{5.106}$$

The above expression suggests the following set of evolution equations for the $J^{(n)}$:

$$\nabla T^{-1} - \alpha_1 j^{(1)} + \beta_1 \nabla \cdot J^{(2)} = \mu_1 J^{(1)},$$

$$\tag{5.107}$$

$$\beta_{n-1} \nabla J^{(n-1)} - \alpha_n j^{(n)} + \beta_n \nabla \cdot J^{(n+1)} = \mu_n J^{(n)},$$

with $\mu_n \geq 0$, as required from the positiveness of the entropy production. Defining τ_n as $\tau_n = \alpha_n/\tau_n$ and writing ∇T^{-1} in terms of ∇T, the set of equations (5.107) can be written as follows

$$\tau_1 j^{(1)} = -(J^{(1)} + \lambda \nabla T) + \frac{\beta_1 \tau_1}{\alpha_1} \nabla \cdot J^{(2)},$$

$$\tau_n j^{(n)} = -J^{(n)} + \frac{\beta_n \tau_n}{\alpha_n} \nabla \cdot J^{(n+1)} + \frac{\beta_{n-1} \tau_n}{\alpha_n} \nabla J^{(n-1)} \quad (n = 2, 3, ..., N).$$

(5.108)

Since $J^{(n)}$ is the flux of $J^{(n-1)}$, this implies, by the very definition of a flux requiring that $\rho j^{(n-1)} = -\nabla \cdot J^{(n)} + ...$, that the coefficient of $\nabla \cdot J^{(n)}$ in the equation for $\rho j^{(n-1)}$ must be equal to -1. As a consequence, $\rho \beta_n = -\alpha_n$ is a relation which reduces considerably the number of independent parameters. Indeed, we have a set of $3N$ coefficients α_n, β_n, and τ_n, but the above relation between β_n and α_n reduces them to $2N$ independent coefficients in the macroscopic theory. Furthermore, the theory of fluctuations predicts that $\alpha_n = k_B \langle \delta J^{(n)} \delta J^{(n)} \rangle^{-1}$, so that we are left with N independent coefficients at the level of the fluctuation theory. In addition, the kinetic theory allows one to express τ_n in terms of molecular interactions, from which it follows that the formalism contains no free parameters.

In the (ω, k)-Fourier space the hierarchy of equations (5.108) may be written as a generalised transport law with (ω, k)-dependent coefficients

$$\tilde{J}^{(1)}(\omega, k) = -ik\lambda(\omega, k)\tilde{T}(\omega, k),$$

(5.109)

where $\tilde{J}^{(1)}(\omega, k)$ and $\tilde{T}(\omega, k)$ are the Fourier transforms of $J^{(1)} \equiv q(r, t)$ and $T(r, t)$ while $\lambda(\omega, k)$ stands for

$$\lambda(\omega, k) = \cfrac{\lambda_0}{1 + i\omega\tau_1 + \cfrac{k^2 l_1^2}{1 + i\omega\tau_2 + \cfrac{k^2 l_2^2}{1 + i\omega\tau_3 + \cfrac{k^2 l_3^2}{1 + i\omega\tau_4}}}},$$

(5.110)

with $l_n^2 = \beta_n^2 (\mu_n \mu_{n+1})^{-1} > 0$. The result (5.110) allows us to define a k-dependent thermal conductivity $\lambda(k)$ in the steady state. Compact expressions for $\lambda(k)$ may be obtained in special cases:

(a) if l_n are given by $l_n^2 = \alpha_{n+1} l^2$, with $\alpha_n = n^2[(2n + 1)(2n - 1)]^{-1}$ (see [5.24]), the continued-fraction expansion for $\lambda(k)$ given by (5.110) may, in the limiting case $\omega = 0$, be written as

$$\lambda(k) = \frac{3\lambda_0}{(lk)^2}\left[\frac{lk}{\tan^{-1}(lk)} - 1\right];$$

(5.111a)

(b) when $l_n^2 = (1/4)l^2$, which represents a model for two-dimensional systems [5.24], it is found that for $\omega = 0$

$$\lambda(k) = \frac{2\lambda_0}{(lk)^2}\left[\sqrt{1+(lk)^2} - 1\right]. \tag{5.111b}$$

Expressions of transport coefficients in the form of a continued-fraction expansion, such as (5.110), are well known in non-equilibrium statistical mechanics [5.25] and they have proved very useful in the analysis of critical phenomena, where non-local effects become important and where the usual polynomial expansions of the generalised transport coefficients in terms of powers of k fail to converge.

Introducing (5.110) into the energy balance equation $\rho c \dot{T} = -\nabla \cdot q$ yields the dispersion relation

$$-i\omega = \cfrac{\chi k^2}{1+i\omega\tau_1 + \cfrac{k^2 l_1^2}{1+i\omega\tau_2 + \cfrac{k^2 l_2^2}{1+i\omega\tau_3 + \cfrac{k^2 l_3^2}{1+i\omega\tau_4}}}}. \tag{5.112}$$

The value of the phase speed $v_p = \omega / \mathrm{Re}\, k$ depends on the order of approximation considered in (5.112). For instance, up to third-order and in the high-frequency limit, one obtains

$$v_{p1}^2(\infty) = \frac{\chi}{\tau_1},$$

$$v_{p2}^2(\infty) = \frac{\chi}{\tau_1} + \frac{l_1^2}{\tau_1\tau_2}, \tag{5.113}$$

$$v_{p3}^4(\infty) = \left(\frac{\chi}{\tau_1} + \frac{l_1^2}{\tau_1\tau_2} + \frac{l_2^2}{\tau_2\tau_3}\right)v_{p3}^2(\infty) - \frac{\chi}{\tau_1}\frac{l_2^2}{\tau_2\tau_3}.$$

For ideal gases, the convergence of (5.113) is seen to be slow, and therefore it is necessary to use an asymptotic expression including an infinite number of higher-order fluxes. A proposed scheme is the following. Let us define $H_n(\omega,k)$ by $H_n(\omega,k) = \lambda_n(\omega,k)/\lambda_0$, where $\lambda_n(\omega,k)$ is the nth-order approximant to $\lambda(\omega,k)$ in (5.110). In the asymptotic limit, it is found that

$$H_\infty(\omega,k) = \frac{1}{1 + i\omega\tau_\infty + l_\infty^2 k^2 H_\infty(\omega,k)}, \tag{5.114}$$

with τ_∞ and l_∞ being the limits of τ_n and l_n for high n. This scheme has proved to be sufficiently accurate for a wide variety of problems in physics [5.26]. Solving (5.114) with respect to $H_\infty(\omega,k)$ results in

$$H_\infty(\omega,k) = \frac{-(1 + i\omega\tau_\infty) \pm \sqrt{(1 + i\omega\tau_\infty)^2 + 4k^2 l_\infty^2}}{2l_\infty^2 k^2}. \tag{5.115}$$

The dispersion equation (5.112) thus becomes $i\omega = -\chi k^2 H_\infty(\omega,k)$ which leads in the high-frequency limit to

$$i\omega = \frac{\chi}{2l_\infty^2}\left(i\omega\tau_\infty \pm \sqrt{4l_\infty^2 k^2 - \omega^2 \tau_\infty^2}\right), \tag{5.116}$$

while the corresponding phase speed is

$$v_{p\infty}^2 = \frac{\chi}{\tau_\infty - (l_\infty^2 / \chi)}. \tag{5.117}$$

By identifying this expression for the phase velocity with the standard expression $v_p = (\chi/\tau_{eff})^{1/2}$, one may introduce a 'renormalized' or 'effective' relaxation time by

$$\tau_{1eff} = \tau_\infty - (l_\infty^2/\chi). \tag{5.118}$$

The interest of this procedure is that it takes account of the presence of all the fluxes although one single flux, the first-order one, is used in the analysis; the key point is that its relaxation time is no longer given by τ_1 but rather by τ_{eff} defined by relation (5.118).

An analogous development for the viscous pressure tensor would yield the following expression for the effective relaxation times of the corresponding variables

$$\tau_{2eff} = \tau_2 - (l_\infty'^2 / v), \tag{5.119}$$

v being the kinematic viscosity, and l_∞' the corresponding correlation length. It may be shown that [5.24]

$$l_\infty^2 = \frac{3k_B T}{4m}\tau_1^2, \qquad l_\infty'^2 = \frac{3k_B T}{4m}\tau_2^2. \tag{5.120}$$

By recalling that $\chi = 5k_BT/(3m)\tau_1$ and $v = (k_BT/m)\tau_2$, it is found from (5.118) and (5.119) that

$$\tau_{1eff} = \tfrac{11}{20}\tau_{1Grad}, \qquad \tau_{2eff} = \tfrac{1}{4}\tau_{2Grad}, \tag{5.121}$$

where τ_{1Grad} and τ_{2Grad} are the respective relaxation times in the thirteen-moment approximation. These values for the effective times are comparable to their respective values obtained by fitting the experimental data for ultrasound velocity which are $\tau_{1exp} = 0.40\tau_{1Grad}$, $\tau_{2exp} = 0.29\tau_{2Grad}$ (see [11.5]).

It can thus be stated that the introduction of the whole set of higher-order fluxes and their elimination through the definition of an effective relaxation time is able to yield results which are in better agreement with the experimental results than a simple theory restricted a priori to the first thirteen moments. The above procedure presents the advantage that it does not ignore a priori all the other fluxes. At high frequencies more and more fluxes contribute to the value of the phase speed of dissipative signals according to (5.113). By considering only the first-order fluxes as independent variables, one obtains nevertheless a satisfactory qualitative understanding of high-frequency phenomena, in contrast with the classical theory of irreversible processes, which predicts unrealistic results. However, to achieve a better quantitative agreement with experiments, the introduction of higher-order fluxes is needed. Furthermore, the possibility of recovering the continued-fraction expansion for transport coefficients from EIT provides a closer connection with microscopic theories than classical irreversible thermodynamics. The ability to gain specific information about the memory functions from thermodynamic arguments is an important feature, because of the extreme difficulty of deriving explicit results from statistical mechanics, where the problem is generally circumvented by mathematical modelling.

One should finally note that since the evolution equations of EIT contain more variables than those of classical descriptions, they require more boundary conditions. This problem has been examined in the kinetic theory of rarefied gases [5.27] and more recently from a macroscopic point of view [5.28].

Problems

5.1 (a) Show how to obtain the results (5.37) and (5.38) from (5.32–33) and (5.36). (b) By following the procedure leading to (5.30–33), determine the non-equilibrium contributions to the entropy and the entropy flux up to the fourth order in ϕ.

5.2 Assuming $f(r,c,t) = f_{eq}(c)[1 + \phi(r,c,t)]$ with $\phi(r,c,t) \ll 1$, derive the linearized form of the Boltzmann collision operator. Study the H-theorem in this approximation.

5.3 The shear viscosities for He and Ar at 273 K and 1 atm are respectively 1.87×10^{-5} kg m^{-1} s^{-1} and 2.11×10^{-5} kg m^{-1} s^{-1}. (a) Calculate the collision time τ_2 for both gases. (b) If $\tau_1 = \frac{3}{2}\tau_2$, evaluate the thermal conductivity of both gases at 273 K and 1 atm, if their respective atomic weights are 4 and 40.

5.4 An expression for the relaxation time is $\tau^{-1} = \sqrt{2}\, n\sigma_0\langle v\rangle$, where n is the number of molecules per unit volume; $\langle v\rangle = (8k_BT/\pi m)^{1/2}$ the mean speed of the molecules, and σ_0 the collision cross-section, which for rigid spheres of diameter d is given by $\sigma_0 = \pi d^2$. (a) Determine the dependence of the viscosity and the thermal conductivity on n and T. (b) Using the data of Problem 5.3, evaluate the atomic diameters of He and Ar. (c) From dimensional arguments, find the dependence of the collision cross-section σ_0 on n and T when the interaction between the particles is described by a potential of the form $V(r) \sim r^{-s}$, with s being a constant. Show that η and λ behave as $T^{(s+4)/2s}$.

5.5 A monatomic gas composed of molecules of atomic diameter 2×10^{-10} m is subjected to a plane Couette flow at 273 K and 1 atm. (a) Determine the relaxation time τ_2 and calculate the shear viscosity at low shear rates. (b) A non-linear analysis of the Boltzmann equation yields for the shear-rate dependent viscosity the expression

$$\eta(\dot\gamma) = \eta_0 \frac{\sinh^{-1}(\tau\dot\gamma)}{\tau\dot\gamma},$$

with τ being a relaxation time which may be identified as τ_2 and η_0 the shear viscosity for low shear rates [see (5.103)]. Evaluate the shear viscosity for $\tau\dot\gamma = 0.5$.

5.6 Show that the kinetic equation for the distribution function in the relaxation-time approximation

$$\frac{\partial f}{\partial t} + c \cdot \nabla f = -\frac{1}{\tau}(f - f_{eq})$$

satisfies an H-theorem. Find the expression of the corresponding entropy production.

5.7 Find the evolution equations of $\mathbf{P}^v$ and q from the kinetic equation for the distribution function in the relaxation-time approximation.

5.8 Grad proposed to expand the non-equilibrium distribution function in Hermite polynomials of the dimensionless molecular velocity $\overline{C} = [m/(k_BT)]^{1/2}C$:

$$f = f_{eq}[1 + \sum_n (1/n!)a_i^{(n)}(r,t)H_i^{(n)}(\overline{C})],$$

where the $H_i^{(n)}$ are tensors of order n with subscripts i denoting a collection of n labels ($i = \{i_1,i_2,...,i_n\}$) and f_{eq} is the local-equilibrium distribution function. The coefficients $a_i^{(n)}$ are also tensors of order n, given by $a_i^{(n)} = (m/\rho)\int f H_i^{(n)}dC$. The first three polynomials are $H_i^{(1)} = \overline{C}_i$, $H_{ij}^{(2)} = \overline{C}_i\,\overline{C}_j - \delta_{ij}$, $H_{ijk}^{(3)} = \overline{C}_i\,\overline{C}_j\overline{C}_k - (\overline{C}_i\,\delta_{jk} + \overline{C}_j\,\delta_{ki} + \overline{C}_k\,\delta_{ij})$. (a) Show that $a_i^{(1)} = 0$, $a_{ij}^{(2)} = (1/p)\,P_{ij}^\nu$, and $a_{ijk}^{(3)} = (1/p)(m/k_BT)^{1/2}S_{ijk}$, with $S_{ijk} = \int \overline{C}_i\overline{C}_j\overline{C}_k f dC$. (b) Show that

$$\sum_{r=1}^{3} a_{irr}^{(3)} = \frac{2}{p}\left(\frac{m}{k_BT}\right)^{1/2} q_i.$$

(c) The Hermite polynomials $H_i^{(n)}$ and $H_j^{(m)}$ are orthogonal, i.e. their product $\int f_{eq}H_j^{(m)}H_i^{(n)}dC$ is zero unless $n = m$ and i is a permutation of j. Show that the non-equilibrium entropy has the form

$$s = s_{eq} - \tfrac{1}{2}k_B\sum_n \alpha_n a_i^{(n)}a_i^{(n)}.$$

Find the form of α_n in terms of the products of $H_i^{(n)}$. (Summation over all permutations of the set of subindices i is assumed in the expression for f and s.). (See R. M. Velasco and L. S. García-Colín, J. Non-Equilib. Thermodyn. **18** (1993) 157.)

5.9 To clarify the asymptotic expressions used in Sect. 5.7 for transport coefficients, (a) show that the continued fraction

$$R = \cfrac{a}{1 + \cfrac{a}{1 + \cfrac{a}{1 + ...}}}$$

tends, in the asymptotic limit of an infinite expansion, to

$$R_\infty = \tfrac{1}{2}\left(\sqrt{1 + 4a} - 1\right).$$

[*Hint:* Note that in this limit, R_∞ must satisfy $R_\infty = a/(1 + R_\infty)$]. (b) From this result, show relation (5.111b). (c) Using the method outlined in this problem, and assuming that all the relaxation times and correlation lengths are equal, prove (5.114) and (5.115).

Chapter 6

Fluctuation Theory

Up to now we have developed a thermodynamic formalism and have derived evolution equations for the dissipative fluxes. In the Part III of this book, these equations are used to interpret experimental data or applied to situations which are beyond the scope of the classical irreversible thermodynamics. In this chapter, we focus our attention on the generalised Gibbs equation, which is the central result of extended irreversible thermodynamics (EIT). Our aim is to show that the generalised entropy plays an essential role in the description of fluctuations around equilibrium states of not only the conserved variables but the dissipative fluxes, too.

The main reason to study the fluctuations is that it provides a link between entropy and probability and that it allows one to compute the non-classical coefficients of the entropy in terms of the equilibrium distribution function. The line of thought involving fluctuations is somewhat intermediate between the macroscopic and the microscopic descriptions: this is the reason why some authors refer to it as a mesoscopic theory. In fact, some other modern approaches to non-equilibrium thermodynamics beyond local equilibrium introduce, as additional variables of the theory, the second moments of the fluctuations out of equilibrium instead of the fluxes (see for instance [7.24]). However, to take into account this alternative possibility one must consider fluctuations around non-equilibrium states, which will be examined in Chap. 7.

Still another reason to consider fluctuations is that their study around equilibrium states has led to one of the most important results of classical non-equilibrium thermodynamics, namely, the Onsager reciprocal relations, the derivation of which is reviewed in Sect. 6.6.

6.1 Einstein's Formula. Second Moments of Equilibrium Fluctuations

The basic question to be answered is: How far is it possible to extend the relation between entropy and probability when the classical entropy is replaced by a generalised one? One way to solve this problem is to have non-equilibrium ensembles in the Gibbsian sense and this has been the line of thinking of some authors. However, the construction of such ensembles in general situations remains an open problem, which is discussed in Chap. 7.

In the present chapter, we follow an intermediate way based on the Einstein relation for the probability of fluctuations. For isolated systems, the probability of fluctuations of thermodynamic variables with respect to their equilibrium reference values is related to the change in entropy ΔS associated with the fluctuation

$$\Delta S = S - S(\text{equilibrium}), \tag{6.1}$$

where S is the total entropy of a system of mass M and volume V. The relation between the probability and the entropy change was proposed by Einstein by inverting the well known Boltzmann–Planck formula for the entropy,

$$S = k_B \ln W, \tag{6.2}$$

with W the probability of the macrostate and k_B the Boltzmann constant. In this way one is led to

$$W \approx \exp(\Delta S / k_B). \tag{6.3}$$

If the fluctuations are small, the quantity ΔS may be expanded:

$$\Delta S \approx (\delta S)_{eq} + \tfrac{1}{2}(\delta^2 S)_{eq}, \tag{6.4}$$

δS and $\delta^2 S$ being the first and second differentials of the entropy. Recalling that, for an isolated system, the entropy is maximum in equilibrium, it follows that $(\delta S)_{eq} = 0$ and $(\delta^2 S)_{eq} \leq 0$. Introduction of these results into (6.3) leads to

$$W \approx \exp\left[\tfrac{1}{2}\delta^2 S / k_B\right]. \tag{6.5}$$

This is the well-known Einstein formula for the probability of fluctuations. This relation is valid for moments up to second order, but it fails for higher-order moments [6.1].

Einstein's relation (6.5) is easily understood for isolated systems. Of course, the total energy and the total volume of an isolated system cannot fluctuate. Consider, however, an isolated system composed of two subsystems in thermal and mechanical contact.

The internal energy and the volume may be distributed between both subsystems in different ways, one of which corresponds to the equilibrium situation defined by the equality of the temperatures and pressures of both subsystems. The fluctuations considered here refer to those in the distribution of internal energy and volume between both subsystems.

However (6.5) is a particular case of a more general expression formulated for non-isolated systems. When the surroundings are incorporated, Einstein [6.2] suggested to express the probability of fluctuations as

$$W \approx \exp(-\Delta \mathcal{A}/k_B T_0), \tag{6.6}$$

where $\mathcal{A}$, the availability, is a new quantity defined as

$$\mathcal{A} = U - T_0 S + p_0 V, \tag{6.7}$$

in which T_0 and p_0 are the constant temperature and pressure of the surroundings. Note that $\mathcal{A}$ is not a property of the system alone but of the system in a given environment. It can be shown that $\Delta \mathcal{A}$ measures the maximum amount of useful work which can be extracted from the system during exchanges with the surroundings. If the system and surroundings are in thermal and mechanical equilibrium, $T = T_0$ and $p = p_0$, the change in $\mathcal{A}$ is given by

$$\Delta \mathcal{A} = \Delta U - T \Delta S + p \Delta V = \Delta G, \tag{6.8}$$

G being the Gibbs function, and (6.6) takes the form

$$W \approx \exp(-\Delta G/k_B T). \tag{6.9}$$

Similarly, for a system at fixed V and T, $\Delta \mathcal{A} = \Delta F$, with F the Helmholtz free energy, and the probability of fluctuations becomes

$$W \approx \exp(-\Delta F/k_B T). \tag{6.10}$$

Expressions (6.9) and (6.10) are valid for large and small fluctuations. Restricting our study to small fluctuations, $\Delta \mathcal{A}$ can be expanded in a series as follows:

$$\Delta \mathcal{A} \approx \delta \mathcal{A}_{eq} + \tfrac{1}{2}(\delta^2 \mathcal{A})_{eq} + ...,$$

where the first-order term is zero at equilibrium because G (or F), at fixed T and p (or fixed V and T), are minimum in equilibrium. This allows us to write

$$W \approx \exp\left[-\tfrac{1}{2}\delta^2 G/k_B T\right], \quad W \approx \exp\left[-\tfrac{1}{2}\delta^2 F/k_B T\right]. \tag{6.11}$$

These equations are analogous to (6.5) and may still be written as (6.5); indeed, it is easy to see that $\delta^2 F$ at constant T is proportional to $\delta^2 S$. To do so, remember that $F = U - TS$; to study the fluctuations of U at constant T, U is the independent variable and S is a function of U. Thus, one has $\delta^2 F = -T\delta^2 S$ and from (6.11) one recovers (6.5) but with $\delta^2 S$ computed at constant T and V. Similarly, it is easy to see that $\delta^2 G = -T\delta^2 S$ which, combined with (6.11), again leads to (6.5) but with $\delta^2 S$ computed at constant T and p. We have dealt explicitly with this variety of situations to emphasize that (6.5) is not only applicable to isolated systems but to many other situations, provided that one keeps in mind that different quantities are to be fixed depending on the external constraints placed on the system.

Now, we shall use the generalised entropy of EIT to obtain information about the fluctuations around equilibrium states [6.3–4]. The second differential of the generalised entropy may be derived directly from the Gibbs equation (2.39), and when this expression is introduced into (6.5) one obtains

$$W(\delta u,\delta v,\delta p^v,\delta q,\delta\overset{\circ}{\mathbf{P}}{}^v) \approx \exp\{(M/2k_B)[(T^{-1})_u(\delta u)^2 + 2(T^{-1})_v\delta u\delta v + (T^{-1}p)_v(\delta v)^2$$

$$-(v\tau_0/\zeta T)\delta p^v\delta p^v - (v\tau_1/\lambda T^2)\delta q\cdot\delta q - (v\tau_2/2\eta T)\delta\overset{\circ}{\mathbf{P}}{}^v : \delta\overset{\circ}{\mathbf{P}}{}^v]\} \qquad (6.12)$$

where the subscripts u and v stand for partial derivatives with respect to u and v respectively. Note that the terms coupling the fluctuations of u and v to those of the fluxes are proportional to the average value of the fluxes, which is zero at equilibrium. In Chap. 7 we will consider fluctuations around non-equilibrium steady states, where these terms do not longer vanish.

To obtain the second moments of the fluctuations from (6.12), it must be recalled that for a multivariant Gaussian distribution function

$$W \approx \exp\left[-\tfrac{1}{2}E_{ij}\delta x_i\delta x_j\right] \qquad (6.13)$$

the second moments are given by

$$\langle\delta x_i\delta x_j\rangle = (E^{-1})_{ij}, \qquad (6.14)$$

where the brackets $\langle...\rangle$ denote the average over the probability distribution (6.13). Since in equilibrium the fluctuations of the classical variables u and v are uncoupled with the fluctuations of the fluxes, as seen in (6.12), the second moments of the fluctuations of u and v may be obtained from (6.12) and (6.14) by taking into account that the matrix corresponding to the second derivatives of the entropy with respect to these variables is

$$E_{uv} = -\frac{M}{k_B}\begin{pmatrix} (T^{-1})_u & (T^{-1})_v \\ (T^{-1})_v & (T^{-1}p)_v \end{pmatrix}. \tag{6.15}$$

By inversing the matrix and introducing the result in (6.14), one obtains

$$\langle \delta u \delta u \rangle = k_B c_p T^2 - 2k_B T^2 pv\alpha + k_B T p^2 v k_T,$$

$$\langle \delta u \delta v \rangle = k_B T^2 v\alpha - k_B T pv k_T, \tag{6.16}$$

$$\langle \delta v \delta v \rangle = k_B T v k_T,$$

where c_p is the specific heat at constant pressure, α the coefficient of thermal expansion, and k_T the isothermal compressibility. After some calculations it may be seen that (6.15) can also be written in the alternative form: $\langle \delta u \delta u \rangle = -(k_B/M)(\partial u/\partial T^{-1})_{T^{-1}p}$, $\langle \delta u \delta v \rangle = -(k_B/M)(\partial v/\partial T^{-1})_{T^{-1}p}$ and $\langle \delta v \delta v \rangle = -(k_B/M)(\partial v/\partial T^{-1}p)_{T^{-1}}$. Similarly, the second moments of the fluctuations of the fluxes are given by

$$\langle \delta q_i \delta q_j \rangle = k_B \lambda T^2 (\tau_1 V)^{-1} \delta_{ij},$$

$$\langle \delta p^v \delta p^v \rangle = k_B \zeta T (\tau_0 V)^{-1}, \tag{6.17}$$

$$\langle \delta P_{ij}^v \delta P_{ij}^v \rangle = k_B \eta T (\tau_2 V)^{-1} \Delta_{ijkl},$$

with δ_{ij} the Kronecker symbol and $\Delta_{ijkl} = (\delta_{ik}\delta_{jl} + \delta_{il}\delta_{jk} - \frac{2}{3}\delta_{ij}\delta_{kl})$.

Expressions (6.17) are worth of further examination. They relate the dissipative coefficients λ, ζ, and η to the fluctuations of the fluxes with respect to equilibrium. These relations may be interpreted in two different ways: either by stating that the dissipative coefficients determine the strength of the fluctuations or saying that the fluctuations determine the dissipative coefficients. The latter point of view has been exploited in the Green–Kubo relations for the dissipative coefficients, which are one of the cornerstones of modern non-equilibrium statistical mechanics. Expressions (6.17) for the second moments of the fluctuations allow us also to compute the non-classical coefficients of the Gibbs equation (3.11) from the equilibrium distribution function and the microscopic expressions of the dissipative fluxes. In Sect. 7.7, it is shown how to generalise these results to fluctuations around non-equilibrium steady states (see also Problem 7.5).

Here we will make evident the connection of (6.17) with the well-known Green–Kubo formulae. To this end, recall the Green–Kubo relations for thermal conductivity and shear viscosity:

$$\lambda = \frac{V}{k_B T^2} \int_0^\infty \langle \delta q_i(0) \delta q_i(t) \rangle \, dt,$$

$$\eta = \frac{V}{k_B T} \int_0^\infty \langle \delta P_{ij}^v(0) \delta P_{ij}^v(t) \rangle \, dt \quad (i \neq j),$$

(6.18)

respectively. In (6.18) the brackets $\langle \ldots \rangle$ stand for equilibrium average, and no summation is implied over repeated indices. At this point, it must be underlined that (6.17–18) are valid for all kinds of dynamics; in particular, if the evolution of the fluctuations of the fluxes is described by the Maxwell–Cattaneo relaxational equations (2.64–66), then, after integration, (6.17–18) reduce to

$$\lambda = \frac{\tau_1 V}{k_B T^2} \langle \delta q_i(0) \delta q_i(0) \rangle,$$

$$\eta = \frac{\tau_2 V}{k_B T} \langle \delta P_{ij}^v(0) \delta P_{ij}^v(0) \rangle,$$

(6.19)

which are nothing else than the results expressed by (6.17).

The Green–Kubo relations have generated a wide interest in connection with the study of the evolution of the dissipative fluxes in statistical mechanics. This point is worth emphasizing, because one of the objectives of EIT is just the determination of evolution equations for the dissipative fluxes, but from a macroscopic point of view. This provides an interesting point of contact between the macroscopic and the microscopic approaches.

6.2 Ideal Gases

In this section we compute the coefficients appearing in the non-classical part of the generalised entropy by using the expressions for the second moments derived in the previous section and the equilibrium distribution function for ideal gases.

As seen in Chap. 5, the kinetic definitions for q and $\mathbf{P}^v$ are

$$q = \int \tfrac{1}{2} m C^2 \mathbf{C} f \, d\mathbf{c}$$

(6.20)

and

$$\mathbf{P}^v = \int \left(m \mathbf{C}\mathbf{C} - \tfrac{1}{3} p \mathbf{U} \right) f \, d\mathbf{c},$$

(6.21)

where f is the distribution function, c the velocity of a molecule, v the mean velocity, C $= c - v$ the molecular velocity relative to the mean motion, m the mass of a molecule, U the identity tensor, and p the thermodynamic pressure.

The fluctuations of q near equilibrium are due to the fluctuations δf of the distribution function and the fluctuations δv of the mean velocity. Up to first-order terms in the fluctuations one has

$$\delta q = \int \tfrac{1}{2}mC^2 C \delta f \, dc - \left[\int \tfrac{1}{2}mC^2 f_{eq} \, dc\right]\delta v - \left[\int mCC f_{eq} \, dc\right]\cdot \delta v. \qquad (6.22)$$

For an ideal gas at equilibrium, the second and third integrals on the right-hand side of (6.22) give respectively $\rho u \delta v$ and $p \delta v$, and hence

$$\delta q = \int \tfrac{1}{2}mC^2 C \delta f \, dc - \rho h \delta v, \qquad (6.23)$$

where $h = u + (p/\rho)$ is the enthalpy per unit mass, which for a classical monatomic ideal gas takes the form $h = \tfrac{5}{2}k_B T/m$.

Since δv is, by definition, taken from

$$\rho \delta v = \int mC \delta f \, dc, \qquad (6.24)$$

the fluctuation of the heat flux given by (6.23) can be written as

$$\delta q = \int \left(\tfrac{1}{2}mC^2 - \tfrac{5}{2}k_B T\right) C \, \delta f \, dc. \qquad (6.25)$$

Using this expression, frequently called the subtracted heat flux, to compute the second moments of the fluctuations, one finds that

$$\langle \delta q \delta q \rangle = \int dc \int dc' \left(\tfrac{1}{2}mC^2 - \tfrac{5}{2}k_B T\right) C \left(\tfrac{1}{2}mC'^2 - \tfrac{5}{2}k_B T\right) C' \langle \delta f(C)\delta f(C')\rangle. \qquad (6.26)$$

For the fluctuations of the non-diagonal components of the viscous pressure tensor, an analogous derivation yields

$$\delta \mathbf{P}^v = \int mCC \delta f \, dc, \qquad (6.27)$$

from which

$$\langle \delta P_{ij}^v \delta P_{kl}^v \rangle = \int dc \int dc' m C_i C_j \, m C_k' C_l' \langle \delta f(C)\delta f(C')\rangle. \qquad (6.28)$$

The expression for the second moments of the fluctuations of the distribution function may be obtained [6.5] from Einstein's formula (6.5). When Boltzmann's definition of entropy (5.27), namely

$$S = -k_B V \int f \ln f \, dc,$$
(6.29a)

where V is the volume of the system, is introduced into (6.5), one obtains

$$\Pr(\delta f) \approx \exp\left[-\frac{1}{2} \frac{V}{f_{eq}(r,c,t)} (\delta f(r,c,t))^2 \right]$$
(6.29b)

From (6.29b), it can be written

$$\langle \delta f(r,c,t) \delta f(r',c',t) \rangle = \frac{1}{V} f_{eq}(r,c) \delta(r-r') \delta(c-c')$$
(6.30)

and consequently (6.26) and (6.28) reduce to

$$\langle \delta q \delta q \rangle = \frac{1}{V} \int dc \left(\tfrac{1}{2} mC^2 - \tfrac{5}{2} k_B T \right)^2 CC f_{eq},$$
(6.31a)

$$\langle \delta P_{ij}^v \delta P_{kl}^v \rangle = \frac{1}{V} \int dc \, m^2 C_i C_j C_k C_l f_{eq}.$$
(6.31b)

Introducing the Maxwell–Boltzmann distribution function,

$$f_{eq} = n \left(\frac{m}{2\pi k_B T} \right)^{3/2} \exp\left(-\frac{mC^2}{2 k_B T} \right),$$
(6.32)

into (6.31a, b) and substituting the corresponding results in (6.17), one obtains for the ratios τ_1/λ and τ_2/η:

$$(\tau_1 v/\lambda T^2) = \tfrac{2}{5}(p^2 T)^{-1}, \qquad (\tau_2 v/2\eta T) = \tfrac{1}{2}(v/pT).$$
(6.33)

The present theory may also be applied to degenerate ideal gases. In this case, the entropy reads [6.6]

$$S = -k_B V \int \left[f \ln \frac{f}{y} \mp y \left(1 \pm \frac{f}{y} \right) \ln \left(1 \pm \frac{f}{y} \right) \right] dc,$$
(6.34)

where the upper sign applies for Bose–Einstein statistics and the lower one for Fermi–Dirac statistics; the quantity y stands for $y = (2s+1)/h^3$ for particles with spin s and mass m different from zero and $y = 2s/h^3$ for particles with vanishing mass. Here, h is the Planck constant. Combining the Einstein formula (6.5) with (6.34), one obtains for the second moments of the fluctuations of the distribution function

$$\langle \delta f(r,c,t)\delta f(r',c',t)\rangle = -\frac{1}{V}\left(\frac{\partial f}{\partial \alpha}\right)_{eq}\delta(r-r')\delta(c-c'),\tag{6.35}$$

where the equilibrium distribution function is given by

$$f_{eq} = \frac{y}{\exp(\alpha + x^2)\mp 1},\tag{6.36}$$

with $x = [\frac{1}{2}m/(k_B T)]^{1/2}C$ and $\alpha = -\mu/k_B T$, μ being the chemical potential. The expressions for q and $\mathbf{P}^v$ are identical with those of the preceding case, but now, of course, the expressions for the enthalpy h and pressure p are those corresponding to degenerate gases.

The results are consequently given in terms of the functions $I_n(\alpha)$ defined as

$$I_n^{\pm}(\alpha) = \int_0^{\infty} \frac{x^n}{\exp(\alpha + x^n)\pm 1}dx\tag{6.37}$$

which satisfy the relation $dI_n/d\alpha = -\frac{1}{2}(n-1)I_{n-2}(\alpha)$. For the quantities $(\tau_1 v/\lambda T^2)$ and $(\tau_2 v/2\eta T)$, it is found that

$$\frac{\tau_1 v}{\lambda T^2} = \left\{\frac{5}{3}\frac{k_B T^{7/2}}{mA}\left[\frac{7}{5}\frac{k_B T}{m}I_6(\alpha) - \frac{5}{3}\frac{k_B T^{5/2}}{mpA}I_4^2(\alpha)\right]\right\}^{-1}\tag{6.38}$$

and

$$\frac{\tau_2 v}{2\eta T} = \frac{3mA}{4k_B T^{7/2}I_4(\alpha)},\tag{6.39}$$

with $A^{-1} = 4\pi\, 2^{1/2}yR^{3/2}m$ and R being the gas constant. These results coincide with those derived by Liu and Müller [2.3] by a different method. In the limit of high values of α, (6.38–39) reduce to (6.33).

The results of this section are in agreement with those of the kinetic theory of ideal gases discussed in Chap. 5. The interest of the theory of fluctuations is outlined by (6.33) and (6.38–39), which show that it is possible to calculate the extra terms of entropy from equilibrium statistical mechanics without appealing to a non-equilibrium kinetic theory.

For obtaining results in the case of dilute real gases it must be taken into account the interaction effects between particles [6.7]. This leads to several corrections to the coefficients $(\tau_1 v/\lambda T^2)$ and $(\tau_2 v/2\eta T)$, which have been evaluated in [6.7] up to the first order in a virial expansion.

6.3 Fluctuations and Hydrodynamic Stochastic Noise

Expressions (6.16) can also be related to the classical formulae of Landau and Lifshitz for hydrodynamic stochastic noise [6.8–9]. These authors assume that this noise arises from fast fluctuations of the dissipative fluxes; by using an ad hoc version of the Onsager–Machlup formulation [6.10], they were able to obtain expressions for the second moments of the noise. Whereas in the usual theory the stochastic noise appears as a purely mathematical entity, in EIT it receives a more precise physical meaning, as shown below.

To make explicit the relation between rapid fluctuations and noise, we calculate the time correlation functions from (6.17) and (2.60–62); it is found that

$$\langle \delta q_i(0)\delta q_j(t)\rangle = k_B \lambda T^2 (\tau_1 V)^{-1}\delta_{ij}\exp(-t/\tau_1), \tag{6.40a}$$

$$\langle \delta P_{ij}^v(0)\delta P_{ij}^v(t)\rangle = k_B \eta T(\tau_2 V)^{-1}\Delta_{ijkl}\exp(-t/\tau_2). \tag{6.40b}$$

In the limit of vanishing relaxation times, we recover the classical formulae of Landau and Lifshitz, since in this limit $\tau^{-1}\exp(-t/\tau) \rightarrow 2\delta(t)$, $\delta(t)$ being the Dirac delta function. Accordingly, in this limit we recover not only the classical deterministic formulation of linear irreversible thermodynamics but its fluctuating hydrodynamics counterpart too [6.11]. In a strict sense, δq and $\overset{0}{\delta P}{}^v$ may be considered as noises only in the limit of vanishing τ. In general, one should include a white noise f in the evolution equations of these variables; for instance, one could write (see Problem 6.6)

$$\delta \dot{q} + \tau^{-1}\delta q = f \tag{6.41a}$$

with

$$\langle f_i(t)f_j(t')\rangle = \frac{2\lambda k_B T^2}{\tau^2}\delta_{ij}\delta(t-t'). \tag{6.41b}$$

When $\tau \rightarrow 0$, δq tends to τf and it may be considered as a true stochastic noise, rather than the fluctuation of a basic variable.

Let us also mention that by writing $q = -\lambda \nabla T + \delta q$ and allowing ∇T to fluctuate, this formalism leads to spatial correlations of the temperature fluctuations (see Problem 6.7) [6.12]; this was proposed by several authors as a possible origin of the $1/f$ noise [6.13] but we do not enter here into this specialized subject.

6.4 The Entropy Flux

The next step is to calculate the coefficients appearing in the entropy flux in terms of the moments of fluctuations [6.14]. The procedure is not straightforward and the final result not so direct and appealing as for the coefficients occurring in the expression of the generalised entropy.

Assume that the fluctuation in the distribution function, $\delta f = f - f_{eq}$, is given by

$$\delta f = f_{eq}(\delta\lambda \cdot \hat{q} + \delta\overset{0}{\lambda}{}^{v} : \overset{\wedge}{\mathbf{P}}{}^{v}), \qquad (6.42)$$

with $\hat{q}$ and $\overset{\wedge}{\mathbf{P}}{}^{v}$ the microscopic operators corresponding to the fluxes and $\delta\lambda$ and $\delta\overset{0}{\lambda}{}^{v}$ parameters which are to be related to the fluctuations of the fluxes. This expression is the simplest one leading to non-zero values for δq and $\delta\overset{0}{\mathbf{P}}{}^{v}$.

According to (6.42), we have the following relations between $\delta\lambda$, $\delta\overset{0}{\lambda}{}^{v}$ and δq, $\delta\overset{0}{\mathbf{P}}{}^{v}$:

$$\delta q = \int \hat{q}\,\delta f\,\mathrm{d}c = \left(\int \hat{q}_1^2 f_{eq}\mathrm{d}c\right)\delta\lambda,$$

$$\qquad (6.43)$$

$$\delta\overset{0}{\mathbf{P}}{}^{v} = \int \overset{\wedge}{\mathbf{P}}{}^{v}\,\delta f\,\mathrm{d}c = 2\left(\int (\hat{P}_{12}^{v})^2 f_{eq}\,\mathrm{d}c\right)\delta\overset{0}{\lambda}{}^{v},$$

wherein we have assumed that only the component q_1 of q and P_{12}^{v} of $\mathbf{P}^{v}$, with $P_{12}^{v} = P_{21}^{v}$, are non-vanishing.

Introduction of (6.42) into expression (6.29a) of entropy and in the corresponding expression for the entropy flux

$$J^{s} = -k_{B}\int f \ln f\, C\,\mathrm{d}c, \qquad (6.44)$$

yields, up to second order in the fluctuations of the fluxes,

$$\Delta S = S - S_{eq} = -\tfrac{1}{2}k_{B}V\int f_{eq}(\hat{q}\cdot\delta\lambda + \overset{\wedge}{\mathbf{P}}{}^{v} : \delta\overset{0}{\lambda}{}^{v})^2\mathrm{d}c, \qquad (6.45a)$$

$$\Delta J^s = J^s - T^{-1}q = -\tfrac{1}{2}k_B \int f_{eq}(\hat{q}\cdot\delta\lambda + \overset{\wedge}{\mathbf{P}}{}^{\nu}:\delta\overset{0}{\lambda}{}^{\nu})^2 C\,d\mathbf{c}. \tag{6.45b}$$

These expressions can be recast in the form

$$\Delta S = -\tfrac{1}{2}k_B V\Big[\int f_{eq}\,\hat{q}_1^2\,d\mathbf{c}\Big]\delta\lambda\cdot\delta\lambda - \tfrac{1}{2}k_B V\Big[2\int f_{eq}(\hat{P}_{12}^{\nu})^2 d\mathbf{c}\Big]\delta\overset{0}{\lambda}{}^{\nu}:\delta\overset{0}{\lambda}{}^{\nu}, \tag{6.46a}$$

$$\Delta J^s = -\tfrac{1}{2}k_B\Big[4\int f_{eq}\,\hat{q}_1\hat{P}_{12}^{\nu}C_2\,d\mathbf{c}\Big]\delta\overset{0}{\lambda}{}^{\nu}\cdot\delta\lambda\,, \tag{6.46b}$$

which, according to (6.43), may be rewritten as

$$\Delta S = -\frac{1}{2}k_B V\Big[\int f_{eq}\,\hat{q}_1^2\,d\mathbf{c}\Big]^{-1}\delta q\cdot\delta q - \frac{1}{2}k_B V\Big[2\int f_{eq}(\hat{P}_{12}^{\nu})^2 d\mathbf{c}\Big]^{-1}\delta\overset{0}{\mathbf{P}}{}^{\nu}:\delta\overset{0}{\mathbf{P}}{}^{\nu}, \tag{6.47}$$

$$\Delta J^s = -k_B\frac{\int C_2\hat{P}_{12}^{\nu}\hat{q}_1 f_{eq}\,d\mathbf{c}}{\int \hat{q}_1^2 f_{eq}\,d\mathbf{c}\int(\hat{P}_{12}^{\nu})^2 f_{eq}\,d\mathbf{c}}\delta\overset{0}{\mathbf{P}}{}^{\nu}\cdot\delta q.$$

By comparing (6.47) with their corresponding macroscopic expressions

$$\Delta S = -\tfrac{1}{2}\rho V(\alpha_1\delta q\cdot\delta q + \alpha_2\delta\overset{0}{\mathbf{P}}{}^{\nu}:\delta\overset{0}{\mathbf{P}}{}^{\nu}), \tag{6.48a}$$

$$\Delta J^s = \beta\delta\mathbf{P}^{\nu^0}\cdot\delta q, \tag{6.48b}$$

we find that

$$\alpha_1 = k_B\Big[\int \hat{q}_1^2 f_{eq}\,d\mathbf{c}\Big]^{-1}, \qquad \alpha_2 = k_B\Big[2\int(\hat{P}_{12}^{\nu})^2 f_{eq}\,d\mathbf{c}\Big]^{-1}, \tag{6.49}$$

and

$$\beta = -k_B\frac{\int C_2\hat{P}_{12}^{\nu}\hat{q}_1 f_{eq}\,d\mathbf{c}}{\int \hat{q}_1^2 f_{eq}\,d\mathbf{c}\int(\hat{P}_{12}^{\nu})^2 f_{eq}\,d\mathbf{c}}. \tag{6.50}$$

after use of the definitions $\alpha_1 = v\tau_1/\lambda T^2$, $\alpha_2 = v\tau_2/2\eta T$ and (6.31). Relations (6.49) are identical to those previously given by (6.19), whereas (6.50) is new and expresses the coefficient β in the entropy flux as a function of fluctuations. Writing the denominator in terms of the dissipative coefficients and using the relations (6.16) for the relaxation times, one obtains the more compact form

$$\beta = -\frac{\tau_1 \tau_2}{k_B \lambda \eta T^3} \int C_2 \hat{P}_{12}^v \, \hat{q}_1 \, f_{eq} \, d\mathbf{c} .$$
(6.51)

This is a particular case of a formula derived by Ernst [6.15] from the kinetic theory by means of projection-operator methods. More general expressions of β in terms of correlation functions of fluxes have been given by several authors [6.16]. Application of (6.51) to ideal – either classical or degenerate – gases is straightforward and yields

$$\beta = -\tfrac{2}{5} pT,$$
(6.52)

a well-known result in kinetic theory, as shown in Chap. 5. The corresponding expression of β for dilute real gases is derived in [6.7].

6.5 Application to a Radiative Gas

It is easy to apply the above formalism to a radiation gas or a mixture of radiation (photons, neutrinos, gravitons) and a material fluid [6.14b]. The latter is assumed to thermalize in a short time, whereas the mean free time of radiation τ is supposed to be much longer. Such systems are of interest in cosmological and astrophysical problems.

In this case, we may interpret q and $\overset{0}{\mathbf{P}}{}^v$ as

$$\delta q = c \, \delta e_{rad}$$
(6.53)

and

$$\delta \overset{0}{\mathbf{P}}{}^v = (1/c^2) \mathbf{cc} \, \delta e_{rad},$$
(6.54)

where c is the velocity vector (whose magnitude is the speed of light) and δe_{rad} is the fluctuation of the radiation energy per unit volume. When the Stefan–Boltzmann law $e_{rad} = aT^4$, with a the black-body constant, is used for a radiative gas, the only non-zero element of the matrix (6.15) is $(T^{-1})_u$ and the first of expressions (6.16) may be written as

$$\langle \delta e_{rad} \delta e_{rad} \rangle = 4 k_B a T^5 V^{-1}.$$
(6.55)

By introducing this result into (6.17) and (6.50), it is found that

$$\lambda = \frac{\tau V}{k_B T^2} \langle \delta q_1 \delta q_1 \rangle = \frac{4}{3} c^2 a T^3 \tau,$$

$$\eta = \frac{\tau V}{k_B T^2} \langle \delta P_{12}^v \delta P_{12}^v \rangle = \frac{4}{5} a T^4 \tau, \qquad (6.56)$$

$$\beta = -\frac{\tau^2 V}{k_B \lambda \eta T^3} \langle \delta q_1 \delta P_{12}^v C_2 \rangle = -\frac{1}{4 a T^4},$$

which coincide with the results derived in the relativistic kinetic theory [17.7]. In the above model, it was supposed that the relaxation times of q and $\overset{0}{\mathbf{P}}{}^{\nu}$ coincide.

Concerning the bulk viscous pressure, one must take into account that $\delta p^v = \Omega \delta e_{rad}$, with Ω defined as $\Omega = [\partial P_{tot}/\partial(\rho e)]_v - \frac{1}{3}$, P_{tot} being the total scalar pressure of the fluid, and ρ its proper density [6.14b]. One then obtains for the bulk viscosity

$$\zeta = (\tau V/k_B T)\langle \delta p^v \delta p^v \rangle = 4\Omega^2 a T^4 \tau. \qquad (6.57)$$

An interesting point provided by the theory of fluctuations is a reduction of the number of independent parameters. In the macroscopic theory developed in Chap. 2, we have eight parameters, namely τ_1, τ_0, τ_2, λ, ζ, η, β, and β'. Equations (6.16) and (6.51) and its analogue for β' provide five relations between these parameters, so that we are left with just three independent parameters, for instance τ_1, τ_0, and τ_2. In this way, the macroscopic extended irreversible thermodynamics plus fluctuation theory give much more information than the classical theory with the same number of free parameters.

6.6 Onsager's Relations

Up to now, we have mainly considered static aspects of the fluctuations. The analysis of the evolution of fluctuations leads to the Onsager–Casimir reciprocal relations [1.1], which are of key importance in non-equilibrium thermodynamics. Here we shall reproduce the classical derivation of Onsager–Casimir relations. Another presentation based on extended thermodynamics will be given in Sect. 14.5.

Assume that the entropy is a function of several variables $A_1,..., A_n$ and denote by α_β their deviation from the average equilibrium value $\langle A_\beta \rangle$, $\alpha_\beta = A_\beta - \langle A_\beta \rangle$. Thus, the entropy arising from the fluctuations α_β is

$$S = S_{eq} - \tfrac{1}{2} G_{\beta\gamma} \alpha_\beta \alpha_\gamma, \qquad (6.58)$$

with

$$G_{\beta\gamma} = -\left(\frac{\partial^2 S}{\partial\alpha_\beta\partial\alpha_\gamma}\right)_{eq}. \tag{6.59}$$

Summation with respect to repeated subindices is understood. It follows from Einstein's relation (6.5) that the second moments of the fluctuations are given by

$$\langle\alpha_\beta\alpha_\gamma\rangle = k_B(G^{-1})_{\beta\gamma}. \tag{6.60}$$

Onsager assumed that the decay of spontaneous fluctuations of the system is described by the same phenomenological laws that describe the evolution of the perturbations produced by external causes. He supposed, furthermore, that such laws are linear, so that the evolution of α_β is governed by

$$\frac{d\alpha_\beta}{dt} = -M_{\beta\gamma}\alpha_\gamma, \tag{6.61}$$

with $\mathbf{M}$ a matrix independent of α_γ.

Onsager [1.1] took as fluxes the time derivatives of the fluctuations α_β and as corresponding thermodynamic forces the derivatives of the entropy with respect to α_β:

$$J_\beta = \frac{d\alpha_\beta}{dt}, \qquad X_\beta = \left(\frac{\partial S}{\partial\alpha_\beta}\right) = -G_{\beta\gamma}\alpha_\gamma. \tag{6.62}$$

With these definitions of J_β and X_β, the time derivative of the entropy is a bilinear form in the fluxes and the forces

$$\frac{dS}{dt} = -G_{\beta\gamma}\alpha_\gamma\frac{d\alpha_\beta}{dt} = J_\beta X_\beta, \tag{6.63}$$

while the evolution equations (6.61) can be casted in the form of linear phenomenological relations between fluxes and forces, namely

$$J_\beta = L_{\beta\gamma}X_\gamma. \tag{6.64}$$

Indeed, it suffices to introduce the definitions (6.62) into (6.64) to obtain

$$\frac{d\alpha_\beta}{dt} = -L_{\beta\gamma}G_{\gamma\eta}\alpha_\eta. \tag{6.65}$$

Comparison with (6.61) leads to the identification

$$\mathbf{L} = \mathbf{M}\cdot\mathbf{G}^{-1}. \tag{6.66}$$

In a steady state, one has for any couple of variables α_β and α_γ

$$\langle \alpha_\beta(0)\alpha_\gamma(t)\rangle = \langle \alpha_\beta(0+\tau)\alpha_\gamma(t+\tau)\rangle. \tag{6.67}$$

Since this equality is valid for any values of t and τ, in particular, for $\tau = -t$, one has

$$\langle \alpha_\beta(0)\alpha_\gamma(t)\rangle = \langle \alpha_\beta(-t)\alpha_\gamma(0)\rangle. \tag{6.68}$$

Assume that the variable A_β has a well-defined time-reversal parity, i.e. that under the transformations $t \to -t$, $p \to -p$, with p the momentum of the particles, it behaves as $\alpha_\beta(-t) = \varepsilon_\beta\alpha_\beta(t)$, with $\varepsilon_\beta = +1$ for even variables, such as mass and energy, and $\varepsilon_\beta = -1$ for odd variables, such as momentum or heat flux. As a consequence, (6.68) becomes

$$\langle \alpha_\beta(0)\alpha_\gamma(t)\rangle = \varepsilon_\beta\varepsilon_\gamma\langle \alpha_\beta(t)\alpha_\gamma(0)\rangle. \tag{6.69}$$

Note that (6.69) implies that equal-time correlations between even and odd variables vanish in equilibrium. Relation (6.69) is of fundamental importance in deriving the Onsager–Casimir relations: it expresses the microscopic reversibility of the system.

Moreover, direct integration of the evolution equation (6.61) yields

$$\alpha_\beta(t) = \exp[-\mathbf{M}t]_{\beta\gamma}\, \alpha_\gamma(0). \tag{6.70}$$

For small values of t one may expand (6.70) and keep only first-order terms in t. Then (6.69) takes the form

$$\langle \alpha_\beta(0)[\delta_{\gamma\eta} - M_{\gamma\eta}t]\alpha_\eta(0)\rangle = \varepsilon_\beta\varepsilon_\gamma\langle[\delta_{\beta\eta} - M_{\beta\eta}t]\alpha_\eta(0)\alpha_\gamma(0)\rangle, \tag{6.71}$$

which yields

$$M_{\gamma\eta}\langle \alpha_\beta(0)\alpha_\eta(0)\rangle = \varepsilon_\beta\varepsilon_\gamma M_{\beta\eta}\langle \alpha_\eta(0)\alpha_\gamma(0)\rangle. \tag{6.72}$$

Expressing the second moments in terms of the matrix $\mathbf{G}$ of the second derivatives of the entropy, (6.72) can be written as

$$M_{\gamma\eta}(G^{-1})_{\beta\eta} = \varepsilon_\beta\varepsilon_\gamma M_{\beta\eta}(G^{-1})_{\eta\gamma}. \tag{6.73}$$

In view of the relation (6.66) between $\mathbf{L}$ and $\mathbf{M}$, and since $\mathbf{G}$ is a symmetric matrix, it follows from (6.73) that

$$L_{\gamma\beta} = \varepsilon_\beta\varepsilon_\gamma L_{\beta\gamma}, \tag{6.74}$$

which are the well known Onsager–Casimir reciprocal relations.

The present analysis has the advantage of exhibiting explicitly the main assumptions which underlie the derivation of Onsager–Casimir relations, namely, linear evolution equations and the hypothesis that the behaviour of fluctuations is described on the average by the macroscopic transport laws. In the presence of a magnetic field B, the operation of time-reversal will only change the internal magnetic field arising from the motion of the particles, which is automatically reversed under time reversal. Because of the Lorentz force, which is proportional to the vector product of the velocities and the magnetic field strength, the external magnetic field B must be inverted in order that the reversal motion follows the same path as in the direct motion. The same comment applies when the system rotates as a whole with angular velocity $\boldsymbol{\omega}$; the sign of the angular velocity must be changed because the Coriolis force is proportional to the vector product of the linear velocities of the particles and the angular velocity of the system. Therefore, in the presence of an external magnetic field B and rotation $\boldsymbol{\omega}$, the Onsager–Casimir relations read

$$L_{\gamma\beta}(B,\omega) = \varepsilon_\beta \varepsilon_\gamma L_{\beta\gamma}(-B,-\omega). \tag{6.75}$$

The importance and the limitations of the Onsager–Casimir relations have been stressed in Chap. 1.

Problems

6.1 When the relaxation time of the distribution function is not a constant but a function of the peculiar (or relative) velocity C, the Green–Kubo relations for the thermal conductivity and the shear viscosity yield

$$\lambda = (k_B T^2)^{-1} V \langle \tau(C) \hat{q}_x(C) \hat{q}_x(C) \rangle,$$

$$\eta = (k_B T)^{-1} V \langle \tau(C) \hat{P}^v_{xy}(C) \hat{P}^v_{xy}(C) \rangle,$$

with $\hat{q}_x = (\frac{1}{2}mC^2 - \frac{5}{2}k_B T)C_x$ and $\hat{P}^v_{xy} = mC_x C_y$. (a) Show that these expressions coincide with those derived from the relaxation-time approximation of the kinetic theory when τ depends on C. (b) Compare the values of λ and of η obtained when τ is assumed to be a constant and when $\tau(C)$ varies as C^{-1}.

6.2 Consider a non-ideal monatomic gas whose particles interact through a potential $\phi(r_{ij})$ which depends only on the distance r_{ij} between the molecules, r_{ij} being the vector from particle i to particle j. The pressure tensor $\mathbf{P}$ is $\mathbf{P} = \mathbf{P}_c + \mathbf{P}_p$, with $\mathbf{P}_c$ and $\mathbf{P}_p$ being kinetic and potential contributions, as obtained from (5.62) and (5.63). The entropy of the system is given by (5.69). (a) Show that

$$\langle \delta P^v_{c12} \delta P^v_{c12} \rangle = k_B T \eta_c / \tau_c,$$

$$\langle \delta P^v_{p12} \delta P^v_{p12} \rangle = k_B T \eta_p / \tau_p,$$

with η_c and η_p the kinetic and potential contributions to the shear viscosity, and τ_c and τ_p the respective relaxation times of $\overset{0}{\mathbf{P}}{}^v_c$ and $\overset{0}{\mathbf{P}}{}^v_p$. (b) Show that

$$\langle \delta P^v_{c12} \delta P^v_{c12} \rangle = k_B p T,$$

$$\langle \delta P^v_{p12} \delta P^v_{p12} \rangle = k_B p T \left(\tfrac{4}{15} Y_1 + \tfrac{1}{15} Y_2 \right),$$

with Y_1 and Y_2 given by

$$Y_1 = (2\pi n/k_B T)\int g_{eq} \phi' r^3 dr, \qquad\qquad Y_2 = (2\pi n/k_B T)\int g_{eq} \phi'' r^4 dr,$$

where g_{eq} is the equilibrium pair-correlation function.

6.3 The internal energy and the entropy of a photon gas at equilibrium are $U = aT^4 V$ and $S = \tfrac{4}{3} aT^3 V$, with $a = \tfrac{8}{15}\pi^5 k_B^4/c^3 h^3$. (a) Using the first of expressions (6.56), show that the entropy of a photon gas under a radiative heat flux q is, up to the second order,

$$S = \tfrac{4}{3} a^{1/4} U^{3/4} V^{1/4} - \tfrac{3}{8} a^{1/4} U^{-5/4} V^{9/4} q^2.$$

(b) Obtain an expression for $\theta - T$ for this system, with θ the non-equilibrium absolute temperature defined in Sect. 3.2.

6.4 The photon mean free path in photon-electron mixtures is given by $\ell = (n\sigma_T)^{-1}$, with n being the number of electrons per unit volume and σ_T the Thomson scattering cross-section, $(\sigma_T = \tfrac{8}{3}\pi e^4/m_e^2 c^4 = 6.65 \times 10^{-29}$ m$^2)$. According to (6.56) and neglecting the collisions of photons with protons, because of their much smaller scattering cross- section, evaluate the photon mean free path, the thermal conductivity, and the shear viscosity of a radiative gas at 3000 K composed of photons and electrons (and protons) with $n = 4 \times 10^3$ m^{-3}.

6.5 In a mixture of radiation and non-relativistic matter, the pressure is $p = \tfrac{1}{3} aT^4 + nk_B T$ and the energy density $\varepsilon = aT^4 + \tfrac{3}{2} nk_B T + nm_0 c^2$, with n being the number of particles per unit volume and m_0 the rest mass of the particles. Taking into account that $(\partial p/\partial \varepsilon)_n = (\partial p/\partial T)_n (\partial T/\partial \varepsilon)_n$, obtain the bulk viscosity of the mixture from expression (6.57). Show that in the ultrarelativistic limit (i.e. vanishing n) the bulk viscosity is zero.

6.6 Assume that $a(t)$ denotes a set of random variables satisfying a Langevin-like equation

$$\dot{a} + \mathbf{H} \cdot a = f$$

with $\mathbf{H}$ a constant matrix and f a white noise such that

$$\langle f(t) \rangle = 0,$$

$$\langle f(t') f^T(t) \rangle = \gamma \delta(t - t').$$

(a) Show that the matrix γ must satisfy the fluctuation-dissipation relation

$$\mathbf{H} \cdot \sigma + \sigma \cdot \mathbf{H}^T = \gamma$$

with $\sigma(t) = \langle \delta a(t) \delta a^T(t) \rangle$. (b) Prove relation (6.41b).

6.7 In a rigid heat conductor, the generalised entropy $s(u,q)$ is written as

$$s(u,q) = s_{eq}(u) - \tau(2\rho\lambda T^2)^{-1} q \cdot q.$$

In a steady state, where $q = -\lambda\nabla T$, it is given as

$$s(u,\nabla u) = s_{eq}(u) - \tau\lambda(2\rho c_v^2 T^2)^{-1} \nabla u \cdot \nabla u,$$

where $u = c_v T$ and c_v is the specific heat capacity at constant volume considered here as a constant. In terms of energy fluctuations, the second differential of the entropy reads

$$\delta^2 s = -\frac{1}{c_v T^2}(\delta u)^2 - \left(\frac{\tau\lambda}{2\rho c_v^2 T^2}\right)\nabla \delta u \cdot \nabla \delta u.$$

Show that

$$\langle \delta u(r) \delta u(r + r') \rangle = k_B c_v T^2 (\xi^2 r')^{-1} \exp(-r'/\xi),$$

where ξ stands for $(\tau\lambda//\rho c_v)^{1/2}$. Note that in the limit $\xi \to 0$ the well-known result $\langle \delta u(r) \delta u(r + r') \rangle = k_B c_v T^2 \delta(r')$ is recovered. (*Hint*: Calculate the second moments of the Fourier components $|\delta u_k|$.)[See H. Sato, Physica B **97** (1979) 194.]

6.8 Show how to obtain (6.52) from (6.51) for a monatomic ideal gas.

6.9 Assume that third-order terms of the form $\overset{0}{\mathbf{P}}{}^v : \overset{0}{q}q$ were included in the entropy, such that

$$s = s_{eq} - \tfrac{1}{2}\alpha_1 q \cdot q - \tfrac{1}{2}\alpha_2 \overset{0}{\mathbf{P}}{}^v : \overset{0}{\mathbf{P}}{}^v - \alpha'' \overset{0}{\mathbf{P}}{}^v : \overset{0}{q}q.$$

(a) Prove that

$$\alpha'' = - k_B^{-2} \alpha_2 \alpha_1^2 \langle \delta q_x \delta P_{xy}^v \delta q_y \rangle.$$

(b) Show that for an ideal monatomic gas this coefficient is

$$\alpha'' = - \tfrac{9}{25} m (n^2 k_B^3 T^4)^{-1}.$$

[See D. Jou and J. Casas-Vázquez, J. Non-Equilib. Thermodyn. **8** (1983) 127.]. An analogous coupling between viscous pressure and diffusion flux is also interesting in the analysis of the structure factor of polymer solutions under shear flow [see M. Criado-Sancho, D. Jou and J. Casas-Vázquez, Physica A **274** (1999) 466.].

Chapter 7

Information Theory

Information theory was first introduced in statistical mechanics by Jaynes in 1957 [7.1] and from then on has found many applications [7.2–5]. Here we discuss the main basic points underlying the theory and apply them to the analysis of the foundations of extended irreversible thermodynamics (EIT). There is a strong analogy between EIT and some formulations of information theory because information is measured by means of a generalised entropy depending on the various (equilibrium and non-equilibrium) constraints acting on the system. Moreover, information theory is not limited to the quadratic order of approximation and opens therefore an avenue to highly non-linear descriptions.

One of the problems in the information theory is the physical identification of the Lagrange multipliers which are introduced for taking account of the constraints. At equilibrium, the nature of these Lagrange multipliers is elucidated by comparing the microscopic expression for the differential of the entropy with the macroscopic Gibbs equation. One problem out of equilibrium is to find the corresponding macroscopic Gibbs equation with an unambiguous identification of the non-classical parameters related to non-equilibrium constraints. EIT is the only theory which, at the present time, provides such a Gibbs equation.

For simplicity, we focus our analysis on steady states, which are the most direct generalisation of equilibrium states. Usually, information is incorporated in the study of the behaviour of the variables to obtain predictions about their most probable future evolution; this leads in a natural way to the formulation of the transport equations [7.4–6]. Here we are more interested in enlarging the concept of Gibbs ensembles to non-equilibrium states, thus providing a further basis for the understanding of the non-equilibrium entropy and the equations of state of EIT.

7.1 Basic Concepts

Consider a system of N particles characterized by their positions and momenta, $\Gamma_N = \{r_1, p_1,..., r_N, p_N\}$. Assume that we have instruments measuring the local mean values $\langle A_i \rangle$ of a set of extensive observables $A_i(\Gamma_N)$. The problem is to obtain the probability density $f_N(\Gamma_N)$ which maximizes the information about the system compatible with some measured quantities. In other words, the objective is to determine the probability density which maximizes the global entropy S defined by

$$S = -k_B(h^{3N}N!)^{-1}\int f_N(\Gamma_N)\ln f_N(\Gamma_N)\mathrm{d}\Gamma_N, \tag{7.1}$$

subject to the constraints

$$(h^{3N}N!)^{-1}\int f_N(\Gamma_N)\mathrm{d}\Gamma_N = 1,$$

$$(h^{3N}N!)^{-1}\int f_N(\Gamma_N)A_i(\Gamma_N)\mathrm{d}\Gamma_N = \langle A_i \rangle \qquad (i = 1, 2, ...), \tag{7.2}$$

with specified values for the mean values $\langle A_i \rangle$ of the observables A_i at any point r. In (7.1–2) $\mathrm{d}\Gamma_N = \mathrm{d}r_1\mathrm{d}p_1...\mathrm{d}r_N\mathrm{d}p_N$ is the volume element in the phase space, h the Planck constant, and k_B the Boltzmann constant.

The choice of a correct set of observables plays an important role in the formalism; an incomplete set of variables leads to an unsatisfactory distribution function. The choice of variables depends both on theoretical arguments and on the degree of experimental accurateness one wants to ascribe. Information theory requires an adequate selection of the set of variables to give rise to satisfactory predictions. The choice of variables in non-equilibrium situations remains a problem which needs further developments [7.7].

According to the well-known Lagrange procedure, to achieve the maximization of S subject to constraints (7.2), one has to maximize the quantity

$$S - (h^{3N}N!)\int f_N\left\{\lambda_0 + \sum_i[A_i(\Gamma_N) - \langle A_i \rangle]\cdot \lambda_i\right\}\mathrm{d}\Gamma_N \tag{7.3a}$$

where the λ_i are the Lagrange multipliers corresponding to the quantities A_i and the middot indicates a scalar product. In inhomogeneous systems, $\langle A_i \rangle$ depend on the position and consequently the Lagrange multipliers λ_i must be also functions of the position. Since $\langle A_i \rangle$ have fixed values and are not subject to variation, maximization of (7.3a) is equivalent to maximize

$$-k_B \int \left[f_N \ln f_N + f_N \lambda_0 + f_N \sum_i \lambda_i \cdot A_i(\Gamma_N) \right] \mathrm{d}\Gamma_N, \qquad (7.3b)$$

In (7.3), λ_0 is the Lagrange multiplier accounting for normalization, and in what follows subscript i starts with $i = 1$. In EIT the quantities A_i are identified with the internal energy, the particle density and the fluxes. For instance, a simple situation would consist of a closed system with internal energy U subjected to a heat flow Q; in this case, the constraints are U and Q (Fig. 7.1).

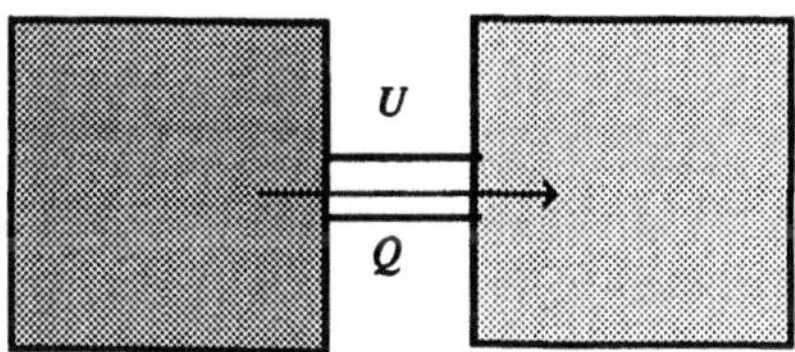

Fig. 7.1. A closed system is placed between two thermal reservoirs at different temperatures. The constraints on the system are the mean values of the energy U and heat flux Q.

Expression (7.3b) is an extremum under the condition that f_N satisfies

$$\frac{\delta}{\delta f_N} \left[f_N \ln f_N + f_N \lambda_0 + f_N \sum_i \lambda_i \cdot A_i(\Gamma_N) \right] = 0. \qquad (7.4)$$

with δ being a functional derivative. This yields

$$f_N = Z^{-1} \exp\left[-\sum_i \lambda_i \cdot A_i(\Gamma_N) \right], \qquad (7.5)$$

where

$$Z = \left(h^{3N} N! \right)^{-1} \int \exp\left[-\sum_i \lambda_i \cdot A_i(\Gamma_N) \right] \mathrm{d}\Gamma_N \qquad (7.6)$$

is a generalised partition function that follows from the normalization condition for f_N.

The Lagrange multipliers are derived from the constraints (7.2). The latter may be written in the compact form

$$-\frac{\partial \ln Z}{\partial \lambda_i} = \langle A_i \rangle, \qquad (7.7)$$

as follows from definition (7.6) of Z and relations (7.2). Introduction of the distribution density (7.5) in the definition (7.1) for the entropy results in

$$S = k_B\left[\ln Z + \sum_i \boldsymbol{\lambda}_i \cdot \langle A_i \rangle\right]. \tag{7.8}$$

In equilibrium, one may take as observable quantity the energy U of the system, $A(\Gamma_N) = \mathcal{H}(\Gamma_N)$, where $\mathcal{H}(\Gamma_N)$ is the microscopic Hamiltonian operator with $\langle \mathcal{H}(\Gamma_N) \rangle = U$. The distribution function (7.5) is then the one corresponding to the canonical ensemble

$$f_N = Z^{-1}\exp\left[-\beta\mathcal{H}(\Gamma_N)\right], \tag{7.9}$$

where we have written $\lambda_1 = \beta$, as it is usual in statistical mechanics. If in addition to the energy the number of particles is chosen as an observable quantity with a specified mean value, the distribution function takes the form

$$f_N = Z^{-1}\exp\left[-\beta\mathcal{H}(\Gamma_N) - \alpha N'\right], \tag{7.10}$$

with N' being the microscopic particle-number operator.

Up to now, the Lagrange multipliers α and β of (7.9) or (7.10) are not assigned a physical meaning. In view to their physical interpretation, let us take the differential expression of (7.8), namely

$$dS = k_B \sum_i \boldsymbol{\lambda}_i \cdot d\langle A_i \rangle + k_B \sum_i \langle A_i \rangle \cdot d\boldsymbol{\lambda}_i + k_B\, d\ln Z = k_B \sum_i \boldsymbol{\lambda}_i \cdot d\langle A_i \rangle. \tag{7.11}$$

The second equality follows from differentiating the expression (7.6) for Z with respect to the Lagrange multipliers, which results in $d\ln Z = -\sum_i \langle A_i \rangle \cdot d\boldsymbol{\lambda}_i$. By identification of (7.11) with the Gibbs equation of classical thermodynamics,

$$dS = T^{-1}dU - \mu T^{-1}dN, \tag{7.12}$$

one is led to the identifications

$$\beta = (1/k_B T), \qquad \alpha = -(\mu/k_B T),$$

with T being the absolute temperature and μ the chemical potential.

In a non-equilibrium state characterized by $\mathbf{P}^v$ and q as additional variables, one should write

$$dS = k_B\beta dU + k_B\alpha dN + k_B\boldsymbol{\lambda}_{\mathbf{P}^v} : d\mathbf{P}^v + k_B\boldsymbol{\lambda}_q \cdot dq. \tag{7.13}$$

Note that the Lagrange multipliers $\lambda_{\mathbf{P}^v}$ and $\lambda_{\mathbf{q}}$ do not have any analogue in classical thermodynamics and therefore cannot be identified in an equilibrium theory. In EIT, such a Gibbs equation is given by (3.11) and after comparison with (7.13) allows us to identify the Lagrange multipliers, up to second order in the fluxes, as

$$\beta = \frac{1}{k_B\theta} \,, \qquad \alpha = -\frac{\mu}{k_B\theta} \,, \qquad \lambda_{\mathbf{P}^v} = -\frac{\tau_2 V}{2k_B\eta T}\mathbf{P}^v \,, \qquad \lambda_{\mathbf{q}} = -\frac{\tau_1 V}{k_B\lambda T^2}\mathbf{q}. \quad (7.14)$$

The main point of interest of the foregoing analysis is the possibility of computing an entropy depending on the constraints acting on the system. In non-equilibrium steady states, typical constraints are the thermodynamic fluxes and forces, which thus may be introduced as variables in expression (7.8) for the entropy. Whether the above fluxes allow to describe the system with sufficient details cannot be answered a priori, but by comparison with experiments.

It is worth noting that an expression for the entropy flux may be obtained in a way similar to that leading to (7.11) [7.12]. The entropy flux is defined as

$$\mathbf{J}^s = -\frac{k_B}{h^{3N}N!}\int f_N \ln f_N \sum_k [\mathbf{v}_k - \langle \mathbf{v}\rangle]\mathrm{d}\Gamma_N, \quad (7.15)$$

with $\mathbf{v}_k$ the velocity of particle k and $\langle \mathbf{v}\rangle$ the average velocity. Introducing into (7.15) the expression (7.5) for f_N and defining by $\mathbf{J}^{A_i}$, the flux of the variables A_i, namely

$$\mathbf{J}^{A_i} = -\frac{1}{h^{3N}N!}\int f_N A_i \sum_i [\mathbf{v}_k - \langle \mathbf{v}\rangle]\mathrm{d}\Gamma_N, \quad (7.16)$$

one finds in a direct way

$$\mathbf{J}^s = k_B \sum_i \lambda_i \cdot \mathbf{J}^{A_i}. \quad (7.17)$$

Note that λ_i has the same tensorial rank as A_i whereas $\mathbf{J}^{A_i}$ is one rank higher, and thus their internal product is indeed a vector. When energy and particle fluxes are considered, the entropy flux (7.17) reduces to the classical expression $\mathbf{J}^s = (1/T)\mathbf{q} - (\mu/T)\mathbf{J}$, but will contain more terms if other fluxes are considered.

It is important to emphasize that the distribution function (7.5) is not the exact one, but an approximate distribution function which gives exact results for the variables taken as constraints, but not for other variables. To have a self-consistent description, one should introduce as constraints all the *slow* variables (i.e. all the variables whose time scale is of order of the time scale of observation) for internal consistency [7.5]. Furthermore, expression (7.5) lacks information about the dissipation and the microscopic dynamics of the system, i.e. about the dynamics of the *fast* variables not included as constraints in the maximum-entropy description. A general approach along these lines is the so-called MaxEnt-NESOM (Non-Equilibrium Statistical Operator Method), which has been applied to several problems, but goes beyond the level of this book; the interested reader is referred to [7.5] and [7.7].

Note that the information-theoretical methods do not specify which variables are necessary for the description of the system. In principle one should include all the variables decaying with a relaxation time comparable to the experimental time [7.5]. Therefore, it is imperious to be careful in using a truncation procedure. Indeed, some systems are described by a hierarchy of variables whose relaxation times differ widely and whose higher-order variables tend to zero very fastly. In other systems, such as ideal gases, instead, all the non-conserved variables decay with almost the same value of the relaxation time, and that would indicate that, in principle, an infinite number of relaxation variables should be taken into account, and not only a finite number of them.

In this chapter we have limited our considerations to non-equilibrium steady states. Dynamical equations may also be obtained by following several approaches, such as those of Robertson [7.6], Zubarev [7.3], the non-equilibrium statistical operator method [7.7], or the generalised projection operator of Ichiyanagi [7.8]. However, the compatibility of the evolution equations with the H-theorem is, in many situations, a subject of controversy [7.7, 9].

7.2 Ideal Gas Under Heat Flux and Viscous Pressure

As a first application consider an ideal monatomic gas out of equilibrium subjected to a heat flux and a viscous pressure. This problem or closely related ones have been dealt with by several authors [7.10–11]. At each position, the mean values of the particle number density n, the energy density per unit volume ρu, the momentum density ρv, the heat flux q, and the components of the pressure tensor $\mathbf{P}$ are supposed to be known. Since the equilibrium pressure p is fixed by u and n, the independent knowledge of $\mathbf{P}$ is a constraint on the viscous part $\mathbf{P}^v$ of the pressure tensor.

We now use the information theory to derive an expression for the non-equilibrium distribution function and the non-equilibrium entropy. We assume a one-particle distribution function f, which is satisfactory for particles moving independently because they do not interact except through instantaneous collisions.

The constraints (7.2) on the distribution function f can be written as

$$n(r) = \int f(r,C)\,\mathrm{d}C, \tag{7.18}$$

$$\rho v(r) = \int mc\, f(r,C)\,\mathrm{d}C, \tag{7.19}$$

$$\rho u(r) = \tfrac{3}{2} n k_B T(r) = \int \tfrac{1}{2} mC^2 f(r,C)\,\mathrm{d}C, \tag{7.20}$$

$$\mathbf{P}^v(r) = \int mCC f(r,C)\,\mathrm{d}C - p\mathbf{U}, \tag{7.21}$$

$$q(r) = \int \tfrac{1}{2} mC^2 C f(r,C)\,\mathrm{d}C, \tag{7.22}$$

with C the peculiar molecular velocity. Since $\mathrm{Tr}\,\mathbf{P}^v = 0$ in (7.21), it is inferred that the viscous pressure is zero.

In the previous section it was shown that the distribution function $f(r,c)$ maximizing the entropy

$$\rho s(r) = -k_B \int f(r,C)\ln f(r,C)\,\mathrm{d}C, \tag{7.23}$$

subject to conditions (7.18–22) has the form

$$f = \exp\left[-\alpha - \beta\tfrac{1}{2}mC^2 - \lambda_v \cdot C - \lambda_{\mathbf{p}^v} : (mCC - p\mathbf{U}) - \lambda_q \cdot \tfrac{1}{2}mC^2 C\right]. \tag{7.24}$$

where $\exp[-\alpha]$ designates the one-particle partition function z and the Lagrange multiplier λ_v has been introduced to ensure the constraint on the value of the relative velocity. The next step is to identify the coefficients α, β, λ_v, $\lambda_{\mathbf{p}^v}$, and λ_q. We assume that the coefficients λ_v, $\lambda_{\mathbf{p}^v}$ and λ_q are small for small values of $\mathbf{P}^v$ and q, an ansatz to be confirmed a posteriori. Expanding the exponential in (7.24) up to the first order in these terms results in

$$f(r,c) = \exp\left[-\alpha - \beta\tfrac{1}{2}mC^2\right]\left[1 - \lambda_v \cdot C - \lambda_{\mathbf{p}^v} : (mCC - p\mathbf{U}) - \lambda_q \cdot \tfrac{1}{2}mC^2 C\right]. \tag{7.25}$$

After substitution of (7.25) in (7.18–22), one obtains, up to first order in the fluxes,

$$\exp[-\alpha] = n\left(\frac{m}{2\pi k_B T}\right)^{3/2}, \qquad \beta = \frac{1}{k_B T}, \tag{7.26}$$

$$\lambda_v = \frac{m}{pk_BT}q, \qquad \lambda_{\mathbf{P}^v} = -\frac{1}{2pk_BT}\mathbf{P}^v, \qquad \lambda_q = -\frac{2m}{5pk_B^2T^2}q. \tag{7.27}$$

Bearing these results in mind, (7.25) becomes

$$f(r,c) = n\left(\frac{m}{2k_BT}\right)^{3/2} \exp\left(-\frac{mC^2}{2k_BT}\right)\left[1 + \frac{1}{2pk_BT}mCC:\mathbf{P}^v\right.$$

$$\left. + \frac{2m}{5pk_B^2T^2}\left(\frac{1}{2}mC^2 - \frac{5}{2}k_BT\right)C\cdot q\right]. \tag{7.28}$$

This is nothing but the expression of the distribution function (5.36) derived within the thirteen-moment approximation, discussed in Chap. 5. The philosophy behind both formalisms is however very different. In Grad's approach, the distribution function is expanded in terms of Hermite polynomials, in such a way that higher-order terms would contain higher-order Hermite polynomials. In information theory, one takes into account the constraints on the system and limits the expansion of the exponential just to the first order. A higher-order expansion would clearly differ from the corresponding one in Grad's theory. Nevertheless, at the present order of approximation, the expressions for the entropy derived from the information theory and Grad's formalism are identical.

The results (7.14), together with the computation of the Lagrange multipliers carried out in (7.27) yield expressions for η and λ which are equivalent to those obtained in the kinetic theory. The macroscopic Gibbs equation thus appears to be a useful complement to non-equilibrium information theory.

7.3 Ideal Gas Under Shear Flow: Non-linear Analysis

The simplest example allowing for a non-linear analysis is the problem of a classical ideal gas under shear (thermal effects are not present) [7.13] with the constraints (7.18–22). For the distribution function which maximizes the entropy, the technique of Lagrange multipliers results in the following expression

$$f = z^{-1}\exp\left[-\frac{1}{2}\left(\sum_i \beta mC_i^2 + \sum_i\sum_j \lambda_{ij}mC_iC_j\right)\right], \tag{7.29}$$

where β and λ_{ij} $(\equiv \lambda_{P_{ij}^v})$ are the Lagrange multipliers corresponding to the constraints on the energy and on the components of the viscous pressure tensor and z is the one-particle partition function. Explicit integration of the partition function z gives [see the mathematical identity (7.78)]

$$z = \frac{(2\pi)^{3/2}}{m^{3/2}\,|\mathbf{M}|^{1/2}}\frac{V}{N}, \tag{7.30}$$

where $|\mathbf{M}|$ is the determinant of the matrix

$$\mathbf{M} = \begin{pmatrix} \beta + 2\lambda_{11} & \lambda_{12} & \lambda_{13} \\ \lambda_{12} & \beta + 2\lambda_{22} & \lambda_{23} \\ \lambda_{13} & \lambda_{23} & \beta + 2\lambda_{33} \end{pmatrix}. \tag{7.31}$$

The Lagrange multipliers are obtained from the constraints, which may be written as

$$\rho u = -\frac{\partial \ln z}{\partial \beta}, \qquad P_{ij}^{v} = -\frac{\partial \ln z}{\partial \lambda_{ij}}. \tag{7.32}$$

If one writes (7.29) in the compact form

$$f = z^{-1}\exp\left[-\tfrac{1}{2}\mathbf{M}:mCC\right], \tag{7.33}$$

and introduces this result in (7.23), it immediately follows that the expression for the entropy is given by

$$S = nk_B\ln z + \tfrac{1}{2}k_B\mathbf{M}:\mathbf{P}. \tag{7.34}$$

Furthermore, taking into account that

$$\mathbf{P} = \int mCCf\,\mathrm{d}C = n\mathbf{M}^{-1},$$

and introducing $U = n\langle\tfrac{1}{2}mC^2\rangle$, one may write

$$\tfrac{1}{2}\mathbf{M}:\mathbf{P} = \beta U + \tfrac{1}{2}\mathrm{Tr}\left[\mathbf{U} - \beta\mathbf{P}n^{-1}\right], \tag{7.35}$$

which reduces to the equilibrium expression when $\mathbf{P} = n\beta^{-1}\mathbf{U}$. By taking into account (7.30), (7.35) and expressing $\mathbf{M}$ in terms of $\mathbf{P}$, the entropy (7.34) can be written as follows

$$S = \text{const} + Nk_B \ln V - \tfrac{3}{2} Nk_B \ln \beta + \beta U$$

$$+ \tfrac{1}{2} Nk_B \left\{ \ln[\det(\mathbf{U} + \mathbf{P}^\nu \beta n^{-1})] - \text{Tr}(\mathbf{P}^\nu \beta n^{-1}) \right\}, \tag{7.36}$$

where $\mathbf{P} = n\beta^{-1}\mathbf{U} + \mathbf{P}^\nu$. In equilibrium, one has $\beta = 1/(k_B T)$ and $\mathbf{P}^\nu = 0$, so that the expression for $\mathbf{P}$ reduces to $\mathbf{P} = nk_B T\mathbf{U}$, as expected. Note that (7.36) is non-linear in $\mathbf{P}^\nu$, and goes beyond the quadratic non-equilibrium approximation.

It is easy to express the Lagrange multipliers λ in terms of $\mathbf{P}^\nu$. Indeed, since $\mathbf{M} = \beta\mathbf{U} + \lambda$ and on the other hand $\mathbf{M} = n\mathbf{P}^{-1} = n(n\beta^{-1}\mathbf{U} + \mathbf{P}^\nu)^{-1}$, it follows that

$$\lambda = \beta(\mathbf{U} + \beta n^{-1}\mathbf{P}^\nu)^{-1} - \beta\mathbf{U}. \tag{7.37}$$

The term within the brackets in the right-hand side of (7.37) may be expanded into a series of $\mathbf{P}^\nu$, and limiting this to the first term, one has

$$\lambda \approx \beta^2 n^{-1}\mathbf{P}^\nu \approx -\frac{1}{nk_B^2 T^2}\mathbf{P}^\nu. \tag{7.38}$$

Introducing this result into the expression (7.11) for the Gibbs equation, written per unit mass, it is found that

$$ds = k_B\beta du + \tfrac{1}{2}k_B\lambda : d\mathbf{P}^\nu = T^{-1}du - \frac{\tau v}{2\eta T}\mathbf{P}^\nu : d\mathbf{P}^\nu. \tag{7.39}$$

From the well-known kinetic theory result $\eta = nk_B T\tau$, one observes that (7.39) coincides with the macroscopic expression (3.11) except for the terms in dv and in dq and dp^ν. This is because one has taken as variables the internal energy u and the viscous pressure $\mathbf{P}^\nu$ but not the volume neither the other fluxes. However, the term in dv can be directly obtained from the differential of (7.36), which yields $\pi/\theta = nk_B = p/T$, as noted in (3.27).

When the system is submitted to a fixed shear viscous pressure P_{12}^ν, corresponding to a plane Couette flow, the non-vanishing Lagrange multipliers are β and λ_{12}, and one has

$$z = \frac{(2\pi)^{3/2}}{m^{3/2}} \frac{V}{N}\left(\beta^3 - \beta\lambda_{12}^2\right)^{-1/2}. \tag{7.40}$$

The Lagrange multipliers may be obtained in terms of u and P_{12}^ν as

$$\beta = \frac{1-y}{2\left[R^2+(1-y)\right]}\frac{N}{U}, \qquad \lambda_{12} = \frac{3R^2+2(1-y)}{2R\left[R^2+(1-y)\right]}\frac{N}{U}, \qquad (7.41)$$

with $R = VP_{12}^{\nu}/U$ and $y = (1+3R^2)^{1/2}$. Note that near equilibrium, i.e. when λ_{12} tends to zero, we see that $\beta = \frac{3}{2}N/U$ and $\lambda_{12} = -\beta^2(V/N)P_{12}^{\nu}$. Thus, when $P_{12}^{\nu} = 0$, one recovers from (7.41) the standard Maxwell–Boltzmann distribution function with $\beta = (k_B T)^{-1}$.

The entropy has now the form of a non-linear non-quadratic expression in the fluxes, namely

$$S = S_{eq} + \frac{Nk_B}{2}\ln\frac{27R^2\left[R^2-(y-1)\right]^2}{2(y-1)^3}, \qquad (7.42)$$

where S_{eq} is the equilibrium entropy.

7.4 Ideal Gas Submitted to a Heat Flux: Non-linear Analysis

Consider an ideal non-relativistic gas at rest under fixed values of energy and heat flux. The distribution function, according to the maximum-entropy formalism, is given by [7.14]

$$f = Z^{-1}\exp\left[-\beta\tfrac{1}{2}mC^2 - \boldsymbol{\gamma}\cdot\left(\tfrac{1}{2}mC^2 - \tfrac{5}{2}\beta^{-1}\right)C\right], \qquad (7.43)$$

where the factor $\frac{5}{2}\beta^{-1}$ guarantees that the mean speed is zero. This distribution diverges for high values of the molecular speed, because the operator for the second term in the exponent is odd in the velocity; to avoid this divergence, let us expand (7.43) up to second order in $\boldsymbol{\gamma}$, so that

$$f = Z^{-1}\exp\left(-\beta\tfrac{1}{2}mC^2\right)\left[1 - \boldsymbol{\gamma}\cdot\left(\tfrac{1}{2}mC^2 - \tfrac{5}{2}\beta^{-1}\right)C + \frac{1}{2}(\boldsymbol{\gamma}\cdot C)^2\left(\tfrac{1}{2}mC^2 - \tfrac{5}{2}\beta^{-1}\right)^2\right]. \quad (7.44)$$

Grad's results are recovered if one truncates the expansion at the first-order in $\boldsymbol{\gamma}$. By keeping up to the second-order terms in $\boldsymbol{\gamma}$, one finds for β and $\boldsymbol{\gamma}$ the following expressions

$$\beta = (k_B T)^{-1}\left[1 + \tfrac{2}{5}\frac{m}{p^2 k_B T}\boldsymbol{q}\cdot\boldsymbol{q}\right], \qquad (7.45a)$$

$$\gamma = -\frac{2}{5}\frac{m}{pk_B^2T^2}q. \qquad (7.45b)$$

The expression for γ coincides with that obtained from Grad's expansion. This is not surprising because (7.43) and (5.36) agree up to the first order. It is worth stressing that the Lagrange multiplier β is different from $(k_BT)^{-1}$ and that it depends on the heat flux. In Grad's approach, where second-order terms in q are omitted, one has $\beta = (k_BT)^{-1}$. This is the reason why the question of the difference between absolute non-equilibrium temperature and local-equilibrium temperature does not appear in kinetic theory, where the latter is always used by definition.

Note furthermore that (7.44) yields different values for the averages of v_x^2, v_y^2, and v_z^2. If the heat flux has the direction of the y axis, as in example (3.25), it is found that

$$\left\langle \tfrac{1}{2}mv_x^2 \right\rangle = \left\langle \tfrac{1}{2}mv_z^2 \right\rangle < \tfrac{1}{2}k_BT \ ; \ \left\langle \tfrac{1}{2}mv_y^2 \right\rangle > \tfrac{1}{2}k_BT , \qquad (7.46)$$

as derived from a different basis in Sect. 3.4.

7.5 Relativistic Ideal Gas Under an Energy Flow

As a further illustration, consider a relativistic ideal gas in a non-equilibrium steady state, with prescribed internal energy U and integrated energy flux Vq, where V is the volume and q the energy flux [7.15]. The distribution function maximizing the entropy reads

$$f = Z^{-1}\exp\left[-\beta\textstyle\sum_i p_ic - \gamma\cdot\sum_i p_icc\right], \qquad (7.47)$$

where p_ic is the microscopic expression for the energy of the ith particle and p_icc is the particle contribution to the energy flow (all particles are supposed to move at the speed of light c). According to (7.6), the partition function is defined as

$$Z = \frac{V^N}{N!h^{3N}}\int d\boldsymbol{p}_1...\int d\boldsymbol{p}_N \ \exp\left[-\beta\textstyle\sum_i p_ic - \gamma\cdot\sum_i p_icc\right]. \qquad (7.48)$$

where the factor V^N comes out from integration over the positions of the particles. In contrast with the classical gas studied in Sect. 7.4, the distribution function (7.47) does not diverge because the dependence of the energy flux in terms of the momentum is of first order, instead of third order as for the classical gas. This allows to obtain an explicit expression for the partition function without introducing any truncation of the exponential. The explicit result for (7.48) is

$$Z = \frac{1}{N!}\left(\frac{8\pi V}{\beta^3 c^3 h^3}\right)^N \left(1 - \frac{c^2 \gamma^2}{\beta^2}\right)^{-2N}.$$
(7.49)

We now determine the Lagrange multipliers β and γ in terms of U and Vq. We obtain from the conditions on the mean energy and the mean flux, respectively,

$$U = \frac{3N}{\beta} + \frac{4N}{\beta}\frac{c^2\gamma^2}{\beta^2 - c^2\gamma^2},$$
(7.50)

$$Vq = -\frac{4Nc^2\gamma}{\beta^2 - c^2\gamma^2}.$$
(7.51)

Note that (7.49) and (7.50) diverge for $\gamma = \pm\beta/c$, which sets a maximum value for γ, a feature which will be commented below. Inversion of (7.50) and (7.51) yields for β (U, Vq, N) and $\gamma(U, Vq, N)$

$$\beta = \frac{3N}{U}\frac{1}{y-1},$$
(7.52)

and

$$\gamma = \frac{3N}{Vq^2}\frac{y-2}{y-1}q,$$
(7.53)

where y stands for

$$y = \left[4 - 3\left(\frac{Vq}{cU}\right)^2\right]^{1/2}.$$
(7.54)

It follows from (7.50–51) that for $Vq = 0$ one has $\gamma = 0$ and $\beta = (k_B T)^{-1}$; in addition, (7.49) reduces to the usual equilibrium expression for the partition function. In general, $Vq \neq 0$, $\beta \neq (k_B T)^{-1}$, and the non-equilibrium temperature θ, is given by $\theta = T(y-1)$. Furthermore, it is seen that β and γ diverge for $y = 1$, i.e. when Vq tends to Uc. This domain of validity of Vq is easy to interpret, because cU is the maximum energy flow which may be expected: it corresponds to the energy U carried at the maximum possible speed, which is precisely c.

Recalling that the entropy is given by $S = k_B(\ln Z + \beta U + \gamma \cdot Vq)$ and taking account of (7.49) and (7.53), it is found that

$$S = S_{eq} + Nk_B \ln\left[\tfrac{1}{16}(y-1)(y+2)^2\right],$$
(7.55)

which shows that the entropy may depend on the energy flux Vq in a non-quadratic way.

It is easy to obtain from (7.47) the corresponding expression for the pressure tensor, which has the general form

$$\mathbf{P} = \pi \mathbf{U} + b\mathbf{qq}, \tag{7.56}$$

where the non-equilibrium pressure π is defined by $\pi/\theta = (\partial S/\partial V)_{Vq,U,N}$ and b is given by the condition that $\mathrm{Tr}\,\mathbf{P} = 3\pi + bq^2 = 3p$, with p the local-equilibrium pressure. Note that it is the integrated energy flux Vq rather than the flux q which is kept constant during differentiation. By differentiating the entropy with respect to V, one obtains for π

$$\pi = nk_B\theta = \frac{1}{3}\frac{U}{V}(y-1), \tag{7.57a}$$

and consequently

$$b = -\frac{1}{q^2}(y-2). \tag{7.57b}$$

Inserting these two expressions into (7.56) leads to

$$\mathbf{P} = \frac{1}{3}\frac{U}{V}\left[(y-1)\mathbf{U} - \frac{1}{q^2}(y-2)\mathbf{qq}\right]. \tag{7.58}$$

Observe that for $y = 2$, the pressure tensor reduces to $\mathbf{P} = \frac{1}{3}(U/V)\mathbf{U} = p\mathbf{U}$ which is the equilibrium pressure tensor. The term in $\mathbf{qq}$ becomes dominant when the energy flux approaches cU (i.e. $y \rightarrow 1$) and plays an important role in radiation hydrodynamics, where the pressure tensor is written as [7.16]

$$\mathbf{P} = \frac{U}{V}\left(\frac{1-\chi}{2}\mathbf{U} + \frac{3\chi-1}{2}\frac{\mathbf{qq}}{q^2}\right), \tag{7.59}$$

χ being the so-called Eddington factor. By comparison of (7.58) and (7.59), one has

$$\chi = \frac{5-2y}{3}. \tag{7.60}$$

This expression for the Eddington factor has also been obtained in [7.16b] on arguments based on a rational formulation of EIT, and by requirements of Lorentz invariance (see Prob. 17.2) [7.16c, d].

Let us briefly comment about the behaviour of the entropy when the energy flux approaches its maximum value. According to (7.55), the entropy is diverging instead of vanishing, as a result of using classical rather than quantum statistics. If one wishes to study electromagnetic radiation, the relevant statistics is that of Bose–Einstein and one should maximize the corresponding expression for the entropy, namely

$$S = -k_B \left(h^{3N} N! \right)^{-1} \int [f \ln f - (1+f)\ln(1+f)]d\Gamma . \tag{7.61}$$

This yields for the distribution function

$$f = [\exp(\beta pc + \boldsymbol{\gamma} \cdot p c \boldsymbol{c}) - 1]^{-1}. \tag{7.62}$$

Here, we do not consider a fixed number of particles because we are dealing with photons whose particle number is not fixed. The calculations of β and $\boldsymbol{\gamma}$ are rather cumbersome [7.17] and the final results are

$$\beta = \frac{1}{2k_B} \left(\frac{aV}{U} \right)^{1/4} \frac{(y+2)^{1/2}}{(y-1)^{3/4}}, \tag{7.63}$$

and

$$\boldsymbol{\gamma} = -\frac{3}{4} \left(\frac{a}{V} \right)^{1/4} \frac{V\boldsymbol{q}}{c^2 U^2} \frac{1}{(y+2)^{1/2}(y-1)^{3/4}}. \tag{7.64}$$

Here, a is the radiation constant appearing in the well known expression $U = aT^4 V$ for the internal energy of radiation at equilibrium. Now expression (7.61) for the entropy becomes

$$\frac{S}{V} = \frac{2}{3} a^{1/4} \left(\frac{U}{V} \right)^{3/4} (y-1)^{1/4}(y+2)^{1/2}, \tag{7.65}$$

which tends to the expected value $S/V = \frac{4}{3} aT^3$ at equilibrium ($Vq = 0$) and which vanishes for $V^2 q^2 \to c^2 U^2$.

7.6 Heat Flow in a Linear Harmonic Chain

Harmonic chains are ideal systems which shed an original light on thermodynamics. Heat flow in harmonic chains coupled at each end to heat reservoirs at different temperatures has been studied in reference [7.18]. In a harmonic chain, the phonon mean free path is infinite, so that the energy flux along it is not proportional to the temperature gradient but to the temperature difference between the reservoirs located at its ends.

To avoid complications associated with the boundary conditions, Miller and Larson [7.18] eliminate the boundaries by considering that chain ends are linked together to form a ring. In this case, the system turns out to be a 'superconductor' of thermal energy, because of its infinite heat conductivity: a heat flux lasts indefinitely, without appealing to boundary reservoirs to sustain it. Such a chain ring is characterized by the constraints

$$\langle \mathcal{H} \rangle = U, \qquad \langle \mathcal{J} \rangle = Q, \tag{7.66}$$

where $\mathcal{H}$ is the Hamiltonian of the system, $\mathcal{J}$ the heat flux operator, U the mean (internal) energy of the chain, and Q the mean heat flux along the ring.

The system consists of a linear chain of N particles, each of mass m. Each particle is connected to its nearest neighbours by Hookean springs with stiffness κ. The Nth particle is connected by a spring to the first particle, so that the chain forms a closed ring. One may choose a system of dimensionless quantities where the mass is expressed in terms of m, time in units of $(m/\kappa)^{1/2}$, and energy in units of $(h/2\pi)(\kappa/m)^{1/2}$. Let q_α be the displacement from equilibrium for each particle α ($\alpha = 1,..., N$) and p_α its conjugate momentum.

The Hamiltonian $\mathcal{H}(q_1, p_1,..., q_N, p_N)$ is given by

$$\mathcal{H} = \tfrac{1}{2}\sum_\alpha p_\alpha^2 + \tfrac{1}{2}\sum_\alpha (q_{\alpha+1} - q_\alpha)^2. \tag{7.67}$$

The microscopic operator $\mathcal{J}(q_1, p_1,..., q_N, p_N)$ for the heat flux is

$$\mathcal{J} = -\tfrac{1}{2}N\sum_\alpha (q_{\alpha+1}p_\alpha - q_\alpha p_{\alpha+1}). \tag{7.68}$$

To obtain this form for $\mathcal{J}$ consider that the system is mentally divided into a right subsystem and a left subsystem. We may identify all the particles with labels $\alpha > \gamma$ (where γ is fixed) as the right subsystem, and those with labels $\alpha \le \gamma$ as the left subsystem. The two subsystems interact via a single coupling spring. Half of the potential energy of the coupling spring is attributed to each subsystem. Let $\mathcal{J}_\alpha$ be the net rate of increase of the energy of the subsystem characterized by $\alpha > \gamma$. It is given by the rate of work performed on it by the coupling spring, $-(q_{\gamma+1} - q_\gamma)p_{\gamma+1}$, plus half the rate of increase of the potential energy of the coupling spring, $\tfrac{1}{2}(q_{\gamma+1} - q_\gamma)(p_{\gamma+1} - p_\gamma)$. This yields

$$\mathcal{J}_\alpha = -\tfrac{1}{2}(q_{\alpha+1} - q_\alpha)(p_{\alpha+1} + p_\alpha). \tag{7.69}$$

Taking the average of (7.69) over all pairs gives back (7.68), because of the cancellations due to the fact that the sums are modulo N. From Hamilton's equations, it is also checked that $\mathcal{J}$ is a constant of the motion.

Now we transform $\mathcal{H}$ and $\mathcal{J}$ in terms of normal coordinates η_α and canonical momenta $\xi_\alpha = d\eta_\alpha/dt$, which reduce the potential energy to a canonical quadratic form. The transformation matrix $\mathbf{A}$ relating q and η, defined by $\eta = \mathbf{A} \cdot q$, is given by [7.19]

$$A_{\alpha\gamma} = (1/N)^{1/2}\left[\sin(2\pi\alpha\gamma/N) + \cos(2\pi\alpha\gamma/N)\right]. \tag{7.70}$$

In terms of the normal coordinates, the Hamiltonian of the system becomes

$$\mathcal{H}(\eta,\xi) = \tfrac{1}{2}\sum_\alpha (\xi_\alpha^2 + \omega_\alpha^2\eta_\alpha^2), \tag{7.71}$$

where ω_α, the angular frequency of the α-th mode, is given by $\omega_\alpha = 2\sin(\pi\alpha/N)$. Taking into account that matrix $\mathbf{A}$ given by (7.70) is orthogonal and symmetric ($\mathbf{A} = \mathbf{A}^{-1}$), the heat flux operator (7.68) reads as

$$\mathcal{J} = -(1/N)\sum_\mu \sin(2\pi\mu/N)\left[\eta_\mu\xi_{N-\mu} - \eta_{N-\mu}\xi_\mu\right], \tag{7.72}$$

with $M = N/2$ if N is even and $M = (1/2)(N - 1)$ if N is odd. We have also used the result $\omega_\mu = \omega_{N-\mu}$, which follows from the definition of ω_α.

The dimension of the phase space of the system that we are studying is $2N$, but by assuming that the centre of mass of the system remains fixed, the dimension is reduced to $2(N - 1)$. The next problem is to find the reduced probability distribution function f maximizing the entropy

$$S = -k_B\left(h^{N-1}\right)^{-1}\int f \ln f \, d\Gamma_{N-1}, \tag{7.73}$$

subject to the constraints (7.66). Here $d\Gamma_{N-1} = d\eta_1 d\xi_1 ... d\eta_{N-1} d\xi_{N-1}$. Note that the factor $N!$ in (7.1) does not appear in (7.73) because the particles, being kept in a given order, are no longer indistinguishable.

The result for f is, as shown in Sect. 7.1,

$$f = Z^{-1}\exp\left(-\beta\mathcal{H} - \gamma\mathcal{J}\right), \tag{7.74}$$

where β and γ are the respective Lagrange multipliers and Z is the partition function

$$Z = \left(h^{N-1}\right)^{-1} \int \exp(-\beta \mathcal{H} - \gamma \mathcal{J}) \, d\Gamma_{N-1}. \tag{7.75}$$

To obtain the partition function, note that both $\mathcal{J}$ and $\mathcal{H}$ can be expressed as a sum over pairs μ and $N - \mu$ of normal coordinates sharing the same characteristic frequency ω_μ, i.e. $\mathcal{J} = \sum_\mu \mathcal{J}_\mu$ and $\mathcal{H} = \sum_\mu \mathcal{H}_\mu$, with

$$\mathcal{H}_\mu = \tfrac{1}{2}\left[\xi_\mu^2 + \xi_{N-\mu}^2 + \omega_\mu^2(\eta_\mu^2 + \eta_{N-\mu}^2)\right],$$

$$\mathcal{J}_\mu = N^{-1} \sin(2\pi\mu/N)(\eta_\mu \xi_{N-\mu} - \eta_{N-\mu}\xi_\mu). \tag{7.76}$$

Consequently, the partition function can be expressed as a product $Z = \prod_\mu z_\mu$ with

$$z_\mu = h^{-2} \int d\eta_\mu d\eta_{N-\mu} d\xi_\mu d\xi_{N-\mu} \exp\left(-\beta \mathcal{H}_\mu - \gamma \mathcal{J}_\mu\right). \tag{7.77}$$

To calculate these integrals one may employ the general result

$$\int dr \exp\left(R \cdot r - \tfrac{1}{2} r \cdot M \cdot r\right) = (2\pi)^{n/2}(|M|)^{-1/2} \exp\left(\tfrac{1}{2} R \cdot M^{-1} \cdot R\right), \tag{7.78}$$

where r and R are n-dimensional vectors and M is a positive definite square matrix. To evaluate z_μ, r is considered to be the four-dimensional vector with components $(\eta_\mu, \eta_{N-\mu}, \xi_\mu, \xi_{N-\mu})$, R the null vector 0, and M the matrix M_μ, given by

$$M_\mu = \beta\omega_\mu \begin{pmatrix} \omega_\mu & 0 & 0 & -y_\mu \\ 0 & \omega_\mu & y_\mu & 0 \\ 0 & y_\mu & \omega_\mu^{-1} & 0 \\ -y_\mu & 0 & 0 & \omega_\mu^{-1} \end{pmatrix}, \tag{7.79}$$

where $y_\mu = y \cos(\pi \mu/N)$ and $y = \gamma(N\beta)^{-1}$. It turns out that $z_\mu = |M_\mu|^{-1/2}$, so that

$$z_\mu = \beta^{-2}\omega_\mu^{-2}\left(1 - y_\mu^2\right)^{-1}. \tag{7.80}$$

The total partition function may be calculated in the thermodynamic limit by noting that $\ln Z = \sum_\mu \ln z_\mu$ may be converted into an integral in the limit of large N since $|\omega_{\mu-1} - \omega_\mu|$ becomes arbitrarily small. One finds asymptotically that

$$\ln Z = \int_0^M \ln z_\mu d\mu = -(N/\pi)\int_0^{\pi/2} d\phi\left[2\ln(2\beta) + 2\ln(\sin\phi) + \ln(1 - y^2\cos^2\phi)\right]$$

$$= -N\ln\left\{\tfrac{1}{2}\beta\left[1 + (1-y^2)^{1/2}\right]\right\}. \tag{7.81}$$

The final result for Z can be written as

$$Z = \left\{\tfrac{1}{2}\beta\left[1 + (1-y^2)^{1/2}\right]\right\}^{-N}. \tag{7.82}$$

The Lagrange multipliers β and γ may be found in terms of U and Q through the constraints (7.66), which in analogy to (7.7) are expressed by

$$U = \langle\mathcal{H}\rangle = -\frac{\partial \ln Z}{\partial\beta}, \qquad Q = \langle\mathcal{I}\rangle = -\frac{\partial \ln Z}{\partial\gamma}. \tag{7.83}$$

When (7.82) is introduced into (7.83), one obtains for β and y

$$\beta = \frac{1+x^2}{\varepsilon(1-x^2)}, \qquad y = -\frac{2x}{(1-x^2)}, \tag{7.84}$$

with $\varepsilon = U/N$ and $x = Q/\varepsilon$. In terms of these quantities, (7.82) becomes

$$Z = \left[\varepsilon(1-x^2)\right]^N. \tag{7.85}$$

For $x = 0$ one recovers the usual equilibrium results, whereas for $x^2 \to 1$ both β and y diverge. Thus the presence of a critical heat flux $Q = \varepsilon$ is outlined from the above considerations.

The entropy and the generalised Lagrange multiplier β deserve special comments. According to (7.73) and (7.74), the entropy may be written as

$$S = k_B(\beta U + \gamma Q + \ln Z). \tag{7.86}$$

In the thermodynamic limit when N tends to infinity, the entropy per particle turns out to be, in view of the explicit form (7.82) of Z,

$$s = \lim_{N\to\infty}\frac{S}{N} = k_B\left[1 + \ln\varepsilon + \ln(1-x^2)\right] = s_{eq} + k_B\ln(1-x^2). \tag{7.87}$$

As expected, the presence of the heat flux modifies the value of s. For small values of the heat flux, (7.87) reduces to

$$s(\varepsilon, Q) = s_{eq} - \frac{k_B}{\varepsilon^2} Q^2 . \tag{7.88}$$

The dependence of this expression on the heat flux Q provides a further corroboration of the basic assertions of extended irreversible thermodynamics, stating that the entropy is a function of the heat flux out of equilibrium, and allows one to explore higher-order terms in the heat flux.

The Lagrange multiplier β can be interpreted in terms of a generalised absolute temperature θ defined as $\theta = (k_B\beta)^{-1}$; from (7.84) one has

$$\theta^{-1} = T^{-1}\left(\frac{1+x^2}{1-x^2}\right) = T^{-1}\left(1 + \frac{2x^2}{1-x^2}\right) \tag{7.89}$$

indicating that the generalised temperature θ differs from the usual local-equilibrium temperature T ($\equiv \varepsilon/k_B$) by terms at least of the order x^2 or Q^2. The same result can be obtained from (7.87) and the definition $\theta^{-1} = (\partial s / \partial \varepsilon)_Q$.

The entropy (7.87) diverges when $x \to 1$, i.e. when the absolute temperature $\theta = 1/k_B\beta$ tends to zero (in this limit the heat flux tends to the maximum value). This is so because of the use of classical statistics rather than quantum statistics. Indeed, when the non-equilibrium temperature θ becomes lower than the Einstein temperature of the lattice, it is necessary to resort to Bose–Einstein's rather than classical statistics [7.20]; therefore the results obtained by Miller and Larson are no longer valid for a heat flux larger than a given value. Consider a harmonic chain with a linearized dispersion relation $\omega = c|k|$, c being the phonon speed and $|k|$ the magnitude of the wavevector. A quantum analysis of the system under the constraint of a fixed energy density ε and fixed energy flux q yields for the distribution function maximizing the entropy

$$f(k; \beta, \gamma) = \left[\exp\left(\beta\hbar c|k| + \gamma\hbar c^2|k|\right) - 1\right]^{-1}, \tag{7.90}$$

where β and γ are the Lagrange multipliers and $\hbar = h/(2\pi)$.

The entropy behaviour in the quantum limit when $\varepsilon_D(\beta \pm \gamma \cdot c) \gg 1$ (with ε_D the Debye energy $\varepsilon_D = \hbar c\pi/l$) is

$$s = \frac{k_B}{2}\left(\frac{\pi}{6\hbar}\right)^{1/2}\left[(\varepsilon c + q)^{1/2} + (\varepsilon c - q)^{1/2}\right]. \tag{7.91}$$

The Lagrange multipliers β and γ are given by

$$\beta = \frac{1}{2}\left(\frac{\pi}{6\hbar}\right)^{1/2}\left[\frac{1}{(\varepsilon c + q)^{1/2}} + \frac{1}{(\varepsilon c - q)^{1/2}}\right]$$

$$= \frac{1}{2k_BT}\left[\frac{1}{(1+x)^{1/2}} + \frac{1}{(1-x)^{1/2}}\right], \tag{7.92}$$

$$\gamma = \frac{1}{2c}\left(\frac{\pi}{6\hbar}\right)^{1/2}\left[\frac{1}{(\varepsilon c + q)^{1/2}} - \frac{1}{(\varepsilon c - q)^{1/2}}\right]$$

$$= \frac{1}{2ck_BT}\left[\frac{1}{(1+x)^{1/2}} - \frac{1}{(1-x)^{1/2}}\right], \tag{7.93}$$

where $T(\varepsilon) \equiv (6\hbar c\varepsilon / k_B^2\pi)^{1/2}$ is the local-equilibrium temperature and $x = q/\varepsilon c$. At equilibrium $q = 0$ and consequently $\gamma = 0$, $\beta = (k_BT)^{-1}$ and (7.90) becomes the equilibrium Bose–Einstein distribution function. The relation between T and ε follows from the quantum equation of state at low T, namely

$$\varepsilon = \frac{k_B^2\pi^2}{3hc}T^2. \tag{7.94}$$

Note that the results (7.84) are not recovered in the classical limit because, in the present problem, the linearized dispersion $\omega = c|k|$ has been used instead of the exact dispersion relation.

It is interesting to note that the expression for the specific heat at constant heat flux defined as $c_q = (\partial\varepsilon/\partial\theta)_q$ is

$$c_q = \frac{4\varepsilon^{3/2}}{k_B\theta^2}\left(\frac{6\hbar c}{\pi}\right)^{1/2}\frac{(1-x^2)^{3/2}}{(1+x)^{3/2} + (1-x)^{3/2}}. \tag{7.95}$$

Note that c_q vanishes in the limit $x \to 1$, i.e. when the non-equilibrium absolute temperature θ tends to zero, and this corresponds to a third-law-like behaviour [7.20]. This provides a generalization of the third law to non-equilibrium steady states: indeed, in equilibrium, θ coincides with the equilibrium temperature so that the vanishing of θ means the vanishing of T; however, in non-equilibrium, even at a non-zero value of T, θ may become zero at sufficiently high values of the heat flux. A similar quantum behaviour at high values of the energy flux was also found by Fröhlich who analysed Bose–Einstein condensation of phonons in non-linear systems at high values of the energy flux [7.21].

7.7 Information Theory and Non-equilibrium Fluctuations

Fluctuations around non-equilibrium steady states are a lively topic [7.22] as they combine dynamical and statistical non-equilibrium contributions. Here, we focus our attention on the statistical aspects. Information theory is particularly useful for studying the fluctuations of the variables around the non-equilibrium steady state. With this objective in mind, let us express the second moments of the macroscopic values of the observables $A_i(\Gamma_N)$ around their average values; it is straightforward to check after differentiation of (7.7) that

$$\langle (A_i(\Gamma_N) - \langle A_i \rangle)(A_j(\Gamma_N) - \langle A_j \rangle) \rangle = \frac{\partial^2 \ln Z}{\partial \lambda_i \partial \lambda_j}. \tag{7.96}$$

Expression (7.96) may be written in terms of the second derivatives of the generalised entropy by taking into account the mathematical relation

$$\sum_k \frac{\partial \langle A_i \rangle}{\partial \lambda_k} \frac{\partial \lambda_k}{\partial \langle A_j \rangle} = \delta_{ij}. \tag{7.97}$$

Using (7.7) and (7.11), (7.97) may be alternatively expressed as

$$-\frac{1}{k_B} \sum_l \frac{\partial^2 \ln Z}{\partial \lambda_i \partial \lambda_l} \frac{\partial^2 S}{\partial \langle A_l \rangle \partial \langle A_j \rangle} = \delta_{ij}, \tag{7.98}$$

from which it follows that $\partial^2 \ln Z / \partial \lambda_i \partial \lambda_l$ is related to the inverse of the derivative $\partial^2 S / \partial \langle A_l \rangle \partial \langle A_j \rangle$; accordingly, the second moments of fluctuations may be written as

$$\langle (A_i(\mu') - \langle A_i \rangle)(A_j(\mu') - \langle A_j \rangle) \rangle = -\frac{1}{k_B} \left[\left(\frac{\partial^2 S}{\partial \langle A \rangle \partial \langle A \rangle} \right)^{-1} \right]_{ij}. \tag{7.99}$$

This shows that the maximum-entropy formalism relates the second moments of the fluctuations of thermodynamic quantities in a non-equilibrium steady state to the second order derivatives of a generalised entropy. Formally, this is the same situation as in equilibrium; there is, however, an important difference, because the entropy function is no longer the classical one but contains as supplementary variables the fluxes acting as constraints on the system.

In the usual approaches of non-equilibrium hydrodynamic fluctuations [7.22], it is assumed that the hydrodynamic noise, owing to fast fluctuations of q and P^v, keeps its local-equilibrium form, as determined from the local temperature, pressure, and velocity. The intuitive argument in support of this assumption is that these fluctuations are so fast that they do not have enough time to be influenced by the non-equilibrium conditions of the system. However, if the time of decay of the fluctuations of the fluxes is not negligibly small, the noise may 'know' that the system is out of equilibrium and may undergo some changes.

Of course, such modifications of the noise are minute for frequencies much lower than the inverse of the relaxation times of the fluxes, as in light scattering in liquids. However, they could be perceptible at frequencies comparable to the inverse of the relaxation times, as observed in light scattering in gases or neutron scattering in liquids.

To obtain an expression for the probability of fluctuations, we define a generalised specific free energy g as

$$g = -k_B \theta \ln Z = -\theta s + \sum_i \lambda_i \cdot \langle A_i \rangle, \tag{7.100}$$

where λ_i are the Lagrange multipliers. In analogy with (6.9), the probability of fluctuations at constant λ_i is given by

$$W \approx \exp\left(-\frac{1}{k_B \theta} \Delta g\right) \tag{7.101}$$

where Δg is the change in g related to the fluctuations. It is now possible to calculate the second moments of the fluctuations in non-equilibrium steady states by using the identifications (7.14) of the Lagrange multipliers λ_i appearing in Eq. (7.13).

As an illustration, we examine the fluctuations in a gas under a temperature gradient; viscous effects are omitted. The respective Lagrange multipliers for the energy and the heat flux in the steady state are $\beta = (k_B \theta)^{-1}$ and $\lambda_q = \tau_1 V (k_B T^2)^{-1} \nabla T$. We recall further that the Lagrange multiplier related to the volume is $\lambda_v = \pi (k_B \theta)^{-1}$, where π is the generalised pressure defined by (2.2). Thus the generalised free energy (7.100) takes the form

$$g = -\theta s + u + \pi v + \frac{\tau_1 v \theta}{T^2} \nabla T \cdot q. \tag{7.102}$$

The first differential of g at constant values of the multipliers is

$$\delta g = -\theta \delta s + \delta u + \pi \delta v + \frac{\tau_1 v \theta}{T^2} \nabla T \cdot \delta q, \tag{7.103}$$

or, taking into account the dependence of s with respect to u, v, and q,

$$\delta g = -\theta \left(\theta^{-1} \delta u + \pi \theta^{-1} \delta v - \frac{\tau_1 v \theta}{\lambda T^2} q \cdot \delta q \right) + \delta u + \pi \delta v + \frac{\tau_1 v \theta}{T^2} \nabla T \cdot \delta q. \tag{7.104}$$

It is immediately seen that this expression vanishes in the steady state where $q = -\lambda \nabla T$. The second differential of g is simply $\delta^2 g = -\theta \delta^2 s$, because the last three terms of (7.102) are linear in u, v, and q, and consequently (7.101) becomes an Einstein–Boltzmann relation in which s is the generalised entropy. From earlier results [see (2.29)] it follows that

$$s = s_{eq} - \tfrac{1}{2} \alpha_1 q \cdot q, \tag{7.105}$$

with $\alpha_1 = (\tau_1 v / \lambda T^2)$, while its second differential is

$$\delta^2 s = \left(\theta^{-1} \right)_u (\delta u)^2 + 2 \left(\theta^{-1} \right)_v \delta u \delta v + \left(\theta^{-1} \pi \right)_v (\delta v)^2 - \alpha_1 \delta q \cdot \delta q$$

$$+ 2\lambda(\nabla T)\alpha_{1v} \cdot \delta v \delta q + 2\lambda(\nabla T)\alpha_{1u} \cdot \delta u \delta q. \tag{7.106}$$

When this result is inserted into (6.5), one obtains an explicit expression for the probability of the fluctuations δu, δv, and δq. It is not our purpose to write the cumbersome general expressions for the second moments, which can be found in [7.23a,b]. Up to second order in $q_0 = -\lambda \nabla T$ it is found that

$$\langle \delta q \delta q \rangle = \frac{k_B \lambda T^2}{\tau_1 v} \left[1 + \frac{c_v T^2}{\alpha_1} \left(\frac{\partial \alpha_1}{\partial u} \right)^2 q_0^2 \right], \tag{7.107a}$$

$$\langle \delta u \delta q \rangle = -\frac{k_B \lambda T^2}{\tau_1 v \Delta} q_0 \left[\left(T^{-1} \right)_v \alpha_{1v} - \left(T^{-1} p \right)_v \alpha_{1u} \right], \tag{7.107b}$$

$$\langle \delta v \delta q \rangle = -\frac{k_B \lambda T^2}{\tau_1 v \Delta} q_0 \left[\left(T^{-1} \right)_v \alpha_{1u} - \left(T^{-1} \right)_u \alpha_{1v} \right], \tag{7.107c}$$

where the subscripts u and v mean the respective partial derivatives and Δ stands for

$$\Delta = \left(T^{-1} \right)_u \left(T^{-1} p \right)_v - \left(T^{-1} \right)_v^2. \tag{7.108}$$

Two points are worth mentioning: (1) The correlations $\langle \delta u \delta q \rangle$ and $\langle \delta v \delta q \rangle$, which vanish at equilibrium because of the opposite time-reversal symmetry of the corresponding quantities, differ from zero out of equilibrium: this is a manifestation of the breaking of the time-reversal symmetry out of equilibrium; (2) non-equilibrium corrections of order q_0^2 appear in the second moments of the heat flux fluctuations and, as a consequence, in the hydrodynamic noise.

We can write (7.107) explicitly for a monatomic ideal gas. One has $(T^{-1}p)_v = -p(Tv)^{-1}$, $(T^{-1})_v = 0$, $(T^{-1})_u = -(c_v T^2)^{-1}$ with $c_v = \frac{3}{2}k_B/m$. Furthermore, in (6.33) we have found that $\alpha_1 = \frac{2}{5}(k_B^2 T^3 n^2)^{-1}$, and therefore that $\alpha_{1u} = -\frac{4}{5}m(n^2 T^4 k_B^3)^{-1}$ and $\alpha_{1v} = \frac{4}{5}m(n k_B^2 T^3)^{-1}$. The final expressions are thus

$$\langle \delta q \delta q \rangle = \frac{k_B \lambda T^2}{\tau_1 v}\left[1 + \frac{25}{8}\pi \ell^2 (\nabla \ln T)^2\right],$$

$$(7.109)$$

$$\langle \delta u \delta q \rangle = -2uN^{-1}\lambda \nabla T, \qquad \langle \delta v \delta q \rangle = 2vN^{-1}\lambda \nabla T,$$

where ℓ is the mean free path given by $(8k_B T/\pi m)^{1/2}\tau_1$. These results are of the same order, but not are exactly equal to the non-equilibrium corrections which may be found from the kinetic theory of gases [7.22]. This is not surprising because of the various approximations which have been introduced.

The above procedure, which relates a generalised thermodynamic potential to non-equilibrium fluctuations, differs from the procedure followed by some authors [7.24] that construct non-equilibrium thermodynamic potentials by starting from knowledge of the non-equilibrium fluctuations, as obtained from suitable dynamical equations, and using then the Einstein relation (6.5) as a definition of the potential. In contrast, we start from a thermodynamic potential and study its consequences for non-equilibrium fluctuations.

The explicit examples studied in Sects. 7.3–6 also exhibit the dependence on the fluxes. For instance, the second moments of the fluctuations of the energy and of the heat flux around their mean values in the harmonic chain ring studied in Sect. 7.6 are given by

$$\langle (\mathcal{H} - U)(\mathcal{H} - U) \rangle = \frac{\partial^2 \ln Z}{\partial \beta^2} = N\varepsilon^2 \frac{1+x^2}{1-x^2}, \qquad (7.110a)$$

$$\langle (\mathcal{J} - Q)(\mathcal{J} - Q) \rangle = \frac{\partial^2 \ln Z}{\partial \gamma^2} = \frac{\varepsilon^2}{2N}\frac{1+4x^2-x^4}{1-x^2}, \qquad (7.110b)$$

$$\langle (\mathcal{H} - U)(\mathcal{J} - Q) \rangle = \frac{\partial^2 \ln Z}{\partial \beta \partial \gamma} = -2N\varepsilon^2 \frac{x}{1-x^2}. \qquad (7.110c)$$

The heat flux influences the second moments of fluctuations, which diverge for a critical value of the heat flux $Q = \varepsilon$. The second moments of the fluctuations of energy and heat flux vanish at equilibrium, because these quantities are of different time-reversal parity. However, the presence of a non-vanishing mean heat flux ($x \neq 0$) leads to a non-vanishing covariance, as predicted by the above macroscopic results.

Problems

7.1 The thermal conductivity of a dielectric solid is $\lambda = \frac{1}{3}\rho c_v c_0^2 \tau_1$, with c_0 the phonon speed, τ_1 the relaxation time due to resistive phonon collisions, c_v the heat capacity per unit mass, and ρ the mass density. In the Debye model c_v is proportional to T^3 at low tem-peratures. Starting from (7.107a), determine the second-order corrections to the second moments of the fluctuations of the heat flux.

7.2 Find the partition function for the distribution (7.44).

7.3 Obtain the second moments of the fluctuations of U and P_{12} in terms of β and λ_{12} for a classical gas under shear studied in Sect. 7.3.

7.4 Obtain the second moments of the fluctuations of the energy and the energy flux in the relativistic gas studied in Sect. 7.5.

7.5 Using the identification for the Lagrange multiplier for the heat flux $\lambda_q = -\tau \nabla \theta^{-1}$ and the expression (7.99) for the second moments of the fluctuations, show that

$$-\frac{\partial q}{\partial \nabla \theta} = \frac{\tau V}{k_B \theta^2}\langle \delta q\, \delta q\rangle_{neq},$$

where $\langle ...\rangle_{neq}$ stands for the average at non-equilibrium steady states. This expresion generalises the fluctuation-dissipation expression (6.19) for non-equilibrium steady states and allows one to compute non-linear thermal conductivities for transport laws of the form $q = -\lambda(\nabla \theta)\nabla \theta$ [for an application of this equation to a radiative gas, see D. Jou and M. Zakari, J. Phys. A **28** (1995) 1585].

7.6 Show that the expressions for β and γ in (7.63) and (7.64) may also be obtained from $\theta^{-1} = (\partial S/\partial U)_{V,J,N}$ and $\gamma = (\partial S/\partial J)_{U,V,N}$.

Chapter 8

Linear Response Theory

In Chap. 6 we discussed the microscopic foundations of extended irreversible thermo-dynamics (EIT) through the fluctuation theory by focusing the attention on the main features of the second moments of the fluctuations. The present chapter deals with more general methods of non-equilibrium statistical mechanics, excerpting in particular the dynamical aspects of fluctuations. First, we start from the Liouville equation and introduce the projection operator technique to relate the memory functions to the time correlation functions of the fluctuations of the fluxes.

This result plays a central role in modern non-equilibrium statistical mechanics and it is also of special interest in EIT, because it emphasizes the role played by the evolution of the fluctuations of the thermodynamic fluxes in specifying the memory functions. The microscopic expressions for such evolution, though compact and elegant, are merely formal, and they are overwhelmingly difficult. Thus, to achieve practical results, one must model such evolution and EIT may be very useful as a thermodynamic tool to propose admissible models for the phenomenological evolution equations of the corresponding fluxes.

The method of projection operators is useful, since it provides a general and explicit method for obtaining the evolution equations of the basic variables. Classically, the set of variables consists of the conserved slow variables (energy, linear momentum, mass) and some order parameters characterizing second-order phase transitions. Here we present a generalisation by adding the dissipative fluxes to the basic set of variables. The projection operator technique is especially helpful when the fluxes become 'slow' variables because their relaxation times become sufficiently long.

8.1 Projection Operator Methods

In this section, we briefly introduce the main concepts and ingredients of the linear response theory [8.1–4]. Its main objective is to provide a microscopic expression for the memory function relating a flux $J(t)$ at a given time t to its respective conjugate force $X(t')$ at previous times t':

$$J(t) = \int_0^\infty \mathbf{K}(t-t') \cdot X(t') dt' ; \tag{8.1}$$

$\mathbf{K}(t - t')$ is the so-called memory function. Note that integration of the relaxational evolution equation of the Maxwell–Cattaneo form, namely,

$$\tau \frac{\partial J}{\partial t} = -(J - \mathbf{L} \cdot X) \tag{8.2}$$

with $\mathbf{L}$ the corresponding phenomenological transport coefficient, yields, for a zero initial flux,

$$J(t) = \int_0^\infty \frac{\mathbf{L}}{\tau} \exp[-(t-t')/\tau] \cdot X(t') dt'. \tag{8.3}$$

By comparing (8.3) and (8.1), one may identify the memory function as

$$\mathbf{K}(t-t') = \frac{\mathbf{L}}{\tau} \exp\left(-\frac{t-t'}{\tau}\right), \tag{8.4}$$

indicating that in the linearized version of EIT, the memory function decays exponentially. In Chap. 5 we have justified equations of the form (8.2), and even more general equations, by starting from the Boltzmann equation. However, the latter is valid for semidilute gases, when binary collisions are the dominant relaxation mechanism. Modern statistical mechanics has tried to formulate more general versions of transport theory, without these restrictions on the density. This is achieved by considering N-particle distribution functions instead of the more restrictive one-particle distribution function, and by using the Liouville equation for the description of its evolution.

In contrast to kinetic theory, we work now in the phase space of the whole system instead than in the configuration space of a single particle and we start from the fundamental Liouville equation

$$\frac{\partial f_N}{\partial t} = -i \mathcal{L}_N f_N , \tag{8.5}$$

where f_N is the N-particle distribution function describing a system of N point particles of mass m and $\mathcal{L}_N$ is the Liouville operator, defined as

$$i\mathcal{L}_N = \sum_{i=1}^{N}\left(\frac{\partial \mathcal{H}_N}{\partial p_i}\cdot\frac{\partial}{\partial r_i} - \frac{\partial \mathcal{H}_N}{\partial r_i}\cdot\frac{\partial}{\partial p_i}\right) = \{\mathcal{H}_N,\ldots\}, \qquad (8.6)$$

with r_i and p_i the position and momentum of the ith particle, and $\mathcal{H}_N$ the Hamiltonian of the system; the braces $\{\,,\,\}$ stand for the Poisson bracket. Liouville equation is essentially a reformulation of the Hamilton equations of motion

$$\dot{r}_i = \frac{\partial H_N}{\partial p_i} \quad , \quad \dot{p}_i = -\frac{\partial H_N}{\partial r_i}. \qquad (8.7)$$

Consider the set of all dynamical properties $A, B, C,\ldots$ of the system and define a scalar product as

$$(A,B) = \int d\Gamma_N f_{Neq}\, AB. \qquad (8.8)$$

At this point, it should be realized that, unlike situations near equilibrium, linear response theory is faced with important open problems when the system is driven far from equilibrium. For instance, to study the evolution of perturbations around a non-equilibrium steady state, it could be asked whether the non-equilibrium steady state distribution function should not be preferred in the definition (8.8) instead of the equilibrium distribution function.

It is easy to show that the Liouvillian operator $\mathcal{L}_N$ is a linear Hermitian operator ($\mathcal{L}_N^\dagger = \mathcal{L}_N$) in the Liouville space defined by the set of all dynamical variables and a metric given by the scalar product (8.8). Indeed, integrating by parts and assuming that f_{Neq} tends sufficiently fast to zero at the boundaries of integration it can be written

$$(\mathcal{L}_N A, B^*)^* = (\mathcal{L}_N B, A^*) \qquad (8.9)$$

and consequently

$$(\exp\,[i\mathcal{L}_N t]A, B^*)^* = (\exp\,[-i\mathcal{L}_N t]B, A^*), \qquad (8.10)$$

where the asterisk denotes the complex conjugate.

One defines a projection operator $\mathcal{P}$ that projects an arbitrary observable B onto a given set of observables A by

$$\mathcal{P}B = (B, A^*)(A, A^*)^{-1}\, A. \qquad (8.11)$$

Note that A can also be considered to be a set of several observables (e.g. mass, linear momentum, energy and so on) instead of a single one, in which case A is a column vector whose components are the set of observables. The idea beyond this projection is to concentrate the attention on the directly observable variables of the set A rather than dispersing the attention on the whole set (in principle infinite) of variables. Therefore, beyond its apparently formal aspect, the idea behind the projection operators has a very physical and intuitive motivation.

The quantity

$$Q = 1 - \mathcal{P} \qquad (8.12)$$

is also a projection operator which projects onto a subspace orthogonal to A. Since by definition of Q, one has $iL_N = i(Q + \mathcal{P})L_N$, and

$$e^{A+B} = e^A + \int_0^1 e^{(A+B)(1-\lambda)} B e^{A\lambda} d\lambda \qquad (8.13)$$

(recall that A and B are matrices which in general do not commute), it follows that

$$\exp[iL_N t] = \exp[iQL_N t] + \int_0^t dt' \exp[iL_N(t-t')]i\mathcal{P}L_N \exp[iQL_N t']. \qquad (8.14)$$

So far with general considerations; our aim is to obtain evolution equations for the physically relevant variables $A(\Gamma)$ which depend on the phase-space coordinates. When the particles follow the trajectories dictated by the Hamilton equations of motion, the evolution of $A(\Gamma)$ is governed by

$$\frac{dA(\Gamma)}{dt} = \sum_{i=1}^{N} \left(\frac{\partial \mathcal{H}_N}{\partial p_i} \cdot \frac{\partial A}{\partial r_i} - \frac{\partial \mathcal{H}_N}{\partial r_i} \cdot \frac{\partial A}{\partial p_i} \right) = iL_N A. \qquad (8.15)$$

Note that dA/dt is not necessarily contained in the subspace of the selected variables A. Integration of (8.15) yields

$$A(t) = \exp[iL_N t] A(0) \qquad (8.16)$$

and differentiation with respect to time gives

$$dA(\Gamma)/dt = \exp[iL_N t] iL_N A(0) = \exp[iL_N t](\mathcal{P} + Q)iL_N A(0). \qquad (8.17)$$

The insertion of the operator $\mathcal{P} + Q$ does not alter the result, because it is the identity operator. From the definition (8.11) of $\mathcal{P}$ it follows that

$$\exp\,[i\mathcal{L}_N t]\,\mathcal{P}\,i\mathcal{L}_N A = (i\mathcal{L}_N A, A^*)(A, A^*)^{-1}\,\exp\,[i\mathcal{L}_N t]\,A = i\Omega A(t), \qquad (8.18)$$

where A stands for $A(0)$ and the operator Ω for

$$i\Omega = (i\mathcal{L}_N A, A^*)(A, A^*)^{-1}. \qquad (8.19)$$

The quantity Ω has the dimensions of the reciprocal of time and gives the frequency at which A rotates in the subspace A, in the absence of the orthogonal variables. Introduction of (8.13) and (8.19) into (8.18) yields

$$\frac{dA(t)}{dt} = i\Omega A(t) + \exp[iQ\mathcal{L}_N t]Q\,i\mathcal{L}_N A(t)$$

$$+ \int_0^t dt'\,\exp[i\mathcal{L}_N(t-t')]i\mathcal{P}\mathcal{L}_N\,\exp[iQ\mathcal{L}_N t']Q\,i\mathcal{L}_N A. \qquad (8.20)$$

To have a more compact expression, set

$$F(t) = \exp\,[iQ\mathcal{L}_N t]\,Q\,i\mathcal{L}_N A, \qquad (8.21)$$

which is called the random force and may also be written as

$$F(t) = Q\exp\,[iQ\mathcal{L}_N t]\,Q\,i\mathcal{L}_N A = QF(t). \qquad (8.22)$$

This means that $F(t)$ is orthogonal to A, which is expressed by

$$(F(t), A^*) = 0, \qquad (8.23)$$

indicating that there is no correlation between A and the random force, an important conclusion to be exploited below.

The term $i\mathcal{P}\mathcal{L}_N F(t)$ appearing in the last integral of (8.20) can be written as

$$i\mathcal{P}\mathcal{L}_N F(t) = i\mathcal{P}\mathcal{L}_N QF(t) = (i\mathcal{L}_N QF(t), A^*)(A, A^*)^{-1}\,A. \qquad (8.24)$$

Since Q and $\mathcal{L}_N$ are both Hermitian, it follows that

$$(i\mathcal{L}_N QF(t), A^*) = -\,(F(t), (Q\,i\mathcal{L}_N A)^*) = -\,(F(t), F^*(0)). \qquad (8.25)$$

By defining the memory function $K(t)$ as

$$K(t) = (F(t), F^*(0))(A, A^*)^{-1} \qquad (8.26)$$

and combining (8.26), (8.8), and (8.18), one obtains the generalised Langevin (or Mori) equation:

$$\frac{dA(t)}{dt} = i\Omega A(t) - \int_0^t dt' K(t') A(t - t') + F(t).$$ (8.27)

Relation (8.26) between the memory function and the random force is known as the fluctuation–dissipation theorem.

From (8.27) it is easy to derive the time correlation function $C(t)$ for the variables A, an interesting quantity in non-equilibrium statistical mechanics because it is measurable from light- and neutron-scattering experiments, and defined by

$$C(t) = (A(t), A^*(0)).$$ (8.28)

Taking the scalar product of (8.27) with A^* and using (8.22), one has

$$\frac{dC(t)}{dt} = i\Omega C(t) - \int_0^t dt' K(t') C(t - t').$$ (8.29)

If (8.27) for the microscopic operator A is averaged over an initial non-equilibrium distribution function, one obtains for the macroscopic value $\langle A \rangle$

$$\frac{d\langle A(t) \rangle}{dt} = i\Omega \langle A(t) \rangle - \int_0^t dt' K(t') \langle A(t - t') \rangle + \langle F(t) \rangle,$$ (8.30)

where $\langle A(t) \rangle$ and $\langle F(t) \rangle$ denote the respective ensemble averages of $A(t)$ and $F(t)$, with $\langle F(t) \rangle$ explicitly given by

$$\langle F(t) \rangle = \int d\Gamma \, f_N(\Gamma, 0) \exp[iQ\mathcal{L}_N t] Q \, i\mathcal{L}_N A(\Gamma).$$ (8.31)

Equations (8.29) and (8.30) are at the heart of the theoretical efforts to describe the experiments in generalised hydrodynamics (see Chap. 12) and they deserve some comments. On the one side, there exists no clear-cut criterion for selecting the variables, except that of the relative slowness of their evolution, which becomes problematic at very high frequencies. Expression (8.26) is worth of attention, because it relates the memory function to the correlation of the noise. In fluctuating hydrodynamics, the noise is attributed to the fast fluctuations of the heat flux and the viscous pressure tensor. Thus, in the light of (8.26) and of (6.18) we could write for the memory functions generalising thermal conductivity and viscosity, the relations

$$\lambda(t - t') = \frac{V}{k_B T^2} \langle \delta q_i(t) \delta q_i(t') \rangle$$ (8.32a)

$$\eta(t-t') = \frac{V}{k_BT}\left\langle \delta P_{ij}^v(t)\delta P_{ij}^v(t')\right\rangle \qquad (8.32b)$$

(in these equations, there is no summation with respect to repeated indices). These memory functions complement the transport equations which, in the spirit of (8.1), read as

$$q(t) = -\int_0^t \lambda(t-t')\nabla T(t')dt'$$

$$(8.33)$$

$$\mathbf{P}^v(t) = -\int_0^t 2\eta(t-t')\mathbf{V}(t')dt'$$

When the relaxation times of the fluctuations of the fluxes are very short, in such a way that the memory functions are non-negligible only for very small values of $t - t'$ the bounds of the integrals may be extended to infinity and the thermodynamic forces may be taken out of the integral with the value corresponding to $t = t'$. Doing so, one recovers from (8.32) and (8.33) the Green–Kubo expressions (6.18).

Notwithstanding the elegance and generality of the procedure, the apparent simplicity of equations (8.32–33) is misleading because of the complexity of the Liouville operator. This is why some simplifications must be introduced, usually in the form of mathematical models for the memory functions, to be able to deal with practical problems. Extended irreversible thermodynamics is useful as it provides some hints for such modellings. In the next section we illustrate the previous method by deriving evolution equations for the dissipative fluxes when the latter are chosen as independent variables.

8.2 Evolution Equations for Simple Fluids

In their analysis of neutron-scattering experiments in liquids (see Chap. 12), Akcasu and Daniels [8.5] go beyond the local-equilibrium description by introducing the heat flux and the viscous pressure tensor among the set of variables and by applying the method presented above to obtain the evolution equations for this enlarged set of variables. We present here the main steps of their formalism, which is of special interest in the context of EIT because it starts from the same selection of basic variables.

The variables in this description are the mass density ρ, the momentum density $J = \rho v$, the energy E, the energy flux Q, and the pressure tensor $\mathbf{P}$. In Fourier space, which is suitable to study the response of the system to external sollicitations, the microscopic operators of the aforementioned variables for a simple fluid are

$$\rho(k) = m \sum_{\alpha} \exp[i k \cdot r_{\alpha}],$$

$$J(k) = \sum_{\alpha} m v_{\alpha} \exp[i k \cdot r_{\alpha}],$$

$$E(k) = \sum_{\alpha} \left[\tfrac{1}{2} m v_{\alpha}^2 + \tfrac{1}{2} \sum_{\beta} \phi(r_{\alpha\beta}) \right] \exp[i k \cdot r_{\alpha}],$$

$$Q(k) = \sum_{\alpha} \left[\tfrac{1}{2} m v_{\alpha}^2 + \tfrac{1}{2} \sum_{\beta} \phi(r_{\alpha\beta}) \right] v_{\alpha} \exp[i k \cdot r_{\alpha}] \tag{8.34}$$

$$+ \tfrac{1}{4} \sum_{\alpha\beta} (v_{\alpha} + v_{\beta}) \cdot r_{\alpha\beta} r_{\alpha\beta} \frac{\phi'(r_{\alpha\beta})}{r_{\alpha\beta}} \frac{1 - \exp[-i k \cdot r_{\alpha\beta}]}{-i k \cdot r_{\alpha\beta}} \exp[i k \cdot r_{\alpha}],$$

$$\mathbf{P}(k) = \sum_{\alpha} \left\{ m v_{\alpha} v_{\alpha} + \tfrac{1}{2} \sum_{\beta} \left(r_{\alpha\beta} r_{\alpha\beta} \frac{\phi'(r_{\alpha\beta})}{r_{\alpha\beta}} \frac{1 - \exp[-i k \cdot r_{\alpha\beta}]}{-i k \cdot r_{\alpha\beta}} \right) \right\} \exp[i k \cdot r_{\alpha}],$$

where $r_{\alpha\beta} = r_{\alpha} - r_{\beta}$ is the relative position of particles α and β, $\phi(r_{\alpha\beta})$ denotes the interaction potential between the particles and $\phi' = d\phi/dr_{\alpha\beta}$; the last terms in $E(k)$, $Q(k)$ and $\mathbf{P}(k)$ take into account the contributions of the intermolecular potential, as it was already done in Sect. 5.5 for the viscous pressure. Expressions (8.34) are useful for calculating the transport coefficients, but here we are more interested in the general form of the evolution equations rather than in computing these coefficients.

Instead of $E(k)$, $\mathbf{P}(k)$, and $Q(k)$, it is convenient to introduce the quantities

$$\mathcal{E}(k) = E(k) - \frac{\left(E(k), \rho^*(k) \right)}{\left(\rho(k), \rho^*(k) \right)} \rho(k), \tag{8.35a}$$

$$\mathbf{P}^v(k) = \mathbf{P}(k) - \frac{\left(\mathbf{P}(k), \rho^*(k) \right)}{\left(\rho(k), \rho^*(k) \right)} \rho(k) - \frac{\left(\mathbf{P}(k), \mathcal{E}^*(k) \right)}{\left(\mathcal{E}(k), \mathcal{E}^*(k) \right)} \mathcal{E}(k), \tag{8.35b}$$

$$q(k) = Q(k) - \frac{\left(Q(k), J^*(k) \right)}{\left(J(k), J^*(k) \right)} J(k). \tag{8.35c}$$

The variable $\mathcal{E}(k)$ may be identified as $\mathcal{E}(k) = \rho c_v(k) T(k)$, with $T(k)$ the absolute temperature and $c_v(k)$ the specific heat at constant volume. The variables $\mathbf{P}^v(k)$ and $q(k)$ correspond, of course, to the viscous pressure and to the heat flux; indeed in (8.35c) the convective energy current is subtracted from the total energy current, and in (8.35b) the equilibrium pressure (a function of T and ρ) is subtracted from the total pressure tensor.

Let us now establish the evolution equations for the 14 components of the state vector, namely

$$A = \left\{ \rho, \mathcal{E}, P^v_{11}, P^v_{22}, P^v_{33}, P^v_{13}, P^v_{23}, P^v_{12}, J_1, J_2, J_3, q_1, q_2, q_3 \right\}. \tag{8.36}$$

The first eight variables are even, whereas the last six are odd functions of the velocity. Hence, the state vector (8.36) may be decomposed into even and odd parts:

$$A = \mathrm{col}\{A^e, 0\} + \mathrm{col}\{0, A^o\}. \tag{8.37}$$

As a consequence, the static correlation matrix $\phi = (A, A^\dagger)$, with $A^\dagger$ the transposed of A, splits into two disjoint submatrices

$$\phi = \begin{pmatrix} \phi^e & 0 \\ 0 & \phi^o \end{pmatrix}, \tag{8.38}$$

with

$$\phi^e = \left(A^e, A^{e\dagger} \right) = \begin{pmatrix} (\rho, \rho^*) & 0 & 0 \\ 0 & (\mathcal{E}, \mathcal{E}^*) & 0 \\ 0 & 0 & (\mathbf{P}^v, \mathbf{P}^{v\dagger}) \end{pmatrix} \tag{8.39}$$

and

$$\phi^o = \left(A^o, A^{o\dagger} \right) = \begin{pmatrix} (J, J^\dagger) & 0 \\ 0 & (q, q^\dagger) \end{pmatrix}. \tag{8.40}$$

The block diagonality of ϕ^e and ϕ^o is a consequence of the choice of the state variables in (8.36).

The frequency matrix $i\Omega = (A, A^\dagger)\phi^{-1}$ defined by (8.19) is here

$$i\Omega = \begin{pmatrix} 0 & (A^e, A^{o\dagger})(\phi^o)^{-1} \\ (A^e, A^{e\dagger})(\phi^e)^{-1} & 0 \end{pmatrix}. \tag{8.41}$$

The usual conservation laws of mass, momentum and energy read

$$\frac{\partial \rho}{\partial t} = i\boldsymbol{k} \cdot \boldsymbol{J}, \tag{8.42}$$

$$\frac{\partial \boldsymbol{J}}{\partial t} = i\boldsymbol{k} \cdot \mathbf{P}, \tag{8.43}$$

$$\frac{\partial E}{\partial t} = i\boldsymbol{k} \cdot \boldsymbol{Q}. \tag{8.44}$$

It follows that the random noise $\boldsymbol{F} = Q\,i\mathcal{L}_N A$ has only nine non-zero components, corresponding to the dissipative fluxes

$$\boldsymbol{F} = \left\{ 0,0,F_{11}^{v},F_{22}^{v},F_{33}^{v},F_{13}^{v},F_{23}^{v},F_{12}^{v},0,0,0,F_{1}^{q},F_{2}^{q},F_{3}^{q} \right\}. \tag{8.45}$$

Accordingly, the memory function matrix $\mathbf{K}(t) = (\boldsymbol{F}(t), \boldsymbol{F}^{*}(0))$ is of the form

$$\mathbf{K}(t) = \begin{pmatrix} 0 & 0 & 0 & 0 \\ 0 & \boldsymbol{\phi}^{vv}(t) & 0 & \boldsymbol{\phi}^{vq}(t) \\ 0 & 0 & 0 & 0 \\ 0 & \boldsymbol{\phi}^{qv}(t) & 0 & \boldsymbol{\phi}^{qq}(t) \end{pmatrix}, \tag{8.46}$$

where $\boldsymbol{\phi}^{vv}$ and $\boldsymbol{\phi}^{qq}$ are 6×6 and 3×3 square matrices, while $\boldsymbol{\phi}^{vq}$ and $\boldsymbol{\phi}^{qv}$ are 3×6 and 6×3 rectangular matrices respectively.

Introducing (8.38), (8.45), and (8.46) into (8.29) results in the following evolution equations of $\mathbf{P}^{v}$ and $\boldsymbol{q}$:

$$\dot{\mathbf{P}}^{v}(t) - \left(\dot{\mathbf{P}}^{v},\boldsymbol{q}^{\dagger}\right)\left(\boldsymbol{q},\boldsymbol{q}^{\dagger}\right)^{-1}\boldsymbol{q}(t) + \int_{0}^{t}dt'\,\boldsymbol{\phi}^{vq}(t-t')\cdot\boldsymbol{q}(t')$$

$$-\left(\dot{\mathbf{P}}^{v},\boldsymbol{J}^{\dagger}\right)\left(\boldsymbol{J},\boldsymbol{J}^{\dagger}\right)^{-1}\boldsymbol{J}(t) + \int_{0}^{t}dt'\,\boldsymbol{\phi}^{vv}(t-t'):\mathbf{P}^{v}(t') = \mathbf{F}^{v}(t) \tag{8.47}$$

and

$$\dot{\boldsymbol{q}}(t) - \left(\dot{\boldsymbol{q}},\mathcal{E}^{*}\right)\left(\mathcal{E},\mathcal{E}^{*}\right)^{-1}\mathcal{E}(t) + \int_{0}^{t}dt'\,\boldsymbol{\phi}^{qq}(t-t')\cdot\boldsymbol{q}(t')$$

$$-\left(\dot{\boldsymbol{q}},\mathbf{P}^{v\dagger}\right)\cdot\left(\mathbf{P}^{v},\mathbf{P}^{v\dagger}\right)^{-1}\mathbf{P}^{v}(t) + \int_{0}^{t}dt'\,\boldsymbol{\phi}^{qv}(t-t'):\mathbf{P}^{v}(t') = \boldsymbol{F}^{q}(t). \tag{8.48}$$

These equations are exceedingly complicated, and therefore the following simplifications are usually introduced:

$$\int_{0}^{t}dt'\,\boldsymbol{\phi}^{vv}(t-t'):\mathbf{P}^{v}(t') = \mathbf{W}^{v}(k)\cdot\mathbf{P}^{v}(t), \tag{8.49a}$$

$$\int_0^t dt'\, \boldsymbol{\phi}^{qq}(t-t') \cdot \boldsymbol{q}(t') = \mathbf{W}^q(k) \cdot \boldsymbol{q}(t), \tag{8.49b}$$

with

$$\mathbf{W}^v(k) = \int_0^\infty dt\, \boldsymbol{\phi}^{vv}(t), \qquad \mathbf{W}^q(k) = \int_0^\infty dt\, \boldsymbol{\phi}^{qq}(t). \tag{8.50}$$

In generalised hydrodynamics, the cross terms in $\boldsymbol{\phi}^{vq}$ and $\boldsymbol{\phi}^{qv}$ are usually neglected. Defining the quantities $\boldsymbol{\Sigma}$, $\boldsymbol{\Lambda}$, $\mathbf{A}$, and $\mathbf{B}$ (which are fourth-rank tensors at the exception of $\boldsymbol{\Lambda}$ which is of second-rank) by

$$\beta V^{-1}\left(\dot{\mathbf{P}}^v, \boldsymbol{J}_k\right) = \mathrm{i}\boldsymbol{k} \cdot \boldsymbol{\Sigma},$$

$$\left(\dot{q}, \mathcal{E}^*\right)\left(\mathcal{E}, \mathcal{E}^*\right)^{-1} = \mathrm{i}\boldsymbol{k} \cdot \boldsymbol{\Lambda}, \tag{8.51a}$$

$$\left(\dot{\mathbf{P}}^v, q^\dagger\right) \cdot \left(q, q^\dagger\right)^{-1} = \mathrm{i}\boldsymbol{k} \cdot \mathbf{A},$$

$$\left(\dot{q}, \mathbf{P}^{v\dagger}\right) \cdot \left(\mathbf{P}^v, \mathbf{P}^{v\dagger}\right)^{-1} = \mathrm{i}\boldsymbol{k} \cdot \mathbf{B}, \tag{8.51b}$$

one obtains from (8.47) and (8.48) the simplified expressions:

$$\dot{\mathbf{P}}^v - \boldsymbol{\Sigma} \cdot \boldsymbol{\varepsilon} + \mathbf{W}^v \cdot \mathbf{P}^v - \mathbf{A} \cdot \mathrm{i}kq = \mathbf{F}^v(t) \tag{8.52}$$

and

$$\dot{q} - \mathrm{i}\boldsymbol{k} \cdot \boldsymbol{\Lambda} + \mathbf{W}^q \cdot \boldsymbol{q} - \mathbf{B} \cdot \mathrm{i}k\mathbf{P}^v = \boldsymbol{F}^q(t), \tag{8.53}$$

where $\boldsymbol{\varepsilon}(k)$ is the rate of deformation tensor, i.e.

$$\boldsymbol{\varepsilon} = \frac{\mathrm{i}}{2\rho}(\boldsymbol{kJ} + \boldsymbol{Jk}). \tag{8.54}$$

Relations (8.52) and (8.53) have the same form as the linear macroscopic equations for the heat flux and the viscous pressure tensor proposed in (2.64–66), $\boldsymbol{\Lambda}$ being the thermal diffusivity tensor and $\mathbf{W}^v$ and $\mathbf{W}^q$ frequency matrices. In some aspects (8.52–53) are less general than (2.64–66), because the coupling between the viscous pressure tensor and the heat flux, which was however present in (8.47–48), has been neglected. The contribution of the projection operator formalism is that it yields expressions for the coefficients in terms of correlations between the variables.

It is worth emphasizing that when the space of the relevant dynamical variables is enlarged to include dissipative fluxes, the projection operator method predicts some results derived within macroscopic extended irreversible thermodynamics. Nevertheless, in the microscopic formalism no attention is paid to the thermodynamic implications of the evolution equations. By taking into account such relation, EIT may provide a useful complement to the analysis of the evolution equations, whose final form must nevertheless be phenomenologically modelled if it is wanted to study actual problems.

8.3 Continued-Fraction Expansions

In Sect. 5.6 we saw that when higher-order fluxes are introduced into the description one is led to a hierarchy of equations which yields, in Fourier space, a continued-fraction expansion of the transport coefficients. In this section we point out that similar results may also be obtained from a microscopic perspective within the framework of the projection-operator techniques, when the noise appearing in evolution equations of the form (8.27) is taken as a new variable for which an evolution equation must be formulated [8.6]. This point of view has some analogy with EIT, where the fluctuations of the dissipative fluxes, which are considered as a noise in classical fluctuating hydrodynamics, are incorporated as additional variables in the theory.

When the noise in the evolution equation of the variable of order $n - 1$ is selected as an additional variable, one obtains a hierarchy of equations of the form

$$\frac{dF_n}{dt} = i\Omega_n F_n(t) - \int_0^t K_n(t')F_n(t-t')dt' + F_{n+1}(t). \tag{8.55}$$

Indeed, the evolution equation for F_0 is

$$\frac{dF_0(t)}{dt} = i\Omega_0 F_0(t) - \int_0^t K_0(t')F_0(t-t')dt' + F_1(t) , \tag{8.56}$$

with [see for instance (8.19) and (8.26)]

$$\Omega_0 = \langle \dot{F}_0 F_0^* \rangle \langle F_0 F_0^* \rangle^{-1} ; \qquad K_0(t) = \langle F_1(t)F_1^* \rangle \langle F_0 F_0^* \rangle^{-1} . \tag{8.57}$$

The time correlation function of F_0 is defined as

$$C(t) \equiv \langle F_0(t)F_0^* \rangle \langle F_0 F_0^* \rangle^{-1} . \tag{8.58}$$

After multiplying (8.56) by F_0^* and taking into account that $\langle F_1(t)F_0^* \rangle = 0$, one is led to

$$\frac{dC(t)}{dt} = i\Omega_0 C(t) - \int_0^t K_0(t')C(t-t')dt' \, , \tag{8.59}$$

the Laplace transform of which is

$$C(s) = \frac{C(0)}{s - i\Omega_0 + K_0(s)} \, . \tag{8.60}$$

To calculate $K_0(s)$ we use the property that $K_0(t)$ is the time correlation function of F_1 and that F_1 satisfies an equation of the form

$$\frac{dF_1(t)}{dt} = i\Omega_1 F_1(t) - \int_0^t K_1(t')F_1(t-t')dt' + F_2(t) \, , \tag{8.61}$$

with

$$\Omega_1 = \langle \dot{F}_1 F_1^* \rangle \langle F_1 F_1^* \rangle \quad \text{and} \quad K_1(t) = \langle F_2(t)F_2^* \rangle \langle F_1 F_1^* \rangle^{-1}. \tag{8.62}$$

Multiplying (8.61) by F_1^* and averaging, one finds, after that $\langle F_2(t)F_1^* \rangle = 0$ is taken into account,

$$\frac{dK_0(t)}{dt} = i\Omega_1 K_0(t) - \int_0^t K_1(t')K_0(t-t')dt'. \tag{8.63}$$

The Laplace transform of (8.63) is

$$K_0(s) = \frac{K_0(0)}{s - i\Omega_1 + K_1(s)}, \tag{8.64}$$

and introduction of this result into (8.60) yields

$$C(s) = \frac{C(0)}{s - i\Omega_0 + \dfrac{K_0(0)}{s - i\Omega_1 + K_1(s)}} \, . \tag{8.65}$$

This procedure may continue indefinitely because $K_n(t)$ is proportional to $\langle F_{n+1}(t)F_{n+1}^* \rangle$, and all the F_n obey evolution equations of the same form, so that the Laplace transform of the correlation function for F_0 may be written as a continued fraction in s of the form

$$C(s) = \cfrac{C(0)}{s - i\Omega_0 + \cfrac{K_0(0)}{s - i\Omega_1 + \cfrac{K_1(1)}{s - i\Omega_2 + \dots}}}, \qquad (8.66)$$

with

$$K_{n-1}(0) = \langle F_n(0)F_n^*(0)\rangle\langle F_{n-1}(0)F_{n-1}^*(0)\rangle^{-1}, \quad i\Omega_n = \langle \dot{F}_n(0)F_n^*(0)\rangle\langle F_n(0)F_n^*(0)\rangle^{-1}.$$

This result is comparable to (5.110) for the continued-fraction expansion of the transport coefficients by recalling that, according to (8.32), the memory function generalising the transport coefficients is given by the time correlation function of the fluctuations of the dissipative fluxes. Other microscopic justifications of continued-fraction expansions may be found in the non-equilibrium statistical operator method [8.7]. Such a continued fraction was, however, obtained by assuming that the only couplings between variables of different orders are those with the variables of the immediately higher and lower orders, i.e. that the time derivative of variable of order r is related only to the variables of orders $r - 1$, r and $r + 1$. In general situations, more general couplings are allowed, and the form of the hierarchy of evolution equations is much more intricate.

Problems

8.1 Assume that the relaxation time τ of the distribution function depends on the peculiar molecular velocity C according to $\tau(C) \sim C^a$, where a is a constant. The respective relaxation times for q and $\mathbf{P}^v$ are then given by

$$\tau_1^{-1} = \langle (d\hat{q}_x/dt)\hat{q}_x\rangle\langle \hat{q}_x\hat{q}_x\rangle^{-1}$$

and

$$\tau_2^{-1} = \langle (d\hat{P}_{xy}^v/dt)\hat{P}_{xy}^v\rangle\langle \hat{P}_{xy}^v\hat{P}_{xy}^v\rangle^{-1}$$

with $\hat{q}_x$ and $\hat{P}_{xy}^v$ the microscopic operators for q_x and P_{xy}^v and $\langle (d\hat{q}_x/dt)\hat{q}_x\rangle = \langle \tau(C)\hat{q}_x(C)\hat{q}_x(C)\rangle$ and $\langle (d\hat{P}_{xy}^v/dt)\hat{P}_{xy}^v\rangle = \langle \tau(C)\hat{P}_{xy}^v(C)\hat{P}_{xy}^v(C)\rangle$. Find τ_1/τ_2 in terms of a and calculate this ratio for $a = 0$ and $a = -1$.

8.2 In general, the non-equilibrium entropy for an ideal monatomic gas should be expanded in terms of all the moments of the velocity distribution function rather than in terms of just the first thirteen moments. To estimate the relaxation times of the higher-order moments, introduce the quantities

$$M^{(n)} = \int C^n f(C)\, dc$$

and compute the corresponding relaxation times according to the expression

$$\tau_n^{-1} = \langle (dM^{(n)}/dt)M^{(n)}\rangle\langle M^{(n)}M^{(n)}\rangle^{-1} = \langle \tau^{-1}(C)C^nC^n\rangle\langle C^nC^n\rangle^{-1}.$$

Determine τ_n for $\tau(C) \sim C^a$, with $a = 0$, $a = -1$, and $a = +1$.

8.3 Let the Laplace transform $C(s)$ of an autocorrelation function be given by $C(s) = [s + K(s)]^{-1}$, with $K(s)$ the Laplace transform of the corresponding memory function. Assume that $K(s)$ may be expanded as a continued fraction of the form $K(s) = K_0/s + K_1/s + K_2/s + K_3...$ Show that the first and second approximations, defined by cutting the expansion at K_1 or at K_2 respectively, lead to the following memory functions:

$$K(t) = K_0 \exp[-K_1 t],$$

and

$$K(t) = K_0 \exp(-bt)a^{-1}(a\cos at + b\sin at),$$

with $b = -(K_2/2)$ and $a^2 = K_1 - K_2^2/4$.

8.4 An alternative derivation of (8.14) starts from the Laplace transform of $G(t) \equiv e^{iLt}$, namely, $\tilde{G}(s) = 1/(s - iL)$. Since $iL = i(Q + P)L$, it follows that

$$\tilde{G}(s) = \frac{1}{s - i(Q+P)L}.$$

By using the identity

$$\frac{1}{A} - \frac{1}{B} = \frac{1}{A}(B - A)\frac{1}{B}$$

check that

$$\tilde{G}(s) = \frac{1}{s - iQL} + \frac{1}{s - iL}\, iPL\, \frac{1}{s - iQL}$$

and, by Laplace transforms, show (8.14).

8.5 (a) Let $P_r(t)$ be the probability that a system is in the state r at time t. Assume that the time evolution of $P_r(t)$ is given by the master equation

$$dP_r/dt = \sum_s \left[W_{rs} P_s(t) - W_{sr} P_r(t) \right],$$

with W_{rs} the transition probability from state s to state r per unit time. For isolated systems, $W_{rs} = W_{sr}$. Show that the master equation satisfies an H-theorem, i.e. define $H = \sum_r P_r \ln P_r$ and show that $dH/dt \leq 0$. (b) Consider a very simple situation with only two states, corresponding to particles travelling to the right (state 1) or to the left (state 2). Thus, $W_{11} = W_{22} = \mathcal{T}$ (the transmission coefficient) and $W_{12} = W_{21} = \mathcal{R}$ (the reflection coefficient). Assume that $P_1 + P_2 = n$, with n the total concentration of particles, and that all the particles are moving at speed v. Show that the entropy corresponding to a non-equilibrium steady state with n particles per unit volume and a non-vanishing flux of particles j is

$$S = -\frac{1}{2} k_B \left(n + \frac{j}{v} \right) \ln \left(n + \frac{j}{v} \right) - \frac{1}{2} k_B \left(n - \frac{j}{v} \right) \ln \left(n - \frac{j}{v} \right).$$

This expression shows clearly the highly non-linear character of the dependence of the entropy on the flux.

Chapter 9

Computer Simulations

Computer simulations have become an important tool in statistical mechanics, since they allow the study of systems in conditions hardly accessible to experimental observations. Furthermore, they are useful to assert, at least in some situations, the very foundations of macroscopic formalisms. In particular, in equilibrium, computer simulations have been especially fruitful in the analysis of the equations of state and phase transitions of systems composed of interacting particles; out of equilibrium, they are very helpful in the calculation of transport coefficients and in the formulation of non-equilibrium thermodynamics beyond the local-equilibrium approximation. However, out of local equilibrium several conceptual problems arise which are not present in equilibrium, as, for instance, the definition of temperature or pressure and their relation with measurements.

The interpretation of the results provided by computer simulations relies on assumptions such as the meaning of temperature (which is usually identified as the kinetic temperature) and the kind of average to be performed to obtain truly significant macroscopic results. As a consequence, a detailed comparison of the results of simulations with experimental observations may shed a critical view on these crucial and subtle matters. Most attention has been focused on the calculation of transport coefficients out of equilibrium in the linear regime, and more recently the non-linear regime has been the subject of active research. In particular, non-equilibrium equations of state and non-linear transport equations have been extensively studied by Evans and collaborators [9.1]; in this chapter we will pay special attention to their contributions and emphasize the aspects which are most closely related to extended irreversible thermodynamics (EIT).

9.1 Computer Simulations of Non-equilibrium Steady States

Let us consider a set of particles described by a Lennard–Jones interaction potential in a non-equilibrium steady state characterized by a constant shear rate $\dot{\gamma}$. Systems under a stationary and homogeneous shear rate are one of the most thoroughly investigated situations in non-equilibrium molecular dynamics. Indeed, it is considerably easier to simulate systems in a velocity gradient than systems in a temperature gradient [9.2]. Dissipation due to viscous pressure produces heat which must be removed to keep the system in a steady state; for this reason, several thermostating procedures have been proposed to maintain the temperature constant. In this section we outline the basic equations used in the molecular analysis and comment briefly on the thermostating problem.

According to Evans and collaborators [9.1], the microscopic equations describing the motion of the molecules under the action of a velocity gradient ∇v are given by

$$\dot{r}_i = \frac{p_i}{m} + (\nabla v)^T \cdot r_i,$$

$$\dot{p}_i = F_i - (\nabla v)^T \cdot p_i - \alpha p_i, \tag{9.1}$$

where F_i is the force acting on the molecule i of mass m_i, and αp_i (with p_i the momentum of the molecule i) is a Gaussian thermostat which removes energy from the system so as to keep the temperature constant. This is achieved by imposing that (see Problem 9.2)

$$\alpha = -\frac{V\mathbf{P}^v : (\nabla v)}{\sum_i (p_i^2 / m_i)}, \tag{9.2}$$

where V is the total volume. This description is not fully realistic, as it implies that heat is removed at the same point as where dissipation is produced, whereas in real situations heat is eliminated across the boundaries of the system. Furthermore, the Gaussian thermostat produces sharply defined kinetic energy and only potential energy is distributed canonically. A better choice for the thermostat would be the so-called Nosé–Hoover thermostat [9.3], which yields results in agreement with the canonical ensemble, not only for the potential but also for the kinetic energy. Since the rate of viscous heating is quadratic in the shear rate, both thermostats will obviously lead to the same results in the linear regime. Several debates about the best choice of a thermostat have arisen in relation to some ordering transitions in dense fluids [9.4].

One of the main objectives of non-equilibrium molecular dynamics is to obtain the transport coefficients of fluids with a higher precision than the usual fluctuation–dissipation techniques based on Green–Kubo expressions (see Chap. 6). Furthermore, numerical simulations are also relevant for checking recent ideas on non-equilibrium temperature and pressure and on fluctuations in non-equilibrium states [9.5]. In their earlier calculations of 1980, Evans and Hanley [9.6] did not calculate the entropy and defined simply the temperature in terms of the average kinetic energy of the particles; however, later, Evans [9.7] computed the entropy and the corresponding temperature for an isoenergetic planar Couette flow at low densities. The large mean free paths in this low-density regime require very long runs to achieve an accuracy comparable to that for dense fluids; as a matter of fact, the results were achieved after 15 million time-step calculation runs. Table 9.1, which shows some of Evans' results, makes evident the difference between the kinetic (local-equilibrium) temperature T and the (non-equilibrium) thermodynamic temperature θ defined as the derivative of the energy with respect to the entropy. It is to be noticed that $\theta < T$, in agreement with the results of EIT (see Sect. 3.2).

9.2 Non-equilibrium Equations of State

Non-equilibrium equations of state for the pressure and the internal energy of fluids under shear have been computed by Hanley and Evans [9.6]. These authors have carried out their calculations for a system of 108 particles with an interaction potential of the form $\phi(r) = 4\varepsilon(\sigma/r)^{12}$ (r is the distance between particles, ε and σ the strength and range of the potential) at fixed values of density and temperature, under a constant shear rate $\dot{\gamma}$. The first result deserving special attention is the shear-rate dependence of the pressure $p(T, V, \dot{\gamma})$ and the internal energy per particle $u(T, V, \dot{\gamma})$,

$$p(T,V,\dot{\gamma}) = p_0(T,V) + p_1(T,V)\dot{\gamma}^{3/2},$$

$$u(T,V,\dot{\gamma}) = u_0(T,V) + u_1(T,V)\dot{\gamma}^{3/2}.$$

(9.3)

In Table 9.1 are shown some numerical values for $p_1(T, V)$ and $u_1(T, V)$. Pressure, energy, shear rate, and temperature are expressed in dimensionless form in terms of the parameters ε and σ. The relations between the dimensional and dimensionless quantities are

$$p = p^*(\varepsilon/\sigma^3), \quad u = u^*\varepsilon, \quad \dot{\gamma} = \dot{\gamma}^*(\varepsilon/m\sigma^2)^{1/2}, \quad T = T^*(\varepsilon/k_B),$$

where the symbols with asterisks denote dimensionless quantities. The problem is that when these quantities are scaled with the parameters of atomic fluids, the typical shear rates used in non-equilibrium molecular dynamics are much higher than those found in current experiments (Problem 9.3). The results are more satisfactory with colloidal suspensions, which may be modelled by the same two-particle interaction potential, but with a relaxation time of the order of milliseconds, much longer than the relaxation times of atomic fluids, of the order of picoseconds. Correspondingly, for a given value of the dimensionless shear rate (which is given by the product of the actual shear rate times the relaxation time) the actual shear rate corresponding to a colloidal suspension is one millionth of the shear rate corresponding to an atomic fluid.

Table 9.1. Values of the dimensionless quantities p_1^* and u_1^* of (9.3) obtained at $\dot{\gamma}^* = 1$ (a density of 0.844 and a temperature of 0.722 correspond to the triple point of the fluid) [9.7]

Temperature	Density	p_1^*	u_1^*
0.72	0.84	2.81	0.58
1.00	0.84	2.20	0.49
1.00	0.73	0.72	0.15
1.75	0.84	1.31	0.31
1.75	0.73	0.44	0.14

The main difference between (9.3) and the results of EIT is the non-analytic dependence of p and u on $\dot{\gamma}^{3/2}$; EIT predicts a dependence on $\dot{\gamma}^2$ [9.8]. However, the contradiction is only apparent: computer simulations have been performed at high values of $\dot{\gamma}$ in order to make easily visible the effects of shear; in contrast, in the present description of EIT it is supposed that $\dot{\gamma}$ remains small. The transition from a $\dot{\gamma}^2$ to a $\dot{\gamma}^{3/2}$ behaviour has been studied in the context of extended thermodynamics by Nettleton [9.9], who introduced the volume fraction of locally dilated spherical regions as a further state variable. At a high shear rate, Nettleton observes a bifurcation in the asymptotic solution with a $\dot{\gamma}^{3/2}$ dependence for p and u and a $\dot{\gamma}^{1/2}$ dependence for the viscosity. In Sect. 9.3 we re-examine this difference in the frame of a more general non-linear version of EIT.

The second point to be emphasized is that p_1 and u_1 in (9.3) satisfy the relation

$$N\left(\frac{\partial u_1}{\partial V}\right)_{T,\dot{\gamma}} = -p_1 + T\left(\frac{\partial p_1}{\partial T}\right)_{V,\dot{\gamma}}. \tag{9.4}$$

Such a relationship was proved analytically for the soft-sphere fluid but not for the Lennard–Jones fluid, in which case the computed expressions for p_1 and u_1 are consistent with (9.4) within an error of about 5%. Relation (9.4) is well known for equilibrium systems, with p and u being the equilibrium pressure and internal energy. What is new and unexpected is that (9.4) is also satisfied by the non-equilibrium contributions of p and u. This led Hanley and Evans to propose a generalised Gibbs relation of the form

$$dU = TdS - pdV + \zeta d\dot{\gamma}, \qquad (9.5)$$

where $\zeta(T, V, \dot{\gamma})$ is a state function reflecting the shear-rate dependence of the thermodynamical potential. Expression (9.4) is easily derived from (9.5), since

$$\left(\frac{\partial U}{\partial V}\right)_{T,\dot{\gamma}} = -p + T\left(\frac{\partial S}{\partial V}\right)_{T,\dot{\gamma}}. \qquad (9.6)$$

Introducing the Maxwell relation $(\partial S / \partial V)_{T,\dot{\gamma}} = (\partial p / \partial T)_{V,\dot{\gamma}}$ into (9.6) and recalling that the internal energy is $U = Nu$, we see that (9.6) is identical to (9.4).

A third point predicted by the numerical results refers to thermodynamic stability. By using the Helmholtz potential

$$dF = -SdT - pdV + \zeta d\dot{\gamma}, \qquad (9.7)$$

and proceeding in close analogy with classical thermodynamics, one obtains the following stability conditions:

$$C_v(T, V, \dot{\gamma}) \geq 0 \qquad \text{(thermal stability)},$$

$$-(\partial p / \partial V)_{T,\dot{\gamma}} \geq 0 \qquad \text{(mechanical stability)}, \qquad (9.8)$$

$$(\partial \zeta / \partial \dot{\gamma})_{T,V} \geq 0 \qquad \text{(shear-rate stability)},$$

$$\left(\frac{\partial p}{\partial V}\right)_{T,\dot{\gamma}} \left(\frac{\partial \zeta}{\partial \dot{\gamma}}\right)_{T,V} \geq \left[\left(\frac{\partial p}{\partial \dot{\gamma}}\right)_{T,V}\right]^2.$$

Such stability conditions are generally not satisfied for any value of the shear rate. For instance, one notes that $(\partial u_1 / \partial T)_{V,\dot{\gamma}}$ may be negative according to Table 9.1, which means that a fluid may be thermally less stable in the presence of a shear. In view of

Table 9.1, $(\partial p / \partial \rho)_{T,\dot{\gamma}}$ is seen to be positive so that, under a shear, mechanical instability is enhanced. The fourth condition in (9.8) is more severe than the second one in the sense that it leads to a value of the critical shear rate lower than the one inferred from the thermal stability condition. Hanley and Evans have interpreted the shear-rate influence on phase transitions as a consequence of the breakdown of solid-like structures of charged suspended colloidal particles when the system is sheared [9.10]. Clearly, numerical simulations strongly suggest that thermodynamics should be extended beyond its classical description.

Table 9.2. Values for the non-equilibrium entropy, the kinetic temperature T and the thermodynamic temperature θ at energy $u = 2.134$ at different densities and different values of the shear rate $\dot{\gamma}$. All the quantities are expressed in units of the parameters of the molecular potential ε, σ, and k_B [9.1]. The uncertainties in the entropy are ± 0.005 and in temperature ± 0.04

ρ	$\dot{\gamma}$	s	T	θ
0.100	0.0	5.917	2.175	2.126
0.100	0.5	5.653	2.171	2.048
0.100	1.0	5.392	2.169	1.963
0.075	0.0	6.213		
0.075	0.5	5.852	2.190	2.088
0.075	1.0	5.499	2.188	1.902

Numerical simulations are also useful to test the theoretical ideas on non-equilibrium temperature and non-equilibrium pressure. In their earlier calculations of 1980, Evans and Hoover [9.6] did not evaluate the entropy and defined the temperature in terms of the average kinetic energy of the particles. In 1989, Evans computed explicitly the entropy of shear states under shear flow [9.7]. This was achieved by considering a system of 32 soft discs interacting through a potential of the form $\phi(r) = \varepsilon(\sigma/r)^{12}$, truncated at $r = 1.5\sigma$. Evans calculated the entropy for an isoenergetic planar Couette flow by using the following expression for the entropy (see Sect. 5.5):

$$\frac{S}{N} = 1 - k_B \ln\left(\frac{n}{2\pi m k_B T}\right) - \frac{1}{2} n k_B \int g(r_{12}) \ln g(r_{12}) dr_{12} \qquad (9.9)$$

with $g(r_{12})$ being the radial distribution function; the entropy was obtained by integrating the relevant distribution functions over the whole simulation volume, by forming histograms for $g(r)$ and $f(p)$. Table 9.2 shows some of Evans' results, which make evident the difference between the kinetic (local-equilibrium) temperature T and the (non-equilibrium) thermodynamic temperature θ defined as the derivative of the energy with respect to the entropy.

Furthermore, Evans has calculated the non-equilibrium pressure, defined as $\pi = -(\partial U / \partial V)_{\dot{\gamma}}$, and has compared it with the kinetic pressure p obtained from the trace of the pressure tensor. Some of his results are reproduced in Table 9.3.

Table 9.3. Values of the local-equilibrium pressure p versus the non-equilibrium pressure π at energy $u = 2.134$ and density $\rho = 0.100$ for different values of the shear rate $\dot{\gamma}$ [9.1]

$\dot{\gamma}$	p	π
0.0	0.244	0.215
0.5	0.245	0.145
1.0	0.247	0.085

It is evident from these results that there are significative differences between the values of θ and T and between those of π and p in the presence of shear rates. Note, furthermore, that $\theta < T$ and $\pi < p$, as predicted by EIT, according to the arguments presented in Sect. 3.5. Evans [9.7] noticed that the numerical data for the thermodynamic pressure π agree with the minimum eigenvalue of the pressure tensor and conjectured that this is a general feature. He argued that, if the entropy is related to the minimum reversible work to accomplish a virtual change, in a non-equilibrium steady state the minimum work, changing the volume by dV, will be that carried out by moving the wall perpendicular to the direction corresponding to the minimum eigenvalue of the pressure tensor. We have found in (3.23) a similar result for a gas submitted to a heat flux, where the pressure becomes lower in the directions perpendicular to the heat vector, and in (3.29) for fluids under shear flow.

9.3 Dependence of the Free Energy on the Shear Rate: Non-linear Approach

We have already stated that the main difference between the results (9.3) of numerical simulations and those of EIT is that the former predicts a non-analytical dependence of p and E on the shear rate $\dot{\gamma}$ while EIT exhibits a dependence on $\dot{\gamma}^2$. Indeed, it is found from (9.3) that the free energy $F(T, V, \dot{\gamma})$ is

$$F(T,V,\dot{\gamma}) = F_{eq}(T,V) + F_1(T,V)\dot{\gamma}^{3/2}, \tag{9.10}$$

in contrast with the $\dot{\gamma}^2$ behaviour found in Chap. 2.

A strict comparison with EIT requires a macroscopic theory valid at very high values of the shear rate. Indeed, computer simulations have been performed at high values of $\dot{\gamma}$ in order to emphasize the effects of shear (Problem 9.3 shows the very unrealistic values of the actual shear rate which correspond to the value $\dot{\gamma}^* = 1$ of the dimensionless shear rate considered in NEMD (Non-Equilibrium Molecular Dynamics) simulations of atomic fluids). In contrast, the quadratic approximation supposes that $\dot{\gamma}$ remains small. Nevertheless, the transition from a $\dot{\gamma}^2$ to a $\dot{\gamma}^{3/2}$ regime may be apprehended qualitatively in a rather simple way in the context of EIT [9.11]. To achieve this goal, let us start from the following expression of the free energy of EIT

$$F(T,V,\dot{\gamma}) = F_{eq}(T,V) + \tfrac{1}{2}\tau V \eta \dot{\gamma}^2, \tag{9.11}$$

where we have used $P_{12}^v = \eta\dot{\gamma}$ to write $F(T,V,\dot{\gamma})$ instead of $F(T,V,P_{12}^v)$. In general, the viscosity η will be a function of $\dot{\gamma}$ and will decrease with increasing $\dot{\gamma}$ (a phenomenon known as shear-thinning) [9.4]. Detailed expressions for such a dependence of $\eta(\dot{\gamma})$ may be obtained from kinetic theory, from numerical simulations, and from maximum-entropy arguments. Here, we will use the latter to emphasize the internal unity of the present text. Taking into account expression (7.40) for the Lagrange multiplier λ_{12} conjugated to the viscous pressure component P_{12}^v in information theory, as well as the identification

$$\lambda_{12} = -\frac{\tau}{\eta k_B T} P_{12}^v = \frac{\tau\dot{\gamma}}{k_B T}, \tag{9.12}$$

one may obtain η as a function of P_{12}^v as

$$\eta = -\frac{P_{12}^v}{\dot{\gamma}} = -\eta_0 \frac{3R^2\left[R^2 + (1-y)\right]}{2\left[R^2 + (2/3)(1-y)\right]}, \tag{9.13}$$

where R and y are given by

$$R = P_{12}^v / Nu\,, \qquad y = (1 + 3R^2)^{1/2}, \tag{9.14}$$

while η_0 is the shear viscosity in the low shear-rate limit. For low values of R, this expression tends to η_0, whereas it tends to 0 when R approaches 1, in such a way that

$\eta(\dot{\gamma})\dot{\gamma}$ tends to the finite value Nu (corresponding to $R = 1$). The behaviour of $\eta(\dot{\gamma})$ is given implicitly by (9.13), but it is difficult to invert this relation analytically. Its behaviour can be grossly modelled by [9.11]

$$\eta = \frac{\eta_0}{\left[1+(a\tau\dot{\gamma})^n\right]^{1/n}}, \qquad (9.15)$$

where a and n are fitting parameters. It must be noted that (9.15) does not follow from first principles; it is rather an heuristic simplification of (9.13), but it captures the essential features of the asymptotic behaviour of the pressure tensor at high and low values of the shear rate. Accordingly, expression (9.11) for the free energy may be written as

$$F(T,V,\dot{\gamma}) = F_{eq}(T,V) + \frac{\tau V}{2}\frac{\eta_0}{\left[1+(a\tau\dot{\gamma})^n\right]^{1/n}}\dot{\gamma}^2. \qquad (9.16)$$

Note that the asymptotic behaviour of F at low $\dot{\gamma}$ is of the form $\dot{\gamma}^2$, whereas for high values of $\dot{\gamma}$ it behaves like $\dot{\gamma}$; therefore, at the intermediate regime F is expected to depend on $\dot{\gamma}$ as $\dot{\gamma}^{3/2}$. A linear behaviour of F (or S) with $\dot{\gamma}$ is moreover observed in NEMD; when the calculations are extrapolated to regions with small shear rate, which are the most difficult to simulate, the possibility of a crossover to a $\dot{\gamma}^2$ dependence is not excluded by the results of NEMD.

9.4 Shear-Induced Heat Flux and the Zeroth Law

It was stressed that numerical simulations may lead to a non-equilibrium temperature dependent on the shear rate, as in EIT. In this last section, we examine the relation between this temperature and the heat flux, which is of interest in connection with temperature measurements. This was pointed out in our discussion about non-equilibrium temperature in Chap. 3. Here, we will consider it from the perspective of NEMD.

Evans and co-workers have studied the flow of a fluid in a Poiseuille configuration [9.12–14] and in a non-homogeneous shear flow produced by a sinusoidal transverse force [9.15]. The purpose was to study the conditions of thermal 'equilibrium' between non-equilibrium steady states characterized by different shear rates, in mutual thermal and mechanical contact. One interesting result is the occurrence of a heat flow between different 'subsystems' even when they are at the same kinetic (i.e. local-equilibrium) temperature. This reinforces the arguments presented in Chap. 3, in which it was claimed that the the heat flux is related to the gradient of the temperature θ rather than to the gradient of the local-equilibrium temperature T. Here, instead of a heat flux, the non-equilibrium parameter is the shear rate. The results of Evans et al. [9.12–14] confirm that layers with identical kinetic temperatures may have shear-rate dependent thermodynamic temperatures, since they exchange heat.

However, as commented in Chap. 3 and in Sect. 9.2, such thermodynamic temperature should decrease with increasing shear rates. In contrast, Evans et al. obtained a heat flux in the opposite sense, i.e. from the layers with higher shear rate to those with lower shear rate. To describe this result in a macroscopic way, they assumed an ad hoc constitutive equation for the heat flux of the form

$$q = -\lambda \nabla T - \xi \nabla \left[(\nabla v) : (\nabla v)^T \right], \tag{9.17}$$

where ξ is a coupling coefficient which describes the 'isothermal' generation of heat flux due to a velocity gradient between parallel plates located at $y = 0$ and $y = d$. For a planar Poiseuille flow with constant viscosity η, the solution for the velocity profile is

$$v_x(y) = -\frac{1}{2\eta} \frac{dp}{dx} (y - d)y, \tag{9.18}$$

and the leading-order terms in the heat equation are

$$\rho c_v \frac{dT}{dt} = \lambda \nabla^2 T + \xi \nabla^2 \dot{\gamma}^2 + \eta \dot{\gamma}^2, \tag{9.19}$$

where $\dot{\gamma} \equiv \partial v_x / \partial y$ as usual. The temperature profile in a steady state now has the form

$$T(y) = -\frac{\xi}{4\lambda\eta^2} \left(\frac{dp}{dx} \right)^2 (d - 2y)^2 - \frac{1}{192\lambda\eta} \left(\frac{dp}{dx} \right)^2 (d - 2y)^4 + \text{const.} \tag{9.20}$$

The effect of the coupling in the first term will dominate when $d - 2y < \sqrt{48\xi/\eta}$ and will change the shape of the temperature profile from quartic to parabolic in the central region of the flow.

Although the modelling (9.17) fits the results of the non-equilibrium molecular dynamics simulations, it is an ad hoc assumption and does not clarify the role of the non-equilibrium temperature. In contrast, an equation analogous to (9.17) may be obtained in the framework of EIT. We may start from the evolution equation (2.69) for the heat flux, namely

$$\tau \dot{q} + q = -\lambda \nabla \theta - \frac{a\lambda}{nk_B} \nabla \cdot \mathbf{P}^v, \tag{9.21}$$

with a being a numerical constant. Furthermore, recall that in Sect. 3.3 we wrote the pressure tensor as

$$\mathbf{P} = \pi \mathbf{U} + \mathbf{P}^v = nk_B \theta \mathbf{U} + \mathbf{P}^v. \tag{9.22}$$

For a plane shear flow, this tensor is given the form

$$\mathbf{P} = \begin{pmatrix} nk_B\theta & 0 & 0 \\ 0 & nk_B\theta & 0 \\ 0 & 0 & nk_B\theta \end{pmatrix} + \begin{pmatrix} b\dot{\gamma}^2 & -\eta\dot{\gamma} & 0 \\ -\eta\dot{\gamma} & b\dot{\gamma}^2 & 0 \\ 0 & 0 & 0 \end{pmatrix}, \tag{9.23}$$

up to the second order in the shear rate. The value of the coefficient b is given by the condition that $\mathrm{Tr}\,\mathbf{P} = 3p = 3nk_B T$, identically. This implies that

$$3nk_B\theta + 2b\dot{\gamma}^2 = 3nk_B T,$$

and therefore

$$b\dot{\gamma}^2 = \frac{3}{2}nk_B(T - \theta). \tag{9.24}$$

The expression (9.23) for $\mathbf{P}$ may be introduced into (9.21) and it follows that, in a steady state, the component of the heat flux in the y direction (perpendicular to the flow) is

$$q_y = -\lambda \frac{\partial \theta}{\partial y} - \frac{a\lambda}{nk_B} \frac{\partial}{\partial y}(b\dot{\gamma}^2). \tag{9.25}$$

This equation shows a coupling between the heat flux and the gradient of the shear rate; however, in contrast to (9.17), it exhibits explicitly the role of θ. Moreover, (9.25) does not exclude the possibility that the heat flows in the direction of $\nabla\theta$. To see that, let us eliminate b between (9.24) and (9.25) so that

$$q_y = -\lambda\frac{\partial\theta}{\partial y} - \frac{3a\lambda}{2}\frac{\partial}{\partial y}(T-\theta) = \left(\frac{3a}{2}-1\right)\lambda\frac{\partial\theta}{\partial y}, \tag{9.26}$$

where in the second equality we have assumed that T is homogeneous. It follows that for $3a > 2$, the coupling between $\mathbf{P}^v$ and q in (9.24) justifies that q_y has the same direction as the gradient of θ instead of the opposite direction.

In EIT, the presence of coupling terms comes out in a rather natural way from the general framework. Furthermore, the existence of this coupling shows that the vanishing of the non-equilibrium temperature gradient in (9.21) does not guarantee a zero heat flux. Therefore, the zeroth principle of thermodynamics must be taken with caution in non-equilibrium situations, because of the different possible couplings between the system being analysed and the system being used as a thermometer. This conclusion is not completely surprising; for instance, in classical irreversible thermodynamics, when both heat and matter may flow, it is possible to combine the temperature gradient and concentration gradient in such a way that the net heat exchange vanishes in spite of the existence of a temperature gradient. Here, the heat flux could be made to vanish by a suitable combination of the gradients of θ and $\dot{\gamma}$.

Problems

9.1 In one of the earliest versions of non-equilibrium molecular dynamics of fluids under shear, the microscopic equations of motion were derived from a so-called DOLLS Hamiltonian [9.17], defined as

$$\mathcal{H} = \sum_i \frac{\mathbf{p}_i\cdot\mathbf{p}_i}{2m} + \sum_{i,j}\phi(r_i,r_j) + \sum_i(\nabla v):\mathbf{p}_i\mathbf{r}_i,$$

where ϕ is the intermolecular potential and the last term describes the action of the flow. (a) By using the Hamiltonian algorithm, i.e.

$$\dot{r}_i = \frac{\partial\mathcal{H}}{\partial p_i}, \qquad\qquad \dot{p}_i = -\frac{\partial\mathcal{H}}{\partial r_i},$$

derive the equations of motion from $\mathcal{H}$ and compare them with (9.1). (b) Check that in contrast with the equations derived from the DOLLS Hamiltonian, (9.1) (also called the SLLOD equations of motion) [9.18] are not derivable from a Hamiltonian.

9.2 In order for the kinetic energy of the system to remain constant one must require that $\sum_i p_i \cdot \dot{p}_i = 0$. (a) Apply this condition to (9.1b) and show that

$$\alpha = \frac{\sum_j F_j \cdot p_j - (\nabla v) : \sum_j p_j p_j}{\sum_i p_i \cdot p_i}.$$

(b) Show that this expression may be rewritten in the form of (9.2).

9.3 In the tables of this chapter, the several quantities (temperature, density, shear rate, etc.) are given in dimensionless form in units of m (mass of the particles) and of the parameters ε and σ describing the strength and range of the intermolecular potential. The values of the dimensionless temperature and number density at the triple point are

$$T^* = \frac{k_B T}{\varepsilon} = 0.722, \qquad n^* = n\sigma^3 = 0.844.$$

The triple point of Ar (molecular mass = 40 g mol^{-1}) corresponds to $T_t = 83.8$ K and $p_t = 6.88 \times 10^5$ N m^{-2}, and the number density corresponding to the liquid phase is $n = 2.6 \times 10^{28}$ m^{-3}. (a) Find ε and σ for Ar. (b) Estimate the actual value of the shear rate corresponding to the dimensionless value $\dot{\gamma}^* = 1$. (c) Evaluate the relaxation time for Ar and compare it with that of a globular polymer with $\sigma = 10^{-4}$ m and molar mass 10^3 kg mol^{-1} and $T_t = 300$ K. Recall that

$$\dot{\gamma} = \gamma * \left(\frac{\varepsilon}{m\sigma^2}\right)^{1/2}, \qquad \tau = \left(\frac{m\sigma^2}{\varepsilon}\right)^{1/2}.$$

9.4 To describe fluid systems under shear flow with shear rate $\dot{\gamma}$, the following generalised free energy per unit mass is used

$$f(T,v,\dot{\gamma}) = f_{eq}(T,v) + \alpha(T,v,\dot{\gamma}),$$

$\alpha(T,v,\dot{\gamma})$ is a function given, for instance, by $\alpha = A v^{-x} T^{-z} \dot{\gamma}^{2(y+1)}$, A is a positive constant and x, y, and z are constant parameters. From the stability conditions

$$\left(\frac{\partial^2 f}{\partial T^2}\right)_{v,\dot{\gamma}} \leq 0; \quad \left(\frac{\partial^2 f}{\partial v^2}\right)_{T,\dot{\gamma}} \geq 0; \quad \left(\frac{\partial^2 f}{\partial \dot{\gamma}^2}\right)_{T,v} \geq 0,$$

$$\left(\frac{\partial^2 f}{\partial v^2}\right)\left(\frac{\partial^2 f}{\partial \dot{\gamma}^2}\right) \geq \left(\frac{\partial^2 f}{\partial v \partial \dot{\gamma}}\right)^2 ,$$

analyse how a shear rate can modify the stability of the system.

9.5 The values obtained by Evans for the internal energy u and the entropy s for density $\rho = 0.100$ in a numerical analysis of a system constituted of 32 soft discs described by a potential $\phi(r) = \varepsilon(\sigma/r)^{12}$, at different values of the shear rate $\dot{\gamma}$, are listed in Table 9.4 [9.7].

Table 9.4. Values of the non-equilibrium entropy s for different values of the internal energy u and of the shear rate $\dot{\gamma}$.

u	$s(\dot{\gamma} = 0.0)$	$s(\dot{\gamma} = 0.5)$	$s(\dot{\gamma} = 1.0)$
1.921	5.812	5.539	5.275
2.134	5.917	5.653	5.392
2.346	6.013	5.492	5.747

Estimate the mean value of the non-equilibrium temperature θ at different shear rates by using the values of the entropy presented in Table 9.4, and the definition $\theta = (\partial U / \partial S)_{\dot{\gamma}}$, and by approximating the derivative by the corresponding incremental ratio.

9.6 According to Evans [9.7], the values for the entropy of the system of 32 soft discs mentioned in the previous problem, at internal energy $u = 2.134$ and for different densities, are given in Table 9.5.

Table 9.5. Values of the non-equilibrium entropy s for different values of the density ρ and of the shear rate $\dot{\gamma}$.

ρ	$s(\dot{\gamma} = 0.0)$	$s(\dot{\gamma} = 0.5)$	$s(\dot{\gamma} = 1.0)$
0.100	5.917	5.653	5.392
0.125	5.686	5.478	5.267
0.075	6.213	5.852	5.499

Obtain the thermodynamic pressure at different shear rates by approximating the derivative $\pi\theta^{-1} = (\partial s/\partial v)_{u,\dot{\gamma}}$ by the corresponding incremental ratio.

9.7 In their analysis of the isothermal shear-induced heat flow discussed in Sect. 9.4, Baranyai et al. [9.15] proposed for the heat flux the constitutive equation (9.17), namely

$$q = -\lambda \nabla T - \xi \nabla\!\left[(\nabla v):(\nabla v)^T\right],$$

with ξ being a coupling coefficient. (a) Show that for a velocity profile of the form $v = (\dot\gamma y, 0, 0)$ the energy balance equation takes the form (9.19). (b) Given the classical expression for the entropy production, namely

$$\sigma^s_{CIT} = q \cdot \nabla T^{-1} - T^{-1}\mathbf{P}^v : (\nabla v),$$

and if the fluid is assumed to be Newtonian, discuss under which conditions expression (9.17) for the heat flux is compatible with positive entropy production.

9.8 The extended Gibbs equation for a viscoelastic fluid in the internal energy representation is

$$dU = \theta dS - \pi dV + \mu dN + \frac{\tau}{2\eta}\mathbf{P}^v : d(V\mathbf{P}^v).$$

a) Show that in a steady state

$$\left(\frac{\partial U}{\partial(V\mathbf{P}^v)}\right)_{S,V,N} = -\tau(\nabla v)^s.$$

(b) In equilibrium thermodynamics, the way to change a variable in $U(S, V, N)$ for the corresponding derivative of U with respect to this variable is a Legendre transform. For instance, when S is replaced by $T = (\partial U/\partial S)_{V,N}$, the Legendre transform is

$$F(T,V,N) = U - \frac{\partial U}{\partial S}S = U - TS.$$

Analogously to the equilibrium situation, if one wishes to use $\tau(\nabla v)^s$ as variable instead of $\mathbf{P}^v$, one should make a Legendre transform as

$$F(\theta,V,N,\tau(\nabla v)^s) \equiv U - \frac{\partial U}{\partial S}S - \frac{\partial U}{\partial(V\mathbf{P}^v)} : V\mathbf{P}^v.$$

Compare this expression with that obtained by direct substitution of $\mathbf{P}^v$ with $-2\eta(\nabla v)^s$ in the free energy $F = F_{eq} + \frac{1}{4}J V\mathbf{P}^v : \mathbf{P}^v$.

(c) Compare the form of the chemical potential derived from the Legendre transform $F(\theta, V, N, \tau(\nabla v)^s)$ and from the direct substitution of $\mathbf{P}^v$ with $-2\eta(\nabla v)^s$ in F. Note that the non-equilibrium contribution has opposite sign in the two expressions.

9.9 By following the statements of Problem 9.8, discuss the Legendre transform from Vq to its conjugate in the generalised Gibbs equation, namely $-\tau\nabla T^{-1}$, with τ being the relaxation time of the heat flux in the Maxwell–Cattaneo equation.

Part III

Selected Applications

Chapter 10

Hyperbolic Heat Conduction

There exists an impressive amount of literature on heat conduction and hyperbolic equations. It is not our purpose to examine this topic exhaustively, but to stress the more significant and illustrative aspects in connection with extended irreversible thermodynamics (EIT). Two main motivations underlie such a vast literature. One of them, of a theoretical nature, refers to the so-called 'paradox' of propagation of thermal signals with infinite speed. The second, more closely related to experimental observations, deals with the propagation of second sound, ballistic phonon propagation and phonon hydrodynamics in solids at low temperatures, where heat transport departs dramatically from the usual parabolic description.

In the classical theory of heat transport, thermal signals obey a parabolic equation. In the linear approximation, this implies that the influence of such a signal is felt instantaneously throughout the whole system, or, otherwise stated, that the thermal signal propagates at infinite velocity. One of the first motivations for EIT was precisely to remove the paradox of infinite speed of propagation.

A macroscopic theory with finite speed of propagation finds its roots in the kinetic theory and in experiments. From an experimental point of view, the search for generalising the Fourier equation was launched in the 1960s by the discovery of second sound and ballistic phonon propagation in some dielectric crystals at low temperature. This observation stimulated also the development of microscopic models of heat conduction supporting the generalised macroscopic transport equations. Such analyses are of interest in solid-state physics because they provide useful and relevant information on phonon scattering processes. However, most works on heat propagation are concerned with the dynamical consequences of the transport equations, without paying much attention to their thermodynamic implications. In Chap. 2 we have already emphasized the need to deal consistently with both dynamical and thermodynamical aspects.

10.1 The Finite Speed of Thermal Signals. Second Sound

The thermodynamical analysis of hyperbolic heat transport (described by the Maxwell–Cattaneo's equation) has been carried out in Sect. 2.1. Here we explore the physical consequences of the presence of relaxational terms in the transport equation (2.4). For simplicity, we consider a rigid solid or an incompressible perfect fluid at rest, which, according to the hypotheses of EIT, is locally characterized by the specific internal energy u per unit mass and the heat flux vector q. The energy balance equation reduces in this case to

$$\rho \frac{\partial u}{\partial t} = -\nabla \cdot q , \tag{10.1}$$

where no energy supply has been considered. The evolution equation for q is assumed to be the equation (2.4),

$$\tau \frac{\partial q}{\partial t} = -(q + \lambda \nabla \theta). \tag{10.2}$$

Consider infinitesimally small thermal disturbances around an equilibrium reference state with vanishing velocity. In this case, θ and T are identical up to second-order corrections and $\partial u / \partial t$ is given by $\partial u / \partial t = c_v \partial T / \partial t$, with c_v the heat capacity per unit mass at constant volume. The evolution equations of T and q are therefore given by:

$$\rho c_v \frac{\partial T}{\partial t} = -\nabla \cdot q \tag{10.3}$$

and

$$\tau \frac{\partial q}{\partial t} = -(q + \lambda \nabla T), \tag{10.4}$$

where for simplicity τ and λ are assumed constant. Introduction of (10.4) into (10.3), results in a hyperbolic equation of the telegrapher type, namely,

$$\tau\frac{\partial^2 T}{\partial t^2} + \frac{\partial T}{\partial t} - \chi\nabla^2 T = 0, \tag{10.5}$$

with $\chi = \lambda/\rho c_v$ being the thermal diffusivity. By assuming plane thermal waves of the form

$$T = T'\exp[i(\omega t - kx)], \tag{10.6}$$

where ω is the (real) frequency, k the (complex) wave number and T' the amplitude, the dispersion relation obtained by substituting (10.6) into (10.5) reads

$$-\tau\omega^2 + i\omega + \chi k^2 = 0. \tag{10.7}$$

It follows that the phase velocity v_p and the attenuation distance α are given by

$$v_p = \frac{\omega}{\mathrm{Re}\,k} = \frac{\sqrt{2\chi\omega}}{\sqrt{\tau\omega + \sqrt{1 + \tau^2\omega^2}}} \tag{10.8a}$$

and

$$\alpha = -\frac{1}{\mathrm{Im}\,k} = \frac{2\chi}{v_p}. \tag{10.8b}$$

At low frequencies ($\tau\omega \ll 1$), it is found that $v_p = (2\chi\omega)^{1/2}$ and $\alpha = (2\chi/\omega)^{1/2}$, which are the results predicted by the classical theory based on Fourier's law. In the high-frequency limit ($\tau\omega \gg 1$), the first-order term derivative $\partial T/\partial t$ in (10.5) is small compared with the two other terms and (10.5) becomes a wave equation whose solutions are known in the literature as second sound. The quantities v_p and α tend to the limiting values $v_{p\infty}$ and α_∞ which are given by

$$v_{p\infty} \equiv U = \sqrt{\frac{\chi}{\tau}}, \qquad \alpha_\infty = 2\sqrt{\chi\tau}, \tag{10.9}$$

where $v_{p\infty}$ can be identified with the speed of propagation of thermal pulses as shown in the next section. The velocity U is usually called the velocity of second sound. It is interesting to note that within the limit $\tau = 0$, which corresponds to the Fourier law, one has $v_{p\infty} = \infty$ and $\alpha_\infty = 0$. Note that when τ and λ diverge but τ/λ remains finite, as in the case of superfluids and solids at low temperature, (10.5) reduces to a wave equation. According to kinetic theory, it is found that for solids in the Debye approximation $\chi/\tau =$

$c_0^2/3$, with c_0 the phonon velocity, and for monatomic gases $\chi/\tau = \frac{5}{3}k_B T/m$. As a consequence, the second sound speed in a solid and a monatomic gas are given respectively by

$$U(\text{solid}) = \frac{1}{\sqrt{3}}c_0 , \qquad U(\text{monatomic gas}) = \left(\frac{5}{3}\frac{k_B T}{m}\right)^{1/2} . \qquad (10.10)$$

The problem of propagation of thermal signals was first dealt with by Cattaneo [10.1] and Vernotte [10.2] (see also [2.12]), who based their analysis on kinetic theory arguments and added a relaxation term to the Fourier equation. This point of view has been used in several contexts, such as in the analysis of waves in thermoelastic media, fast explosions, and second sound in solids. Wide-ranging bibliography can be found in [2.10] and [10.3–5]. However, all is not well with the Maxwell–Cattaneo equation, which is insufficient in several aspects: it does not describe accurately the broadening of the second-sound signals, it does not incorporate the phenomenon of ballistic phonon propagation (i.e. the propagation of phonons without any collision) nor the dependence of the second sound on temperature, and it does not deal with phonon hydrodynamics. A more general equation is therefore proposed and discussed in Sect. 10.3.

It should also be noticed that the telegrapher equation (10.5) does not conserve the positive character of the solutions, i.e. it may yield negative values for T even when the initial profile for T is positive everywhere. This unpleasant feature may be avoided by imposing not only an initial positive profile for T but also a restriction on the heat flux, namely, that $q \leq \rho u U$ (see Problem 10.8), which is one of the requirements of thermodynamic stability obtained in Chap. 3 (see also Problem 3.5). This emphasizes once more the necessity of using a generalised entropy.

It should be observed that it is not necessary to introduce relaxation terms into the heat equation to obtain a finite velocity for the propagation of thermal signals. This feature is also predicted by taking a non-linear diffusion equation with a thermal conductivity depending on the temperature, for instance $\lambda \approx T^n$ with $n > 0$ [10.6]. However, experiments show that, for fluids or metals at intermediate and high temperatures, this dependence is not realistic.

Rather than the conceptual problem of infinite speed of propagation, a more interesting situation from a practical point of view is the study of second sound in dielectric solids at low temperature. Bibliographical reviews on this topic can be found for instance in [10.3–4]. The presence of second sound in solids, suggested by Peshkov in 1947 and by Wards and Wilks in 1951, was analysed in terms of the Maxwell–Cattaneo equation by Chester [10.7], who obtained for the second sound the result (10.10).

Theoretical interpretations from a microscopic point of view were proposed by Guyer and Krumhansl [10.8], Ackerman and Guyer [10.3], and Mikhail and Simons [10.9].

To end this section with some explicit data, let us mention that some measurements of the speed of second sound have been carried out on NaF and Bi samples [10.10]. These measurements allow one to determine the value of τ from the first of relations (10.9). It was shown by Coleman and Newman [10.11] that the temperature dependence of the measured speed is well fitted by an empirical law of the form

$$U(T)^{-2} = A + BT^n. \tag{10.11}$$

For NaF, second sound is observed in the temperature range $10.0 < T < 18.5$ K, and the values for n, A, and B are [10.11b] $n = 3.1$, $A = 9.09 \times 10^{-12}$ s^2/cm^2, and $B = 2.22 \times 10^{-15}$ s^2/cm^{2}K^n, U being given in cm/s. In Bi crystals, second sound was observed for $1.4 < T < 4.0$ K, and n, A, and B are $n = 3.75$, $A = 9.07 \times 10^{-11}$ s^2/cm^2, and $B = 7.58 \times 10^{-13}$ s^2/cm^{2}K^n. An order of magnitude for U is then 21×10^4 cm/s for NaF at 17.5 K and of 7.8×10^4 cm/s for Bi at 3.4 K. An extensive analysis of heat pulse experiments and their thermodynamic interpretation can be found in Dreyer and Struchtrup review [10.12].

10.2 Heat Pulses

The response of a system to a perturbation having the form of an energy pulse is of practical interest. Such a situation is met when a material sample is suddenly heated at one of its boundaries for a short time, by means of a laser pulse, an electrical discharge, or a fast exothermic chemical reaction (see Fig. 10.1). This is typical, for instance, in very short (picosecond) irradiation of small samples in semiconductor processing, laser-induced nuclear fusion, catalysis of fast reactions, explosions, or the collapse of supernovae [10.14]. The energy transfer in such fast processes cannot be described with Fourier's law and requires a more involved description.

To study the response of the system subject to a general perturbation due to an energy supply term $g(r,t)$, we rewrite the energy balance equation (10.3) as

$$\rho c_v \frac{\partial T}{\partial t} = -\nabla \cdot \boldsymbol{q} + g(r,t). \tag{10.12}$$

Here, $g(r,t)$ could be, for instance, the power supplied by a laser pulse per unit of mass of the sample. When (10.12) is combined with (10.4), one obtains instead of (10.5)

$$\tau\frac{\partial^2 T}{\partial t^2} + \frac{\partial T}{\partial t} - \chi\nabla^2 T = \frac{1}{\rho c_v}\left[g + \tau\frac{\partial g}{\partial t}\right].$$

(10.13)

It is important to note that the time derivative of the supply term appears on the right-hand side of (10.13), and that such a term has no counterpart in the classical theory of heat.

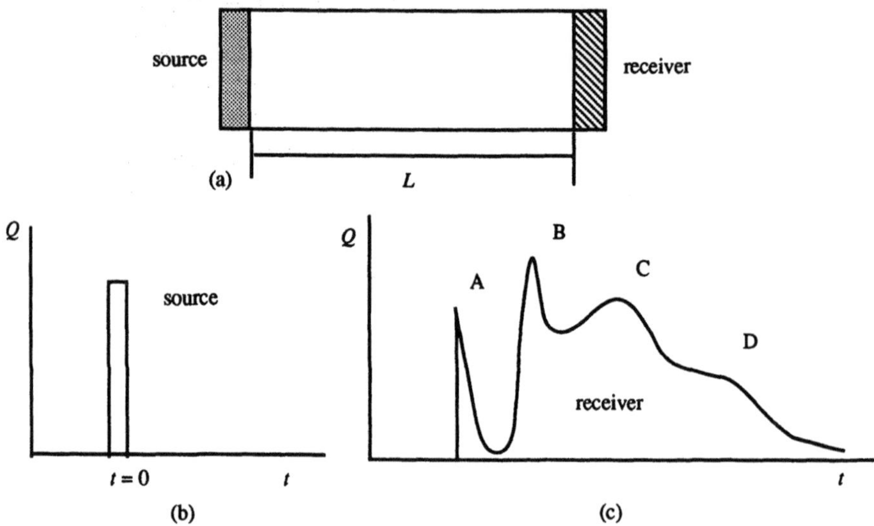

Fig. 10.1. (a) Schematic representation of heat pulse experiments. (b) Heat input at the source. (c) Heat output at the receiver. The signal at the receiver is composed of: ballistic phonons with the longitudinal sound speed (peak A); ballistic phonons with transverse sound speed (peak B); second sound (peak C) and a tail (D) due to the diffusive heat propagation. A, B, C, and D can be described within the extended irreversible thermodynamic formalism whereas only D is accounted for by Fourier's law.

The reader not interested in the mathematical method of resolution may skip straight to (10.31). A useful method for determining the temperature field is to introduce the Green function $G(r,t;r_0,t_0)$, which is the solution of

$$\tau\frac{\partial^2 G}{\partial t^2} + \frac{\partial G}{\partial t} - \chi\nabla^2 G = \delta(r - r_0)\delta(t - t_0).$$

(10.14)

The temperature shift ΔT with respect to its reference value T is given in terms of G by

$$\Delta T(r,t) = \int dt_0 \int dr_0\, G(r,t;r_0,t_0)\frac{1}{\rho c_v}\left[g(r_0,t_0) + \tau\frac{\partial g(r_0,t_0)}{\partial t_0}\right].$$

(10.15)

We first proceed to calculate the Green function. In one dimension (solutions for two and three dimensions may be found in [10.13a]), G only depends on $x' = x - x_0$, and $t' = t - t_0$. In what follows, we drop the primes and simply write $G(x,t)$, which we express as

$$G(x,t) = \int_{-\infty}^{\infty} dk\, e^{ikx}\, \hat{G}(k,t), \tag{10.16}$$

where $\hat{G}(k,t)$ denotes the Fourier transform of $G(x,t)$. Introduction of (10.16) into (10.14) yields

$$\tau \frac{d^2\hat{G}}{dt^2} + \frac{d\hat{G}}{dt} + \chi k^2 \hat{G} = \frac{1}{2\pi}\delta(t). \tag{10.17}$$

To calculate $\hat{G}$ we first consider the solutions of the homogeneous part of (10.17). As in the previous section, they are given by $\exp[-i\omega_+ t]$ and $\exp[-i\omega_- t]$, with ω_+ and ω_- the solutions of the dispersion equation (10.7), i.e.

$$\omega_\pm = \frac{-i \pm \sqrt{4\tau^2 U^2 k^2 - 1}}{2\tau}, \tag{10.18}$$

with U given in (10.9). The solution of the inhomogeneous equation (10.17) satisfying the condition $G(k,t) = 0$ for $t < 0$ is

$$\hat{G}(k,t) = \frac{iU^2}{2\pi\chi} \frac{\exp(-i\omega_+ t) - \exp(-i\omega_- t)}{\omega_+ - \omega_-} H(t). \tag{10.19}$$

Here $H(t)$ is the Heaviside unit step function, defined as $H(t) = 0$ for $t < 0$ and $H(t) = 1$ for $t > 0$. Recall that $dH(t)/dt = \delta(t)$. In virtue of (10.16) one has

$$G(x,t) = \frac{iU^2}{2\pi\chi} H(t) \int_{-\infty}^{\infty} dk \frac{\exp(-i\omega_+ t) - \exp(-i\omega_- t)}{\omega_+ - \omega_-} \exp(ikx). \tag{10.20}$$

Let us first consider the integral involving $\exp(-i\omega_+ t)$:

$$\frac{1}{2U}\exp[-(t/2\tau)] \int_c dk \frac{\exp\left[i\left(kx - \sqrt{k^2 - (2\tau U)^{-2}}\, Ut\right)\right]}{\sqrt{k^2 - (2\tau U)^{-2}}}. \tag{10.21}$$

As contour of integration c take a line in the upper half-plane of k parallel to the real axis of k [10.13]. When $r > Ut$, the contour may be closed by a semicircle in the upper half-plane and the integral is zero, since there are no singularities in this region. When $r < Ut$, the contour extends along the negative imaginary axis. Taking into account that

$$\frac{1}{2\pi}\int_{-\infty+i\varepsilon}^{\infty+i\varepsilon} d\omega \frac{\exp\left[-b\sqrt{a^2-\omega^2}\right]}{\sqrt{a^2-\omega^2}} = J_0\left[a\sqrt{t^2-b^2}\right]H(t-b), \qquad (10.22)$$

with $J_0(x)$ the Bessel function of order zero, one obtains for (10.21)

$$-\frac{\pi i}{U}\exp[-t/2\tau]J_0[i\xi]H(Ut-x), \qquad (10.23)$$

where

$$\xi = \frac{1}{2}\sqrt{\left(\frac{t}{\tau}\right)^2-\left(\frac{r}{U\tau}\right)^2}. \qquad (10.24)$$

The contribution from the integral involving $\exp(-i\omega\_t)$ is

$$-\frac{1}{2U}\exp[-(t/2\tau)]\int_c dk \frac{\exp\left[i\left(kx+\sqrt{k^2-(2\tau U)^{-2}}\,Ut\right)\right]}{\sqrt{k^2-(2\tau U)^{-2}}} \qquad (10.25)$$

and gives

$$-\frac{\pi i}{U}\exp[-t/2\tau]J_0[i\xi]\{1-H(Ut+x)\}. \qquad (10.26)$$

Introduction of (10.23) and (10.26) into (10.20) leads to

$$G(x,t) = \frac{U}{2\chi}\exp[-t/2\tau]I_0[\xi]H(Ut-|x|), \qquad (10.27)$$

where I_0 is the modified Bessel function of the first kind defined by $I_p(\xi) = -(-i)^p J_p(i\xi)$. Since its asymptotic expression for large ξ is

$$I_p(\xi) \approx (2\pi\xi)^{-1/2}e^\xi, \qquad (10.28)$$

it is easily seen that when $\tau \to 0$ (i.e. when U diverges), (10.27) tends to

$$G(x,t) = \frac{1}{2\sqrt{\pi \chi t}} \exp\left(-\frac{x^2}{4\chi t}\right), \tag{10.29}$$

which is the well-known solution of the diffusion equation.

Assume now that the heat supply is a concentrated energy pulse of the form

$$g(x_0,t_0) = g_0\, \delta(x_0)\delta(t_0), \tag{10.30}$$

with $g_0 = \int g(x,t)dx$, so that g_0 represents the total energy supplied to the system. After introducing (10.30) into (10.15), together with (10.27) and the result $dI_0(\xi)/d\xi = I_1(\xi)$, one obtains for the temperature response to a heat pulse

$$\Delta T(x,t) = \frac{g_0}{4\lambda}\exp\left(-\frac{t}{2\tau}\right)\Big\{2I_0(\xi)U\tau\delta(Ut-|x|)$$

$$+\left[I_0(\xi)+\frac{t}{\tau\xi}I_1(\xi)\right]H(Ut-|x|)\Big\}. \tag{10.31}$$

Similarly, the heat flux could be obtained by integration of the Maxwell–Cattaneo equation once the temperature profile is known. It turns out that [10.13b]

$$q(x,t) = \frac{g_0}{4\tau}\exp\left(-\frac{t}{2\tau}\right)\Big\{2I_0(\xi)U\tau\delta(Ut-|x|)+\frac{r}{2U\tau\xi}I_1(\xi)H(Ut-|x|)\Big\}. \tag{10.32}$$

Solution (10.31) describes the response to a heat pulse. The main differences with the classical diffusion equation whose solution is given by

$$\Delta T = \frac{g_0}{2\sqrt{\pi \chi t}}\exp\left(-\frac{x^2}{4\chi t}\right) \tag{10.33}$$

are summarized below.

1. In the classical description, the response to a heat pulse is instantaneously felt in all space, because $\Delta T(x,t)$ is non-zero at any position in space for $t > 0$. In contrast, $\Delta T(x,t)$, given by (10.31), vanishes for $x > Ut$, so that a perturbation front propagates with speed $U = (\chi/\tau)^{1/2}$, while the positions beyond the front are unaltered by the action of the pulse.

2. In (10.31) there are two main terms: the first, related to $\delta(Ut - x)$, is a reproduction of the initial pulse damped by the exponential factor and propagating with speed U. The second term, related to $H(Ut - x)$, corresponds to the wake, which, for sufficiently long

times $Ut \gg x$, tends to the usual diffusive solution, as may be seen from (10.29). Such a structure is observed in the propagation of heat pulses in solids at low temperatures and in the structure of the X-ray peaks coming from X-ray bursters in astrophysics.

3. The description based on a generalised hyperbolic equation predicts the propagation of a front peak which is absent in the parabolic model equation, meaning that the latter underestimates the high temperature peak which may be present in the wave front. Although both situations are characterized by the same total energy, the distribution of thermal energy is very different: in the parabolic case it is distributed in the whole spatial region, whereas there is a sharp concentration of energy in the hyperbolic case (see Fig. 10.2).

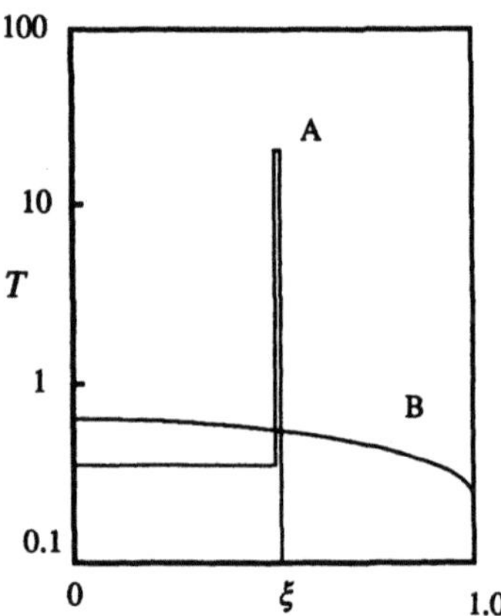

Fig. 10.2. Schematic comparison of hyperbolic and parabolic heat transport: response to a thermal pulse with finite width at short times. A: hyperbolic solution; B: parabolic solution. The position is expressed in terms of ξ (see (10.24)). The graph represents the temperature distribution at time $t = 0.5\tau$.

10.3 Beyond the Maxwell–Cattaneo Equation

A more sophisticated model than the Maxwell–Cattaneo equation was proposed by Guyer and Krumhansl [10.8], who based their description of heat transport on phonon kinetic theory. Accordingly, the phonons undergo two types of collisions, either R or N collisions. R collisions are resistive collisions of phonons with defects of the lattice and (or) the boundaries of the crystal and so-called Umklapp phonon–phonon processes; R collisions conserve energy but not momentum and have a characteristic time τ_R. The second type of scattering is referred to as normal N processes; they do not only conserve energy but also momentum, and their characteristic time will be denoted by τ_N. Fourier's law is only valid at low frequencies, or in terms of relaxation times, when $1/\tau_R$ tends to infinity, which corresponds to a very high frequency of R collisions. Cattaneo's equation is well suited to describe heat transport processes in the range $1/\tau_N \to \infty$ and $1/\tau_R \to 0$, i.e. when the frequency of N phonon–phonon collisions is large compared to that of R collisions, or $\tau_N \ll \tau_R$. However, experimental results concerning second-sound propagation suggest that in the temperature range $(0 < T < 20\ \mathrm{K})$ where second sound is observed in crystals, the frequency of normal and resistive processes may be of the same order and it is even not excluded that situations where $\tau_N \gg \tau_R$ may be found. In these regimes, it was soon recognised that non-local effects play an important role. Since non-local terms were introduced in Chap. 2, we shall take advantage of this opportunity to explore their physical meaning [10.15]. The characteristic phenomena in the regime dominated by non-local terms are referred to as 'phonon hydrodynamics', because of the similarities between the equations for the heat fluxes and the equations of hydrodynamics.

10.3.1 Generalised Maxwell–Cattaneo Models

In Sect. 5.7, the role of the higher-order fluxes was emphasized. Here we take into account the flux of the heat flux, described by a second-order tensor $\mathbf{Q}$, and write instead of the Maxwell–Cattaneo equation (10.2) the following expression:

$$\tau_1 \frac{\partial \boldsymbol{q}}{\partial t} = -(\boldsymbol{q} + \lambda \nabla T) + \nabla \cdot \mathbf{Q}. \tag{10.34}$$

The tensor $\mathbf{Q}$, assumed to be symmetric, may be split in the usual form $\mathbf{Q} = Q\mathbf{U} + \mathbf{Q}^s$, the scalar Q being one-third of the trace of $\mathbf{Q}$ and $\mathbf{Q}^s$ the deviatoric part. According to the results of Sect. 5.7 [see (5.108)] the evolution equations for Q and $\mathbf{Q}^s$ may be written as

$$\tau_0 \dot{Q} = -Q + \beta' \nabla \cdot \boldsymbol{q}, \tag{10.35}$$

$$\tau_2 (\mathbf{Q}^s)^{\cdot} = -\mathbf{Q} + 2\beta'' (\overset{0}{\nabla \boldsymbol{q}})^s \tag{10.36}$$

Note that when the relaxation times are neglected, these expressions parallel the Newton–Stokes equations, with $\boldsymbol{q}$ playing the role of the fluid velocity. This is characteristic of the hydrodynamic regime of phonon flow, which is indeed described by these equations.

The thermodynamics underlying the model (10.34–36) is easily derived by following the same procedure as in Sect. 2.3. It is left as an exercise [10.15] to show that the following results are compatible with (10.34–36) and the positiveness of entropy production, the latter being given by

$$\sigma^s = \frac{1}{\lambda T^2}\boldsymbol{q}\cdot\boldsymbol{q} + \frac{1}{2\lambda T\beta''}\mathbf{Q}^s : \mathbf{Q}^s + \frac{1}{\lambda T\beta'}Q^2 \geq 0. \tag{10.37}$$

It follows that $\lambda > 0$, $\beta'' > 0$ and $\beta' > 0$. The corresponding Gibbs equation and entropy flux have the forms

$$\mathrm{d}s = \frac{1}{\theta}\mathrm{d}u - \frac{\tau_1 v}{\lambda T^2}\boldsymbol{q}\cdot\mathrm{d}\boldsymbol{q} - \frac{\tau_2 v}{2\lambda T^2\beta''}\mathbf{Q}^s : \mathrm{d}\mathbf{Q}^s - \frac{\tau_0 v}{\lambda T^2\beta'}Q\mathrm{d}Q, \tag{10.38}$$

and

$$\boldsymbol{J}^s = \frac{1}{\theta}\boldsymbol{q} + \frac{1}{\lambda T^2}\mathbf{Q}^s\cdot\boldsymbol{q} + \frac{1}{\lambda T^2}Q\boldsymbol{q}. \tag{10.39}$$

As a consequence of the convexity property, the relaxation times are positive.

Assuming that the relaxation times τ_0 and τ_2 are negligibly small, and substituting (10.35–36) into (10.34), one obtains for the evolution of the heat flux, in the linear approximation,

$$\tau_1\frac{\partial\boldsymbol{q}}{\partial t} = -(\boldsymbol{q} + \lambda\nabla T) + \beta''\nabla^2\boldsymbol{q} + \left(\beta' + \tfrac{1}{3}\beta''\right)\nabla(\nabla\cdot\boldsymbol{q}). \tag{10.40}$$

Expression (10.40) may be compared with the results of Guyer and Krumhansl [10.8]. Starting from a linearized Boltzmann equation in the Callaway approximation in three dimensions, these authors established the following equation for the heat flux at low temperatures:

$$\frac{\partial\boldsymbol{q}}{\partial t} + \frac{1}{\tau_R}\boldsymbol{q} + \frac{1}{3}\rho c_v c_0^2\nabla T = \frac{1}{5}c_0^2\tau_N\left[\nabla^2\boldsymbol{q} + 2\nabla(\nabla\cdot\boldsymbol{q})\right], \tag{10.41}$$

where c_v is the heat capacity per unit mass at constant volume and c_0 is the Debye velocity. Comparison between (10.40) and (10.41) yields the following identifications:

$$\tau_1 = \tau_R, \quad \frac{\lambda}{\tau_R} = \frac{1}{3}\rho c_v c_0^2, \quad \beta'' = \frac{1}{5}c_0^2\tau_N\tau_R, \quad \beta' = \frac{1}{3}c_0^2\tau_N\tau_R. \tag{10.42}$$

Thus, the relaxation time τ_1 of the heat flux may be identified as the characteristic time of the resistive collisions τ_R. The second expression in (10.42) for λ/τ_R is a well-known result and is frequently used to determine the value of the relaxation time.

To examine the conditions under which second sound may be observed, we introduce (10.36) into (10.34) and combine it with the energy balance equation. Writing the final result in Fourier space yields the following dispersion relation for longitudinal thermal waves:

$$\rho c_v i\omega = \frac{-\lambda k^2}{1 + i\omega\tau_R + \dfrac{\left(\frac{4}{3}\beta'' + \beta'\right)k^2}{1 + i\omega\tau_2}}, \tag{10.43}$$

with $\tau_0 = \tau_2$. The dispersion relation for transverse waves is analogous to (10.43) but with β'' instead of $\frac{4}{3}\beta'' + \beta'$. For $\omega\tau_R \gg 1 \gg \omega\tau_2$ and ignoring non-local effects, (10.43) leads to the phase velocity

$$v_p = \left(\chi/\tau_R\right)^{1/2}. \tag{10.44}$$

When (10.44) is combined with the second equation in (10.42) one recovers the result $v_p = c_0/\sqrt{3}$, c_0 being the phonon speed. This is the regime usually referred to as second sound, and it may be attained in a low-temperature regime when τ_2 and τ_R are of different orders of magnitude. At higher frequencies, when both $\omega\tau_2 \gg 1$ and $\omega\tau_R \gg 1$, the phase velocity obtained from (10.43) is

$$v_p = \left(\frac{\chi}{\tau_R} + \frac{l^2}{\tau_R\tau_2}\right)^{1/2}, \tag{10.45}$$

with $l^2 = \frac{4}{3}\beta'' + \beta'$ for longitudinal waves and $l^2 = \beta''$ for transverse waves. This regime is typical of ballistic propagation of phonons, i.e. a flow in which phonons travel through the whole crystal without suffering any collision. Whereas the speed of the second sound is $\left(\frac{1}{3}\right)^{1/2} c_0$, the velocity of ballistic phonons should be equal to c_0. Indeed, (10.45) combined with (10.42) yields the speed $\left(\frac{14}{15}\right)^{1/2} c_0$ for longitudinal ballistic phonons if it is assumed that $\tau_2 = \tau_N$; and $\left(\frac{8}{15}\right)^{1/2} c_0$ for transverse ballistic phonons [10.12]. These results may be improved by introducing a higher number of fluxes: Dreyer and Struchtrup [10.12] show that $v_p \approx c_0$ when more than 30 moments are used; alternatively, one may use an infinite number of moments, define an effective relaxation time τ_2 and follow the procedure proposed in Sect. 5.7.

The non-local effects not only modify the speed of propagation with respect to that predicted by the Maxwell–Cattaneo equation (10.2), but they yield a rounding up and a widening of the pulse profile, which instead of the vertical parallelogram shown in Fig. 10.1 takes a Gaussian profile which becomes wider as a function of time.

10.3.2 Phonon Hydrodynamics

Besides second sound and ballistic propagation of phonons, phonon hydrodynamics is useful to gain information about the physics of phonon motion. Poiseuille flow of phonons, first detected in solid He^4, may be observed in a cylindrical heat conductor of radius R when the mean free paths $\ell_N = c_0\tau_N$ and $\ell_R = c_0\tau_R$ satisfy $\ell_N\ell_R >> R^2$ and $\ell_N <<$ R. In this case, (10.41) reduces in a steady state ($\partial q/\partial t = 0$, $\nabla \cdot q = 0$) to

$$\nabla T - \frac{9}{5}\frac{\tau_N}{\rho c_v}\nabla^2 q = 0. \tag{10.46}$$

This form is similar to the equation describing Poiseuille flow of Newtonian fluids provided that q, ∇T, and $\frac{9}{5}(\tau_N/\rho c_v)$ are regarded as the velocity v, the pressure gradient ∇p, and the viscosity η, respectively. In cylindrical coordinates, (10.46) reads, for an imposed temperature gradient, and defining q as the axial component of q,

$$\frac{1}{r}\frac{d}{dr}\left(r\frac{dq}{dr}\right) = \frac{5}{3}\frac{\rho c_v}{\tau_N}\nabla T, \tag{10.47}$$

so that

$$q(r) = -\frac{5}{12}\frac{\rho c_v}{\tau_N}\left(R^2 - r^2\right)\nabla T, \tag{10.48}$$

and the total heat flux across the cylinder is

$$Q = 2\pi\int_0^R dr\,r\,q(r) = -\frac{5\pi}{24}\frac{\rho c_v}{\tau_N}R^4\nabla T. \tag{10.49}$$

In contrast to the Fourier regime, where the total rate of heat transfer across the cylinder for a given temperature gradient is proportional to R^2, in the Poiseuille phonon flow it is proportional to R^4. This difference allows one to distinguish which of the two regimes is actually satisfied. By defining an apparent thermal conductivity as

$$\lambda_{ap} = -\frac{Q}{\pi R^2 \nabla T}, \tag{10.50}$$

it turns out from (10.49) and (10.42) that

$$\lambda_{ap} = \frac{5}{24} \frac{\rho c_v}{\tau_N} R^2. \tag{10.51}$$

This result is important as it provides a way to evaluate the normal collision times τ_N, by measuring λ_{ap}, while according to the third of expressions (10.42), the resistive collision time τ_R will be determined by the measurement of the thermal conductivity.

10.3.3 Two-Temperature Models

Up to now, we have considered relaxational effects described by one single temperature. However, many systems consist of several subsystems, each of which being assigned its own temperature. These subsystems are, for instance, the electrons and the lattice in a metal submitted to a short-pulse laser heating, where the electron temperature is much higher than the lattice temperature for a short time. This situation may be described by the following evolution equations for the electron and lattice temperatures T_e and T_l respectively [10.5a]:

$$c_e \frac{\partial T_e}{\partial t} = \nabla \cdot (\lambda \nabla T_e) - C(T_e - T_l), \tag{10.52}$$

$$c_l \frac{\partial T_l}{\partial t} = C(T_e - T_l). \tag{10.53}$$

The constant C describes the electron–phonon coupling, which accounts for the energy transfer from the electrons to the lattice, and c_e and c_l are the specific heats of the electrons and lattice per unit volume respectively. When the solution of equation (10.53), namely, $T_e = T_l + (c_l/C)\partial T_l/\partial t$ is introduced into (10.52), one is led to

$$\nabla^2 T_l + \frac{c_l}{C} \frac{\partial \nabla^2 T_l}{\partial t} = \frac{c_l + c_e}{\lambda} \frac{\partial T_l}{\partial t} + \frac{c_e c_l}{\lambda C} \frac{\partial^2 T_l}{\partial t^2}. \tag{10.54}$$

Such an equation can also be obtained by eliminating q between the energy balance equation and Guyer–Krumhansl equation under the following identifications: $3\tau_N/5 = c_l/C$, $3/\tau_R c^2 = (c_l + c_e)/\lambda$, $3/c^2 = c_e c_l/\lambda C$.

It is still interesting to note that an equation of the form of (10.54) may be derived from the following constitutive equation, known as the dual-phase-lag equation [10.5]

$$\tau_1 \dot{q} + q = -\lambda \left(\nabla T + \tau' \frac{\partial \nabla T}{\partial t} \right), \tag{10.55}$$

and which takes into account relaxation effects in both the heat flux and the temperature gradient. However, such a procedure yields a parabolic equation, while by starting with the heat flux and the flux of the heat flux as independent variables one is led to a hyperbolic equation which contains the parabolic equation (10.54) in the limit of a vanishing relaxation time of the flux of the heat flux.

Other systems where a two-temperature description is convenient are, for instance, heterogeneous systems where the liquid and solid phases are at different temperatures; polyatomic gases, wherein one can ascribe different temperatures for translational and internal degrees of freedom; and liquid helium II, where different temperatures may be assigned to the normal and superconductor fluids. These situations, and the derivation of the suitable generalised equations for heat transfer, have been thoroughly reviewed by Tzou within the framework of the dual-phase-lag formalism [10.5a].

10.4 Second Sound Under a Heat Flow

The experiments on second sound in solids at low temperature mentioned in Sect. 10.3 were performed by perturbing samples initially at equilibrium. The non-classical term in the generalised temperature expression (3.14) may have a non-negligible influence on the speed of waves in systems in non-equilibrium steady states and deserve special attention [10.11].

The evolution of a thermal perturbation is described by the energy balance equation (10.1) and by the evolution equation (10.2) for the heat flux. After taking into account the form of (3.14) for θ, (10.2) may be written, up to second order in q, as follows:

$$\tau \frac{\partial q}{\partial t} = -q - \lambda \nabla T + \frac{1}{2} \lambda T^2 \nabla \left(\frac{\partial \alpha}{\partial u} q \cdot q \right), \tag{10.56}$$

with $\alpha = \tau v / \lambda T^2$. Linearizing (10.56) around the reference state characterized by the heat flux q_0, and combining it with (10.1), one is led to the following equation for the propagation of thermal perturbations in the one-dimensional case up to the first order in q_0

$$\rho\frac{\partial^2 T}{\partial t^2} + \frac{\rho\lambda T^2}{\tau}q_0\frac{\partial\alpha}{\partial u}\frac{\partial^2 T}{\partial x\partial t} - \frac{\lambda}{\tau c_v}\frac{\partial^2 T}{\partial x^2} + \frac{\rho}{\tau}\frac{\partial T}{\partial t} = 0. \tag{10.57}$$

It follows from (10.57) that the speed of propagation of thermal pulses, or thermal waves in the high-frequency limit, is not the same in all directions (see Fig. 10.3). The speed v_+ in the same direction of the heat flux q_0 and the speed v_- in the opposite direction are found to be

$$v_+ = U\left(\sqrt{\phi^2+1} - \phi\right), \qquad v_- = U\left(\sqrt{\phi^2+1} + \phi\right), \tag{10.58}$$

with U given by (10.9) and $\phi = -(\lambda T^2/2\tau U)(\partial\alpha/\partial u)q_0$. The difference in both speeds v_- and v_+, up to the first order in q_0, is given by

$$\Delta v = v_- - v_+ = 2U\phi = -\frac{\lambda T^2}{\tau}\left(\frac{\partial\alpha}{\partial u}\right)q_0. \tag{10.59}$$

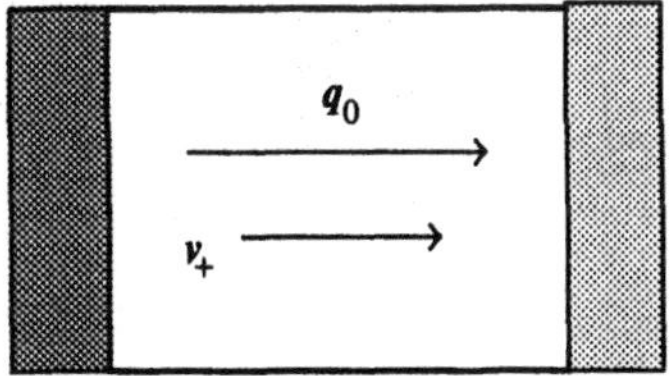

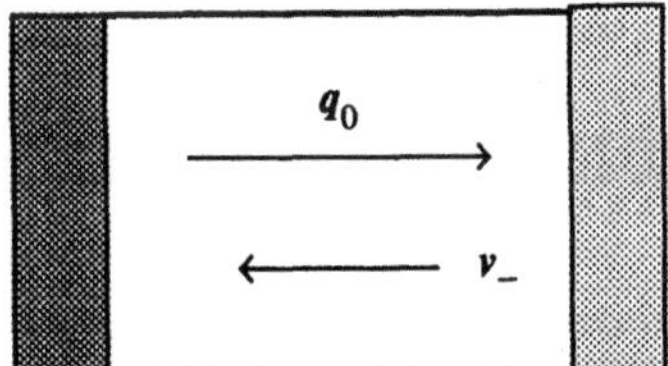

Fig. 10.3. The dotted and the dark parts of the figure indicate respectively the source and the receiver of the heat pulse. The latter is usually a superconducting bolometer and the former a fast heater. The system is out of equilibrium under a mean heat flux q_0. In the figure at the left, the pulse is in the same direction as the main heat flux, while in the figure at the right it is in the opposite direction.

Provided that $\partial\alpha/\partial u < 0$, a small heat pulse propagating in a heat conductor will travel more slowly in the direction of the heat flow than in the opposite direction. Inequality $\partial\alpha/\partial u < 0$ is commonly satisfied in a phonon gas in d dimensions, for which one has [5.17]

$$-\frac{\partial\alpha}{\partial u} = (d+2)\frac{\tau}{\rho\lambda c_v T^3}. \tag{10.60}$$

From (10.59) and (10.60), one obtains for the ratio $\Delta v/U$ in d dimensions

$$\frac{\Delta v}{U} = -(d+2)c_0\tau\nabla\ln T, \qquad (10.61)$$

where τ is the mean time between resistive collisions. For NaF near 18 K, τ and c_0 are $\tau = 7 \times 10^{-8}$ s and $c_0 = 3 \times 10^5$ cm/s [10.4]. A reasonable value for the temperature gradient can be obtained by taking a sample 5 mm in length the sides of which are kept at 10 K and 5 K respectively, so that $\nabla\ln T = 0.8$ cm^{-1}. In these conditions, the ratio $\Delta v/U$ is of the order of 7%, which is not negligible and is susceptible to being tested experimentally.

The speed of thermal pulses through systems in equilibrium states provides information about the ratio $\lambda/c_v\rho\tau$. Since c_v, ρ, and λ are well known and tabulated, one can determine τ. As a consequence, $\alpha = \tau/\rho\lambda T^2$ may be calculated at several temperatures and according to (10.59) this suffices to make predictions about the difference of speeds of thermal pulses under a heat flux. Comparison with experimental data would provide interesting informations on the non-classical terms of the generalised temperature. Note that by using the Guyer–Krumhansl (10.41) rather than the Maxwell–Cattaneo equation (10.2), additional contributions to the difference of speed waves will arise (see Problem 10.11).

10.5 Heat Conduction in a Rotating Rigid Cylinder

Let us now briefly come back to the transformation of time derivatives examined in Chap. 2 (in particular see (2.46)). To this end, consider a rigid cylinder rotating around its axis of symmetry, which is held at a constant temperature T_1, higher than the external temperature T_0. With the transformation (2.46) in mind, the Maxwell–Cattaneo equation of q reads as

$$\tau\dot{q} = -(q + \lambda\nabla T) - \tau a\mathbf{W}\cdot q, \qquad (10.62)$$

where $\mathbf{W}$ is the antisymmetric part of the velocity gradient. For a rigid rotation with angular velocity $\boldsymbol{\omega}$ around the z axis, $v = \boldsymbol{\omega}\times r = -\omega y\mathbf{i} + \omega x\mathbf{j}$, and thus $\mathbf{W}$ takes the explicit form

$$\mathbf{W} = \begin{pmatrix} 0 & \omega & 0 \\ -\omega & 0 & 0 \\ 0 & 0 & 0 \end{pmatrix},$$

and accordingly

$$\mathbf{W}\cdot\boldsymbol{q} = -\omega\times\boldsymbol{q}. \qquad (10.63)$$

In the non-inertial cylindrical reference frame rotating with the cylinder, (10.62) becomes

$$\tau\left(\frac{\partial\boldsymbol{q}}{\partial t} + \boldsymbol{v}\cdot\nabla\boldsymbol{q} + \omega\times\boldsymbol{q}\right) = -(\boldsymbol{q}+\lambda\nabla T) + \tau a\,\omega\times\boldsymbol{q}, \qquad (10.64)$$

where $\partial\boldsymbol{q}/\partial t$ is the rate of change of $\boldsymbol{q}$ as measured by an observer at rest with respect to the rotating cylinder. Since the body is rigid, its radial velocity is zero; furthermore, because of the axial symmetry of the problem, the components q_r and q_θ of the heat flux are independent of the azimuthal angle θ, so that the term $\boldsymbol{v}\cdot\nabla\boldsymbol{q}$ vanishes while (10.64) simplifies as

$$\tau\frac{\partial\boldsymbol{q}}{\partial t} = -\boldsymbol{q} - \lambda\nabla T + \tau(a-1)\,\omega\times\boldsymbol{q}, \qquad (10.65)$$

and in the steady state, for which $\partial\boldsymbol{q}/\partial t = 0$,

$$\boldsymbol{q} = -\lambda\nabla T + \tau(a-1)\,\omega\times\boldsymbol{q}. \qquad (10.66)$$

In terms of the radial and tangential components of $\boldsymbol{q}$, this equation yields

$$q_r - \tau\omega(1-a)q_\theta = -\lambda\frac{\partial T}{\partial r}, \qquad (10.67a)$$

$$q_\theta + \tau\omega(1-a)q_r = 0, \qquad (10.67b)$$

which can also be solved in terms of the temperature gradient as follows:

$$q_r = -\frac{\lambda}{\Delta}\frac{\partial T}{\partial r}, \qquad q_\theta = \frac{\lambda}{\Delta}\tau\omega(1-a)\frac{\partial T}{\partial r}, \qquad (10.68)$$

with $\Delta = 1 + \tau^2\omega^2(1-a)^2$. It is seen that, unless $a = 1$, the components of the heat flux will depend on the angular velocity of the frame. According to the principle of frame-indifference, mentioned in Chaps. 1 and 2, a should be equal to 1 to avoid this frame dependence. However, by taking into account the Coriolis force acting on the molecules of the cylinder, it may be shown through kinetic theory arguments that $a = -1$ [1.30–31], indicating that material indifference cannot be taken as a universal truth (see Problem 10.4).

It follows from (10.68) that the only effect of the rotation on the radial temperature distribution would be a reduction of the apparent thermal conductivity λ by a factor $1/\Delta$. However, at room temperature the relaxation time is of the order of 10^{-10} s in metals and the maximum value for ω in an ultracentrifuge is about 10^4 s^{-1}; therefore such an effect would be difficult to detect, but it could be of interest in heat transfer in neutron stars, where relaxation times may be much longer.

10.6 Non-linear Heat Transfer: Flux Limiters

Up to now we have insisted on the finite speed of thermal signals. A direct consequence is that, for a given energy density, the heat flux cannot reach arbitrarily high values but will be bounded by a maximum saturation value, of the order of the energy density times the maximum speed. Typical situations arise, for instance, in radiative heat transfer, where the speed of photons is the speed of light c, and where the maximum heat flux is $aT^4 c$ [10.24, 25], or in plasma physics, where the speed of electrons is of the order of $(k_B T/m)^{1/2}$ and the energy density is proportional to $k_B T$, so that the maximum heat flux will be of the order of $k_B T(k_B T/m)^{1/2}$ [10.26]. Such bounds on heat transfer imply a drastic reduction of the heat flux with respect to the values predicted by Fourier's law at high temperature gradients and play a very important role, for instance, in laser–plasma interaction in laser-induced nuclear fusion, or in the collapse of stars.

This saturation effect cannot be described by the classical Fourier's law, but may be interpreted by introducing an effective thermal conductivity $\lambda (T, \nabla T)$ that depends not only on the temperature but also on the temperature gradient as

$$q = -\lambda(T,\nabla T)\nabla T. \tag{10.69}$$

For instance, a typical dependence giving rise to a limited heat flux is

$$q = -\frac{\lambda_0(T)}{\sqrt{1+a(\ell\nabla \ln T)^2}}\nabla T, \tag{10.70}$$

with ℓ the mean free path and a a numerical factor. This law reduces to Fourier's law for small temperature gradients and yields a saturation value in the limit of very high temperature gradients.

The form of $\lambda\,(T,\,\nabla T)$ has been the subject of many works in radiation hydrodynamics, and several expressions have been proposed. One of the best known is

$$\lambda(T,x) = 3\lambda_0(T)x^{-1}\left(\coth x - x^{-1}\right) \tag{10.71}$$

derived by Levermore [10.25] from a modified diffusion model for photons, in which $x = l'\nabla T/T, l'$ being a length of the order of the mean free path, while $\lambda_0(T)$ is the thermal conductivity near equilibrium. For small values of the temperature gradient $(x \to 0)$, one has $\lambda \to \lambda_0$, whereas for high values of ∇T one finds $\lambda \to 3\lambda_0 Tl'(\nabla T)^{-1}$, and the corresponding saturation value of the heat flux is $q_{max} = 3\lambda_0 T/l'$. Since $\lambda_0 = \frac{4}{3}aT^3c^2\tau$ (see the first of equations in (6.56)) and taking $l' = 4c\tau$, one finds finally that the maximum allowable value of q is $q_{max} = aT^4c$.

The use of flux limiters is necessary for practical purposes in astrophysical problems, but expression (10.71) turns out to be too cumbersome. Therefore several simpler approximations have been proposed such as

$$\lambda = \lambda_0 \frac{6+3x}{6+3x+x^2}. \tag{10.72}$$

Extended irreversible thermodynamics has been applied to the analysis of flux limiters in two particular circumstances. On the one side, Anile et al. [10.27] have shown that a flux limiter especially interesting in radiation theory is that obtained by assuming that there is a reference frame where the observer may see the radiation in equilibrium, because in this case the entropy of the gas in motion coincides with that derived in extended irreversible thermodynamics. A Lorentz transform yields then expressions for both the Eddington factor and the flux limiter. In this case, the flux-limiter $\lambda(x)$ has no closed analytic form, but it is given by eliminating β from the relations [10.25]

$$\lambda = \frac{3(1-\beta^2)^2}{(3+\beta^2)^2}, \qquad x = \frac{4\beta(3+\beta^2)}{(1-\beta^2)^2}, \tag{10.73}$$

where βc is the speed of the Lorentz frame in which the radiation is in equilibrium.

Another interesting analysis carried out by Jou and Zakari [10.28] is based on the fluctuation–dissipation expression for the thermal conductivity in non-equilibrium steady states derived in Problem 7.10 and on the maximum entropy description for ultrarelativistic gases presented in Sect. 7.5; the result cannot be expressed in closed form but the numerical values are very good agreement with those given by (10.73). Results for (10.71) and (10.73) are shown in Fig. 10.4.

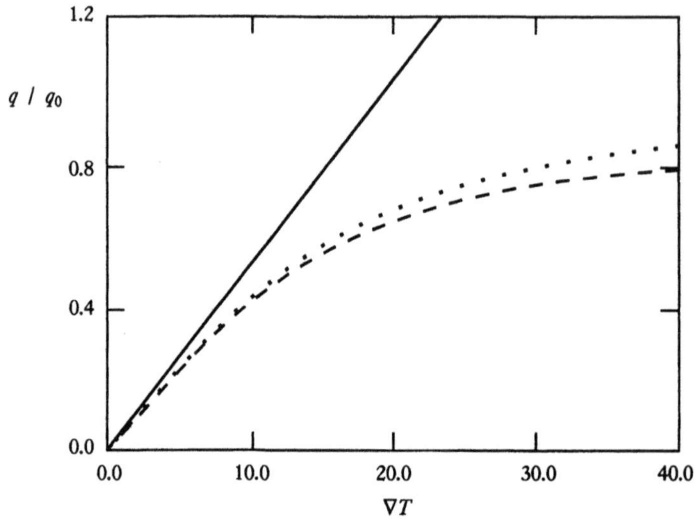

Fig. 10.4. The straight line represents Fourier's law. The curved lines correspond to flux-limited non-linear laws, which give for the heat flux a saturation value q_0. The upper dotted line corresponds to (10.71) and the lower broken line to (10.73).

10.7 Other Applications

We end this chapter by mentioning several situations where the hyperbolic equations of heat transfer have found successful practical applications:

1. *High local heating rate.* An example is the short-pulse laser heating of metals, which is used in the fabrication of microstructures, synthesis of advanced materials, measurement of thin-film properties and the analysis of structure transformations. The duration of these short laser pulses ranges from nanoseconds to femtoseconds. These heating rates are comparable with the thermalization time required by electrons to exchange energy with the lattice and to the relaxation time needed by electrons to change their state. This problem has been studied extensively by Qiu and Tien [10.29]. Other examples of high heating rates are high-speed grinding, high-speed friction or fast explosions.

2. *Fast motion of the heating source.* The rate of heating increases with the speed of the moving heat source. The various consequences of this fast motion have been ex-

amined extensively by Tzou [10.30]. The swinging phenomenon of temperature during the transition of the thermal Mach number (speed of the source divided by the speed of thermal waves) from the subsonic to the supersonic ranges, the physical mechanisms of thermal shock formation, and the local heating induced by dynamic crack propagation in the transonic regime are three examples of problems whose effects cannot be depicted by parabolic diffusion equations but which have been satisfactorily accounted for by hyperbolic equations.

3. *Fast moving interfaces*. Fast motions of interfaces, at a speed higher than the diffusion speed, are found in rapid solidification. Such rapid motions produce significant deviations from local equilibrium at the solid-liquid interfaces, leading, for example, to solute trapping and interfacial undercooling below the equilibrium temperature. These phenomena have been investigated by Sobolev [10.31] (Problem 10.12).

Many other subjects have been explored, both from a theoretical and a practical point of view. For instance, observations of thermal waves are reported in processed meat, applications of the telegrapher equation to IC chips and thermal laser stereo-lithography have been investigated, and general mathematical properties of the telegrapher equations have been studied in the linear and non-linear range,

4. *Superfluids*. The classical theory of superfluid helium II is the two-fluid model proposed by Landau [10.32], which regards the system as composed of a normal component with normal viscosity and nonzero entropy and a superfluid component with zero entropy and zero viscosity. The latter component, which is absent above the lambda transition temperature, was originally considered to be composed by the particles in the condensed Bose–Einstein state, and the normal fluid by the particles in excited states. The expression for the heat flux in this theory has the form $q = -\lambda \, \nabla T + Ts\rho_s(v^{(n)} - v^{(s)})$, with $v^{(n)}$ and $v^{(s)}$ the barycentric velocities of the normal and the superfluid components, s the entropy of the normal fluid and ρ_s the density of the superfluid component. The thermal conductivity is extremely high; the heat flux is not entirely determined by ∇T but its dominant contribution comes from the relative motion of the two components.

The two-fluid model is able to describe many features of liquid helium. However, it is not completely satisfactory because the two components cannot have independent existence and because the superfluid component cannot be strictly interpreted as the Bose–Einstein condensed phase, due to the intense interactions in the liquid. For these reasons, several authors have tried to describe superfluids by adding to the conventional fluid theory a vectorial degree of freedom, to take into account the relative motion of the excitations with respect to the bulk of the fluid. EIT offers a rather natural framework for this kind of description, as it takes the heat flux as an independent variable and, therefore, the dynamics of the relative motion of the excitations may be described by the dynamics of the heat flux. Analyses of the dispersion relation and of the contribution of the heat flux to the pressure tensor, and detailed comparisons of this model with the two-fluid model have been carried out in the literature [10.33–36].

For a wide bibliography on these and other related topics see [2.10] and the www.uab.es/dep-fisica/eit website.

Problems

10.1 The thermal conductivity of a dielectric solid is given by $\lambda = \frac{1}{3}\rho c_v c_0^2 \tau_1$, with c_0 being the phonon speed, τ_1 the relaxation time due to resistive phonon collisions, c_v the specific heat per unit mass at constant volume, and ρ the mass density. In the Debye model for d-dimensional systems, c_v is proportional to T^d at low temperatures. (a) Calculate the speed of the second sound in terms of c_0. (b) Assume that the solid sample is in a non-equilibrium steady state characterized by a heat flux q; when the sample is perturbed, heat waves may move either in the sense of q or in the opposite sense. Determine the difference Δv between the speed of thermal waves along and against q. [*Hint*: Use (10.58).]

10.2 Show that for ideal gases the expression (10.10) for the speed of thermal waves may be written as

$$U^2 = \frac{5}{2}\left(\frac{c_p}{c_v} - 1\right)\frac{k_B T}{m}.$$

Compare U with the speed of sound for monatomic and diatomic gases. (*Hint*: Use the results for λ obtained in Chap. 5.)

10.3 (a) Show that the entropy production corresponding to the Guyer–Krumhansl equation (10.41) is

$$T\sigma^s = \frac{3}{\tau_R \rho c_v c_0^2}\, q \cdot q + \frac{3\tau_N}{5\rho c_v}\left[(\nabla q):(\nabla q)^T + 2(\nabla \cdot q)(\nabla \cdot q)\right].$$

(b) Show that the stationary heat flux that satisfies the Guyer–Krumhansl equation is that corresponding to the minimum entropy production with the constraint $\nabla \cdot q = 0$, i.e. show that the Euler–Lagrange equations that come from

$$\delta \int (T\sigma^s - \gamma \nabla \cdot q)\, dV = 0,$$

with respect to variations of q and γ are the steady-state equations corresponding to $\partial u/\partial t = 0$ and $\partial q/\partial t = 0$ provided one identifies the Lagrange multiplier γ as twice the absolute equilibrium temperature [G. Lebon and P. C. Dauby, Phys. Rev. A **42** (1990) 4710.]

10.4 In the relaxation-time approximation, the solution of the Boltzmann equation for a heat-conducting disk subject to a radial temperature gradient dT/dr is

$$f = f_{eq} - f_{eq}\frac{\tau}{k_B T}\left(\frac{1}{2}mc^2 - 2k_B T\right)c_r\,\frac{d\ln T}{dr},$$

with τ being the relaxation time and c_r the radial component of the molecular velocity. (a) Show that in a frame rotating counterclockwise with angular velocity ω, the solution of the Boltzmann equation is, up to the first order in ω,

$$f = f_{eq} - f_{eq}\frac{\tau}{k_B T}\left(\frac{1}{2}mc^2 - 2k_B T\right)(c_r - 2\omega\tau c_\theta)\frac{d\ln T}{dr}.$$

c_θ being the axial component of the molecular velocity. (b) Find the ratio q_θ/q_r between the axial and radial components of q and compare with expressions (10.67). [See W. G. Hoover et al., Phys. Rev. A **24** (1981) 2109.]

10.5 Sieniutycz and Berry [Phys. Rev. A **43** (1991) 2807] have proposed for heat-conducting fluids in convective motion the following Lagrangian:

$$L(\rho,\rho s,v,v_s) = \tfrac{1}{2}\rho v^2 + \tfrac{1}{2}\rho g s^2 v_s^2 - \rho u(\rho,s),$$

with v being the barycentric velocity, v_s a velocity of entropy diffusion defined as $v_s = J^s/\rho s$, and g a coefficient. (a) Obtain the generalised mechanical momentum $J = \partial L/\partial v$ and thermal momentum $I = \partial L/\partial v_s$. (b) Show that the corresponding Hamiltonian per unit volume defined by

$$\mathcal{H}(\rho,\rho s,I,J) = v\cdot\frac{\partial L}{\partial v} + v_s\cdot\frac{\partial L}{\partial v_s} - L$$

can be written as

$$\mathcal{H}(\rho,\rho s,I,J) = \rho u(\rho,\rho s) + \tfrac{1}{2}\rho^{-1}J^2 + \tfrac{1}{2}\rho g^{-1}(\rho s)^{-2}I^2.$$

(c) Compare this expression with the expression of the generalised $\rho u(\rho,s,q)$ (for

$v = 0$) obtained from the generalized Gibbs equation used in this book and identify g. Show that for a Boltzmann monatomic gas $g = \frac{2}{3}m^2/k_B^2$.

10.6 An energy ε per unit length is suddenly released at the axis of a conducting cylinder. The time evolution of the system is described by the energy balance equation expressed by

$$\rho c_v \frac{\partial T}{\partial t} = -\frac{1}{r}\frac{\partial(rq)}{\partial r},$$

and the Maxwell–Cattaneo equation

$$\frac{\partial q}{\partial t} = -\frac{q}{\tau} - \frac{\lambda}{\tau}\frac{\partial T}{\partial r},$$

with q the radial component of the heat flux. In metals at low temperature λ is a constant whereas $\tau = aT^{-1}$. Write $T(r, t) = at^{-1}f(\xi)$ and $q(r,t) = (\lambda\rho c_v)^{1/2}at^{-3/2}g(\xi)$, with ξ the dimensionless quantity $\xi = r[(\lambda/\rho c_v)t]^{-1/2}$. (a) Show that $g = \frac{1}{2}\xi f$ and that $f(\xi)$ is implicitly given by

$$\xi^2 = 8 + \left(\frac{f-2}{f}\right)^{1/2}\left\{C - 8\ln\left[f^{1/2} + (f-2)^{1/2}\right]\right\},$$

with C being an integration constant related to ε. (b) Show that at $\xi = 2\sqrt{2}$ there is a discontinuity in the solution, where f drops from $f(\xi = 2\sqrt{2}) = 2$ to $f = 0$ for $\xi > 2\sqrt{2}$. Find the speed of the front [H. E. Wilhelm and S. H. Choi, J. Chem. Phys. **63** (1975) 2119.]

10.7 (a) Show that when the thermal conductivity λ depends on T according to $\lambda = \lambda_0 T^n$, the evolution equation for T in cylindrical coordinates is

$$\frac{\partial T}{\partial t} = a\frac{1}{r}\frac{\partial}{\partial r}\left(T^n r \frac{\partial T}{\partial r}\right),$$

with $a = \lambda_0/\rho c_v$. (b) When $t = 0$, an energy ε per unit length is suddenly released along the axis of the cylinder. Assume

$$T(r,t) = (Q/at)^{1/(n+1)}f(\xi),$$

with ξ being the non-dimensional combination $\xi = r(aQ^n t)^{-1/[2(n+1)]}$, where $Q = \varepsilon/\rho c_v$, and show that for $n > 0$ and ρc_v taken as a constant the solution of the equation for T is

$$f(\xi) = \left(\frac{n(\xi_0^2 - \xi^2)}{4(n+1)} \right)^{1/n} H(\xi_0 - \xi),$$

with $H(x)$ the Heaviside function. [See H. E. Wilhelm and S. H. Choi, J. Chem. Phys. **63** (1975) 2119.] (c) Find ξ_0 from the initial condition

$$2\pi \int_0^{\xi_0} f(\xi)\xi \, d\xi = 1$$

and calculate the position of the front $R(t)$ such that $T = 0$ for $r > R(t)$.

10.8 Assume an unidimensional system with an initial temperature profile given by $T(x, 0) = T_0 + \delta T_0 \cos kx$, with T_0 being a uniform value. (a) Obtain the expression for the temperature profile $T(x, t)$ by using the telegrapher equation for T. Note that $T(x, t)$ may reach negative values even though the initial $T(x, 0)$ is everywhere positive. (b) Show that, in the high-frequency limit, a sufficient condition for $T(x, t)$ being always positive is that the initial heat flux $q(x, 0)$ is smaller everywhere than the critical value $q_{crit} = \rho u (\chi / \tau)^{1/2}$ [M. Criado-Sancho and J. E. Llebot, Phys. Lett. A **177** (1993) 323].

10.9 (a) Compare the values of the heat flux obtained from the Fourier law with that obtained from the non-linear expression (10.71) for $x = 0.1$, $x = 1$ and $x = 5$. (b) Show that the maximum difference between (10.72) and (10.73) is of the order of 7% for $x \approx 2.5$.

10.10 (a) Use expression (5.100) to obtain the non-linear thermal conductivity $\lambda(q)$. (b) Show that for high values of ∇T, q behaves as $|q| \approx a^{-1} \ln [2\lambda_0 a \nabla T]$, with $a = (\tau_E / n k_B \lambda_0 T^2)^{1/2}$ (see Sect. 5.6). Thus, note that (5.100) does not describe a saturation of the heat flux, in spite of the fact that it describes a substantial reduction of the non-linear heat flux as compared with the linear theory.

10.11 In Sect. 10.4 we have studied the consequences of the non-equilibrium temperature on the speed of thermal pulses under a heat flux q_0. Assume, instead, that T is the local-equilibrium temperature and that Guyer–Krumhansl equation (10.40) is used. Show that the speed v of thermal pulses along and against the direction of q_0 are

$$v_\pm = v_c \left[-\phi \pm \sqrt{1 + \phi^2 + \gamma T^2 (\partial u / \partial T)} \right]$$

with $\phi = v_c a q_0$, $v_c = \sqrt{\lambda / \tau_1 (\partial u / \partial T)}$, $a = -\frac{1}{2} \partial (\tau_1 / \lambda T^2) / \partial T$ and $\gamma = \frac{4}{3} (\lambda T^2)^{-1} \beta'' \tau_2^{-1} +$.

$+(\lambda T^2)^{-1}\beta'\tau_0^{-1}$. Compare this result with (10.58). [See A. Valenti et al. J. Phys. Condens. Matter **9** (1997) 3117].

10.12 A two-temperature model used to describe ultrafast melting and solification in pulse-laser irradiated materials [see e.g. S. L. Sobolev, Phys. Lett. A **197** (1995) 243] assumes a fast relaxing mode (related to heat-conducting vibrational modes) and a slow relaxing one (related to structural rearrangements). Their respective temperatures and heat capacities are T_1, C_1 and T_2, C_2. Consider a system composed of a liquid and a solid parts separated by an interface at $x = z$. The equations for T_1 and T_2 for the liquid phase are assumed to be

$$C_{1L}\frac{\partial T_1}{\partial t} = \nabla\cdot(\lambda_L\nabla T_1) - \alpha_L(T_1 - T_2) + W_L$$

$$C_{2L}\frac{\partial T_2}{\partial t} = \alpha_L(T_1 - T_2)$$

with W a heat source term and α a heat transfer coefficient between the fast and the slow modes, and analogously for the solid phase (with an S subscript instead of L).

(a) Show that after expressing T_2 in terms of T_1 one obtains

$$\frac{\partial T_1}{\partial t} + \tau\frac{\partial^2 T_1}{\partial t^2} = a\frac{\partial^2 T_1}{\partial x^2} + l^2\frac{\partial^3 T_1}{\partial t\partial x^2} + \frac{W}{C_1+C_2} + \frac{\tau}{C_1}\frac{\partial W}{\partial t},$$

with $\tau = C_1 C_2[\alpha(C_1 + C_2)]^{-1}$, $l^2 = \lambda C_2[\alpha(C_1 + C_2)]^{-1}$ and $a = \lambda(C_1 + C_2)^{-1}$.

(b) Show that in the steady-state regime in the frame moving with the interface, this equation writes as

$$l^2 v\frac{d^3 T}{dx^3} + (a - \tau v^2)\frac{d^2 T}{dx^2} - v\frac{dT}{dx} + \frac{W}{C_1+C_2} + \frac{\tau v}{C_1}\frac{dW}{dx} = 0$$

with v being the velocity of the interface, and find the form of the profile, assuming that $T(x) = \sum_{i=1}^{3} A_i \exp(\mu_i x)$ and $W = 0$. (c) From the matching conditions at the interface and after some cumbersome analyses it may be shown that the interface overheating, defined as $\delta T = T_1(0) - T_2(0)$ is given by [S. L. Sobolev, Phys. Lett. A **197** (1995) 243]

$$\delta T = T_2(0)[(v/a\mu_1) - 1]$$

where μ_1 is the highest value of the μ_i. Show that this tends to 0 for slow interfaces and reaches a maximum for $v \approx (a/\tau)^{1/2}$. This maximum may be rather high if $C_2 \gg C_1$.

Chapter 11

Waves in Fluids

The presence of relaxation terms and higher-order gradients in the dynamical equations makes it very difficult to solve them because they require the introduction of supplementary boundary and initial conditions compared to classical thermo-hydrodynamics. Instead of trying to solve the whole mathematical problem, it is more fruitful to concentrate on the analysis of waves in infinite and semi-infinite media, which is, furthermore, a situation of great experimental interest.

As the relaxation times of the fluxes are usually very small, it is rather difficult to observe their effects in transient processes. However, such effects may be made perceptible in repetitive phenomena, like sound and ultrasound propagation when the frequency becomes comparable to the inverse of the relaxation times of the fluxes. Another situation of interest is provided by shock waves, where the gradients are very high.

Ordinary fluids are described by five independent variables (mass, energy, and the three components of velocity), and therefore they are characterized by five hydrodynamic modes. Additional independent variables must be incorporated at high frequencies and thus the number of independent modes will increase accordingly. In this chapter we shall first discuss the hydrodynamic modes in the classical theory and afterwards we shall analyse the main modifications brought in by extended thermodynamics.

11.1 Hydrodynamic Modes in Simple Fluids

Consider a fluid described by the hydrodynamic balance equations (1.1–3) and the classical Fourier–Stokes–Newton constitutive equations. The combined set of equations can be written as

$$\frac{\partial \rho}{\partial t} + \nabla \cdot (\rho v) = 0, \tag{11.1}$$

$$\frac{\partial v}{\partial t} + (v \cdot \nabla)v + \frac{c^2}{\gamma}\left(\frac{1}{\rho}\nabla\rho + \alpha\nabla T\right) - v\left[\nabla^2 v + \nabla(\nabla \cdot v)\right] - v'\nabla(\nabla \cdot v) = 0, \tag{11.2}$$

and

$$\frac{\partial T}{\partial t} + v \cdot \nabla T + \frac{\gamma-1}{\alpha}\nabla \cdot v - \frac{1}{\rho c_v}\left(\lambda\nabla^2 T + \Phi_\eta\right) = 0, \tag{11.3}$$

with $v = \eta/\rho$, $v' = \zeta/\rho$ being the kinematic viscosities, $c^2 = (\partial p/\partial\rho)_s$ the Laplace sound velocity, $\alpha = -\rho^{-1}(\partial\rho/\partial T)_p$ the coefficient of thermal expansion, $\gamma = c_p/c_v$ the ratio of the specific heats at constant pressure and volume, and Φ_η the frictional heating due to the viscous effects. These equations have been obtained by assuming that λ, η, and ζ are constant and that the pressure and the internal energy are functions of the density and temperature, i.e. $p = p(\rho,T)$, $u = u(\rho,T)$.

The choice of the most convenient independent variables depends on the type of problem to be dealt with. In most problems, one selects as variables the mass density ρ, temperature T, and velocity field v. In view of the applications to be studied in this section, it is preferable to choose the entropy s, the pressure p, the longitudinal current $w = \rho_0\nabla \cdot v$, and the two components of the transverse current, $\mu_i = (\nabla\times v)_i$ $(i = 1, 2)$. We denote by s_0, p_0, $w = 0$, and $\mu_1 = \mu_2 = 0$ the reference state and we superpose infinitesimally small perturbations δs, δp, δw, $\delta\mu_1$, and $\delta\mu_2$. When the variables (ρ, T, v) are changed into (s, p, w, μ_1, μ_2), the set (11.1–3) changes, up to the first order, into

$$\left[\frac{\partial}{\partial t} - (\gamma-1)\chi\nabla^2\right]\delta p + c^2\delta w - \rho_0\frac{\gamma-1}{\alpha}\chi\nabla^2\delta s = 0, \tag{11.4}$$

$$\left[\frac{\partial}{\partial t} - \left(\frac{4}{3}\eta + \zeta\right)\frac{1}{\rho_0}\nabla^2\right]\delta w + \nabla^2\delta p = 0, \tag{11.5}$$

$$\left[\frac{\partial}{\partial t} - \chi\nabla^2\right]\delta s + \frac{\alpha\chi}{\rho_0}\nabla^2\delta p = 0, \tag{11.6}$$

and

$$\left[\frac{\partial}{\partial t} - v\nabla^2\right]\delta\mu_1 = \left[\frac{\partial}{\partial t} - v\nabla^2\right]\delta\mu_2 = 0, \tag{11.7}$$

with $\chi = \lambda(\rho c_p)^{-1}$ the thermal diffusivity.

To solve this set of coupled differential equations it is convenient to use a double Fourier transform, defined as

$$\delta\tilde{\varphi}_k(\omega) = \int_0^\infty dt\, e^{-i\omega t} \int_{-\infty}^\infty dr\, e^{ik\cdot r}\delta\varphi(r,t).\qquad(11.8)$$

Application of transform (11.8) to (11.4–7) leads to

$$\mathbf{M}(k,\omega)\cdot C(\omega) = C(0),\qquad(11.9)$$

with

$$\mathbf{M}(k,\omega) = \begin{pmatrix} i\omega+(\gamma-1)\chi k^2 & \rho_0 c^2 & \rho_0\alpha^{-1}(\gamma-1)\chi k^2 & 0 & 0 \\ -\rho_0^{-1}k^2 & i\omega+v_l k^2 & 0 & 0 & 0 \\ \rho_0^{-1}\alpha\chi k^2 & 0 & i\omega+\chi k^2 & 0 & 0 \\ 0 & 0 & 0 & i\omega+vk^2 & 0 \\ 0 & 0 & 0 & 0 & i\omega+vk^2 \end{pmatrix},$$

$$C(\omega) = \begin{pmatrix} \delta\tilde{p}_k(\omega) \\ \delta\tilde{w}_k(\omega) \\ \delta\tilde{s}_k(\omega) \\ \delta\tilde{\mu}_{1k}(\omega) \\ \delta\tilde{\mu}_{2k}(\omega) \end{pmatrix}, \qquad C(0) = \begin{pmatrix} \delta\tilde{p}_k(0) \\ \delta\tilde{w}_k(0) \\ \delta\tilde{s}_k(0) \\ \delta\tilde{\mu}_{1k}(0) \\ \delta\tilde{\mu}_{2k}(0) \end{pmatrix},$$

where $v_l = (\frac{4}{3}\eta + \zeta)/\rho_0$ is the so-called longitudinal viscosity.

The structure of the hydrodynamic matrix $\mathbf{M}(k, \omega)$ corresponds to two acoustic modes (δp and δw), an entropy mode (δs) and two shear modes (μ_1, μ_2). The first two modes are propagative, i.e. their phase velocity is different from zero, whereas the last three are diffusive, i.e. their phase velocity vanishes. In the following we will focus our attention on the modifications introduced by extended irreversible thermodynamics (EIT).

11.2 Transverse Viscoelastic Waves

The modes related to the transverse components of the velocity are described in classical hydrodynamics by (11.7), and they are purely diffusive. In EIT, such modes become propagative at high frequencies. The situation is parallel to that of thermal modes, examined in Chap. 10: they are purely diffusive in the classical theory of heat conduction but they become propagative in the extended theory.

Assume that the wavevector $\mathbf{k}$ of the propagation is directed along the x axis, and consider the transverse component y of the velocity. The equation describing the evolution of the y-component of the velocity (the z-component will have an analogous description) is the linear momentum balance equation

$$\rho_0 \frac{\partial v_y}{\partial t} = -\frac{\partial P_{xy}^v}{\partial x}, \tag{11.10}$$

coupled to the evolution equations for the heat and viscous pressure (2.64–66). In the present problem, the latter reduce to

$$\tau_1 \frac{\partial q_y}{\partial t} + q_y = \lambda T^2 \beta \frac{\partial P_{xy}^v}{\partial x}, \tag{11.11}$$

and

$$\tau_2 \frac{\partial P_{xy}^v}{\partial t} + P_{xy}^v = -\eta \frac{\partial v_y}{\partial x} + \beta \eta T \frac{\partial q_y}{\partial x}. \tag{11.12}$$

The dispersion relation for these three equations reads

$$k^2 \left[\eta + i\omega\eta\left(\tau_1 + \lambda\beta T^3\right) \right] + i\omega\rho_0 \left[1 + i\omega(\tau_1 + \tau_2) - \omega^2\tau_1\tau_2 \right] = 0. \tag{11.13}$$

At low frequencies ($\omega\tau \ll 1$), one recovers the classical result

$$k = \frac{\omega}{\sqrt{2}} \frac{\rho_0}{\eta}(1 - i), \tag{11.14}$$

whereas in the high-frequency limit ($\omega\tau \gg 1$) one has

$$k^2 \approx \omega^2 \frac{\rho_0}{\eta} \frac{\tau_1\tau_2}{\tau_1 + \lambda\beta^2 T^3}, \tag{11.15}$$

to which corresponds a wave with phase velocity

$$v_p = \frac{\omega}{\mathrm{Re}\,k} = \left(\frac{\eta}{\rho_0\tau_2} \right)^{1/2} \left(1 + \frac{\lambda\beta^2 T^3}{\tau_1} \right)^{1/2}. \tag{11.16}$$

When β and τ_1 are vanishing, with β going to zero faster than τ_1, the phase speed is simply $v_p = (\eta/\rho_0\tau_2)^{1/2}$. Just as for heat waves, this phase speed is infinite for vanishing τ_2.

Joseph and co-workers [11.1] have succeeded in measuring shear-wave speeds in liquids. In Table 11.1 are displayed some of their results for the wave speed, the shear viscosity, the density, and the relaxation time; the latter was obtained from the expression $\tau_2 = \eta(\rho_0 v_p^2)^{-1}$, obtained by putting $\beta = 0$ in the viscoelastic model (11.16).

Table 11.1. Values of the average shear-wave speed, zero-shear viscosity, mass density, and relaxation time for various fluids at $T = 298$ K [11.1]

Fluid	v_p (10^{-2} m/s)	η (Pa.s)	ρ_0 (kg/m^3)	τ (10^{-2}s)
4.5% PEO[1]	86.4	31	1263	3.3
5.0% PEO	125.7	45	1165	2.4
0.5% PAM[2]	23.9	1.4	1117	2.2
2.0% PAA[3]	167.9	98	1315	2.6

[1]Here PEO stands for Poly (Ethylene Oxide) WSR-301 solution in water. The molar weight of the polymer is 4×10^6 and the weight percentage in the solution is given in the first column.

[2]PAM is Poly (Acrylamide) solution in ethylene glycol with Al_2O_3.

[3]PAA is Poly (Acrylic Acid) solution in ethylene glycol of molar weight 4×10^6.

11.3 Ultrasound Propagation in Monatomic Gases

Ultrasound propagation in monatomic gases was one of the main tools used to study the properties of gases before the development of new techniques based on light and neutron scattering. Ultrasound propagation in monatomic gases has been examined from the point of view of the kinetic theory of gases to explain experimental results by Greenspan [11.2] and Meyer and Sessler [11.3]. A macroscopic interpretation in terms of EIT was proposed by Carrassi and Morro [11.4], Anile and Pluchino [11.5], and others, who based their analyses on the relaxational equations (2.64–66). These works were generalised by Lebon and Cloot [11.6], who introduced non-local higher-order terms.

Consider a perturbation around a reference equilibrium state with pressure p_0, density ρ_0, temperature T_0, and velocity zero, so that $\rho = \rho_0 + \delta\rho$, $p = p_0 + \delta p$, $T = T_0 + \delta T$, and $v = 0 + \delta v$. The linearized form of the balance equations is

$$\frac{\partial \delta\rho}{\partial t} + \rho_0 \nabla \cdot \delta v = 0, \tag{11.17a}$$

$$\rho_0 \frac{\partial \delta v}{\partial t} = -\nabla \delta p - \nabla \cdot \delta \mathbf{P}^v, \tag{11.17b}$$

$$\rho_0 \frac{\partial \delta u}{\partial t} = -p_0 \nabla \cdot \delta v - \nabla \cdot \delta q. \tag{11.17c}$$

The perturbation of the internal energy in terms of the perturbation in density and temperature can be written as

$$\delta u = c_v \delta T + \frac{1}{\rho_0}\left(\frac{p_0}{\rho_0} - \frac{c_p - c_v}{\alpha}\right)\delta \rho, \tag{11.18}$$

where non-linear terms in the fluxes have been omitted.

Consider one-dimensional perturbations of the form $\delta \phi = \delta \phi' \exp[i(\omega t - kx)]$, with amplitude $\delta \phi'$, real frequency ω, and complex wavevector k. Inserting these into (11.17), one obtains

$$i\omega \delta \rho' - ik\rho_0 \delta v'_x = 0,$$

$$-ik\delta p' - ik\delta P^{v}_{xx}{}' + i\omega \rho_0 \delta v'_x = 0, \tag{11.19}$$

$$i\omega \rho_0 c_v \delta T' - ik\delta q'_x - \frac{\rho_0}{\alpha}\left(c_p - c_v\right)ik\delta v'_x = 0,$$

where δT can be expressed in terms of $\delta \rho$ and δp by means of $\alpha \rho_0 \delta T = (\gamma/c^2)\delta p - \delta \rho$.

11.3.1 The Classical Theory

In this section we reproduce the classical theory, and for pedagogical reasons we have given the details of the calculations. However, the reader interested only in the results may go directly to relations (11.23) and (11.27).

For a monatomic ideal gas described by the Newton and Fourier equations with zero bulk viscosity one has

$$\delta P^{v}_{xx}{}' = \frac{4}{3}ik\eta \delta v'_x, \qquad \delta q'_x = ik\lambda \delta T'. \tag{11.20}$$

Combination of (11.19) and (11.20) yields

$$\mathbf{A}_c \cdot \mathbf{B}_c = 0, \tag{11.21}$$

where

$$\mathbf{A}_c = \begin{pmatrix} i\omega & -ik\rho_0 & 0 \\ 0 & i\omega\rho_0 + \frac{4}{3}\eta k^2 & -ik \\ -\dfrac{1}{\rho_0 c_p c}(i\omega\rho_0 c_v + \lambda k^2) & -\dfrac{\rho_0}{\alpha}(c_p - c_v)ik & \dfrac{1}{\rho_0 c^2 \alpha}(i\omega\rho_0 c_v + \lambda k^2) \end{pmatrix},$$

$$\mathbf{B}_c = \begin{pmatrix} \delta\rho' \\ \delta v_x' \\ \delta p' \end{pmatrix}, \qquad \mathbf{0} = \begin{pmatrix} 0 \\ 0 \\ 0 \end{pmatrix}.$$

The corresponding dispersion relation given by the vanishing of the determinant of $\mathbf{A}_c$ is

$$\frac{1}{c\rho_0 c_p}k^4\left(\frac{-ic}{\omega} + \frac{4}{3}\frac{\eta\gamma}{c\rho_0}\right) + k^2\left[\left(1 + \frac{i\omega}{c^2\rho_0}\right)\left(\frac{4}{3}\eta + \frac{\lambda\gamma}{c_p}\right)\right] - \frac{\omega}{c} = 0. \qquad (11.22)$$

This is the well-known Kirchhoff equation for acoustic waves. For small values of k and ω one finds for the phase velocity v_p and the absorption coefficient Γ'

$$v_p = \frac{\omega}{\operatorname{Re} k} = c,$$

$$(11.23)$$

$$\Gamma' = -\operatorname{Im} k = \frac{\omega^2}{2\rho_0 c^3}\left[\frac{4}{3}\eta + \frac{\lambda}{c_p}(\gamma - 1)\right].$$

These are the classical results for sound propagation and absorption at low frequencies. At high frequencies, the asymptotic solution of (11.22) behaves as

$$k^2 \approx a + i[b(\omega/c) + O(1/\omega)], \qquad (11.24)$$

where

$$a = \left[\frac{(1-\gamma)l_2 + \gamma l_1 \pm (l_2 - \gamma l_1)}{\gamma l_2} \pm \frac{(2-\gamma)l_1 - l_2}{l_2 - \gamma l_1}\right]\frac{1}{2\gamma l_1 l_2}, \qquad (11.25a)$$

$$b = \frac{1}{2\gamma l_1 l_2}[l_2 + \gamma l_1 \pm (l_2 - \gamma l_1)], \qquad (11.25b)$$

with $l_2 = (4/3)(\eta/c\rho_0)$ and $l_1 = \lambda/(\rho_0 c_p c)$. Thus, in the high-frequency limit, the imaginary part of k^2 dominates, so that $k^2 = ib(\omega/c)$. Since $\sqrt{i} = (1/\sqrt{2})(1+i)$, one is led to

$$\mathrm{Re}\,k \approx \mathrm{Im}\,k \approx \left[\frac{l_2 + \gamma l_1 \pm (l_2 - \gamma l_1)}{8\gamma l_1 l_2}\right]^{1/2} \omega^{1/2}. \tag{11.26}$$

Consequently, the phase velocity reads

$$v_p = \frac{\omega}{\mathrm{Re}\,k} = 2\left[\frac{\gamma l_1 l_2}{l_2 + \gamma l_1 \pm (l_2 - \gamma l_1)}\right]^{1/2} \omega^{1/2}, \tag{11.27}$$

which is unbounded as ω tends to infinity. This remains true for the absorption coefficient $\Gamma' = -\,\mathrm{Im}\ k$.

11.3.2 The Extended Theory

For a vanishing bulk viscous pressure, the one-dimensional equations (2.64) and (2.66) can be expressed as

$$\tau_1 \frac{\partial q_x}{\partial t} + q_x = -\lambda \frac{\partial T}{\partial x} + \beta\lambda T^2 \frac{\partial P^v_{xx}}{\partial x},$$

$$\tag{11.28}$$

$$\tau_2 \frac{\partial P^v_{xx}}{\partial t} + P^v_{xx} = -\frac{4}{3}\eta\frac{\partial v_x}{\partial x} + \frac{4}{3}\eta T\beta \frac{\partial q_x}{\partial x}.$$

These equations, together with (11.19), yield

$$\mathbf{A}_e \cdot \mathbf{B}_e = \mathbf{0}, \tag{11.29}$$

where

$$\mathbf{A}_e = \begin{pmatrix} i\omega & -ik\rho_0 & 0 & 0 & 0 \\ 0 & i\omega\rho_0 & -ik & -ik & 0 \\ -i\omega c_v/\alpha & -ik\rho_0(c_p - c_v)/\alpha & i\omega c_p/\alpha c^2 & 0 & -ik \\ 0 & -\frac{4}{3}ik & 0 & 1+i\omega\tau_2 & \frac{4}{3}ik\eta\beta T \\ ik\lambda/\rho_0\alpha & 0 & ik\lambda/c^2\rho_0\alpha & ik\lambda\beta T^2 & 1+i\omega\tau_1 \end{pmatrix},$$

$$\mathbf{B}_e = \begin{pmatrix} \delta\rho' \\ \delta v'_x \\ \delta p' \\ \delta P^{v\,\prime}_{xx} \\ \delta q'_x \end{pmatrix}, \qquad \mathbf{0} = \begin{pmatrix} 0 \\ 0 \\ 0 \\ 0 \\ 0 \end{pmatrix}.$$

We now neglect for simplicity the cross terms on the right-hand side of (11.28), i.e. we assume that $\beta = 0$. The dispersion relation corresponding to this simplified version of (11.29) turns out to be

$$l_1\left[\gamma l_2 - i\frac{c}{\omega}(1+i\omega\tau_2)\right]k^4 + \left\{f(\omega) + i\frac{\omega}{c}[(1+i\omega\tau_1)l_2 + (1+i\omega\tau_2)\gamma l_1]\right\}k^2$$

$$-\left(\frac{\omega}{c}\right)^2 f(\omega) = 0, \tag{11.30}$$

with $f(\omega) = (1 + i\omega\tau_1)(1 + i\omega\tau_2)$. When $\tau_1 = \tau_2 = 0$, one recovers the Kirchhoff equation (11.22). If (11.30) is multiplied by $(1 - i\omega\tau_1)(1 - i\omega\tau_2)$ to make the last term a real one, we may write it as

$$Ak^4 + Bk^2 - C = 0, \tag{11.31}$$

with

$$A = \left(a_1 + a_1'\omega^2\right) - i(1/\omega)\left(a_2 + a_2'\omega^2\right), \qquad B = \left(1 + b_1'\omega^2 + b_1''\omega^4\right) + i\left(b_2 + b_2'\omega^2\right),$$

$$C = c_1\omega^2 + c_1'\omega^4 + c_1''\omega^6. \tag{11.32}$$

The coefficients in (11.32) are

$$a_1 = cl_1[\gamma(l_2/c) - \tau_1], \qquad a_1' = -cl_1\tau_1^2[\gamma(l_2/c) + \tau_2], \qquad a_2 = cl_1,$$

$$a_2' = cl_1[\tau_2^2 + \gamma(l_2/c)(\tau_1 + \tau_2)], \qquad b_1' = \tau_1^2 + \tau_2^2 + (l_2/c)\tau_2 + \gamma(l_1/c)\tau_1,$$

$$b_1'' = \tau_1\tau_2[\tau_1\tau_2 + (l_2/c)\tau_1 + \gamma(l_1/c)\tau_2], \qquad b_2 = (l_2 + \gamma l_1)c^{-1},$$

$$b_2' = \left(l_2\tau_1^2 + \gamma l_1\tau_2^2\right)c^{-1}, \qquad c_1 = c^{-2}, \qquad c_1' = \left(\tau_1^2 + \tau_2^2\right)c^{-2}, \qquad c_1'' = \tau_1^2\tau_2^2 c^{-2}.$$

At high frequencies, (11.31) leads after rather lengthy calculations to

$$k^2 = \frac{\omega^2}{2a_1'^2}\left\{a_1'\left[-b_1''\pm\left(b_1''^2 + 4a_1'c_1''\right)^{1/2}\right]\right.$$

$$\left.+\frac{i}{\omega}\left[\left(a_1'b_2' + a_2'b_1''\right)\left[\left(b_1''^2 + 4a_1'c_1''\right)^{1/2} \mp b_1''\right] \mp 2a_1'a_2'c_1''\right]\left(b_1''^2 + 4a_1'c_1''\right)^{-1/2}\right\}. \tag{11.33}$$

This relation shows that at high frequencies the imaginary part of k^2 is negligible with respect to the real part.

Within the same high-frequency limit, one obtains two waves with constant phase velocities given by

$$v_p = \frac{\omega}{\mathrm{Re}\,k} = \frac{(2a_1')^{1/2}}{\left[-b_1'' \pm \left(b_1''^2 + 4a_1'c_1''\right)^{1/2}\right]^{1/2}}. \qquad (11.34)$$

These expressions may be calculated explicitly by using kinetic theory results; accordingly, $\lambda = (5/2)(k_B^2 Tn/m)\tau_1$, $\eta = p\tau_2$, $\gamma = 5/3$, $c_v = (3/2)k_B/m$, $\tau_1 = (3/2)\tau_2$, $p = nk_BT$, and the Laplace sound velocity $c^2 = (5/3)k_BT/m$. With these values it is found that

$$\lim_{\omega \to \infty} v_p^{(1)} = 1.64c, \qquad \lim_{\omega \to \infty} v_p^{(2)} = 0.72c, \qquad (11.35)$$

where superscripts 1 and 2 denote the two wave velocities. In the low-frequency limit, the corresponding values for the phase velocity are $\lim_{\omega \to 0} v_p^{(1)} = c$, and $\lim_{\omega \to 0} v_p^{(2)} = 0$. The results are shown in Fig. 11.1 [11.7]. The first wave is the usual acoustic wave, with velocity equal to c at small frequencies and $1.64c$ at high frequencies, in contrast with the classical prediction of an infinite speed in this limit. The second wave corresponds to a diffusive wave which is strongly damped at low frequencies and is therefore not observed. This is no longer true however at high frequencies, at which the absorption coefficients $\Gamma'^{(2)}$ and $\Gamma'^{(1)}$ remain finite and almost constant when ω is varied.

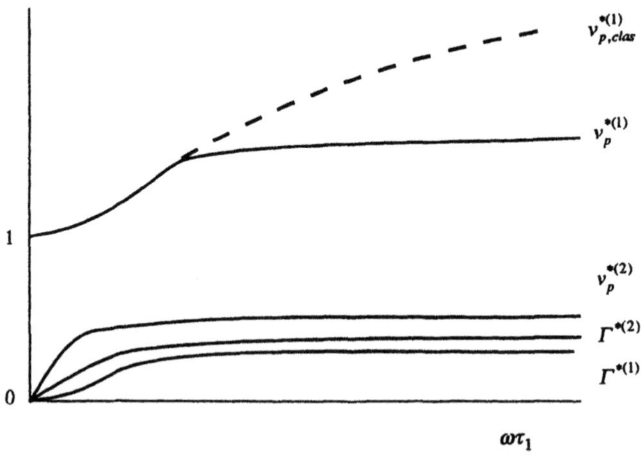

Fig. 11.1. Non-dimensional phase speeds $v_p^*(= v_p/c)$ and absorption coefficients $\Gamma^*(= \tau_1 c\,\Gamma')$ of the acoustic wave 1 and the extra wave 2 as a function of the non-dimensional frequency $\omega\tau_1$ for EIT. The dashed curve represents the phase speed for the classical theory.

A numerical comparison between the respective phase velocities in the classical Kirchhoff and the relaxational theory may be found in Table 11.2, where values of c/v_p are given as functions of $\Lambda k/2\pi$; $\Lambda = \eta(\rho_0 c)^{-1}$ is the mean free path of the molecules. The numerical values in the relaxational case have been calculated by assuming that $\tau_1 = \tau_2 = \Lambda/c\pi$ [11.4]. It is evident that the classical theory deviates appreciably from the experimental results as $\Lambda k/2\pi$ increases, whereas the introduction of the relaxation times makes the comparison much more accurate in the domain of short wavelengths.

Table 11.2. Numerical values of c/v_p as a function of $\Lambda k/2\pi$ [11.4]

$\Lambda k/2\pi$	0.25	0.50	1.00	2.00	4.00	7.00
(c/v_p)(Kirchhoff)	0.40	0.26	0.19	0.13	0.10	0.07
(c/v_p)(relaxational)	0.52	0.43	0.44	0.47	0.48	0.49
(c/v_p)(experimental)	0.51	0.46	0.50	0.46	0.46	0.46

The degree of agreement with the experimental data is very sensitive to the choice of the values of τ_1 and τ_2 [11.4]. It is true that from a macroscopic point of view one is free to choose τ_1 and τ_2 as phenomenological parameters maximizing the agreement with experiments, but the kinetic theory of gases (Chap. 5) and the fluctuation theory (Chap. 6) predict definite relations between τ_1 and λ and between τ_2 and η, so that the macroscopic freedom is only apparent. When the kinetic results are taken into account, as in (11.35), the predicted values of the high-frequency speed do not agree with the experimental ones. This is not surprising, because the relaxation times of heat flux and of viscous pressure are of the order of the collision time. This means that, at frequencies comparable to the inverse of the relaxation times, it is necessary to include in the macroscopic description not only the dissipative fluxes but also higher-order fluxes, as explained in Sect. 5.7. This considerably improves the agreement with experimental data.

By doing so, and letting the relaxation time of the new extra variables tend to zero, it was shown by Lebon and Cloot [11.6] that (2.64) and (2.66) are replaced by

$$\tau_1 \dot{q} = -(q + \lambda \nabla T) + \beta \lambda T^2 \nabla \cdot \mathbf{P}^v + \lambda_1 \nabla \cdot (\nabla q),$$

$$\tau_2 \dot{\mathbf{P}}^v = -(\mathbf{P}^v + 2\eta \mathbf{V}) + 2\eta \beta T (\nabla q) + \eta_1 \nabla \cdot (\nabla \mathbf{P}^v),$$

(11.36)

where λ_1 and η_1 are two supplementary coefficients. The model involves seven unknown parameters $\tau_1, \tau_2, \lambda, \eta, \beta, \lambda_1$, and η_1. Instead of selecting τ_1 and τ_2 as free parameters, as done earlier by Carrassi and Morro [11.4] and Anile and Pluchino [11.5], Lebon and Cloot prefer to choose λ_1 and η_1, because of the uncertainty about the theoretical value of these two quantities. In view of the kinetic theory, the other parameters are related to λ_1 and η_1 by $\lambda = (15/4)[k_B/(Tm)]^{1/2}p\eta_1$, $\beta = -2/(5pT\sqrt{\eta_1})$, $\eta = p\eta_1(k_BT/m)^{-1/2}$, $\tau_1 = (3/2)\eta_1(k_BT/m)^{-1/2}$, and $\tau_2 = \eta_1(k_BT/m)^{-1/2}$, in the case of monatomic Maxwell molecules. Instead of (11.30) one obtains now the more complicated dispersion relation

$$\rho_0 A_2 A_3 + \left(\frac{k}{\omega}\right)^2 \left[i\rho_0\omega\left(\frac{\lambda}{c_v}A_2 + \frac{1}{3}\eta A_3\right) + \rho_0^2 A_2 A_3 \left(\frac{c^2}{\chi}\right)\left(1 + \frac{k_B}{mc_v}\right)\right] - \left(\frac{k}{\omega}\right)^4$$

$$\times \left[\frac{\omega\lambda c^2}{c_v}\left(\frac{\eta\omega}{3c^2} + \frac{i}{\gamma}\rho_0 A_1\right) - \frac{2\rho_0 c^2}{c_p}\lambda^2\eta\beta T\omega^2 + \frac{2\rho_0}{3}\eta^2\beta T\omega^2\left(\frac{k_B}{mc_v} - \frac{2\eta\beta T\omega}{A_2}\right)\right] = 0$$

$$(11.37)$$

with $A_1 = 1 - i\omega\tau_1 + \lambda_1 k^2$, $A_2 = 1 - i\omega\tau_2 + \eta_1 k^2$, and $A_3 = A_1 + (4/3)(\eta\lambda\beta^2 T^3/A_2)k^2$. This equation has been solved numerically [11.6] and compared with the experimental data by Meyer and Sessler [11.3] and by Greenspan [11.2]. In Figs. 11.2 and 11.3, c/v_p and the dimensionless absorption coefficient Γ^* are shown as a function of the dimensionless frequency $\omega^* = \tau^*\omega$ with $\tau^* = (2\eta/p\gamma)$. It is noticed that the dimensionless wave speed v_p/c attains a finite value (approximately 2) at high frequency ($\omega^* > 10$).

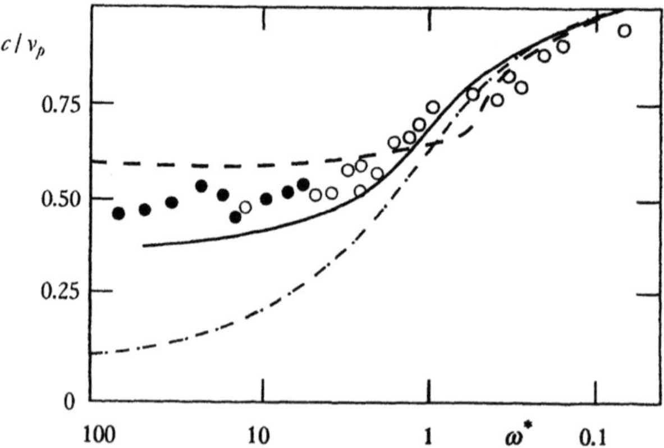

Fig. 11.2. The quantity c/v_p versus dimensionless frequency ω^*. Experimental data: *black circles*, Meyer and Sessler; *white circles*, Greenspan. Theoretical results: — · — · — Navier–Stokes; — — — Anile and Pluchino; ——————— Lebon and Cloot.

As for the damping, it is seen that the absorption coefficient increases at small frequencies up to a maximum value at $\omega^* = 1$ and then decreases slowly for high frequencies. As pointed out by Woods and Troughton [11.8], at high values of ω^* there may be a contribution to the absorption arising from diffusion in the piezoelectric receiver, so that the experimental result for the absorption factor should be considered as an upper limit to the actual value.

In brief, it is observed that, whereas the Navier–Stokes approach provides a good modelling at low frequencies, it is definitively not adequate at high frequencies, say $\omega^* > 2$. By using extended irreversible thermodynamics in its simplest version, i.e. with the heat flux and the viscous pressure tensor as the only extra variables, Anile and Pluchino [11.5] obtained a more satisfactory result for the phase speed. Unfortunately, the results for the absorption coefficients are even worse than those based on the Navier–Stokes theory as soon as ω^* becomes larger than unity.

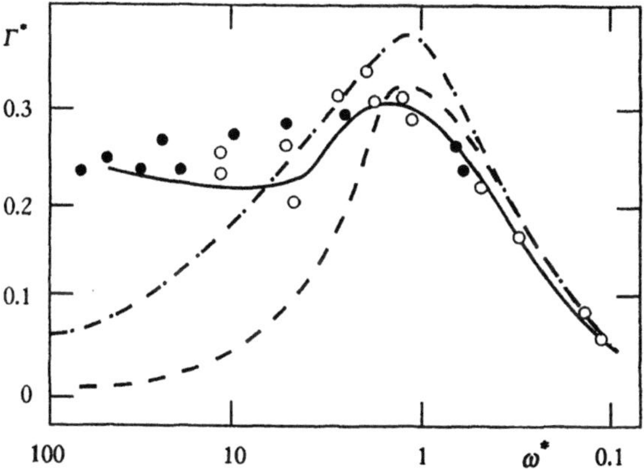

Fig. 11.3. The dimensionless absorption coefficient $\Gamma^* = \tau^* c\,\Gamma'$ versus dimensionless frequency ω^* (= $\tau^* \omega$). Experimental data: *black circles*, Meyer and Sessler; *white circles*, Greenspan. Theoretical results: — · — · — Navier-Stokes; — — — — Anile-Pluchino; ——————— Lebon-Cloot.

Better results are obtained with the strongly non-local model with $\lambda_1 \neq 0$ and $\eta_1 \neq 0$. As shown in Fig. 11.2, the theoretical curve for the phase velocity matches reasonably well the experimental observations by Greenspan and by Meyer and Sessler, even in the high-frequency range ($\omega^* > 2$). Moreover, an excellent agreement between theory and experiment is achieved for the absorption coefficient over the whole range of frequencies (Fig. 11.3). It follows from the previous analysis that, in order to match the high-frequency experimental data, it is necessary to include, in addition to the usual heat flux and viscous pressure tensor, supplementary variables of the form of fluxes of the fluxes.

Here we are concerned with simple monatomic gases. Other causes of relaxational effects are found when the molecules have internal degrees of freedom, or when a chemical reaction takes place in the gas. Such relaxational effects have been observed, and analysed by some authors [11.9–10]. Sound propagation in the framework of EIT in elastic dielectrics and metals has been studied by Kranys [11.11] and Jou et al. [11.12].

11.4 Shock Waves

In shock waves, the density, pressure, and normal velocity are discontinuous across a thin layer. In fact, such a discontinuity layer is not a geometrical surface, but a material surface with a small but non-vanishing thickness, of the order of the mean free path, across which the mentioned quantities exhibit a strong variation. The relaxation terms may thus play a considerable role in the analysis of the structure of shock waves.

11.4.1 General Equations

We consider a monatomic ideal gas, with zero bulk viscosity, in a steady one-dimensional flow. The basic equations describing the structure of a plane shock wave in a frame moving with the wave (see Fig. 11.4) are

$$v\frac{\mathrm{d}\rho}{\mathrm{d}x} + \rho\frac{\mathrm{d}v}{\mathrm{d}x} = 0,$$

$$\rho v\frac{\mathrm{d}v}{\mathrm{d}x} + \frac{\mathrm{d}P_{xx}^v}{\mathrm{d}x} + \frac{\mathrm{d}p}{\mathrm{d}x} = 0,$$

$$\rho v\frac{\mathrm{d}u}{\mathrm{d}x} + \frac{\mathrm{d}q}{\mathrm{d}x} + p\frac{\mathrm{d}v}{\mathrm{d}x} + P_{xx}^v\frac{\mathrm{d}v}{\mathrm{d}x} = 0, \tag{11.38}$$

$$q + \lambda\frac{\mathrm{d}T}{\mathrm{d}x} + \tau_1 v\frac{\mathrm{d}q}{\mathrm{d}x} - \beta\lambda T^2\frac{\mathrm{d}P_{xx}^v}{\mathrm{d}x} = 0,$$

$$P_{xx}^v + \frac{4}{3}\eta\frac{\mathrm{d}v}{\mathrm{d}x} + \tau_2 v\frac{\mathrm{d}P_{xx}^v}{\mathrm{d}x} - \frac{4}{3}\beta\eta T\frac{\mathrm{d}q}{\mathrm{d}x} = 0,$$

where $q = q_x$ and $v = v_x$. The first three equations are the mass, momentum and energy balance equations, respectively, whereas the two last ones come from (2.64) and (2.66), respectively. Note that the material time derivatives reduce in the steady situation to their convective part, for instance, $\mathrm{d}q/\mathrm{d}t = v(\mathrm{d}q/\mathrm{d}x)$ and so on. These equations must be supplemented with the state equations

$$u = c_v T = \frac{c^2}{\gamma(\gamma-1)}, \qquad p = nk_B T = \frac{\rho}{\gamma}c^2, \tag{11.39}$$

in which we have written u and p in terms of the Laplace sound velocity c and the adiabatic coefficient γ, because these expressions are the most usual ones in the analysis of shock waves.

11.4.2 The Classical Approach

The first three equations of (11.38) can be immediately integrated and yield

$$\rho v = M^*,$$

$$\rho v^2 + P_{xx}^v + p = P^*, \tag{11.40}$$

$$\rho v \left[u + \tfrac{1}{2}v^2\right] + \left(P_{xx}^v + p\right)v + q = Q^*,$$

where M^*, P^*, and Q^* are integration constants.

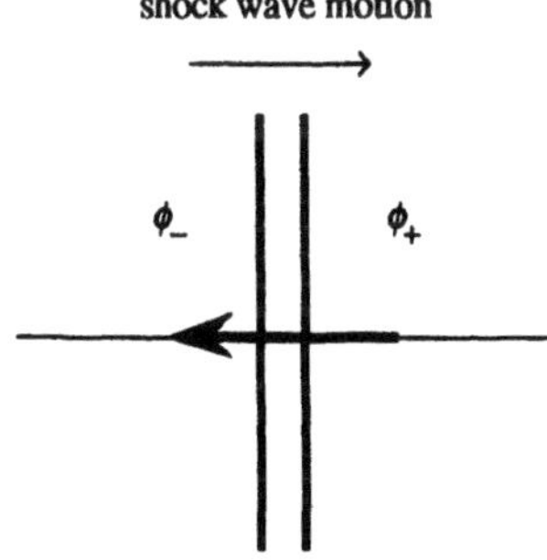

Fig. 11.4. In the reference frame where the shock wave is at rest, the gas flows from the zone at equilibrium (ϕ_+) to the zone perturbed by the shock motion (ϕ_-). In a frame at rest in the gas, the shock wave goes from the left to the right.

Our next task is to relate the values of ρ, v, p, and T behind and ahead of the shock wave (ρ_-, v_-, p_-, and T_-, and ρ_+, v_+, p_+, and T_+ respectively). Far from the shock wave, the dissipative fluxes in (11.38) are negligible, because at long distances equilibrium is reached. The corresponding velocities of the gas with respect to the shock wave are $v_- = M^*/\rho_-$, $v_+ = M^*/\rho_+$. Introducing these expressions into the second equation in (11.40) we can write

$$p_+ + \frac{M^{*2}}{\rho_+} = p_- + \frac{M^{*2}}{\rho_-}, \tag{11.41}$$

so that

$$M^{*2} = \frac{p_- - p_+}{\rho_+^{-1} - \rho_-^{-1}}. \tag{11.42}$$

One may then write the third of equations (11.40) as

$$u_+ + \frac{p_+}{\rho_+} + \frac{1}{2}\left(\frac{M^*}{\rho_+}\right)^2 = u_- + \frac{p_-}{\rho_-} + \frac{1}{2}\left(\frac{M^*}{\rho_-}\right)^2. \tag{11.43}$$

Substitution of M^* by its expression (11.42) yields

$$u_+ - u_- + \tfrac{1}{2}\left(\rho_+^{-1} - \rho_-^{-1}\right)(p_+ + p_-) = 0. \tag{11.44}$$

For given values of p_+ and ρ_+, the latter equation links ρ_- with p_-. Such a relation is known as the shock adiabatic line or Hugoniot adiabatic line: it relates the thermodynamic quantities on both sides of the shock layer. For an ideal gas described by the equations of state (11.39), (11.44) takes the form

$$\frac{\rho_-}{\rho_+} = \frac{(\gamma+1)p_- + (\gamma-1)p_+}{(\gamma-1)p_- + (\gamma+1)p_+}. \tag{11.45}$$

The ideal gas relation $T_+/T_- = (p_+\rho_-/p_-\rho_+)$ then leads to

$$\frac{T_+}{T_-} = \frac{p_+}{p_-}\frac{(\gamma+1)p_- + (\gamma-1)p_+}{(\gamma-1)p_- + (\gamma+1)p_+}. \tag{11.46}$$

The constant M^*, according to (11.42) and (11.45), is given by $M^{*2} = (\rho_-/2)[(\gamma-1)p_- + (\gamma+1)p_+]$. Furthermore, it follows from $v_-\rho_- = v_+\rho_+ = M^*$ that the velocity of the gas with respect to the shock wave may be written as

$$v_-^2 = \frac{1}{2\rho_-}[(\gamma-1)p_- + (\gamma+1)p_+], \tag{11.47a}$$

$$v_+^2 = \frac{1}{2\rho_-}\frac{[(\gamma+1)p_- + (\gamma-1)p_+]^2}{(\gamma-1)p_- + (\gamma+1)p_+}. \tag{11.47b}$$

From these expressions and (11.39), one can write the relations v_-/v_+, p_+/p_- and T_+/T_- in terms of the Mach number $\mathcal{M} = v_-/c_-$, where c_- is the sound velocity behind the discontinuity layer

$$\frac{v_-}{v_+} = \frac{(\gamma+1)\mathcal{M}^2}{(\gamma-1)\mathcal{M}^2+2},$$

$$\frac{p_+}{p_-} = \frac{2\gamma\mathcal{M}^2-\gamma+1}{\gamma+1}, \tag{11.48}$$

$$\frac{T_+}{T_-} = \frac{\left(2\gamma\mathcal{M}^2-\gamma+1\right)\left[(\gamma-1)\mathcal{M}^2+2\right]}{(\gamma+1)^2\mathcal{M}^2}.$$

These expressions are well known [11.13; 6.8] and are independent of the dissipative fluxes.

11.4.3 The EIT Approach

The dissipative terms play an important role in the determination of the spatial structure of the shock layer, i.e. on the spatial dependence of the thermodynamic quantities and the velocity across the layer (Fig. 11.4). Several authors have studied the role of the relaxation terms on shock waves [11.14–15; 5.2]. Here we follow closely the analysis of Anile and Majorana [11.14].

It is customary to use the dimensionless variables $\omega = (M^*/P^*)v$ and $\phi = p/P^*$. The combination of the set (11.38) with the equations of state (11.39) leads to the following spatial dependence of ω and ϕ:

$$a_{11}\frac{d\omega}{dx} + a_{12}\frac{d\phi}{dx} = A_1 D_1,$$

$$a_{21}\frac{d\omega}{dx} + a_{22}\frac{d\phi}{dx} = A_2 D_2, \tag{11.49}$$

where

$$a_{11} = \tfrac{5}{2}\phi^2 + r_1\omega^2 - [r_1(\gamma-1)^{-1} + r_2]\omega\phi - r_1\omega,$$

$$a_{12} = -r_1(\gamma-1)^{-1}\omega^2 + \left(\tfrac{5}{2}-r_2\right)\omega\phi,$$

$$a_{21} = \tfrac{8}{9}\left[1-\tfrac{2}{3}r_2(\gamma-1)^{-1}\right]\phi + \left(\tfrac{16}{45}r_2 - r_3\right)\omega - \tfrac{16}{45}r_2,$$

$$a_{22} = -r_3\omega - \tfrac{16}{45}r_2(\gamma-1)^{-1}\omega,$$

$$A_1 = \tfrac{5}{2}(M^*/P^*)(k_B p/\lambda m), \qquad A_2 = \tfrac{2}{3}(M^*/P^*)(p/\eta),$$

$$r_1 = \tfrac{5}{2}(\tau_1/\rho\lambda T)p^2, \qquad r_2 = -\tfrac{5}{2}p\beta T, \qquad r_3 = \tfrac{2}{3}(p\tau_2/\eta),$$

$$D_1 = (\gamma-1)^{-1}\omega\phi - a^* - \tfrac{1}{2}\omega^2 + \omega, \qquad D_2 = \omega + \phi - 1, \qquad a^* = M^*Q^*/P^{*2},$$

with k_B the Boltzmann constant and m the mass of the molecules. Note that a^* may be written in terms of the properties of the equilibrium state upstream to the shock layer as $a^* = \gamma^2\mathcal{M}^2 (1 + \gamma\mathcal{M}^2)^{-2}[(\gamma-1)^{-1} + \tfrac{1}{2}\mathcal{M}^2]$.

When the determinant D of the homogeneous part of (11.49) is different from zero, i.e. $D = a_{11}a_{22} - a_{12}a_{21} \neq 0$, the set (11.49) may be written as

$$\frac{d\omega}{dx} = \frac{\Delta_1}{D}, \qquad \frac{d\phi}{dx} = \frac{\Delta_2}{D}, \tag{11.50}$$

with $\Delta_1 = A_1 D_1 a_{22} - A_2 D_2 a_{12}$ and $\Delta_2 = A_2 D_2 a_{11} - A_1 D_1 a_{21}$. The equilibrium states located far away from the shock layer are given by $\Delta_1 = \Delta_2 = 0$, or, equivalently, by $D_1 = D_2 = 0$, i.e. they are the solution of

$$\frac{\phi\omega}{\gamma-1} - a^* - \frac{\omega^2}{2} + \omega = 0, \qquad \omega + \phi - 1 = 0. \tag{11.51}$$

The set (11.51) has two real positive solutions $P_- = (\omega_-, \phi_-)$ and $P_+ = (\omega_+, \phi_+)$ whenever $\tfrac{1}{2} < a^* < \tfrac{1}{2}\gamma^2(\gamma^2-1)^{-1}$. These solutions are given by

$$\omega_+ = \frac{\gamma+\varepsilon}{\gamma+1}, \qquad \phi_+ = \frac{1-\varepsilon}{\gamma+1},$$

$$\omega_- = \frac{\gamma-\varepsilon}{\gamma+1}, \qquad \phi_- = \frac{1+\varepsilon}{\gamma+1}, \tag{11.52}$$

with

$$\varepsilon = \sqrt{\gamma^2 - 2(\gamma^2-1)a^*} = -\gamma\frac{1-\mathcal{M}^2}{1+\gamma\mathcal{M}^2}.$$

Note that the results (11.52) are the same as the classical ones (11.48). The quantity ε parametrizes the strength of the shock: for $\varepsilon \to 0$, the shock is weak, whereas for $\varepsilon \to 1$, the shock is strong.

An asymptotic expansion may be obtained for weak shocks. Setting $y = \varepsilon x$ and expanding the solutions $\omega(y,\varepsilon)$ and $\phi(y,\varepsilon)$ of (11.50) in terms of ε, one has

$$\omega(y,\varepsilon) = \gamma(\gamma + 1)^{-1} + \varepsilon\omega_0(y) + \varepsilon^2\omega_1(y),$$

$$\phi(y,\varepsilon) = (\gamma + 1)^{-1} + \varepsilon\phi_0(y) + \varepsilon^2\phi_1(y). \tag{11.53}$$

After introducing these expressions into (11.49) and bearing in mind that $a^* = \frac{1}{2}(\gamma^2 - \varepsilon^2) \times (\gamma^2 - 1)^{-1}$, one obtains at the first order in ε

$$\varepsilon a_{11}\frac{d\omega_0}{dy} + \varepsilon a_{12}\frac{d\phi_0}{dy} = A_1\left\{\frac{\gamma(\omega_0 + \phi_0)}{\gamma^2 - 1} + \varepsilon\left[\frac{\omega_0\phi_0}{\gamma - 1} - \frac{\omega_0^2}{2} + \frac{1}{2(\gamma^2 - 1)}\right]\right\}, \tag{11.54a}$$

$$\varepsilon a_{21}\frac{d\omega_0}{dy} + \varepsilon a_{22}\frac{d\phi_0}{dy} = A_2(\omega_0 + \phi_0). \tag{11.54b}$$

Elimination of ε results in

$$\left[a_{11} - \frac{A_1\gamma}{A_2(\gamma^2 - 1)}a_{21}\right]\frac{d\omega_0}{dy} + \left[a_{12} - \frac{A_1\gamma}{A_2(\gamma^2 - 1)}a_{22}\right]\frac{d\phi_0}{dy}$$

$$= A_1\left[\frac{\omega_0\phi_0}{\gamma - 1} - \frac{\omega_0^2}{2} + \frac{1}{2(\gamma^2 - 1)}\right]. \tag{11.55}$$

Furthermore, inspection of (11.54b) shows that $\phi_0 = -\omega_0 + O(\varepsilon)$, because the left-hand side is of the order of ε, so that (11.55) reduces to

$$\left[\frac{5(1 - \gamma)}{2(1 + \gamma)^2} - \frac{8A_1\gamma}{9A_2(\gamma^2 - 1)(1 + \gamma)}\right]\frac{d\omega_0}{dy} = A_1\left[-\frac{1 + \gamma}{\gamma - 1}\frac{\omega_0^2}{2} + \frac{1}{2(\gamma^2 - 1)}\right]. \tag{11.56}$$

This equation has the form

$$\frac{d\omega_0}{dy} = -\left(A'\omega_0^2 - B'\right), \tag{11.57}$$

whose solution is

$$\omega_0(y) = (B'/A')^{1/2} \tanh(B'y), \tag{11.58}$$

from which it follows that

$$\omega_0(y) = -\phi_0(y) = -(\gamma+1)^{-1}\tanh(B'y), \tag{11.59}$$

with $B'^{-1} = (2/M^*)(\gamma+1)^{-1}[(\gamma-1)(\lambda'/c) + \frac{4}{3}\gamma\eta']$, $A' = (\gamma+1)^2\varepsilon^{-2}B'$, and λ', η' the corresponding transport coefficients evaluated for $\phi = (\gamma+1)^{-1}$. Note that the relaxation coefficients r_1, r_2, and r_3 do not appear in B'. This means that for weak shocks the width $1/B'$ and the structure of the shock layer are not explicitly influenced by the relaxation terms up to order ε. Such terms appear, however, at the higher-order approximations. At these orders, numerical methods are needed to obtain explicit values for the shock structure. Furthermore, as will be seen, the relaxation terms put some limits on the Mach numbers for which a regular structure may exist.

Anile and Majorana [11.14] integrated numerically the set of equations (11.38–39) for argon and evaluated the shock width, defined in general by

$$L = \left(\omega_- - \omega_+\right)\left(\frac{d\omega}{dx}\right)_{max}^{-1}. \tag{11.60}$$

Note that up to first order in ε the width of weak shocks is $L = 2B'^{-1}$, according to (11.59). For more general situations, it is usual to introduce the dimensionless ratio ℓ_-/L, with ℓ_- the upstream mean free path [11.14].

Table 11.3. Ratio of mean free path ℓ_- to shock width L at different Mach numbers $\mathcal{M}$ for three dynamical models: model NSF ($r_1 = r_2 = r_3 = 0$) is the classical Navier–Stokes–Fourier model; models GI ($r_1 = r_2 = 1$, $r_3 = 2/3$) and GII ($r_1 = r_3 = 2/(5\pi)$, $r_2 = 0$) are introduced in the text

$\mathcal{M}$	model NSF	model GI	model GII
1.1	0.0346	0.0346	0.0347
1.3	0.0983	0.0981	0.0985
1.5	0.1527	0.1528	0.1534
1.7	0.1972	———	0.1989
1.9	0.2326	———	0.2357

Three special choices of the parameters r_1, r_2, and r_3 appearing in (11.49) deserve special attention. The case $r_1 = r_2 = r_3 = 0$ corresponds to the classical Newton–Stokes–Fourier theory. From Grad's kinetic model, one has $\lambda = \frac{5}{2}(k_B p/m)\tau_1$, $\eta = \frac{2}{3}p\tau_2$, and $\beta = -\frac{2}{5}(pT)^{-1}$, so that $r_1 = 1$, $r_2 = 1$, and $r_3 = \frac{2}{3}$ (model GI). The third model assumes that $r_1 = \frac{2}{5}\pi^{-1}$, $r_2 = 0$, and $r_3 = r_1$ (model GII), which are the values obtained by Carrassi and Morro [11.4] by comparing the linear dispersion relation derived from (11.31) with the experimental data in the high-frequency limit of the phase speed for sound waves. The results are shown in Table 11.3.

It can be seen that the shock width obtained from models GI and GII agrees well with that computed from the Navier–Stokes–Fourier model for all Mach numbers for which a regular solution exists. By a regular solution is meant a smooth profile without any singularity in the structure of the shock layer, and therefore it is not possible to discriminate between the classical and the extended models in this domain.

However, the occurrence of a critical Mach number in the EIT description means that a regular unique solution can only exist if $D(\omega_-,\phi_-)$ and $D(\omega_+,\phi_+)$ have the same sign, with the determinant $D(\omega,\phi) = a_{11}a_{22} - a_{12}a_{21}$ being constructed from (11.49). If this condition is not met, there exists the possibility of finding an intermediate couple ω', ϕ' such that $D(\omega', \phi') = 0$, for which the right-hand side of (11.50) diverges. For ideal monatomic gases ($\gamma = \frac{5}{3}$) the determinant $D(\omega,\phi)$ reads

$$D(\omega,\phi) = -\left\{\left[\tfrac{5}{2}\phi^2 + r_1\omega^2 - \left(\tfrac{3}{2}r_1 + r_2\right)\omega\phi - r_1\omega\right]\left(r_3 - \tfrac{8}{15}r_2\right)\right.$$

$$\left. +\left[\left(\tfrac{5}{2} - r_2\right)\phi - \tfrac{3}{2}r_1\omega\right]\left[\tfrac{8}{45}(5 - 3r_2)\phi + \left(\tfrac{16}{45}r_2 - r_3\right)\omega - \tfrac{16}{45}r_2\right]\right\}. \qquad (11.61)$$

In model GI, taking into account that $\omega = 1 - \phi$, we can write (11.61) as

$$D(1 - \phi, \phi) = -(1 - \phi)(9.2\phi^2 - 7.2\phi + 1). \qquad (11.62)$$

According to solutions (11.52), the allowed domain of variation of ϕ is $\frac{3}{8}(1 - \varepsilon) \leq \phi \leq \frac{3}{8}(1 + \varepsilon)$, because $\gamma = \frac{5}{3}$. Thus $D(\omega, \phi)$ will be positive over all the domain until it includes the point $\phi = 0.1805$, for which (11.62) vanishes. The corresponding value of ε is $\varepsilon = 0.519$ which in view of (11.52c) yields a critical Mach number $\mathcal{M}_c = 1.65$.

In model GII, (11.62) takes the form

$$D(1 - \phi, \phi) = -(1 - \phi)(3.094\phi^2 - 0.577\phi - 0.016). \qquad (11.63)$$

Now, both $D(\omega_-, \phi_-)$ and $D(\omega_+, \phi_+)$ are negative for $\mathcal{M}$ lower than $\mathcal{M}_c = 2.07$, in which case $D(\omega_-, \phi_-)$ becomes positive. A regular shock structure exists in this case for $\mathcal{M} < 2.07$. Observe, finally, that in the Navier–Stokes–Fourier theory, (11.61) reduces to

$$D(1 - \phi, \phi) = -\tfrac{20}{9}(1 - \phi)\phi^2, \qquad (11.64)$$

which is always negative. This is why the NSF model displays a regular shock wave structure whatever the value of $\mathcal{M}$. Regular shock wave structures have been observed experimentally for $\mathcal{M}$ up to 4, and in computer simulations for $\mathcal{M}$ up to 10.

The main feature of these results is that the proposed versions of EIT predict a regular solution only for sufficiently low values of the Mach number, namely for $\mathcal{M} < 1.65$ in model GI and for $\mathcal{M} < 2.07$ in model GII. Models incorporating higher-order fluxes raise the critical Mach number: for instance, $\mathcal{M}_c = 1.887$ for 21 moments and $\mathcal{M}_c = 2.809$ for 51 moments [11.16, 18]. Some authors [11.17] argue that each characteristic speed of the system, and not only the maximum one, would produce a singularity. However, numerical calculations including higher-order moments indicate that the structure remains regular for speeds lower than the maximum characteristic one [11.18]. Whether the inclusion of an infinite number of moments will yield a finite or an infinite value for the critical Mach number, as well as the role of non-linear terms, remain open questions [11.16–18].

Problems

11.1 Carrassi and Morro took $\tau_1 = \tau_2 = \Lambda/c\pi$ in order to fit the experimental data for the ultrasonic speed presented in Table 11.2. (a) Compare these values for τ_1 and τ_2 with those obtained from the kinetic theory of gases, for which $\lambda = \frac{5}{2}(k_B^2 Tn/m)\tau_1$ and $\eta = p\tau_2$. (b) Obtain the value of the high-frequency limit of the phase speed v_p for $\tau_1 = \tau_2 = \Lambda/c\pi$ and compare with the values given by (11.35).

11.2 (a) Show that if one imposes the requirement that plane waves must propagate with speed lower than the speed of light c, the inequality $\beta^2 \leq c^2 \alpha_1 \alpha_2$ must be obeyed. (b) Check that this inequality is true for the coefficients listed in Table 17.1 for relativistic gases. [*Hint*: Obtain the high-frequency limit of the perturbations of δq_x, δP_{xz}^v for a wave vector k in the z direction according to (11.11–12).]

11.3 The criterion that wave speeds should be lower than the speed of light c is used to develop constraints on possible models for the equations of state of nuclear matter. As a simple illustration, we consider the equations of state in the low-temperature limit:

$$\rho = n[m_0 + E_0(n)] , \qquad p = n^2[dE_0(n)/dn] ,$$

with $E_0(n)$ being the ground-state energy per nucleon, m_0 the nucleon rest mass, n the number of particles per unit volume, and ρ the mass-energy density. (a) Calculate the adiabatic sound speed $v_s = [(\partial p/\partial \rho)_s]^{1/2}$ for this system. (b) One usual equation for $E_0(n)$ is

$$E_0(n) = E_0(n_0) + \frac{K}{18nn_0}(n - n_0)^2 ,$$

with $E_0(n_0) = -16$ MeV, $K = 210$ MeV and $n_0 = 0.160$ fm^{-3}. Find the range of values of n/n_0 for which this equation satisfies $v_s \leq c$. [T. S. Olson and W. A. Hiscock, Phys. Rev. C **39** (1989) 1818.]

11.4 The experimental results for ultrasonic absorption in metals have been described phenomenologically by taking for the viscosity η of the electron gas the expression $\eta = \frac{1}{5} nm\, v_F^2 \tau$, with v_F being the Fermi velocity, n the electron number density, and m the electron mass. Assume for $\mathbf{P}^v$ the evolution equation

$$\tau \dot{\mathbf{P}}^v + \mathbf{P}^v = -2\eta \mathbf{V} + A\nabla^2 \mathbf{P}^v \;,$$

where A is a coefficient which has the dimensions of (length)^2. In contrast with classical gases, where the relaxational effects are more important than the non-local ones, in Fermi gases the latter are more important and the former may be neglected. (a) Using the balance laws of mass, momentum and energy and the corresponding simplified equation for $\mathbf{P}^v$, show that the inverse attenuation time $\alpha_t = -\,\mathrm{Im}\,\omega$ of longitudinal waves is

$$\alpha_t = \tfrac{2}{3}\eta\omega^2(\rho_0 c^3)^{-1}(1 - Ak^2) \;.$$

(b) Compare with the second-order form of the general microscopic expression

$$\frac{2\alpha\rho c\tau}{nm} = \frac{1}{3}\frac{(kl)^2 \tan^{-1}(kl)}{kl - \tan^{-1}(kl)} - 1$$

and obtain the value of coefficient A [D. Jou, F. Bampi and A. Morro, J. Non-Equilib. Thermodyn. **7** (1982) 201.]

11.5 The profile of the pressure p across the layer of a weak shock may be written as

$$p - \frac{p_+ + p_-}{2} = \frac{p_+ - p_-}{2}\tanh\!\left(\frac{x}{D}\right),$$

with

$$D = \frac{8av^2}{(p_+ - p_-)(\partial^2 v/\partial p^2)_s}\;,$$

where v the specific volume and $a = (2\rho c^3)^{-1}[(4/3)\eta + (\lambda/c_p)(\gamma - 1)]$, c_p being the heat capacity at constant pressure per unit mass. Obtain the relation between the width of the shock layer $2D$ and the mean free path ℓ_- for a monatomic perfect gas for $\mathcal{M} = 1.2$. (*Hint*: Use the following results: (1) $\eta = \frac{1}{3}nm\langle v\rangle\ell$, with $\langle v\rangle = [8k_BT/(\pi m)]^{1/2}$; (2) $\lambda =$

$\frac{5}{2}\eta c_v$, with c_v being the heat capacity at constant volume per unit mass; (3) for a perfect gas $(\partial^2 v/\partial p^2)_s = (\gamma + 1)v\gamma^{-2}p^{-2}$; and (4) the relation p_+/p_- in (11.48).)

11.6 The viscous pressure in a one-dimensional flow is given by $P^v_{xx} = \frac{4}{3}\eta(\partial v_x/\partial x)$. One may estimate the viscous pressure in the shock as $P^v_{xx} = \frac{4}{3}\eta[(v_+ - v_-)/(2D)]$, with $2D$ the thickness of the shock layer. (a) By using the value of D derived in the previous problem, determine the ratio P^v_{xx}/p for a shock wave with $\mathcal{M} = 1.2$. (b) Using an analogous procedure, evaluate the heat flux across a shock wave with $\mathcal{M} = 1.2$.

Chapter 12

Generalised Hydrodynamics

In ordinary fluids, the relaxation times of the fluxes are usually very small. One could therefore ask whether the relaxational effects considered in extended irreversible thermodynamics (EIT) are observable. The answer is affirmative: it is possible to obtain information on such effects either from experiments or numerical simulations. Computer simulations were analysed in Chap. 9; in this chapter, we discuss the experimental aspects.

Experimental information has been obtained from light- and neutron-scattering, which allows not only to measure the relaxation times of the fluxes, but also to determine their frequency and wavelength dependences through a recent discipline known as generalised hydrodynamics. In this approach, the relaxation times, which play a central role in EIT, enjoy the status of relevant experimental quantities. On the other hand, computer simulations have opened the road to situations not directly accessible to experiments.

One of the most unexpected results of the light and neutron experiments and related computer simulations is that the behaviour of a fluid at microscopic scales may still be described by means of hydrodynamical equations, provided that the transport coefficients are assumed to be functions of the frequency and the wavevector. This has a special relevance in the context of EIT, as it shows that the extension of hydrodynamic concepts to a regime of high frequencies and short wavelengths is relevant. Extended thermodynamics appears as the thermodynamic framework consistent with generalised hydrodynamics, in the same way that classical irreversible thermodynamics subtends classical hydrodynamics.

12.1 Density and Current Correlation Functions

Spontaneous microscopic fluctuations are present in any macroscopic system. It is generally assumed that their decay parallels the decay of external perturbations. This means that the analysis of fluctuations may provide as much information on the dynamics of the system as the study of external perturbations.

Velocity fluctuations are intimately related to the time correlation function of the density and the momentum of the system. In general terms, a time correlation function is defined as the thermodynamic average of the product of two dynamical variables, each of them expressing at any time and any position in space the spontaneous deviation of a fluid property from its equilibrium average. The time correlation function of the density is important to investigate the non-equilibrium properties of fluids, because it contains most of the relevant information about the dynamics of the system. In that respect, the importance of the density time-correlation function for transport properties is comparable to that of the partition function for equilibrium properties in statistical mechanics.

The time correlation function of the density may be directly derived from light and neutron inelastic spectroscopy or from computer simulations. The ability to perform experiments or simulations at high-frequency and short-wavelength regimes has motivated a very strong interest for that field [12.1–2]. In the above-mentioned experiments, it is possible to work at frequencies so high that they become comparable to the inverse of the relaxation times of fluxes in liquids. Therefore, generalised hydrodynamics constitutes a stimulating basement for checking and discussing the hypotheses and results of EIT.

As mentioned earlier, one of the basic ideas behind generalised hydrodynamics is to keep as far as possible the structure of the classical hydrodynamics transport equations by allowing the usual transport coefficients to depend on the frequency and the wavelength; in physical space, these coefficients take the form of memory functions reflecting the spatial and temporal non-locality of the transport equations. To evaluate these transport coefficients and the corresponding memory functions is among the most challenging problems in modern non-equilibrium statistical mechanics.

The central quantities for a fluid are, from the point of view of generalised hydrodynamics, the density correlation function and the momentum or current correlation function. They are respectively defined as

$$C_{nn}(r',t',r'',t'') = V\langle \delta n(r',t')\delta n(r'',t'')\rangle,$$

$$C_{v_i v_j}(r',t',r'',t'') = V\langle \delta[\rho v_i(r',t')]\delta[\rho v_j(r'',t'')]\rangle,$$

$$(12.1)$$

with $\delta A(r,t) = A(r,t) - \langle A(r,t) \rangle$, A being a dynamical variable such as n, ρ, or v, and $\langle A \rangle$ standing for the equilibrium average; V is the total volume and n the number of particles per unit volume. At equilibrium, which means a homogeneous and stationary state, expressions in (12.1) may only depend on the relative distance $r = |r' - r''|$ and the time interval $t = t' - t''$.

We introduce the Fourier transforms of (12.1):

$$S(k,t) = \int dr \exp[ik \cdot r] C_{nn}(r,t), \tag{12.2a}$$

$$J_{ij}(k,t) = \int dr \exp[ik \cdot r] C_{v_i v_j}(r,t). \tag{12.2b}$$

From symmetry requirements, the current time correlation function (12.2b) may be written as

$$J_{ij} = \frac{k_i k_j}{k^2} J_l(k,t) + \left(\delta_{ij} - \frac{k_i k_j}{k^2} \right) J_t(k,t), \tag{12.3}$$

where $J_l(k, t)$ and $J_t(k, t)$ are the longitudinal and transverse current correlation functions, respectively, and $k^2 = k \cdot k$, with k being the wavevector.

Furthermore, one denotes the Laplace transform of (12.2) as

$$\tilde{S}(k,\omega) = \int_{-\infty}^{\infty} dt \exp[-i\omega t] S(k,t),$$

$$\tag{12.4}$$

$$\tilde{J}_{ij}(k,\omega) = \int_{-\infty}^{\infty} dt \exp[-i\omega t] J_{ij}(k,t),$$

where $\tilde{S}(k,\omega)$ and $\tilde{J}_{ij}(k,\omega)$ are known respectively as the spectral density of the density and velocity fluctuations.

From the continuity equation (1.12) it follows that

$$\frac{\partial n_k(t)}{\partial t} = ik \cdot j_k(t), \tag{12.5}$$

where $n_k(t)$ and $j_k(t)$ denote the Fourier transform of the number density n and the current density nv, respectively. From (12.5) one obtains the basic relation between $\tilde{S}(k,\omega)$ and $\tilde{J}_{ij}(k,\omega)$, namely

$$\tilde{J}_l(k,\omega) = \frac{\omega^2}{k^2} \tilde{S}(k,\omega). \tag{12.6}$$

Although the correlation functions $\tilde{J}_l(k,\omega)$ and $\tilde{S}(k,\omega)$ contain the same information, some spectral characteristics are displayed more clearly by considering one quantity rather than the other, as will be outlined below. The limiting value of $\tilde{S}(k,\omega)$ for $\omega = 0$ is the so-called static structure factor $\tilde{S}(k)$, a frequently used quantity in the study of equilibrium properties of fluids.

12.2 Spectral Density Correlation

12.2.1 The Classical Hydrodynamical Approximation

Let us start with the classical hydrodynamical description. The spectral density of the density fluctuations at low ω and k is easily derivable from the expressions of the linearized hydrodynamical equations (11.1–3). After application of Laplace and Fourier transforms to the balance equations of mass, momentum and energy (11.1–3), one obtains

$$s\delta\tilde{\rho}_k(s) + \mathrm{i}k \cdot \tilde{j}_k(s) = \delta\tilde{\rho}_k(0),$$

$$(s + v_l k^2)\tilde{j}_k(s) + \frac{c^2}{\gamma}\mathrm{i}k\left[\delta\tilde{\rho}_k(s) + \alpha\tilde{g}_k(s)\right] = \tilde{j}_k(0), \qquad (12.7)$$

$$(s + \gamma\chi k^2)\tilde{g}_k(s) + \frac{\gamma - 1}{\alpha}\mathrm{i}k \cdot \tilde{j}_k(s) = \tilde{g}_k(0),$$

where s stands for $\mathrm{i}\omega$, a usual notation in generalised hydrodynamics and which cannot in this chapter be confused with the entropy, $g(s)$ is the Laplace–Fourier transform of $\rho_0\delta T(r,t)$ and v_l is the longitudinal viscosity, given by $v_l = \left(\frac{4}{3}\eta + \zeta\right)\rho^{-1}$. Eliminating $\tilde{j}_k(s)$ and $\tilde{g}_k(s)$ from (12.7), one is led to

$$\delta\tilde{\rho}_k(s) = \frac{N(k,s)}{M(k,s)}, \qquad (12.8)$$

where

$$M(k,s) = \det\begin{vmatrix} s & \mathrm{i}k_x & \mathrm{i}k_y & \mathrm{i}k_z & 0 \\ \gamma^{-1}c^2\mathrm{i}k_x & s + v_l k^2 & 0 & 0 & \gamma^{-1}\alpha c^2\mathrm{i}k_x \\ \gamma^{-1}c^2\mathrm{i}k_y & 0 & s + v_l k^2 & 0 & \gamma^{-1}\alpha c^2\mathrm{i}k_y \\ \gamma^{-1}c^2\mathrm{i}k_z & 0 & 0 & s + v_l k^2 & \gamma^{-1}\alpha c^2\mathrm{i}k_z \\ 0 & \alpha^{-1}(\gamma - 1)\mathrm{i}k_x & \alpha^{-1}(\gamma - 1)\mathrm{i}k_y & \alpha^{-1}(\gamma - 1)\mathrm{i}k_z & s + \gamma\chi k^2 \end{vmatrix},$$

and

$$N(k,s) = \det \begin{vmatrix} \delta\tilde{\rho}_k(0) & ik_x & ik_y & ik_z & 0 \\ \tilde{j}_{kx}(0) & s+v_l k^2 & 0 & 0 & \gamma^{-1}\alpha c^2 ik_x \\ \tilde{j}_{ky}(0) & 0 & s+v_l k^2 & 0 & \gamma^{-1}\alpha c^2 ik_y \\ \tilde{j}_{kz}(0) & 0 & 0 & s+v_l k^2 & \gamma^{-1}\alpha c^2 ik_z \\ \tilde{g}_k(0) & \alpha^{-1}(\gamma-1)ik_x & \alpha^{-1}(\gamma-1)ik_y & \alpha^{-1}(\gamma-1)ik_z & s+\gamma\chi k^2 \end{vmatrix}.$$

$$(12.9)$$

From this result, it is easy to construct the correlation function $\langle \delta\tilde{\rho}_k^*(0)\delta\tilde{\rho}_k(s)\rangle$ by taking into account that cross correlations involving $j_k(0)$ and $g_k(0)$ do not appear, as ρ, T, and v are here statistically independent variables. A superscript asterisk denotes the complex conjugate. In this way one finds that

$$\langle \delta\tilde{\rho}_k^*(0)\delta\tilde{\rho}_k(s)\rangle = \frac{(s+v_l k^2)(s+\gamma\chi k^2)+(1-\gamma^{-1})(ck)^2}{(s-s_0)(s-s_-)(s-s_+)}\langle \delta\tilde{\rho}_k^*(0)\delta\tilde{\rho}_k(0)\rangle, \quad (12.10)$$

where s_0, $s_\pm$ are the roots of the determinant $M(k, s)$, i.e. solutions of

$$z^3 + z^2(\gamma a + b) + z(1+\gamma ab) + a = 0, \qquad (12.11)$$

with $a = \chi k/c$, $b = v_l k/c$, and $z = s/ck$, where c is the Laplace sound velocity and k is the magnitude of wavevector k. The quantities a and b are small for typical light-scattering experiments in an ordinary liquid, where $k \approx 10^5$ cm^{-1} and $c \approx 10^5$ cm/s. Up to the first order in a and b, which means the order in k^2, the solutions of (12.11) are given by

$$z_0 = -a, \qquad z_\pm = \pm i - (1/2)[b + (\gamma-1)a], \qquad (12.12)$$

or, in terms of s

$$s_0 = -\chi k^2, \qquad s_\pm = \pm ick - \Gamma k^2, \qquad (12.13)$$

where $\Gamma = \frac{1}{2}[v_l + (\gamma-1)\chi]$ is related to the sound absorption coefficient $\Gamma' = -\,\mathrm{Im}\,k$ by $\Gamma = \Gamma' c/k^2$.

It is then easy to calculate the spectral density of the density fluctuations, defined as

$$\frac{\tilde{S}(k,\omega)}{\tilde{S}(k)} = 2\,\mathrm{Re}\frac{\langle \delta\tilde{\rho}_k^*(0)\delta\tilde{\rho}_k(s=i\omega)\rangle}{\langle \delta\tilde{\rho}_k^*(0)\delta\tilde{\rho}_k(0)\rangle}; \qquad (12.14)$$

with $\tilde{S}(k)$, the static structure factor, given by $\tilde{S}(k) = \langle \delta\tilde{\rho}_k^*(0)\delta\tilde{\rho}_k(0)\rangle$. Using (12.10) and (12.13), one obtains for (12.14)

$$\frac{\tilde{S}(k,\omega)}{\tilde{S}(k)} = \frac{\gamma-1}{\gamma}\frac{2\chi k^2}{\omega^2+(\chi k^2)^2} + \frac{1}{\gamma}\left[\frac{\Gamma k^2}{(\omega+ck)^2+(\Gamma k^2)^2} + \frac{\Gamma k^2}{(\omega-ck)^2+(\Gamma k^2)^2}\right]$$

$$+\frac{k}{\gamma c}[\Gamma+(\gamma-1)\chi]\left[\frac{\omega+ck}{(\omega+ck)^2+(\Gamma k^2)^2} - \frac{\omega-ck}{(\omega-ck)^2+(\Gamma k^2)^2}\right]. \quad (12.15)$$

This result is important because it shows the structure of the spectrum which can be measured by light-scattering spectroscopy of simple monatomic liquids. In most cases, the spectrum is well approximated by the first three terms in (12.15), which are a sum of three Lorentzians (Fig. 12.1). The central (Rayleigh) peak arises from fluctuations at constant pressure and corresponds to the thermal diffusivity mode. The other two peaks, shifted in frequency by $-ck$ and $+ck$, are the Stokes and anti-Stokes components of the Brillouin doublet, and they correspond to acoustic modes. The amplitude of the last two terms in (12.15) is several orders of magnitude smaller than the amplitude of the Lorentzians. A measurement of the position and widths of the Rayleigh and the Brillouin peaks allows one to determine the sound velocity c, thermal diffusivity χ and sound absorption factor Γ. Note, finally, that because of (12.6) the longitudinal current correlation will lack the central Rayleigh peak owing to the ω^2 factor. The ratio of the integrated intensity of the Rayleigh peak, I_R, and the integrated intensities of the Brillouin peaks, I_B, is known as the Landau–Placzek ratio. Direct integration of (12.15) yields $I_R(2I_B)^{-1} = \gamma - 1$, so that this ratio provides a measurement of γ.

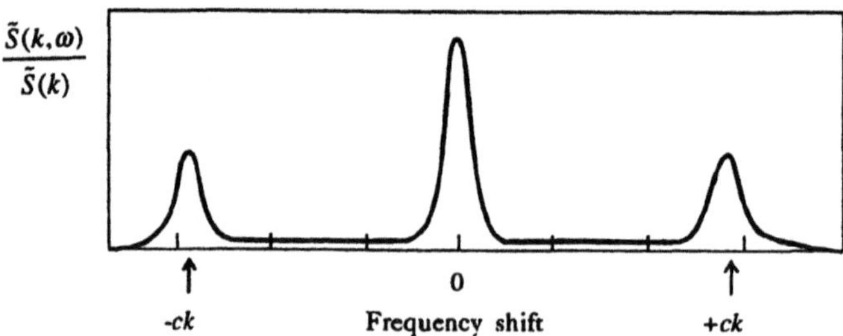

Fig. 12.1. Spectrum of the density fluctuations in a simple fluid. The central (Rayleigh) peak corresponds to the thermal diffusivity mode and the two lateral (Brillouin) peaks, shifted in frequency $\pm ck$, correspond to the acoustic modes. The height and width of the respective peaks are given by equation (12.15).

The frequency range of light-scattering experiments in liquids exceeds considerably that of ultrasonic experiments. Nevertheless, at still higher frequencies, like those found in neutron-scattering experiments in liquids or light-scattering experiments in gases, the above analysis based on classical hydrodynamics is no longer valid. New features appear, like, for instance, new propagation modes or dispersion effects in sound propagation and sound absorption. In the next sections we shall show that these features are well accounted for by EIT.

12.2.2 The EIT Description

In EIT, the linearized hydrodynamical equations (11.1–3) must be complemented by the Maxwell–Cattaneo equations (2.60–62) for the fluxes, which are uncoupled from the set (11.1–3). By substituting solutions of the form $\exp(-st)$ in (2.60–62), it is readily checked that the viscosity and the heat diffusivity coefficients take the form

$$\nu_l(s) = \frac{4}{3}\eta\frac{1}{1+s\tau_2} + \zeta\frac{1}{1+s\tau_0}, \qquad \chi(s) = \chi\frac{1}{1+s\tau_1}, \tag{12.16}$$

where τ_0, τ_1, and τ_2 are the respective relaxation times of the bulk viscous pressure, the heat flux and the viscous pressure tensor. After inserting these expressions into the dispersion relation (12.10), the resulting equation exhibits six roots, three more than the usual three longitudinal modes, and no simple analytic solution exists.

To explore the consequences of (12.16), consider a simple model in which $\tau_1 = \tau_2 = 0$. Such a model is useful for the description of thermal relaxation in molecular fluids, where there is a coupling between internal and translational degrees of freedom. This coupling becomes important when the energy fluctuations in the internal degrees of freedom relax on a time scale comparable to the hydrodynamic characteristic times $(\Gamma k^2)^{-1}$ and $(\chi k^2)^{-1}$. In such a case the first of expressions (12.16) can be written as [12.2]

$$\nu_l(s) = \nu_l + \frac{c_\infty^2 - c_0^2}{1+s\tau_0}\tau_0, \tag{12.17}$$

where c_∞ and c_0 stand for the infinite and zero-frequency sound speed, respectively, and ν_l is the frequency-independent part of $\nu_l(s)$ associated with translational motion whereas the second term in (12.17) describes the contribution of the internal degrees of freedom to $\nu_l(s)$. The frequency domain for which $\nu_l(s = i\omega)$ is the most sensitive to the frequency dependence is for $\omega \approx 10^{-10}$s. When (12.17) is introduced into (12.11), the dispersion relation, under the restrictions $\nu_l k/c_0 \ll 1$ and $\chi k/c_0 \ll 1$, reads as

$$(s+\chi k^2)(s+i\omega_s + \Gamma_s k^2)(s+c_0^2 c_s^{-2}\tau_0^{-1}) = 0 \tag{12.18}$$

with

$$\omega_s \equiv c_s k = c_0 k \left[B + \left(B^2 + (c_0 k \tau_0)^{-2} \right)^{1/2} \right]^{1/2},$$

$$B = \frac{1}{2}\left[\left(\frac{c_\infty}{c_0}\right)^2 - \left(\frac{1}{c_0 \tau_0 k}\right)^2 \right],$$

$$\Gamma_s = \Gamma + \frac{1}{2}\chi\left[1 - \left(\frac{c_0}{c_s}\right)^2 \right] + \frac{1}{2}\left(1 - \gamma\chi\tau_0 k^2\right)\frac{c_\infty^2 - c_0^2}{1 + \omega_s^2 \tau_0^2}\tau_0.$$

Making use of (12.9) and (12.17), we see that the spectral distribution may be expressed as a sum of four Lorentzians, namely

$$\frac{\tilde{S}(k,\omega)}{\tilde{S}(k)} = \left(1 - \gamma^{-1}\right)\frac{2\chi k^2}{\omega^2 + (\chi k^2)^2} + R_C\frac{2(c_0/c_s)^2\tau_0^{-1}}{\omega^2 + (c_0^2 c_s^{-2}\tau_0^{-1})^2}$$

$$+ R_B\left[\frac{\Gamma_s k^2}{(\omega + c_s k)^2 + (\Gamma_s k^2)^2} + \frac{\Gamma_s k^2}{(\omega - c_s k)^2 + (\Gamma_s k^2)^2} \right], \tag{12.19}$$

with

$$R_C = \left\{ \left[1 - \left(\frac{c_s}{c_0}\right)^2 \right]\left[\left(1 - \gamma^{-1}\right)c_0^2 k^2 + \left(\frac{c_0}{c_s}\right)^4\tau_0^{-2} \right] + \left(c_\infty^2 - c_0^2\right)k^2 \right\}\left[\omega_s + \left(\frac{c_0}{c_s}\right)^4\tau_0^{-2} \right]^{-1},$$

$$R_B = \left\{ \left[1 - \left(\frac{c_s}{c_0}\right)^2\left(1 - \gamma^{-1}\right) \right]\left[\omega_s^2 + \left(\frac{c_0}{c_s}\right)^4\tau_0^{-2} \right] + \left(c_\infty^2 - c_0^2\right)k^2 \right\}\left[\omega_s + \left(\frac{c_0}{c_s}\right)^4\tau_0^{-2} \right]^{-1}.$$

A first difference between (12.15) and (12.19), except for neglecting the last two terms of (12.15), is a change in position and width of the Brillouin peaks, given now respectively by c_s and Γ_s instead of c and Γ. But more important is the presence of a new central peak of width τ_0^{-1}. Such a peak is difficult to detect, since it appears as a broad background between the Rayleigh and the Brillouin peaks. Note that $c_s > c$ and $\Gamma_s > \Gamma$, so that the Brillouin peaks are pushed away from the Rayleigh line and become wider than in the absence of relaxational effects. Observe finally that when $c_\infty = c_0 = c$, expression (12.19) reduces to the classical result (12.15), except for the last two terms of (12.15), which have been neglected in (12.19) for simplicity.

12.3 The Transverse Velocity Correlation Function: the EIT Description

Although not directly observable by means of light- or neutron-scattering experiments, the transverse velocity correlation function is worth studying from a fundamental point of view. On the one hand, its analysis is much simpler than that of the longitudinal velocity correlation function and, on the other hand, much effort has been devoted to its study through computer simulations and molecular dynamics.

The transverse velocity correlation function is defined as

$$C_t(r,t,r',t') = \langle \delta v_t(r,t)\delta v_t(r',t')\rangle, \tag{12.20}$$

where $\delta v_t = v_t - \langle v_t\rangle$ is the fluctuation of the component of the velocity transverse to the wavevector of the perturbation. From now on, to fix the ideas, we consider the wavevector in the x direction and the transverse velocity in the y direction. The Fourier transform of $C_t(r-r', t-t')$ will be denoted $J_t(k, t)$. The equation satisfied by $J_t(k, t)$ is derived from the linearized equation of motion (11.10), which for the transverse velocity component is given by

$$\rho_0 \frac{\partial v_y}{\partial t} = -\frac{\partial P_{iy}^v}{\partial x_i}, \tag{12.21}$$

wherein the evolution of $\mathbf{P}^v$ (in absence of bulk effects) is governed by (2.66), namely

$$\tau_2 \frac{\partial P_{ij}^v}{\partial t} + P_{ij}^v + 2\eta V_{ij} - 2\beta^2\eta\lambda T^3 \frac{\partial^2 P_{ij}^v}{\partial x_k \partial x_k} = 0. \tag{12.22}$$

Combination of (12.21) and (12.22) and application of a Fourier transform leads to

$$\tau_2 \frac{\partial^2 J_t}{\partial t^2} + \left(1 + l^2 k^2\right)\frac{\partial J_t}{\partial t} + \nu k^2 J_t = 0, \tag{12.23}$$

with $l^2 = 2\eta\lambda T^3\beta^2$. A Laplace transform with respect to time allows to write (12.23) as

$$\frac{\tilde{J}_t(k,0)}{\tilde{J}_t(k,s)} = \frac{s + (1 + l^2 k^2)\tau_2^{-1}}{s^2 + (1 + l^2 k^2)\tau_2^{-1}s + \nu\tau_2^{-1}k^2}. \tag{12.24}$$

When $\tau_2 = 0$ and $l = 0$, one recovers the transverse correlation function of classical hydrodynamics. When $l = 0$ and $\tau_2 \neq 0$, one obtains the results of the Maxwell viscoelastic model. Expression (12.24) may be formulated in terms of the frequency by taking into account that $\tilde{J}_t(k,\omega) = 2\,\mathrm{Re}\big[\tilde{J}_t(k,s=i\omega)\big]$ and noting that $J_t(k,\,t=0) = v_0^2 = k_B T/m$. The result is

$$\tilde{J}_t(k,\omega) = \frac{2v_0^2\tau_2^{-2}vk^2(1+l^2k^2)}{\left\{\omega^2 - \left[\dfrac{vk^2}{\tau_2} - \dfrac{(1+l^2k^2)^2}{2\tau_2^2}\right]\right\}^2 + \left[\dfrac{vk^2}{\tau_2} - \dfrac{(1+l^2k^2)^2}{4\tau_2^2}\right]\left[\dfrac{(1+l^2k^2)^2}{\tau_2^2}\right]}. \tag{12.25}$$

When $l = 0$, this expression reduces to

$$\tilde{J}_t(k,\omega) = \frac{2v_0^2vk^2}{\tau_2^2\omega^4 + (1-2v\tau_2 k^2)\omega^2 + (vk^2)^2}, \tag{12.26}$$

which is represented in Fig. 12.2. When $\tau_2 = 0$ and $l = 0$, (12.25) simplifies to

$$\tilde{J}_t(k,\omega) = \frac{2v_0^2vk^2}{\omega^2 + (vk^2)^2}. \tag{12.27}$$

The most relevant property of (12.25) in comparison with (12.27), obtained from the classical Navier–Stokes theory, is the occurrence of a maximum of $\tilde{J}_t(k,\omega)$ at a non-zero frequency given by

$$\omega_m = \left[\frac{vk^2}{\tau_2} - \frac{(1+l^2k^2)^2}{2\tau_2^2}\right]^{1/2}. \tag{12.28}$$

The condition for observing such a maximum is that k exceeds the critical value k_c defined by $2v\tau_2 k_c^2 = (1+l^2k_c^2)^2$ and corresponding to $\omega_m = 0$. In the low-k limit, the maximum of $\tilde{J}_t(k,\omega)$ is attained at zero frequency (Fig. 12.2), which means a purely diffusive mode. A maximum at $\omega \neq 0$ reflects the property of propagation of shear waves with speed k/ω_m and is a typical feature of elastic solids rather than viscous fluids. Molecular dynamics computations on argon confirms the presence of such a resonant value for $k > k_c = 0.63\mathring{A}^{-1}$.

Another comment concerns the ratio v/τ_2 that appears in the previous formula. Recall that for an ideal gas, we have found in (5.45b) that

$$\frac{v}{\tau_2} = \rho^{-1}p. \tag{12.29}$$

For a liquid, this result must be replaced by [12.2, 4]

$$\frac{\nu}{\tau_2} = \rho^{-1}G_\infty(k), \tag{12.30}$$

where $G_\infty(k)$ is the wavenumber-dependent high-frequency shear modulus. Since (12.30) is exactly the same expression as that obtained in the theory of viscoelasticity, we may identify τ_2 with the relaxation time of viscoelastic materials.

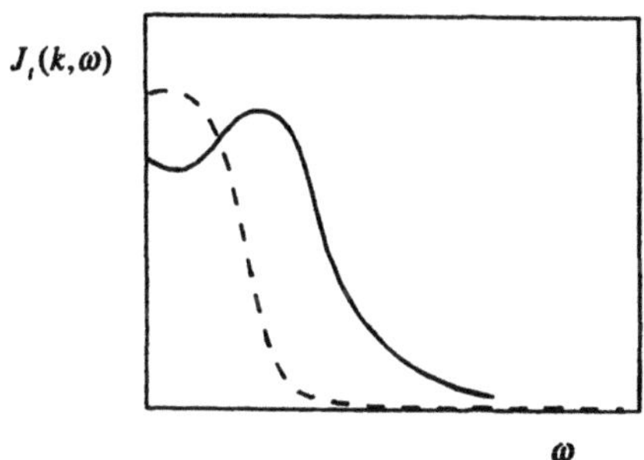

Fig. 12.2. Transverse velocity correlation function $J_t(k,\omega)$ in terms of the frequency for a given k:- - - - classical Navier-Stokes theory (12.27); ————— extended theory (12.26).

The present macroscopic analysis may also be compared with the microscopic Mori–Zwanzig theory presented in Chap. 8. This formalism gives an exact equation for the time evolution of the correlation function in terms of a memory function, namely

$$\frac{\partial J_t(k,t)}{\partial t} = -\int_0^t dt'\, K(k,t-t')J_t(k,t') ; \tag{12.31}$$

$K(k,t)$ may be written as $K(k,t) = k^2 K'(k,t)$, with $K'(k,t)$ a generalised shear-viscosity function which depends on k and t. Usually the memory function is modelled mathematically because it is impossible to derive its explicit expression from the projection operator techniques. In the extended thermodynamic description [12.3], the generalised shear viscosity $K'(k,t)$ may be obtained explicitly by taking the Laplace transform of (12.21) and by comparing with

$$P_{iy}^\nu(k,s) = -ikK'(k,s)v_y(s). \tag{12.32}$$

This procedure yields

$$K'(k,s) = \frac{\eta}{\tau_2}\,\frac{1}{s+(1+l^2k^2)\tau_2^{-1}}, \tag{12.33a}$$

from which it follows that

$$K(k,t) = \frac{\eta k^2}{\tau_2}\exp\left[-(1+l^2k^2)\tau_2^{-1}t\right] . \tag{12.33b}$$

This expression may be written in terms of a k-dependent relaxation time $\tau_2(k)$ defined by

$$\tau_2(k) = \frac{\tau_2}{1+l^2k^2} , \tag{12.34}$$

in the form

$$K(k, t) \sim \exp[-t/\tau_2(k)] . \tag{12.35}$$

An expression of the same form as (12.34) has been used by Chung and Yip [12.6] to fit the computer simulation data for argon-like fluids. It was found that for argon at 85 K, $\tau_2 = 0.26 \times 10^{-12}$ s and $l = 4 \times 10^{-9}$ cm. It is worth emphasizing that (12.33) is precisely the exponential model for the memory function used in generalised hydrodynamics; however, in the present context it is not introduced ad hoc but arises naturally from the postulates of EIT. A more detailed model for $\tau_2(k)$ is discussed in Sect. 12.5.

The wavelength dependence of the relaxation time was also the subject of detailed investigations, because interesting experimental information is gained about it from the position of the maximum of the spectral function $\tilde{J}_l(k,\omega)$. Alley and Alder [12.5] have used computer simulations to study the generalised shear viscosity coefficient $\eta(k, s)$ in a fluid of hard spheres. They have shown that for a zero frequency and small wave number, molecular dynamical data may be fitted by an expression of the form

$$\eta(k,0) = \frac{\eta}{1+a^2k^2} , \tag{12.36}$$

where a, which incorporates the effect of spatial inhomogeneities, is a measure of the deviation from the Stokes friction law. It is worth to notice that expression (12.36) is equivalent to the EIT result (12.33a) for $s = 0$.

12.4 The Longitudinal Velocity Correlation Function: the EIT Description

The longitudinal velocity correlation function is defined in a way similar to (12.20) and is measurable because of its coupling with density fluctuations. In an isothermal system, neglecting relaxational effects of the heat flux, and introducing (2.64) into (2.65) and (2.66), one is led to the following evolution equations for the bulk and shear viscous pressures:

$$\tau_0 \dot{p}^v = -(p^v + \zeta \nabla \cdot v) + \lambda \zeta T^3 \beta'^2 \nabla^2 p^v + \lambda \zeta T^3 \beta' \beta'' \nabla(\nabla \cdot \overset{0}{\mathbf{P}}{}^v), \qquad (12.37)$$

$$\tau_2 (\overset{0}{\mathbf{P}}{}^v)^{\boldsymbol{\cdot}} = -(\overset{0}{\mathbf{P}}{}^v + 2\eta \overset{0}{\mathbf{V}}) + 2\lambda \eta T^3 \beta''^2 \nabla(\nabla \cdot \overset{0}{\mathbf{P}}{}^v) + 2\lambda \eta T^3 \beta' \beta'' (\nabla \overset{0}{\nabla} p^v)^s. \quad (12.38)$$

These expressions when introduced in the longitudinal part of the linearized momentum balance equation give

$$\rho_0 \frac{\partial v_l}{\partial t} = -\nabla(p + p^v) - (\nabla \cdot \overset{0}{\mathbf{P}}{}^v)_l. \qquad (12.39)$$

In the absence of temperature fluctuations, one has $\nabla p = (\rho k_T)^{-1} \nabla \rho$, where k_T is the isothermal compressibility. If the cross terms $\beta' \beta''$ are neglected [12.3], one obtains for $\tilde{J}_l(k,s)$ (the longitudinal correlation function)

$$\frac{\tilde{J}_l(k,0)}{\tilde{J}_l(k,s)} = s + \gamma c_T^2 \frac{k}{s} + \frac{k^2}{\rho_0} \left[\frac{\zeta}{\tau_0(k)s + 1} + \frac{4}{3} \frac{\eta}{\tau_2(k)s + 1} \right], \qquad (12.40)$$

where c_T stands for the isothermal sound velocity and $\tau_i(k)$ for $\tau_i(0)[1 + l_i^2 k^2]^{-1}$, with $l_0^2 = \lambda \zeta T^3 \beta'^2$ and $l_2^2 = 2\eta \lambda T^3 \beta''^2$.

When $\tau_0 = \tau_2 = 0$, i.e. in the Navier–Stokes regime, (12.40) reduces to the classical result

$$\frac{\tilde{J}_l(k,0)}{\tilde{J}_l(k,s)} = s + \gamma c_T^2 \frac{k^2}{s} + v_l k^2. \qquad (12.41)$$

The spectral distribution corresponding to (12.40) reads

$$J_l(k,\omega) = \frac{2v_0^2 \omega^2 k^2 A}{\left[\omega^2 - \gamma c_T^2 k^2 - \omega^2 k^2 (A_0 \tau_0(k) + A_2 \tau_2(k)) \right]^2 + (\omega k^2 v_l)^2}, \qquad (12.42)$$

with v_l defined by (12.16) and $A = A_0 + A_2$ given by

$$A_0 = \frac{\zeta}{\rho_0 \tau_0} \frac{\tau_0(k)}{1 + \omega^2 \tau_0^2(k)}, \qquad A_2 = \frac{4}{3} \frac{\eta}{\rho_0 \tau_2} \frac{\tau_2(k)}{1 + \omega^2 \tau_2^2(k)}.$$

In the limiting case $\tau_0 = \tau_2 = l = 0$, expression (12.42) simplifies to

$$J_l(k,\omega) = \frac{2v_0^2\omega^2 k^2 v_l}{(\omega^2 - \gamma c_T^2 k^2)^2 + (\omega k^2 v_l)^2}. \tag{12.43}$$

The most relevant feature of (12.42) compared to (12.43) is the presence of dispersion effects leading to a wavevector-dependent phase speed

$$v_p^2 = \frac{\omega^2}{k^2} = \gamma c_T^2 + A_0 \omega^2 \tau_0(k) + A_2 \omega^2 \tau_2(k). \tag{12.44}$$

At low frequencies ($\omega\tau_0 \ll 1$ and $\omega\tau_2 \ll 1$) one finds back the classical result $v_p^2 = \gamma c_T^2$, whereas in the high-frequency limit one gets

$$v_{p\infty}^2 = \frac{\zeta}{\rho_0 \tau_0} + \frac{4}{3}\frac{\eta}{\rho_0 \tau_2}. \tag{12.45}$$

Furthermore, the damping function is now

$$\Gamma(k,\omega) = \frac{1}{2}\left(\frac{\omega}{v_p}\right)^2 A, \tag{12.46}$$

which reduces to the classical expression for $\Gamma(k, \omega)$ at low frequencies but does not vanish at high frequencies, at variance with the classical approach.

It is of interest to connect these results with the memory function formalism. By analogy with (12.30), we introduce a memory function through

$$\frac{\partial J_l(k,t)}{\partial t} = -\int_0^t dt'\, K_l(k,t-t')J_l(k,t'). \tag{12.47}$$

The memory function corresponding to the EIT equations (12.37–38) is

$$K_l(k,t) = k^2\left[\gamma c_T^2 + \frac{\zeta}{\rho_0 \tau_0}\exp\left(-\frac{t}{\tau_0(k)}\right) + \frac{4}{3}\frac{\eta}{\rho_0 \tau_2}\exp\left(-\frac{t}{\tau_2(k)}\right)\right], \tag{12.48}$$

which shows explicitly that, in absence of heat flow, there are two exponentially decaying modes associated with the bulk and shear viscosities. This equation is far simpler than the corresponding one proposed in generalised hydrodynamics, although the general idea that $K_l(k, t)$ is governed by a two-relaxation-times model is well accepted [12.2].

By including thermal conductivity, expression (12.40) will contain a supplementary term and will take the form [12.3b]

$$\frac{\tilde{J}_l(k,0)}{\tilde{J}_l(k,s)} = s + \gamma c_T^2 \frac{k^2}{s} + \frac{k^2}{\rho_0}\left[\frac{\zeta}{\tau_0(k)s+1} + \frac{4}{3}\frac{\eta}{\tau_2(k)s+1}\right] + \frac{c_T^2 k^2(\gamma-1)}{s+\chi k^2}, \quad (12.49)$$

which corresponds to a memory function

$$K_l(k,t) = k^2\left\{\gamma c_T^2 + \frac{\zeta}{\rho_0\tau_0}\exp\left[-\frac{t}{\tau_0(k)}\right]\right.$$

$$\left. + \frac{4}{3}\frac{\eta}{\rho_0\tau_2}\exp\left[-\frac{t}{\tau_2(k)}\right] + c_T^2(\gamma-1)\exp(-\chi k^2 t)\right\}. \quad (12.50)$$

Some authors [8.5, 12.6] have proposed microscopic models where the dissipative fluxes (heat flux, viscous pressure) enter as independent variables (see Sect. 8.3). It is found that in the linear approximation these fluxes satisfy the phenomenological equations (2.64–66), with microscopic expressions for the coefficients. Although such models reproduce the predictions of EIT, it should be borne in mind that they are not backed by a thermodynamic theory, which is the main concern of the present book.

12.5 Influence of Higher-order Fluxes

In Sect. 5.6 we discussed the role of higher-order fluxes in EIT and showed to what extent their inclusion modifies the values of the speed of wave propagation. When introduced into the context of generalised hydrodynamics, these higher-order fluxes will influence the position of the peaks and, furthermore, may give additional peaks in the spectra.

Velasco and García-Colín [12.7] have computed the generalised transport coefficients as a function of the frequency and the wavelength, in the 13-, 20- and 26-moment approximations for hard spheres. Their results depend on the number of moments, but they exhibit a qualitatively similar behaviour.

In a series of papers, Weiss and Müller [12.8] have gone beyond these lower-order approximations and have analysed very carefully the spectrum of light scattering in gases in the framework of EIT, by including more and more higher-order fluxes. Their approach is essentially based on the Boltzmann equation for Maxwellian molecules and a closure is achieved by maximizing the entropy.

The number of peaks in the density spectrum depends on the number of higher-order fluxes. It turns out that, to describe the experimental data, one needs a large number of fluxes, for instance, to reproduce accurately the experimental data by Clark in xenon some 300 moments are needed. Under these conditions, additional higher-order moments no longer modify the results.

Another aspect of higher-order fluxes refers to the dependence of the relaxation time on the wave vector. For instance, within the limit $kl \gg 1$, the gas behaves like a system of independent free particles. In this case, the position distribution function at time t of particles having left the origin at time $t = 0$ is given by

$$f(r,t) = \int dr\, f_{eq}(p)\,\delta[r - (p/m)t], \tag{12.51}$$

with p being the momentum of the particles. When the Maxwellian distribution function is introduced into this expression, one obtains after several calculations (see Problem 12.4),

$$J_t(k,t) = \exp\left[-\tfrac{1}{2}(kv_0 t)^2\right], \tag{12.52}$$

where $v_0 = (k_B T/m)^{1/2}$ is the root-mean-square thermal speed. Therefore, one may identify $2^{-1/2}kv_0$ as the inverse of a relaxation time corresponding to a Gaussian decay of the form

$$J_t(k,t) = \exp\left[-\left(\frac{t}{\tau}\right)^2\right]. \tag{12.53}$$

Then, one has

$$\tau = \frac{2^{1/2}}{kv_0}, \tag{12.54}$$

from which it follows that the relaxation time varies as k^{-1} in the high-wavevector limit. However, the relaxation time given by (12.34) behaves in this high-wavevector limit as k^{-2}. To solve this contradiction, let us include higher-order fluxes in the description; in this case, the k dependence of the relaxation time will take the form

$$\tau_2(k) = \frac{\tau_2(0)}{1 + \dfrac{l^2 k^2}{1 + \dfrac{l^2 k^2}{1 + \dots}}}, \qquad (12.55)$$

which is a generalisation of (12.34). In the asymptotic limit, this fraction may be written as (see Problem 5.10)

$$\tau(k) = \frac{\tau_0}{l^2 k^2} \frac{\sqrt{1 + 4l^2 k^2} - 1}{2}, \qquad (12.56)$$

which for high values of k exhibits the required k^{-1} dependence. This shows how the introduction of a large number of higher-order fluxes may modify the scaling laws describing the behaviour of $\tau(k)$ from k^{-2} at the lowest-order approximation to k^{-1} in the asympotic limit of an infinite number of fluxes.

Problems

12.1 The wavelength of visible light ranges from 450 nm to 700 nm. (a) Determine $\omega\tau$ and kl for visible light in a typical monatomic gas (molecular radius $= 2 \times 10^{-10}$ m) at 273 K and 1 atm. (b) Evaluate $\omega\tau$ and kl in a liquid where $\tau = 2 \times 10^{-13}$ s and $l = 4 \times 10^{-11}$ m. (c) Find the energy of scattered neutrons such that $kl = 1$ in this liquid.

12.2 One of the consequences of higher-order hydrodynamics is that the shift of Brillouin peaks is related to the hypersound velocity rather than to the sound velocity. Consider, for instance, the following situation. The frequency shift corresponding to the Brillouin peaks in water for scattering at 90° for light of 0.6328 μm (this is the red light of a He-Ne laser) is 4.33×10^9 Hz. Obtain the hypersonic velocity of sound and compare it with the sound velocity 1491 m/s. (*Hint*: The frequency shift for scattering at an angle θ is given by $\omega = 2\omega_0 n(v/c)\sin(\theta/2)$, with v the hypersound velocity, c the speed of light, and n the index of refraction of the medium.)

12.3 Starting from expression (12.25) for the transverse velocity correlation function, find the critical wave number k_c at which propagating shear waves could be observed in a monatomic perfect gas, for instance, He at 273 K and 1 atm (shear viscosity of $\mathrm{He}^4 = 1.87 \times 10^{-5}$ kg m^{-1} s^{-1}; recall that $\eta = \frac{1}{3} nm\langle v\rangle l$ and $\tau = l/\langle v\rangle$).

12.4 Within the limit $kl \gg 1$, a gas behaves like a system of independent free particles. In this case the transverse velocity correlation function is given by (12.51). Show that when the Maxwellian distribution function is introduced into this expression one obtains

$$J_t(r,t) = (\pi v_0^2 t^2)^{-3/2} \exp\left[-(r/v_0 t)^2\right]$$

By using the Fourier transform show (12.52).

12.5 Compare the behaviour of the k-dependent viscosity found from the following second-order expansions: (a) an expansion in the fluxes (as in the Grad method)

$$\tau_2 \frac{\partial \mathbf{P}^v}{\partial t} + \mathbf{P}^v = -2\eta \mathbf{V} + \nabla \cdot \overset{0}{\mathbf{J}}{}^v,$$

$$\overset{0}{\mathbf{J}}{}^v = -\eta''\left\langle \nabla \mathbf{P}^v \right\rangle,$$

with $\overset{0}{\mathbf{J}}{}^v$ the flux of the viscous pressure tensor, η'' a phenomenological coefficient and $<...>$ the completely symmetrized part of the corresponding third-order tensor.

(b) a gradient expansion, as in the Burnett approach

$$\tau_2 \frac{\partial \mathbf{P}^v}{\partial t} + \mathbf{P}^v = -2\eta \mathbf{V} + \ell^2 \nabla^2 \mathbf{V},$$

(c) Compare them with (12.36). Note that (b) yields an unstable behaviour for high values of k, since the generalized viscosity becomes negative. [For a wide discussion of the instabilities arising in the Burnett approach to generalized hydrodynamics see I. V. Karlin, J. Phys. A: Math. Gen. **33** (2000) 8037].

Chapter 13

Non-classical Diffusion

Although under normal circumstances the classical Fick law gives an excellent description of mass diffusion, it is however not appropriate for describing some particular features, like those concerned with inertial effects or couplings between mass transport and viscous pressure. Inertial effects are of importance in some applications which have been studied in the framework of stochastic processes, as correlated random walks, a topic to be discussed in Sect. 13.2.

Furthermore, as seen in Sect. 13.3, a generalised diffusion equation of the telegrapher type with non-local contributions yields a good description of Taylor's dispersion (dispersion of a solute under a velocity gradient) in the whole range of time scales, in contrast with the classical description which is only satisfactory in the long-time limit. Moreover, the coupling between diffusion and viscous pressure (a coupling between quantities of different tensorial order which is excluded by the classical theory of non-equilibrium thermodynamics) leads to several non-Fickian effects which are relevant in glasses and polymer solutions; it is shown in Sect. 13.4 that the theoretical scheme proposed by EIT is well suited for analysing such effects. Finally, Sect. 13.5 is devoted to the analysis of reaction-diffusion processes, where hyperbolic diffusion is combined with chemical reactions. All these aspects have considerable practical importance: to mention only a few examples, Taylor's dispersion plays a relevant role in the dispersion of contaminants in rivers and estuaries or of hematocrite in veins and arteries, shear-induced diffusion is at the basis of chromatographic techniques of separation of macromolecules of different mass and, on the other hand, it is important in macromolecular processing where homogeneity is required.

13.1 Extended Thermodynamics of Diffusion

In this section we derive the basic equations of mass transport in the framework of EIT, with especial emphasis on relaxational effects. In a multicomponent system, the concentration of the constituents must be included among the set of independent variables. As seen in Chap. 1, in the absence of chemical reactions, the evolution equations for the mass fractions $c_k = \rho_k/\rho$ are

$$\rho \dot{c}_k = -\nabla \cdot \boldsymbol{J}_k \qquad (k = 1, 2, ..., n), \tag{13.1}$$

with $\boldsymbol{J}_k$ the diffusion flux of component k with respect to the barycentric motion. Since $\Sigma_k c_k = 1$ and $\Sigma_k \boldsymbol{J}_k = 0$, only $n - 1$ of the mass fractions and diffusion fluxes are independent variables.

For an n-component mixture, the classical Gibbs equation is written as

$$ds = T^{-1}du + T^{-1}pdv - T^{-1}\sum_k \mu_k dc_k. \tag{13.2}$$

According to the postulates of extended irreversible thermodynamics, the diffusion fluxes $\boldsymbol{J}_k$ are considered as independent variables. For the sake of simplicity, we shall not include the heat flux and the viscous pressure tensor as independent variables. The introduction of these fluxes is straightforward and does not raise fundamental difficulties [13.1]. The generalised Gibbs equation is now given by

$$ds = T^{-1}du + T^{-1}pdv - T^{-1}\sum_k \mu_k dc_k - T^{-1}v\sum_k \alpha_k \boldsymbol{J}_k \cdot d\boldsymbol{J}_k. \tag{13.3}$$

From the balance equations for u, v, and c_k, (1.22), (1.11) and (13.1) respectively, one obtains, neglecting external forces, the material time derivative of the entropy

$$\rho \dot{s} = -\nabla \cdot \left(\frac{1}{T}\boldsymbol{q} - \frac{1}{T}\sum_k \mu_k \boldsymbol{J}_k \right) - \frac{1}{T^2}\boldsymbol{q} \cdot \nabla T$$

$$-\frac{1}{T}\sum_k \boldsymbol{J}_k \cdot \left[T\nabla(\mu_k T^{-1}) + \alpha_k \dot{\boldsymbol{J}}_k \right]. \tag{13.4}$$

Note that, except for the last term on the right-hand side, this expression is the same as in the classical theory of irreversible processes.

Restricting our attention to isothermal and isobaric diffusion in isotropic fluids, the entropy production derived from (13.4) simply reads as

$$\sigma^s = -\frac{1}{T}\sum_k \mathbf{J}_k \cdot \left[(\nabla\mu_k)_{T,p} + \alpha_k \dot{\mathbf{J}}_k\right]. \tag{13.5}$$

This expression suggests to express the constitutive equations in the linear approximation as

$$\alpha_i \dot{\mathbf{J}}_i + (\nabla\mu_i)_{T,p} = -\sum_k L_{ik}\mathbf{J}_k, \tag{13.6}$$

with L_{ik} a positive definite matrix; it is important to stress that the summation rule over repeated indices is not of application in this section, as they concern components and not Cartesian coordinates. The evolution equations for $\mathbf{J}_i$ can be written as

$$\alpha_i \dot{\mathbf{J}}_i = -\sum_k L_{ik}\mathbf{J}_k - (\nabla\mu_i)_{T,p} \qquad (i,\,k = 1, 2, \dots, n-1). \tag{13.7}$$

The analysis of the cross terms in (13.7) will be carried out later on. In the particular case of diffusion in a binary system, expression (13.7) reduces to

$$\alpha \dot{\mathbf{J}}_1 = -a\nabla c_1 - L\mathbf{J}_1, \tag{13.8}$$

where a stands for $(\partial\mu/\partial c_1)_{T,p}$, the quantities μ and α are respectively given by $\mu = \mu_1 - \mu_2$ and $\alpha = \alpha_1 + \alpha_2$, while $\mathbf{J}_1$ is the diffusion flux of component 1 with respect to the barycentric motion. After dropping index 1, (13.8) can be rewritten as

$$\tau \dot{\mathbf{J}} + \mathbf{J} = -\rho D\nabla c, \tag{13.9}$$

with D the diffusion coefficient and τ the relaxation time of $\mathbf{J}$. Comparison of (13.9) with (13.8) leads us to identify $a/L = \rho D$ and $\alpha/L = \tau$ or $\alpha = (\tau a/D)$. With this notation, the Gibbs equation (13.3) becomes

$$ds = T^{-1}du + T^{-1}p\,dv - T^{-1}\mu\,dc - \frac{\tau a v}{\rho D T}\mathbf{J}\cdot d\mathbf{J}. \tag{13.10}$$

The last term in the right-hand side of (13.10) is analogous to that found previously for the generalised entropy, with the flux multiplied by the relaxation time and divided by the transport coefficient. Such a generalised Gibbs equation may also be written by taking the relative velocity of the components as an independent variable [13.2, 1.5].

Since T is uniform, it is useful to write (13.10) in a slightly different form involving the free energy $f = u - Ts$. Integration of (13.10) leads, for the free energy f, to the expression

$$f(T,v,c,\mathbf{J}) = f_{eq}(T,v,c) + \tfrac{1}{2}\alpha v \mathbf{J} \cdot \mathbf{J}. \qquad (13.11)$$

In the steady state, or for slow variations of $\mathbf{J}$, for which $\mathbf{J} = -\rho D \nabla c$, expression (13.11) can be written as

$$f(T,v,c,\nabla c) = f_{eq}(T,v,c) + \tfrac{1}{2} k_B T l^2 (\nabla c) \cdot (\nabla c), \qquad (13.12)$$

where $l = (\alpha D^2/v k_B T)^{1/2}$ represents a correlation length. A free energy depending on the concentration gradient was proposed by Ginzburg and Landau; it is the basis of the Landau–Lifshitz analysis of spatial correlations of density fluctuations, of the Cahn–Hilliard theory of spinodal decomposition and of the Ginzburg–Landau–Wilson Hamiltonian description of critical points. However, it must be kept in mind that (13.12) is less general than (13.11), which remains true even for unsteady processes.

When the evolution equation for the flux (13.9) is introduced into the balance law (13.1) and the quantities ρ, τ, and D are constant one is led to the telegrapher equation

$$\tau \frac{\partial^2 c}{\partial t^2} + \frac{\partial c}{\partial t} = D \nabla^2 c, \qquad (13.13)$$

which has been used in many applications, as shown in the next sections.

13.2 Telegrapher's Equation and Stochastic Processes

An equation of the form of (13.13) can also be derived on microscopic grounds by assuming a correlated random walk, i.e. by a random walk in which the jump directions in any two consecutive intervals are correlated [13.3]; it may also be obtained from a dichotomous model in which particles switch between two different states, and also from a random walk with a continuous distribution function of pausing times [13.4] or from differential stochastic equations with telegrapher's noise [13.4]. The correlated random walk was introduced by Fürth in 1917, to model diffusion in biological and physical problems. The same technique was developed in 1921 by G. I. Taylor to analyse turbulent diffusion; in 1953 it was popularized by the classical work of Goldstein [13.3], and since then it has been widely studied [13.4].

13.2.1 Correlated Random Walk

Assume that at the initial time $t = 0$, a set of particles are present at $x = 0$. At time $t = t_0$, half of the particles jump a distance d to the right and half to the left. From now on, at each interval t_0, every particle jumps a distance d, with a probability p of jumping in the same direction as the previous jump and a probability $q = 1 - p$ of jumping in the opposite direction. Note that in a normal random walk, the probability of jumping to the right or to the left is the same, i.e. $p = 1/2$, so that there is no correlation between the speed direction at successive jumps. We now show that this correlation leads to the telegrapher equation. Assume that $\gamma(n, v)$ is the fraction of particles that at time $t = nt_0$ are at position $x = vd$. We denote by $\alpha(n, v)$ and $\beta(n, v)$ the fraction of particles which are reaching position vd from the left and from the right, respectively. As a consequence, one has

$$\gamma(n, v) = \alpha(n, v) + \beta(n, v). \tag{13.14}$$

Since the particles which at time $(n + 1)t_0$ occupy the position vd were at a previous time at $v - 1$ or $v + 1$, it is easy to check that

$$\alpha(n + 1, v) = p\alpha(n, v - 1) + q\beta(n, v - 1) = p\gamma(n, v - 1) - c\beta(n, v - 1),$$
$$\beta(n + 1, v) = p\beta(n, v + 1) + q\alpha(n, v + 1) = p\gamma(n, v + 1) - c\alpha(n, v + 1), \tag{13.15}$$

with $c = p - q$ the correlation between two successive jumps. Conversely, the particles which at $(n - 1)t_0$ were at vd, will arrive at $t = nt_0$ either at $v + 1$ or $v - 1$, so that

$$\gamma(n, v - 1) = \alpha(n, v + 1) + \beta(n, v - 1). \tag{13.16}$$

From (13.14–16) it follows that

$$\gamma(n + 1, v) = p[\gamma(n, v - 1) + \gamma(n, v + 1)] - c\gamma(n - 1, v). \tag{13.17}$$

This result shows that the fraction of particles which at time $(n + 1)t_0$ are at place vd depends not only on the distribution of particles at the previous time nt_0, as in the usual Markovian random walk, but also on the distribution at $(n - 1)t_0$.

To transform (13.17) into a differential equation, we write $t_0 \to \Delta t$, $d \to \Delta x$, $\Delta x = v\Delta t$ (with v the flight speed of the particle during the jump), $c = 1 - (\Delta t/\tau)$, for $\Delta t < \tau$ and $c = 0$ for $\Delta t > \tau$, and $p = (1 + c)/2 = 1 - (2\tau)^{-1}\Delta t$. This expression for c follows from considering that within the short-time limit, the correlation between the speeds at two successive jumps tends to 1, and that for $\Delta t > \tau$ the correlation is lost; the quantity τ is a constant giving information about the loss of correlation with increasing time. If

$w(t,x)$ is the probability density of the distribution of particles, the previous identifications allow to rewrite (13.17) as

$$w(t+\Delta t,x) = \left(1 - \frac{1}{2\tau}\Delta t\right)[w(t,x-\Delta x) + w(t,x+\Delta x)]$$

$$-\left(1 - \frac{1}{\tau}\Delta t\right)w(t-\Delta t,x). \tag{13.18}$$

Expanding this expression up to second order in Δt and setting finally $\Delta t = \tau$ yields

$$\tau\frac{\partial^2 w}{\partial t^2} + \frac{\partial w}{\partial t} = \tau v^2 \frac{\partial^2 w}{\partial x^2}, \tag{13.19}$$

which is a telegrapher equation with relaxation time τ and diffusion coefficient τv^2. Here there is no drift term because we have assumed that there is no external force causing a systematic motion of the particles in a given direction.

It is also interesting to expand (13.18) at higher order instead of limiting it to the second order. Taking into account that $w(t+\Delta t) = \exp[\Delta t(\partial_t)]$ with $\partial_t = \partial/\partial t$, and analogously for $w(x+\Delta x)$, it results that

$$2[\cosh(\tau\partial_t)-1]w + \frac{\tau}{\Delta t}[\cosh(d\partial_x) - \exp(-\tau\partial_t)]w = 2[\cosh(d\partial_x)-1]w, \tag{13.20}$$

where $\cosh(a)$ and $\exp(-a)$ stand for the corresponding Taylor expansions of these functions in powers of a. One may also write (13.20) in the form

$$\sum_i a_i \frac{\partial^i w}{\partial t^i} = \sum_j b_j \frac{\partial^{2j} w}{\partial x^{2j}}, \tag{13.21}$$

where the explicit expressions for the coefficients a_i and b_j can be directly calculated from (13.20) (Problem 13.5). Expression (13.21) is different from the usual Kramers–Moyal expansion, where only the first-order time derivative is kept on the left-hand side, while an infinite expansion in spatial derivatives appears in the right-hand side. Here, space and time are kept on equal footing, a necessary requisite to capture the effects of memory. As shown by Rosenau [13.5], the spectrum of the telegrapher equation reproduces satisfactorily the spectrum of the original discrete process of the correlated random walk for all wavelengths. This is remarkable because in general the short-wavelength limit of a continuum model differs widely from the corresponding discrete microscopic model. Indeed, the well known Fokker–Planck equation, which is a good description of a completely uncorrelated random walk for long times, differs very much from it in the short-wavelength regime. The reason for the much more satisfactory behaviour of the telegrapher equation is that it preserves the characteristic speed of the walker, d/t_0, in contrast with the Fokker–Planck equation, where this information is lost.

13.2.2 *H*–theorem for Telegrapher Type Equation

A natural question, from the thermodynamic point of view, is whether an H-theorem is satisfied for kinetic equations of the telegrapher type [13.6a]. Consider, for instance, an equation of the form

$$\tau\frac{\partial^2 f}{\partial t^2} + \frac{\partial f}{\partial t} = \nabla \cdot D_0\big[k_B T \nabla f + (\nabla U_{ext})f\big]. \tag{13.22}$$

This relation generalises the Smoluchowski equation describing the dispersion of non-interacting Brownian particles under an external potential density $U_{ext}(x)$. Here, $f(x,t)$ is the probability distribution function of the particles, D_0 the inverse of the friction coefficient and τ a relaxation time. By analogy with kinetic theory, we define the entropy S by

$$S = -k_B \int f(x,t)\big[\ln f(x,t) - 1\big]\mathrm{d}x. \tag{13.23}$$

Since (13.22) is valid in a system at constant temperature T, it is more convenient to use the free energy $F = U - TS$, with $U = \int \mathrm{d}x\, f(x,t)u(x)$ rather than the S function itself. The evolution of F is given by

$$\frac{\mathrm{d}F}{\mathrm{d}t} = \int \frac{\partial f}{\partial t}\big(k_B T \ln f + u(x)\big)\mathrm{d}x. \tag{13.24}$$

Substituting $\partial f/\partial t$ drawn from (13.22) in (13.24), one obtains

$$\frac{\mathrm{d}F}{\mathrm{d}t} = -\tau \int \mathrm{d}x\, \frac{\partial^2 f}{\partial t^2}\big[k_B T \ln f + u(x)\big] - D_0 \int \mathrm{d}x\, \frac{1}{f}\big[k_B T \nabla f + (\nabla u(x))f\big]^2, \tag{13.25}$$

as is easily seen by integration by parts and assuming that the flux vanishes at the boundaries. The second term in the right-hand side of (13.25) is negative, because the friction coefficient is positive, but the first one has no definite sign. When the relaxation time τ is zero, (13.22) reduces to the Smoluchowski equation and $\mathrm{d}F/\mathrm{d}t$ is definite negative, implying an irreversible behaviour. However, $\mathrm{d}F/\mathrm{d}t$ is not always negative when τ is different from zero, so that it may be claimed that generally the H-theorem is not satisfied.

By analogy with the macroscopic formulation of extended irreversible thermodynamics, we introduce a generalised entropy S which depends not only on f but also on the probability flux j, defined by way of the usual conservation equation

$$\frac{\partial f}{\partial t} + \nabla \cdot j = 0. \tag{13.26}$$

The generalised entropy S will now take, instead of (13.23), the form

$$S = -k_B \int dx \left[f \ln f + \alpha(f) j^2 \right], \tag{13.27}$$

where $\alpha(f)$ is an undefined coefficient depending on f which will be identified below. The evolution equation for the generalised free energy $F = U - TS$ obtained by using this generalised entropy S is now given by

$$\frac{dF}{dt} = \int dx \left\{ (\partial f / \partial t)[U_{ext}(x) + k_B T \ln f] + k_B T \alpha' j^2 (\partial f / \partial t) + 2 k_B T \alpha j . (\partial j / \partial t) \right\}, \tag{13.28}$$

where α' stands for $d\alpha (f)/df$. After integration by parts one obtains

$$\frac{dF}{dt} = \int dx \left\{ j . \left[\nabla \left(U_{ext}(x) + k_B T \ln f + k_B T \alpha' j^2 \right) + 2 k_B T \alpha (\partial j / \partial t) \right] \right\}. \tag{13.29}$$

To preserve the negative character of dF/dt, it is observed that j cannot be independent of the term inside the brackets; the simplest relation for j is then

$$j = -D \left[\nabla \left(U_{ext}(x) + k_B T \ln f + k_B T \alpha' j^2 \right) + 2 k_B T \alpha (\partial j / \partial t) \right], \tag{13.30}$$

D being a scalar function of x and t. Introducing (13.30) into (13.29) and assuming that the term in j^2 may be neglected in comparison to $(\partial j / \partial t)$, one recovers (13.22) provided the coefficient α is identified as $\alpha = \tau(2 k_B T D_0 f)^{-1}$. This result is important as it shows that it is possible to formulate an H-theorem for the kinetic telegrapher equation at the condition of generalising in a suitable way the Boltzmann expression for the entropy. A formulation of the H-theorem for a generalised master equation involving relaxational terms was also proposed by Vlad and Ross [13.6c].

Applications of the telegrapher equation in stochastic problems are numerous; let us mention diffusion of light in turbid media, scattering and absorption of laser radiation from biological tissues, heat transport, etc. For example, a telegrapher equation was used to describe the behaviour of photons in stellar atmospheres [13.7]. The double-peak structure resulting from the response of (13.13) to a delta-function source as analysed in Sect. 10.2 may explain that the temporal luminosity profiles observed in many X-ray bursters are composed of a precursor peak and a main diffusion peak.

13.3 Taylor's Dispersion

Taylor's dispersion refers to longitudinal dispersion of a solute in a solvent which flows along a rectilinear duct [13.8]. In 1953, Taylor showed that the combined action of a velocity gradient and transverse molecular diffusion leads at asymptotic long times to a longitudinal diffusion as shown in Fig. 13.1, with an enhanced diffusion compared to normal molecular diffusion. The effective diffusion coefficient D is given by $D = D_m + D_T$, D_m being the molecular diffusion coefficient in a fluid at rest and D_T the Taylor diffusion coefficient, which is of the form

$$D_T = A\overline{U}^2 d^2 / D_m, \tag{13.31}$$

with $\overline{U}$ the mean velocity, d the width of the duct and A a numerical coefficient which depends on the details of the flow (for instance, for a plane Poiseuille flow between parallel plates separated by a distance d, $A = 1/210$ and for a cylindrical Poiseuille flow in a duct of radius d, $A = 1/48$).

Taylor's dispersion is important in many practical situations, as dispersion of pollutants in rivers and estuaries, diffusion in arteries and veins (in most applications, the turbulent diffusion coefficient should be used instead of the molecular diffusion coefficient). Formula (13.31) is also frequently used for calculating D_m from laboratory measurement of D_T, a technique which is faster and more accurate than the direct measurement of D_m. However, since Taylor's formula is only valid at asymptotic long times, one must wait for long times before it is applicable; in the practice, this means that the Taylor regime is only observable in very long ducts. Therefore, many authors have tried to extend Taylor's description towards shorter times [13.9] but the study of the intermediate time regime turns out to be extremely cumbersome. Instead of using a detailed hydrodynamic approach, Camacho [13.10] proposes an approach based on the relaxational Maxwell–Cattaneo equation for the Taylor diffusion flux. Indeed, it is noted that at high frequencies or short times, Eq. (13.13) is reversible, whereas at low frequencies or long times it is irreversible, thus providing a simple framework to describe a transition between a reversible ballistic-like regime to a diffusion regime.

Camacho's formalism captures in an elegant way the essential features of the longitudinal dispersion of the solute along all the time span, and agreement with molecular-dynamics simulations is very satisfactory.

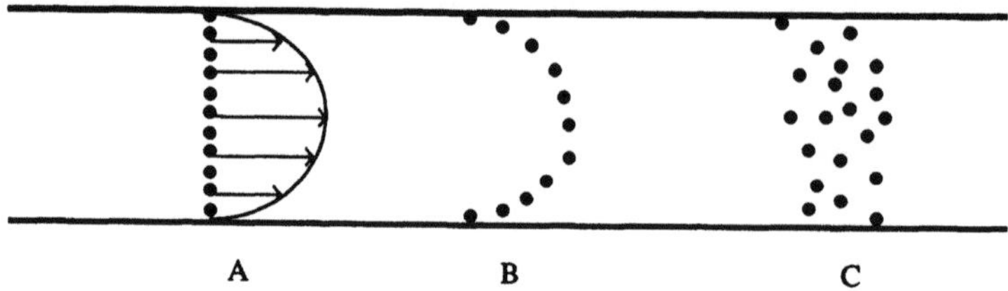

Fig. 13.1. Taylor's dispersion: in A the particles of the solvent are aligned perpendicular to the flow, whose parabolic velocity profile is shown; in B, the particles have moved with the fluid; in C, the combination of the inhomogeneous velocity field and the transversal diffusion of the particles yields a longitudinal diffusion much enhanced with respect to the molecular diffusion in the fluid at rest. The stage C is known as Taylor's dispersion.

In this approach, the Taylor diffusion flux is described by an equation of the form

$$J_T + \tau_T \frac{\partial J_T}{\partial t} + \tau_T \beta \overline{U} \frac{\partial J_T}{\partial x} = -D_T \frac{\partial c}{\partial x} + l_T^2 \frac{\partial^2 J_T}{\partial x^2}. \tag{13.32}$$

where the x axis is taken in the direction of the flow. Here, J_T is the diffusion flux characterizing Taylor's dispersion; the terms including the relaxation time τ_T and the correlation length l_T are similar to those already considered in Sect. 10.3. Besides, Eq. (13.32) contains a new term (the term with coefficient β) describing a transient anisotropic dispersion due to the fact that, at intermediate times, the solute dispersion is different downflow and upflow. The derivation of (13.32) may be found in [13.10].

It must be noticed that the coefficients τ_T, l_T, β, and D_T are not arbitrary. They have been derived [13.10] from an analysis of the behaviour of the Fourier modes of the density, $c_n(x, t)$, the velocity v_n and the fluxes $J_n = \frac{1}{2}c_n(x, t)v_n$, which are related to the Taylor flux by $J_T = \sum_n J_n$. It was found that

$$\tau_T = \sum_{n=1}^{\infty} v_n^2 \tau_n \left(\sum_{n=1}^{\infty} v_n^2 \right)^{-1}, \quad l_T^2 = \tau_T D_m, \quad D_T = \sum_{n=0}^{\infty} \frac{1}{2} v_n^2 \tau_n, \tag{13.33a}$$

$$\beta = \frac{1}{4\overline{U}D_T} \sum_{m,n=1}^{\infty} (v_{m-n} + v_{m+n}) v_n v_m \tau_m, \tag{13.33b}$$

where the relaxation time spectrum τ_n (giving the relaxation times of the nth mode of the diffusion flux J_n) depends itself on the flux. For a plane Poiseuille flow, the results are $D_T = U^2 d^2/(210 D_m)$, $\tau_T = d^2/(42 D_m)$ and $\beta = -1/5$ [13.10].

When Eq. (13.32) is introduced in the solute balance equation (13.1) one obtains

$$\tau_T \frac{\partial^2 c}{\partial t^2} + \frac{\partial c}{\partial t} + \tau_T \beta \bar{U} \frac{\partial^2 c}{\partial x \partial t} - (\tau_T D_m + l_T^2) \frac{\partial^3 c}{\partial x^2 \partial t}$$

$$= D \frac{\partial^2 c}{\partial x^2} + \tau_T \beta \bar{U} D_m \frac{\partial^3 c}{\partial x^3} + l_T^2 D_m \frac{\partial^4 c}{\partial x^4}, \tag{13.34}$$

wherein the various coefficients are defined in (13.33).

The results derived from this equation have been compared with a numerical simulation of a Poiseuille flow between parallel plates. The philosophy behind such a simulation is simple, and was first performed by Rigord [13.11]. One starts by considering two parallel lines separated by a distance d and several points (N) which represent the solute particles distributed uniformly at $x = 0$. At every time step t_s, each particle undergoes two motions: a convective motion plus a Brownian motion; it travels a length $l_c(y) = v(y)t_s$, which corresponds to Poiseuille convection of the solvent flow with $v(y) = (3/2)U(1 - 4y^2/d^2)$ and a length l_d, independent of y, in a random direction. As the system evolves, one computes the average $\langle (\Delta x)^2 \rangle$ of the square of the displacements with respect to the mean convective motion at different times, and represents the ratio $(\Delta x)^2(t)/2t = D^*(t)$ versus the penetration length of the solute as a whole, given by $L = \langle x(t) \rangle$. The quantity D^* represents an 'effective dispersion coefficient', since it has the dimensions of a diffusion coefficient and it tends asymptotically to D in the long-time limit. Other simulations consist of letting the system evolve during a time t_i and suddenly reverse the velocity field and represent again $D^*(t)$ as a function of $\langle x(t) \rangle$. The latter are called echo dispersion simulations whereas the former ones are called transmission dispersion simulations. The results of the theoretical predictions based on (13.34) are compared with those of simulations in Fig. 13.2. The agreement is seen to be very satisfactory for all the time span and for a wide range of velocities (always in the laminar regime). Thus, it can be concluded that the evolution equation (13.32) for the flux captures all the essential details of the interaction between the velocity flow and the longitudinal diffusion for all time regimes.

In the transmission simulations one distinguishes three regions: a horizontal line at short penetration lengths, not shown in Fig. 13.2, which corresponds to the domain of molecular diffusion characterized by D_m, an intermediate regime, the almost diagonal line with squares, and a horizontal line at long penetration lengths, corresponding to fully developed Taylor dispersion, characterized by $D_m + D_T$. Regarding echo simulations, it is noticed that for very short inversion times (curve 1) the effective dispersion coefficient approaches the value D_m. This corresponds to a partially reversible behaviour where almost all the particles go back to their original initial position after the inversion, except for a small diffusion which has slightly increased the width of the original profile. In contrast, in the long time limit (curves 4–6), the echo curves are almost flat: the inversion of the velocity field will not bring the particles back to their initial distribution because the dispersion is fully irreversible.

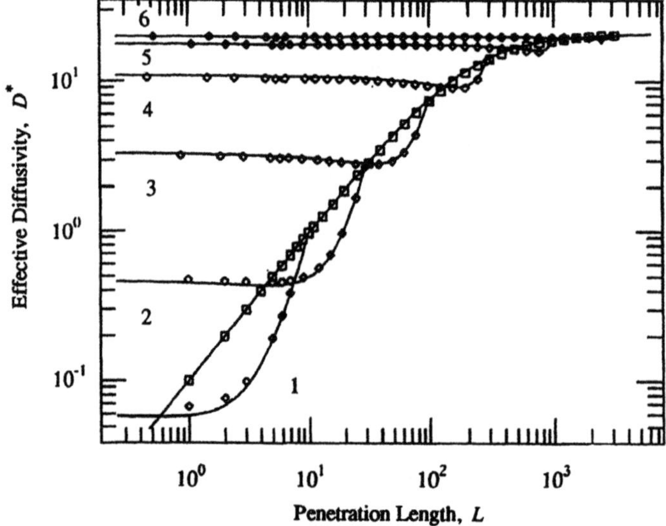

Fig. 13.2. Comparison between the theoretical results predicted from (13.34) (solid lines) and the numerical simulations of [13.10b]. Squares and diamonds correspond to transmission and echo simulations respectively, as explained in the text. The numbered curves correspond to different reversal times in echo simulations.

More details are given in [13.10], where an explicit expression of the non-equilibrium entropy and its relation to the evolution equations for the diffusion flux is derived in agreement with the predictions of extended irreversible thermodynamics.

13.4 Non-Fickian Diffusion in Polymers

Departures from the Fick law are observed in several diffusion processes, particularly those involving polymers [13.12]. Diffusion plays a rate-controlling role in many industrial and biological processes. In particular, it is of interest in dry-spinning of fibres, in coating of substrates via deposition and subsequent drying of polymer solutions, and in the design of drug-delivery systems. Long relaxation times have also a great influence on diffusive processes in glasses and justify that special emphasis be put on the study of relaxational effects in diffusion.

Let us turn our attention to some phenomenological observations. The most common experiments where non-Fickian effects are perceptible are sorption, desorption and permeation of a small solvent in a film or membrane of glassy polymer. The initial concentration of the permeant is suddenly increased (sorption) or decreased (desorption) around the sample, or at one side of it (permeation). The experimental quantities of interest are the mass uptake $m(t)$ per unit surface as a function of time, in sorption and desorption experiments, or the volume throughput $Q(t)$ per unit area of membrane, in permeation.

The mass uptake $m(t)$ is defined as the mass of solvent absorbed by the film per unit area, i.e.

$$m(t) = M\int_0^l \mathrm{d}x\big[c(x,t) - c_0\big], \tag{13.35}$$

where M is the molar mass of the solvent and $c(x, t)$ its concentration, in moles per unit volume, across the film, of thickness l. In permeation, the volume throughput per unit area is

$$Q(t) = \int_0^t \mathrm{d}t' J(t'), \tag{13.36}$$

with J the volume flux of the solvent. Both $m(t)$ and $Q(t)$ have been widely studied in the context of the classical diffusion theory [13.13]. Especially interesting for our discussion are the short-time limit behaviour of $m(t)$ and the long-time behaviour of $Q(t)$, which in the classical Fickian theory are respectively given by [13.13]

$$\frac{m(t)}{m(\infty)} = \left(\frac{16Dt}{\pi l^2}\right)^{1/2}, \tag{13.37a}$$

$$Q(t) \approx \frac{D(c_1 - c_0)}{l}\left(t - \frac{l^2}{6D}\right) \text{ for } (t \to \infty), \tag{13.37b}$$

with c_1 and c_0 the concentrations of the solvent on each side of the membrane. It follows that $m(t)/m(\infty)$ behaves as $t^{1/2}$ at short times and that the time lag in permeation, i.e. the time at which $Q(t)$ vanishes, behaves as l^2 and is given by $t^* = l^2/(6D)$. It must be kept in mind that these expressions refer to differential sorption and permeation, i.e. they correspond to small changes in concentration, in order to avoid non-linearities arising from the sensitive dependence of D on c.

Expressions (13.37a, b) are usually satisfied far from the glass transition temperature. Close to the transition, and below it, non-Fickian features appear and manifest themselves through several phenomena: case-II and super-case-II diffusion, which are discussed below, two-stage sorption, sigmoidal sorption, and pseudo-Fickian behaviour.

The physical cause responsible for these peculiar behaviours is usually interpreted as the coupling of viscous stresses and diffusion. During diffusion of small molecules in a polymer solution, this coupling, resulting from the swelling due to the solvent, produces a relative motion between neighbouring chains of the polymer, whose mutual friction will emerge in a viscous stress. Since this stress is coupled to diffusion, it is natural to take it into account in the same way as in Chap. 2 viscous pressure was coupled with heat conduction. Paralleling the developments of Chap. 2, we simply rewrite (2.64–66) with the heat flux replaced by the diffusion flux.

The resulting equations are

$$\tau_1 \dot{J} = -(J + D\nabla c) + \beta'' \tilde{D} T \nabla \cdot \overset{0}{\mathbf{P}}{}^v + \beta' \tilde{D} T \nabla p^v, \tag{13.38}$$

$$\tau_0 \dot{p}^v = -(p^v + \zeta \nabla \cdot v) + \beta' T \zeta \nabla \cdot J, \tag{13.39}$$

and

$$\tau_2 (\overset{0}{\mathbf{P}}{}^v)^{\cdot} = -(\overset{0}{\mathbf{P}}{}^v + 2\eta \overset{0}{\mathbf{V}}) + 2\beta'' T \eta (\nabla^0 J)^s, \tag{13.40}$$

where β' and β'' arise from the coupling of diffusion and viscous effects in the expression of the entropy flux, which, in absence of heat effects, is given by

$$J^s = -\mu T^{-1} J + \beta'' \overset{0}{\mathbf{P}}{}^v \cdot J + \beta' p^v J. \tag{13.41}$$

The last two terms on the right-hand side of (13.41) are non-classical and analogous to those introduced in (2.11); the quantity $\tilde{D}$ in (13.38) stands for $\tilde{D} = D(\partial \mu / \partial c)^{-1}$. For simplicity, and in accordance with the literature, we restrict ourselves to the one-dimensional problem, with only c, J_x, and P_{xx}^v as variables, and ignore the bulk effects ($p^v = 0$); moreover, we assume that $\tau_0 = \tau_2, \beta' = \beta'' = \beta$, and that the velocity gradients vanish. After these simplifications, (13.38–40) reduce to

$$\tau_1 \dot{J}_x + J_x = -D \frac{\partial c}{\partial x} + \beta \tilde{D} T \frac{\partial P_{xx}^v}{\partial x}, \tag{13.42a}$$

$$\tau_2 \dot{P}_{xx}^v + P_{xx}^v = T\beta\eta_l \frac{\partial J_x}{\partial x},\tag{13.42b}$$

with $\eta_l = \frac{4}{3}\eta + \zeta$ the longitudinal viscosity.

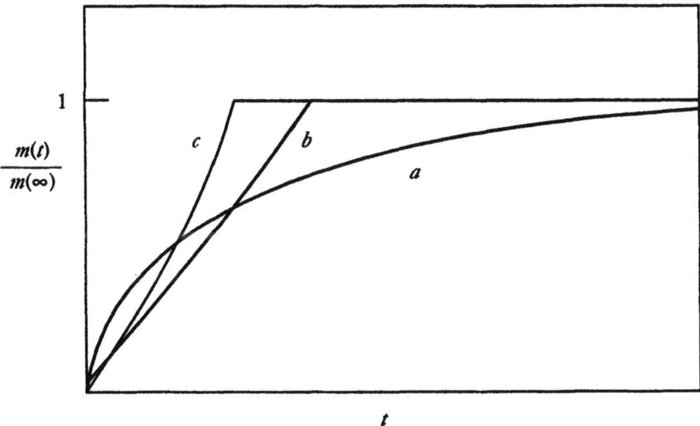

Fig. 13.3. Fractional mass uptake $m(t)/m(\infty)$ as a function of time t for Fickian diffusion (a), case-II diffusion (b), and super-case-II diffusion (c).

Alfrey, Gurnee, and Lloyd observed for the first time in 1965 [13.12] that when the solvent molecules penetrate into the polymer, a sharp advancing boundary is produced between the inner glassy region and the outer swollen gel. The boundary moves with a constant velocity in case-II diffusion, or it accelerates in super-case-II diffusion. The short-time expression for the mass uptake may be expressed generally as $m(t)/m(\infty)$ $\sim t^n$, with $n = 1$ for case-II, $n > 1$ for super-case-II, and $n = 1/2$ for the usual Fickian diffusion (Fig. 13.3). Although a well-defined boundary could also appear in classical Fick diffusion with a sufficiently steep variation of the diffusion coefficient with the concentration, such a boundary would advance as $t^{1/(2+n)}$ (if D depends on c as c^n) and not as t. Case-II diffusion has been observed in solutions of methanol in polymethyl methacrylate (PMMA) in the range 0°C–15°C, in solutions of alkanes in polystyrene in the range 25°C–50°C, in solutions of benzene in epoxy resins. Super-case-II diffusion has been observed for ethanol, propanol and butanol in thin films of PMMA [13.14].

Case-II diffusion can be simply described by means of a Maxwell–Cattaneo equation as obtained from (13.42a) with $\beta = 0$. This has been done by Neogi [13.14c] in the context of non-equilibrium thermodynamics with internal variables. The fractional mass uptake in this model for times longer than l^2/D is given by

$$\frac{m(t)}{m(\infty)} = \frac{2}{l}\sqrt{\frac{D}{\tau}}\,t \qquad t < \frac{l}{2}\sqrt{\frac{\tau}{D}},$$

$$\frac{m(t)}{m(\infty)} = 1 \qquad t > \frac{l}{2}\sqrt{\frac{\tau}{D}}. \tag{13.43}$$

This corresponds to two fronts of solvent advancing from the left and from the right towards the centre of the membrane with speed $(D/\tau)^{1/2}$, which is the speed predicted by the Maxwell–Cattaneo equation. Each front must travel a distance $l/2$ before arriving at the centre. When the fronts collide, they may occasionally clash and give rise to oscillations which die away rapidly. Some characteristic speeds of the fronts in case-II diffusion are given in Table 13.1.

Table 13.1. Speeds of fronts in case-II diffusion [13.14, 15]

System	Speed (m/s)	T (°C)
n-pentane in biaxially oriented polystyrene	1.4×10^{-6}	30
	1.3×10^{-6}	35
n-pentane in cast annealed polystyrene	0.8×10^{-6}	30
	0.7×10^{-6}	35
benzene in epoxy resins	5.9×10^{-8}	70
methanol in PMMA	1.0×10^{-9}	24
	1.0×10^{-7}	62

In super-case-II, the front accelerates. This is observed in very thin membranes of the order of 1 mm thickness. The acceleration of the front has been explained by the fact that the differential swelling stress increases when the residual thickness of the core is reduced. Super-case-II diffusion cannot be interpreted from the simple Maxwell–Cattaneo model, because the Maxwell–Cattaneo equation does not include the coupling between stress and diffusion described by the last term on the right of (13.42a).

The set of equations (13.42) is richer than the Maxwell–Cattaneo law. To show this by means of a simple example, we neglect τ_1 in the evolution equation (13.42a) of the diffusion flux and keep the relaxational effects only for the viscous pressure. Under this assumption, (13.42a) may be written as

$$J_x = -\tilde{D}\frac{\partial}{\partial x}\left(\mu - T\beta P_{xx}^v\right). \tag{13.44}$$

It is worth comparing this equation with the basic idea underlying the Thomas–Windle theory [13.15a], one of the most successful models of super-case-II diffusion. These authors assume that the stress P_{xx}^v affects the chemical potential in the same way as an 'osmotic' pressure; in other words, the diffusion originates in the gradient of a non-equilibrium chemical potential given by

$$\mu_i(T,p+P_{xx}^v,c_i) = \mu_{i,eq}(T,p,c_i) + \int_p^{p+P_{xx}^v} dp'\frac{\partial\mu_i}{\partial p'} = \mu_{i,eq}(T,p,c_i) + v_i P_{xx}^v, \quad (13.45)$$

with v_i the partial volume of species i per unit mass. Comparison of (13.45) and (13.44) yields $-T\beta = v_1$, with v_1 the partial volume per unit mass of the solvent.

To close the set of equations, a supplementary relation between P_{xx}^v and J_x is needed. This relation is provided by (13.42b), written in the form

$$\tau_2\dot{P}_{xx}^v + P_{xx}^v = -\eta_l v_1\frac{\partial J_x}{\partial x}. \quad (13.46)$$

In their theory, Thomas and Windle take $\tau_2 = 0$, but this is not consistent because it yields zero transport when η_l is very high, in contrast with experimental observations. To circumvent this difficulty, Durning and Tabor [13.15b] used (13.46) with τ_2 different from zero. Combining (13.44) and (13.46), they obtain for the diffusion flux

$$J_x = -D\frac{\partial c}{\partial x} - D''\frac{\partial}{\partial x}\int_{-\infty}^t dt'\exp[-(t-t')/\tau_2]\frac{\partial c}{\partial t'}, \quad (13.47)$$

after using the mass balance equation $\partial c/\partial t = -\partial J_x/\partial x$ and identifying D'' as $D'' = v_1^2\tilde{D}\rho\eta_l/\tau_2$. Durning and Tabor analysed the consequences of (13.47) in classical and oscillatory sorption. When suitable functions for $D(c)$ and $\eta_l(c)$ are used in combination with (13.47), a good agreement with experimental results is achieved [13.15b]. An equation like (13.47) has also been proposed in the context of glassy diffusion [13.15b]. All these examples show clearly that EIT provides a useful thermodynamic framework for non-Fickian diffusion.

13.5 Hyperbolic Reaction-Diffusion Systems

Up to now, we have considered that the only process influencing concentration changes is mass transport, eventually coupled with viscous shearing. In some systems, concentration changes may also be due to chemical reactions. Assume, for instance, that a function $F(n)$, generally non-linear, describes the change of particles concentration n per unit time and volume due to chemical reactions. The balance equation is

$$\frac{\partial n}{\partial t} + \nabla \cdot J = F(n),\tag{13.48}$$

where J is the flux of particles, usually given by Fick's law $J = -D\nabla n$. Here, instead, we assume that J satisfies a relaxational equation of the Maxwell–Cattaneo type (13.9). By combining (13.48) and (13.9), one obtains in the one-dimensional case the following non-linear hyperbolic reaction-diffusion equation (HRD) commonly called reaction-telegrapher equation

$$\tau\frac{\partial^2 n}{\partial t^2} + \frac{\partial n}{\partial t} = D\frac{\partial^2 n}{\partial x^2} + F(n) + \tau\frac{\partial F}{\partial t}.\tag{13.49}$$

The presence of the relaxation time τ does not only influence diffusion (first term on the left-hand side) but well the chemical reaction (two last terms on the right-hand side), in such a way that not only $F(n)$ but also its time derivative has an influence on the evolution of concentration distribution.

Many practical applications of HRD equations can be found in natural systems [13.16]. For instance, in the spreading of an epidemics, τ is related to the incubation time, which may be comparable or even longer that the time corresponding to the diffusive propagation of the corresponding micro-organisms between neighbouring people. A further example is the propagation of forest fires: τ is the mean ignition time, i.e. the average time needed to set fire from a burning tree to a green tree [13.18]. Another illustration is found in human migrations, where τ is the time scale between two successive migrations; experimental values obtained from archaeological data for the spread of the neolithic transition in Europe indicate that $\tau = 25$ years [13.19]. Hyperbolic Lotka–Volterra equations are used to describe the interaction between farmers and hunter-gatherers in the expansion of agricultural communities; in this case, τ is the mean generation time and its value is 12.5 years [13.17]. A next example of relaxational effects is the self-inductance of the axon membrane, which contributes with a finite relaxation time to the propagation of action potential in nerves. HRD equations may also be derived from kinetic theory arguments [13.19]. Thermodynamic properties derived from the kinetic model indicate that the entropy is a function of the concentration and the flux of particles.

Mathematically, HRD equation (13.49) has two homogeneous solutions which are obtained by setting $F(n) = 0$. The resulting solutions are connected by a heteroclinic orbit in the phase space which represents the irreversible process joining two equilibrium states. When reactive processes are coupled to diffusion and both occur on the same time scale, the dynamics of the system is described by a wave front. The speed v of the wave front in the long time limit is found to be constant unlike parabolic reaction-diffusion equations, where the speed of the wavefront is unbounded. In hyperbolic reaction-diffusion processes, the speed has always upper and lower bounds [13.16, 18]:

$$v_L \le v \le (D/\tau)^{1/2}. \tag{13.50}$$

Here we show, by means of a simple example, how to determine the upper bound [13.20]. Take a piecewise linear reaction term of the form

$$F(n) = \Omega\left[-n + H(n - \alpha)\right], \tag{13.51}$$

with Ω the inverse of a reaction time, α a given value of the concentration, defining the model and H the Heaviside step function. Shape-preserving wavefronts correspond to similarity solutions of the form $n(x, t) = n(x - vt)$, with v the front speed. Assuming that $n > \alpha$ for $x - vt > 0$ and $n < \alpha$ for $x - vt < 0$, one may replace $H(n - \alpha)$ by $H(x - vt)$, whose Fourier transform is

$$\hat{H}(k,t) = \frac{e^{-ikvt}}{ik}. \tag{13.52}$$

The Fourier transform of (13.49) is then given by

$$\tau\frac{\partial^2 \hat{n}(k,t)}{\partial t^2} + (1 + \Omega\tau)\frac{\partial \hat{n}(k,t)}{\partial t} = -Dk^2\hat{n}(k,t) - \Omega\hat{n}(k,t) + \Omega\left(\frac{1}{ik} - \tau v\right)e^{-ikvt}. \tag{13.53}$$

In terms of the dimensionless quantities $k^* = k\sqrt{D/\Omega}$, $x^* = x\sqrt{\Omega/D}$, $a = \Omega\tau$ and $v^* = v/\sqrt{\Omega D}$, the solution for the density profile may be finally expressed, after some lengthy calculations, as [13.20]

$$n(x^*) = \frac{1}{2} + H(x^*)\left[\frac{1}{2} + \left(-\frac{1}{2} + \frac{v^*(1-a)}{2\sqrt{v^{*2}(1-a)^2 + 4}}\right)e^{m_- x^*}\right]$$

$$+ H(-x^*)\left[-\frac{1}{2} + \left(\frac{1}{2} + \frac{v^*(1-a)}{2\sqrt{v^{*2}(1-a)^2 + 4}}\right)e^{m_+ x^*}\right] \tag{13.54}$$

where

$$m_{\pm} = \frac{-v^*(1+a) \pm \sqrt{v^{*2}(1-a)^2 + 4}}{2(1-av^{*2})}.$$

$$(13.55)$$

The conditions that n must tend to 1 for positive high values of x^* and to zero for large (but negative) values of x^* require, respectively, that $m_- < 0$ and $m_+ > 0$. These restrictions are only fulfilled when $|v^*| \leq 1/\sqrt{a}$, which, in terms of the wave speed yields $v \leq \sqrt{D/\tau}$, as given by (13.50). This upper limit for the speed of the front diverges in the parabolic case, when τ vanishes. The upper bound coincides with that obtained for the telegrapher equation in the high frequency limit. Linearization around the equilibrium solutions, variational techniques, asymptotic analysis and numerical simulations have been used to determine these bounds in more general problems [13.16, 18].

Problems

13.1 *Anomalous diffusion.* In usual diffusion, the second moments of the displacement in the long-time limit behave as $\langle(\Delta x)^2\rangle \approx t$. In several situations, one finds instead of $\langle(\Delta x)^2\rangle \approx t^{2/\delta}$. If $\delta < 2$ this behaviour is called superdiffusive and if $\delta > 2$ it is called sub-diffusive [J. P. Bouchaud and A. Georges, Phys. Rep. **195** (1990) 127 and M. Sahimi, Rev. Mod. Phys. **65** (1993) 1393]. Anomalous diffusion, which we have not included in the text, is an active field of research. This behaviour may arise if the stochastic motions are characterized either by a wide distribution of waiting times, or of jumping lengths, in contrast to the usual Brownian motion which is characterized by a very narrow distribution of waiting times and jumping lengths. (a) Show that a behaviour $\langle(\Delta x)^2\rangle \approx t^3$ may arise if one assumes an equation of the form $\partial c/\partial t = D(x, t)\partial^2 c/\partial x^2$ with $D(x, t) \approx x^a t^b$ with $3a + 2b = 4$. (b) Generalise this result for $\langle(\Delta x)^2\rangle \approx t^{2/\delta}$.

13.2 Another way to obtain anomalous diffusion is to start from a generalised equation of the form $\partial c/\partial t = D\partial^2 c^{q-1}/\partial x^2$. (a) Show that the solution of this equation with an initial delta condition has the form $c(x,t) = t^{-1/q} f(\xi)$ with $\xi = t^{-1/q} x$ and thus yields $\langle\Delta x^2\rangle \approx t^{2/(q-1)}$. (b) Show that for $q = 2, f(\xi) = K \exp[-\xi^2/4D]$, that for $q > 2, f(\xi) = B(A^2 - \xi^2)^{1/(q-2)}$, and for $q < 2, f(\xi) = B(A^2 + \xi^2)^{-1/(2-q)}$ with A and B constants which depend on the value of q.

13.3 *Two-layer model and the telegrapher equation.* The so-called two-layer model consists of a system whose particles jump at random between two states, 1 and 2, having associated velocities $v_1 = v$ and $v_2 = -v$ respectively along the x axis. Assume that the rate R of particle exchange between the two states per unit time and length is proportional to the difference of the probability densities P_1 and P_2, i.e. $R = r(P_1 - P_2)$. Show that the evolution equations for the total probability density $P = P_1 + P_2$ and for the probability flux $J = (P_1 - P_2)v$ are respectively

$$\frac{\partial P}{\partial t} + \frac{\partial J}{\partial x} = 0,$$

$$\tau \frac{\partial J}{\partial t} + J = -D \frac{\partial P}{\partial x},$$

with $\tau = 1/2r$ and $D = v^2\tau$ [see C. van den Broeck, Physica A **168** (1990) 677; J. Camacho and M. Zakari, Phys. Rev. E **50** (1994) 4233].

13.4 *Taylor's dispersion.* Assume that particles whose molecular diffusion coefficient in water is $D_m = 10^{-9}$ m^2/s are introduced as a thin transversal line in a plane Poiseuille flow between two parallel plates which are 2 cm apart, and with a mean speed of 1 cm/s. (a) Evaluate the Taylor diffusion coefficient. (b) The time necessary to reach the Taylor dispersion regime is of the order of the time needed for a particle to diffuse (by molecular diffusion) across half the width of the channel; estimate the time necessary to reach the Taylor's regime and the corresponding length of the tube. (c) Evaluate the width of the region occupied by the solute, i.e. $\langle (\Delta x)^2 \rangle$ at $t = 10^3$s according to the data in Fig. 13.2 and compare it to the value obtained from Taylor's expression for the diffusion coefficient.

13.5 Find the development (13.21) from (13.20) and obtain the explicit expressions for the coefficients appearing in (13.21), which are a_i (i odd) $= (i!)^{-1}(\Delta t)^{i+1}$, a_i(i even) $= (i!)^{-1} (\Delta t)^i [2\tau - \Delta t]$, $b_j = [(2j)!]^{-1}(\Delta x)^{2j}[2\tau - \Delta t]$.

13.6 *Fast solidification fronts.* Rapid solidification of alloys is a situation of technological interest where the relaxational aspects of diffusion must be taken into acccount when the velocity v of the solidifcation front separating the solid and liquid phases is comparable or higher than the diffusion speed $v_D = (D/\tau_D)^{1/2}$, where D is the diffusion coefficient and τ_D the relaxation time of the diffusion flux [P. K. Galenko and D. A. Danilov, J. Crystal Growth **216** (2000) 512, Phys. Lett. A **272** (2000) 207]. As a simple illustration of he influence of the finite value of v_D (in contrast with the classical parabolic situation where $\tau_D = 0$ and v_D diverges), consider $k(v)$, the partition cofficient of the solute, defined by the ratio of concentrations of solute in the solid and the liquid phases. It is given by

$$k(v) = \frac{k_e\left(1 - \dfrac{v^2}{v_D^2}\right) + \dfrac{v}{v_{DI}}}{1 - \dfrac{v^2}{v_D^2} + \dfrac{v}{v_{DI}}} \quad \text{for } v < v_D$$

and $k(v) = 1$ for $v > v_D$, where v_{DI} is the diffusion speed at the solidification front and k_e the equilibrium partition coefficient. Estimate the difference between k_e and $k(v)$ in a 3% Ag–Cu alloy and a 10% Ni–Fe alloy where

Alloy	k_e	v_D (ms^{-1})	v_{DI} (ms^{-1})
Ag–Cu	0.44	10	10
Ni–Fe	0.21	37	25

for wavefronts at $v = 5$ ms^{-1} and 15 ms^{-1} (the velocity v of the wavefront depends on the degree of undercooling with respect to the melting point).

13.7 Assume the hyperbolic reaction-diffusion equation (13.49), with a source term given by $F(n) = \Omega f(n)$, where Ω is the inverse of a reaction time. Use the dimensionless variables proposed in Sect. 13.5, namely, $t^* = \Omega t$, $x^* = x(\Omega/D)^{1/2}$ and $a = \Omega \tau$, and consider $\partial f/\partial t = f'(n)\partial n/\partial t$ with $f'(n) = \partial f/\partial n$. Assume a concave positive reaction term with $f(n = 0) = f(n = 1) = 0$ and a traveling wavefront $n(x - vt)$, with v the speed of the front. (a) Show that the front will satisfy

$$(1 - av^2)\frac{\partial^2 n}{\partial z^2} + v[1 - af'(n)]\frac{\partial n}{\partial z} + f(n) = 0,$$

with $z = x - vt$. (b) Linearize this equation around $n = 0$ and assume solutions of the form $n \cong \exp(\lambda z)$. Show that the positiveness of n requires that

$$v \geq v_L = \frac{2\sqrt{f'(0)}}{1 + af'(0)}.$$

(c) Obtain this lower bound for the logistic source term $f(n) = n(1 - n)$. (V. Méndez, J. Fort and J. Farjas, Phys. Rev. E **60** (1999) 5231).

Chapter 14

Electrical Systems

This chapter is devoted to the analysis of electrical phenomena. After establishing the generalised equations for electrical transport, we analyse in the first section the second moments of the fluctuations in non-equilibrium steady states, and discuss the crossed terms which link fluxes and forces. In the second section, we revisit Onsager's reciprocal relations and observe that the formalism of extended irreversible thermodynamics (EIT) is closer to Onsager's original treatment than classical irreversible thermodynamics (CIT).

In the next sections, we discuss two topics of practical interest: charge transport in submicronic electronic devices and dielectrical relaxation in fluids. Recent descriptions of charge transport in submicronic electronic devices have opened a promising field of application for EIT. Indeed, although the carrier transport can be described by means of the Boltzmann equation, to solve it is a very difficult task and, furthermore, it contains more information than needed in practical applications. Therefore, it is common in practice to consider a reduced number of moments, which are directly related with density, charge flux, internal energy, energy flux and so on, and which are measurable and controllable variables. This kind of approach is referred to as a hydrodynamical model and EIT is very helpful in determining which truncations of the hierarchy of the evolution equations for the moments are compatible with thermodynamics. Finally, we briefly discuss another problem of practical interest in materials sciences: dielectric relaxation in liquids in the range of high frequencies. From a theoretical point of view, this analysis is illuminating as it serves to illustrate the relations between EIT and the thermodynamics with internal variables.

14.1 Electrical Systems: Evolution Equations

We consider here electric conduction in a rigid metallic sample. We assume that the electric current is due to the motion of electrons with respect to the lattice. The independent variables for this problem in CIT are the internal energy per unit mass u and the charge per unit mass, z_e; in EIT, the electric current i is selected as an additional independent variable. For the system under study, the balance equations of charge and internal energy read

$$\rho \dot{z}_e = -\nabla \cdot i, \tag{14.1a}$$

$$\rho \dot{u} = -\nabla \cdot q + i \cdot E, \tag{14.1b}$$

respectively, with E the electric field and $i \cdot E$ the Joule heating term. In order to find the evolution equation for i we proceed by analogy with Sect. 13.1.

Ignoring heat transport for the moment, the generalised Gibbs equation takes the form [14.1]

$$ds = T^{-1}du - T^{-1}\mu_e dz_e - \alpha i \cdot di, \tag{14.2}$$

with μ_e being the chemical potential of electrons and α a phenomenological coefficient independent of i. Equation (14.2) leads, by following the same procedure as in Sect. 13.1, to the evolution equation for i:

$$\tau_e \frac{di}{dt} = -(i - \sigma_e E'), \tag{14.3}$$

where $E' = E - T\nabla(T^{-1}\mu_e)$, τ_e is the relaxation time, and σ_e the electrical conductivity, provided that α in (14.2) is identified as $\alpha = \tau_e (\rho \sigma_e T)^{-1}$. The generalised entropy is now given by

$$\rho s = \rho s_{eq} - \frac{\tau_e}{2\sigma_e T} i \cdot i. \tag{14.4}$$

Equation (14.3) generalises the usual Ohm law $i = \sigma_e E$ and is often used in plasma physics and in the analysis of high-frequency currents [14.2] but without any reference to its thermodynamic context. Expression (14.3) has also been obtained in the kinetic theory in the linearized relaxation-time approximation.

Relation (14.3) can similarly be derived from the simplest model of ionic conduction in a dilute solution. Consider a system with a number density n of ions of charge q, mass m and apparent radius r moving in a fluid of viscosity η; the current density is then given by $i = nqv$, v being the drift velocity of ions. By supposing that the number of ions is much smaller than the number of neutral solvent particles, the barycentric velocity of the system is practically zero. Assuming that the resistive force is given by Stokes' law, one recovers (14.3), with

$$\sigma_e = nq^2(6\pi\eta r)^{-1} \quad \text{and} \quad \tau_e = m(6\pi\eta r)^{-1}.$$

When these results are introduced in expression (14.4), one obtains

$$\rho s = \rho s_{eq} - \tfrac{1}{2}nmv^2 T^{-1}, \tag{14.5}$$

which indicates that the effect of the current is to diminish the equilibrium entropy by an amount equal to the kinetic energy of ions divided by the absolute temperature.

If instead of particles with the same radius, they have different radii distributed according to a law $f(r)$, then the current density takes the form

$$i(t) = \int_0^\infty \mathrm{d}r f(r) q v(r,t), \tag{14.6}$$

with q the charge of each particle and v the drift velocity of particles of radius r. After sudden suppression of the electrical field E, the current i will decay due to the viscosity of the solvent. According to the Stokes law, the decrease of i is given by

$$i(t) = \int_0^\infty \mathrm{d}r f(r) q v_{ss}(r) \exp[-t/\tau(r)], \tag{14.7}$$

where $v_{ss} = qE(6\pi\eta r)^{-1}$ is the steady-state drift velocity and $\tau(r)$ is the relaxation time of the particles, given by $\tau(r) = m(r)(6\pi\eta r)^{-1} \approx r^2$. Writing v_{ss} in terms of the steady state flux $i(0) = \sigma_e E$, relation (14.7) becomes

$$i(t) = i(0)\int_0^\infty \mathrm{d}r f(r) q^2 (6\pi\eta r \sigma_e)^{-1} \exp[-t/\tau(r)]. \tag{14.8}$$

Before proceeding further, we recall some mathematical results. Let us define a function of time $\chi(t)$ by

$$\chi(t) = \int_0^\infty \mathrm{d}x g(x) \exp[-t/\tau(x)],$$

with $\tau(x) = \tau_0 x^s$ and $g(x) \approx x^m \exp(-cx)$, where s, m and c are numerical constants. Application of the saddle-point method shows that at sufficiently long times, $\chi(t)$ behaves as

$$\chi(t) \approx \exp\left[-(t/\tau_0)^{1/(1+s)}\right]. \tag{14.9}$$

Transposing this result to (14.8) with $\tau(r) \approx r^2$ and $f(r)$ given by $f(r) \approx r^m \exp(-cr)$, it is found that $i(t)$ decays as

$$i(t) = i(0)\exp\left[-(t/\tau)^{1/3}\right]. \tag{14.10}$$

The interesting result is that $i(t)$ does not decay as a single exponential, but rather as a stretched exponential owing to the presence of the exponent $1/(1 + s)$. Such a form of decay is known under the name of the Kohlrausch or Williams–Watts law. This behaviour is not unusual in physics; it generally arises in phenomena described by a hierarchy of variables with relaxation times obeying a scaling law [14.3]. The above example was treated to make explicit that EIT is not restricted to simple exponentials, but that it can also cope with more general situations. The corresponding generalised entropy is now given by

$$s = s_{eq} - \frac{1}{2T} \int_0^\infty dr\, f(r) \frac{\tau(r)}{\sigma_e(r)} q v(r) \cdot q v(r). \tag{14.11}$$

Another interesting consequence of expression (14.3) can be drawn from the following example. Consider an alternating electric field $E = E_0 \cos(\omega t)$ acting on the system. The electric current solution of (14.3) has the form

$$i = i_1 \cos \omega t + i_2 \sin \omega t, \tag{14.12}$$

with $i_1 = \sigma_e[1 + \omega^2 \tau^2]^{-1} E_0$ and $i_2 = \sigma_e \omega \tau_e[1 + \omega^2 \tau^2]^{-1} E_0$. In CIT the entropy production is given by $\sigma_{CIT}^s = T^{-1} i \cdot E$, and in our example by

$$\sigma_{CIT}^s = \frac{\sigma_e}{T} \frac{\cos^2 \omega t + \omega \tau_e \cos \omega t \sin \omega t}{1 + \omega^2 \tau_e^2} E_0^2. \tag{14.13}$$

The denominator is always positive but not necessarily the numerator. At low frequencies, the first term in the numerator overcomes the second one and the classical entropy production is positive, but at high frequencies the second term dominates, and

σ_{CIT}^{s} becomes negative in each cycle in a given time interval. Note that the average of (14.13) over a cycle is positive. Thus, at sufficiently high frequencies, the classical expression for the second law is instantaneously violated, but not on the average. In contrast, the entropy production as obtained from EIT is always positive, since its expression is

$$\sigma_{EIT}^{s} = (T\sigma_{e})^{-1} i^{2},$$

which is unconditionally positive.

The second moments of the fluctuations of the electric current density i are easily derived from the generalised entropy (14.2). Its second differential, at constant electron density, is

$$\delta^{2}s = -\left(\frac{1}{c_{v}T^{2}} + \frac{1}{2}\frac{\partial^{2}\alpha}{\partial u^{2}}i_{0}^{2}\right)(\delta u)^{2} - \alpha\delta i \cdot \delta i - 2\frac{\partial\alpha}{\partial u}i_{0} \cdot \delta u\delta i, \qquad (14.14)$$

where c_{v} is the specific heat capacity at constant volume, i_{0} the mean electric current, and α stands for $\tau_{e}v(T\sigma_{e})^{-1}$. Introduction of (14.14) into the Einstein formula (6.5) for the probability of fluctuations yields the second moments of the fluctuations in presence of a mean electric current i_{0}:

$$\langle \delta i_{x}\delta i_{x}\rangle = \frac{k_{B}T\sigma_{e}}{\tau_{e}v}\left[1 + \frac{c_{v}T^{2}}{\alpha}\left(\frac{\partial\alpha}{\partial u}\right)^{2}i_{0}^{2}\right]. \qquad (14.15)$$

In (14.15) it is assumed that i_{0} is directed along the x axis. In equilibrium, $i_{0} = 0$ and one obtains

$$\langle \delta i_{x}\delta i_{x}\rangle = \frac{k_{B}T\sigma_{e}}{\tau_{e}v}, \qquad (14.16)$$

which is the Green–Kubo expression for the electrical conductivity in the case of an exponential decay of the fluctuations of the current. This result is analogous to expressions (6.19) for the fluctuations of the heat flux and the viscous pressure. Equation (14.16) is also related to the well-known Nyquist formula for current fluctuations, as it gives the fluctuations in terms of the electrical resistance.

Relation (14.16) allows one to express σ_{e} in terms of τ_{e}. Indeed, the microscopic operator for the density current is $i = \int eCf dC$ with C being the peculiar velocity and e the electron charge, from which follows that the second moments of i are, both in classical and in Fermi–Dirac statistics, $\langle \delta i_{x}\delta i_{x}\rangle = ne^{2}k_{B}T(mv)^{-1}$. One thus recovers the well-known Drude relation, $\sigma_e = (ne^2/m)\tau_e$, from which follow that $\alpha = (Tn^2e^2)^{-1}$. When this result is introduced into relation (14.15), together with the expressions $du = c_v dT$ and $i_0 = \sigma_e E$, one finds that

$$\langle \delta i_x \delta i_x \rangle = \frac{k_B T \sigma_e}{\tau_e v}\left[1 + \frac{Te^2}{m^2 c}\left(\frac{E\tau}{T}\right)^2\right]. \tag{14.17}$$

The non-equilibrium correction can be evaluated by recalling that the heat capacity per unit mass for an electron gas in Fermi–Dirac statistics is $c = \pi^2 k_B^2 T(2m\varepsilon_F)^{-1}$, with ε_F the Fermi energy. Defining a mean free path ℓ as $\ell = \tau_e v_F$, with v_F the Fermi velocity $v_F = (2\varepsilon_F/m)^{1/2}$, one may write (14.17) as

$$\langle \delta i_x \delta i_x \rangle = \frac{k_B T \sigma_e}{\tau_e v}\left[1 + \pi^{-2}\left(\frac{eE\ell}{k_B T}\right)^2\right]. \tag{14.18}$$

This expression is comparable to the result obtained from the kinetic theory of gases [14.4], which predicts a coefficient 0.156 instead of π^{-2} ($= 0.101$) in front of the non-equilibrium corrections.

The generalised Gibbs equation (14.2) has also been used to analyse the statistics of open and closed pores in simple models of biological membranes; this is an interesting topic in biophysics [14.5] and is explicitly treated in the second of references in [14.5] (see Problem 14.1). Furthermore, more elaborated versions of EIT including higher-order fluxes have been applied to transport problems in semiconductors, as shown in Sect. 14.3.

14.2 Cross Terms in Constitutive Equations: Onsager's Relations

Amongst the most important features of classical non-equilibrium thermodynamics are the Onsager reciprocal relations; as repeatedly recalled, they cannot be obtained from purely macroscopic arguments alone, but only with the help of supplementary hypotheses, such as the time-reversal symmetry of equilibrium correlation functions and linear regression of fluctuations. It is natural to ask whether the formalism of EIT is able to shed a new light on Onsager's relations. As expected, it turns out that Onsager's results remain outside the scope of macroscopic thermodynamics but, notwithstanding this observation, the EIT formulation is closer to the original Onsager derivation than that of classical non-equilibrium thermodynamics [14.7].

Consider a rigid and isotropic body, crossed by a heat flux q and an electric flux i. The generalised Gibbs equation has the form

$$ds = T^{-1}du - \mu_e T^{-1}dz_e - (\rho T)^{-1}(\alpha_{11}q + \alpha_{12}i) \cdot dq$$

$$- (\rho T)^{-1}(\alpha_{21}q + \alpha_{22}i)di. \tag{14.19}$$

In virtue of the balance equations (14.1a,b) for u and z_e, one obtains for the entropy balance

$$\rho \frac{ds}{dt} + \nabla \cdot \left(\frac{1}{T}q - \frac{\mu_e}{T}i \right) = q \cdot \left(\nabla T^{-1} - \frac{\alpha_{11}}{T}\frac{dq}{dt} - \frac{\alpha_{21}}{T}\frac{di}{dt} \right)$$

$$+ i \cdot \left(\frac{E}{T} - \nabla \frac{\mu_e}{T} - \frac{\alpha_{12}}{T}\frac{dq}{dt} - \frac{\alpha_{22}}{T}\frac{di}{dt} \right), \tag{14.20}$$

from which follow immediately the expressions of the entropy flux J^s and the entropy production σ^s. The entropy production, given by the right-hand side of (14.20), has still the structure of a bilinear form. To obtain the simplest evolution equations for q and i compatible with a definite positive entropy production, one assumes linear relations between the thermodynamic forces and the fluxes q and i. This results in

$$\nabla T^{-1} - \frac{\alpha_{11}}{T}\frac{dq}{dt} - \frac{\alpha_{21}}{T}\frac{di}{dt} = \mu'_{11}q + \mu'_{12}i,$$

$$\tag{14.21}$$

$$\frac{E}{T} - \nabla \frac{\mu_e}{T} - \frac{\alpha_{12}}{T}\frac{dq}{dt} - \frac{\alpha_{22}}{T}\frac{di}{dt} = \mu'_{21}q + \mu'_{22}i,$$

with $\mu'_{11} \geq 0$, $\mu'_{22} \geq 0$, and $\mu'_{11}\mu'_{22} \geq \frac{1}{4}(\mu'_{12} + \mu'_{21})^2$, as a consequence of $\sigma^s > 0$.

To show that the matrix of the coefficients $\boldsymbol{\mu}'$ is symmetric, we start from Onsager's original result (Sect. 6.6) stating that if the evolution equations are given by $d\beta/dt = \mathbf{L} \cdot (\partial S/\partial \beta)$, then $\mathbf{L}$ is a symmetric matrix. Let us assume that ∇T^{-1} and $ET^{-1} - \nabla(\mu_e T^{-1})$ vanish in (14.21), so that they refer to fluctuations near an equilibrium state. Then, (14.21) can be presented in the form

$$\begin{pmatrix} dq/dt \\ di/dt \end{pmatrix} = -T(\boldsymbol{\alpha}^T)^{-1} \cdot \boldsymbol{\mu}' \cdot \begin{pmatrix} q \\ i \end{pmatrix}. \tag{14.22a}$$

This expression may be rewritten in terms of the derivatives of the generalised entropy (14.19) with respect to q and i as

$$\begin{pmatrix} dq/dt \\ di/dt \end{pmatrix} = \rho T^2 \left[(\alpha^T)^{-1} \cdot \mu' \cdot \alpha^{-1} \right] \cdot \begin{pmatrix} \partial S/\partial q \\ \partial S/\partial i \end{pmatrix}. \tag{14.22b}$$

According to Onsager's results, the matrix $\mathbf{L} = (\alpha^T)^{-1} \cdot \mu' \cdot \alpha^{-1}$ is symmetric. Since α is symmetric, because it is the matrix of the second derivatives of s, it follows that $\mu' = \alpha \cdot \mathbf{L} \cdot \alpha$ is itself symmetric. In contrast, the matrix τ of the relaxation times, given by $\tau = (T\mu')^{-1} \cdot \alpha$ is generally not symmetric.

In non-equilibrium steady states, (14.21) may be compared with the usual phenomenological equations [1.5] expressing the coupling between thermal and electrical effects and given by

$$\nabla T^{-1} = \frac{1}{\lambda T^2} q - \frac{\Pi + \mu_e}{\lambda T^2} i,$$

$$\tag{14.23}$$

$$E - \nabla \mu_e = \frac{\varepsilon}{\lambda} q - \left(\varepsilon \frac{\Pi + \mu_e}{\lambda} - r \right) i,$$

where λ is the thermal conductivity at zero electric current, ε the differential thermoelectric power or Seebeck coefficient, Π the Peltier coefficient, and r the isothermal electrical resistivity. By comparison of (14.21) and (14.23), one is led to the identifications

$$\mu'_{11} = (\lambda T^2)^{-1}, \qquad\qquad \mu'_{21} = (\varepsilon T - \mu_e)(\lambda T^2)^{-1},$$

$$\mu'_{12} = -(\Pi + \mu_e)(\lambda T^2)^{-1}, \qquad \mu'_{12} = rT^{-1} + (\mu_e - T)(\Pi + \mu_e)(\lambda T^2)^{-1}.$$

From the symmetry of μ' it follows that $-(\Pi + \mu_e) = \varepsilon T - \mu_e$, so that

$$\varepsilon T = -\Pi, \tag{14.24}$$

which is the well-known second Thomson relation.

It is interesting to emphasize that the argument leading to the symmetry of the phenomenological matrix μ' is parallel to that presented by Onsager himself, who postulated that thermodynamic fluxes are time derivatives of state variables, whilst the forces are derivatives of the entropy with respect to state variables. Since in EIT the quantities q and i are basic state variables, it is clear that (14.22) relates the time derivatives of the basic variables with $\partial S/\partial q$ and $\partial S/\partial i$, just like in the Onsager original derivation, where the time derivatives of the basic variables (here, the thermodynamic fluxes) are linearly related to the derivatives of the entropy with respect to these variables (the thermodynamic forces). In CIT, the transposition of Onsager's arguments from the microscopic to the macroscopic level raises some difficulties, since the heat flux and the pressure tensor cannot be directly identified as time derivatives of state variables.

14.3 Hydrodynamical Models of Transport in Semiconductors and Plasmas

Electrical transport can be described by means of a Boltzmann equation for the charge carriers. However, to solve directly this equation is mathematically a very hard task. A possible simplification consists in restricting to a few practically relevant lowest-order moments of the distribution function. This kind of approach is called the hydrodynamical approach [14.8].

14.3.1 Transport in Semiconductors

One challenge in the analysis of transport phenomena is the study of charge transport in submicronic semiconductor devices, which is essential for the optimization of their functioning and design.

The evolution equations for the moments are directly obtained from the Boltzmann equation. In the case of electrons in the conduction band of the semiconductor, it takes the form

$$\frac{\partial f}{\partial t} + v(k) \cdot \nabla f - eE \cdot \nabla_k f = Q, \qquad (14.25)$$

with $f(x,k,t)$ being the distribution function and k the electron momentum, comprised in the first Brillouin zone, $v(k) = \nabla_k \varepsilon$ is the electron group velocity and Q the collision term. In the effective-mass approximation [14.9], the energy $\varepsilon(k)$ is given by $\varepsilon(k) = \frac{1}{2}k^2/m^*$ and $v(k) = k/m^*$, m^* being the effective electron mass (for instance, in silicon $m^* = 0.26m_e$ with m_e being the electron mass). Multiplying (14.25) by several products of the components of k, and integrating one obtains a hierarchy of equations for the different moments. Besides the mass conservation equation for the zeroth-order moment, one finds [14.9]

$$\frac{\partial}{\partial t}(nv_i) + \frac{\partial \theta_{ij}}{\partial x_j} + \frac{neE_i}{m^*} = Q_i, \qquad (14.26)$$

$$\frac{\partial \theta_{<ij>}}{\partial t} + \frac{\partial \theta_{<ij>r}}{\partial x_r} + \frac{2neE_{<i}v_{j>}}{m^*} = Q_{<ij>}, \qquad (14.27)$$

in which θ_{ij}, Q_i and Q_{ij} stand for the moments of f and of Q with respect to k, namely $\theta_{ij} = (1/m^{*2})\int dk\, f k_i k_j$, $Q_i = (1/m^*)\int dk\, Q k_i$, $Q_{ij} = (1/m^*)\int dk\, Q k_i k_j$ and so on, and the symbol $<\cdots>$ denotes the completely symmetric and traceless part of the corresponding tensor. Furthermore, one obtains for the energy $W = \int dk\, f\varepsilon(k)$ and the energy flux $S_i = \int dk\, f\varepsilon(k)v(k)$ the following equations:

$$\frac{\partial W}{\partial t} + \frac{\partial S_i}{\partial x_i} + neE_i v_i = Q_w, \tag{14.28}$$

$$\frac{\partial S_i}{\partial t} + \frac{\partial S_{ij}}{\partial x_j} + e\left[E_j\theta_{ij} + (W/m^*)E_i\right] = Q_i', \tag{14.29}$$

where $Q_w = \int dk\, Q\varepsilon(k)$, $Q_i' = (1/m^*)\int dk\, Q\varepsilon(k)k_i$, and $S_{ij} = (1/m^{*2})\int dk\, f\varepsilon(k)k_i k_j$. These equations are complemented by the Poisson equation which relates the electrical potential to the charge density.

Depending on the choice of variables and the level at which the hierarchy is truncated, one obtains different hydrodynamical models. A simple one is the so-called drift-diffusion model [14.8], which includes the number density of electrons and holes among the set of independent variables, but not their energies W. A more sophisticated but popular model is Baccarani–Wordeman's one [14.8], wherein the energy of electrons and holes are taken as independent variables, but not the heat flux, which is assumed to be given by the Fourier law.

These models present severe drawbacks; to optimize the description, a sound analysis of other possible truncations is highly desirable. Let us go back to the scheme (14.26–29), based on the variables n, v_i, W, S_i, and $\theta_{<ij>}$ (both for electrons and for holes); to close the problem one needs constitutive equations expressing S_{ij} and the collision terms Q_w and Q_i' in terms of the basic variables. Extended irreversible thermodynamics provides a way to determine which kind of truncations are compatible with thermodynamics. Application of EIT to the analysis of submicronic devices has been performed by Anile et al. [14.9]. We summarize here their main results concerning the closure problem, the modelling of the collision terms and the heat flux, but to avoid lengthy developments we will skip all the details.

We first introduce some definitions. We decompose the electron momentum k according to $k = m^*(u + c)$, with u being the average velocity and c the peculiar electron velocity, while $\hat{\theta}_{ijr}$ represents the moments of the distribution function with respect to c. The part of S_{ij} independent of u is $\hat{\theta}_{ijr}$ and the part of the energy flux S_i independent of u is the heat flux q_i. By applying the methods of EIT, the following closure relations have been obtained [14.9]:

$$\hat{\theta}_{<ijk>} = \frac{1}{5}\left(q_i\delta_{jk} + q_j\delta_{ki} + q_k\delta_{ij}\right) + \frac{2m^*}{15nk_BT}\left(q_i\hat{\theta}_{jk} + q_j\hat{\theta}_{ki} + q_k\hat{\theta}_{ij}\right)$$

$$-\frac{2m^*}{15nk_BT}q_l\left(\hat{\theta}_{<li>}\delta_{jk} + \hat{\theta}_{<lj>}\delta_{ki} + \hat{\theta}_{<lk>}\delta_{ij}\right),$$

$$(14.30)$$

$$\hat{\theta}_{ijrr} = \left[\frac{5n}{2m^*}(k_BT)^2 + \left(\sigma - \frac{2}{5nk_BT}\right)\mathbf{q}\cdot\mathbf{q}\right]\delta_{ij} + 2\sigma q_iq_j + \frac{7k_BT}{2}\hat{\theta}_{<ij>} + \frac{m^*}{n}\hat{\theta}_{<il>}\hat{\theta}_{<jl>}.$$

Here, a definite value of the unknown parameter σ is provided by the maximum-entropy approach, or alternatively it may be determined by comparison with Monte Carlo computer simulations or by experiments. The best value for σ is -0.1321, up to values of E of the order of 10^5 V/cm.

The constitutive equations for the collision terms will be obtained from the Baccarani–Wordeman model, and take the form

$$Q_i = -\frac{nv_i}{\tau_p}, \quad Q_i' = -\frac{S_i}{\tau_q}, \quad Q_w = -\frac{W-W_0}{\tau_w}, \quad Q_{<ij>} = -\frac{\theta_{<ij>}}{\tau_s}, \quad (14.31)$$

where τ_p, τ_q, τ_w and τ_s are the corresponding relaxation times. It turns out that if τ_p and τ_q are independent of the fluxes (what is assumed by Baccarani and Wordeman), Onsager's relations are only satisfied when the two relaxation times are equal. However, Monte Carlo simulations show that this assumption is only valid at high energy, and that it is strongly violated at low energy. In order to recover the Onsager relations at high and low energies, Anile et al. redefine τ_p and τ_q as

$$\frac{1}{\tau_p} = a + b\frac{S}{J}, \qquad \frac{1}{\tau_q} = a' + b'\frac{S}{J}, \qquad (14.32)$$

with S and J the magnitude of the energy flux S and the electric flux J respectively. The Onsager relations are then satisfied if $a' + \frac{5}{2}b'k_BT = \frac{5}{2}k_BT[a + \frac{7}{2}bk_BT]$.

Finally, in contrast with most theories, where the Fourier law is assumed a priori, Anile et al. [14.9] derive from the evolution equation (14.29) the following constitutive equation for the heat flux in the steady state:

$$\frac{q_i}{\tau_q} = -\frac{5nk_B^2T}{2m}\frac{\partial T}{\partial x_i} + \frac{5}{2}k_BTJ_i\left(\frac{1}{\tau_p} - \frac{1}{\tau_q}\right). \qquad (14.33)$$

The second term in the right-hand side is not classical but plays a great role in semi-conductor devices, as it accounts for the strong decrease of the effective thermal conductivity; this effect is important to reproduce some details of the electron velocity profiles in submicro-electronic devices. It may be noted that the extra term vanishes when $\tau_p = \tau_q$. An alternative way to describe this reduction is to appeal to flux-limiters as in Sect. 10.8.

A way to check the quality of the truncation is to compare the predictions of the hydrodynamic models with Monte Carlo simulations when applied to a typical device such as a $n^+ - n - n^+$ ballistic diode (Fig. 14.1).

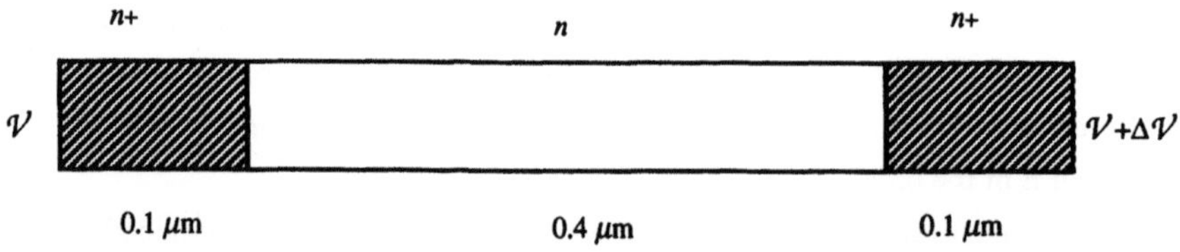

Fig. 14.1. A n^+–n–n^+ silicon diode. The doping density in the region n^+ is higher than in the region n .

Assume that this diode consists of a 0.1 μm n^+ region, a 0.4 μm n region and another 0.1 μm n^+ region, at 300 K and with a doping density $n = 5 \times 10^{23}$ m^{-3} in the channel and $n^+ = 10^{26}$ m^{-3} in the 0.1 μm regions. The results of Monte Carlo simulations are generally used as a benchmark to check the results provided by hydrodynamic descriptions. In particular, for silicon at room temperature (and considering only intravalley scattering with acoustic phonons and inelastic intervalley scattering with optical phonons), Anile et al. find $\tau_w = 3.82 \times 10^{-12}$ s, $a = 42.8$ ps^{-1}, $b = 1.07$ (eVps)$^{-1}$, $a' = 4.322$ ps^{-1}, and $b' = 0.0045$ eVps^{-1} (ps stands for picosecond). With these values, the hydrodynamic model provides results which are in good agreement with Monte Carlo simulations as reflected by Fig. 14.2. Nevertheless, it is worth stressing that, compared to the Monte Carlo simulation, the hydrodynamic models are much more economical with regard to the computing time consumption. To obtain one point in the Monte Carlo simulation, it takes several hours, in comparison with the few minutes necessary to get the whole curve for the profile in the hydrodynamical model.

Another important topic in microelectronic devices of nanometric size is the contribution of ballistic electrons to charge transport, because the size of the device becomes comparable to the mean-free path. The ballistic transport of electrons and holes was observed in GaAs in 1985, and since that time, many efforts have been displayed to take advantage of this characteristics to increase the speed and improve the efficiency of the devices [14.10].

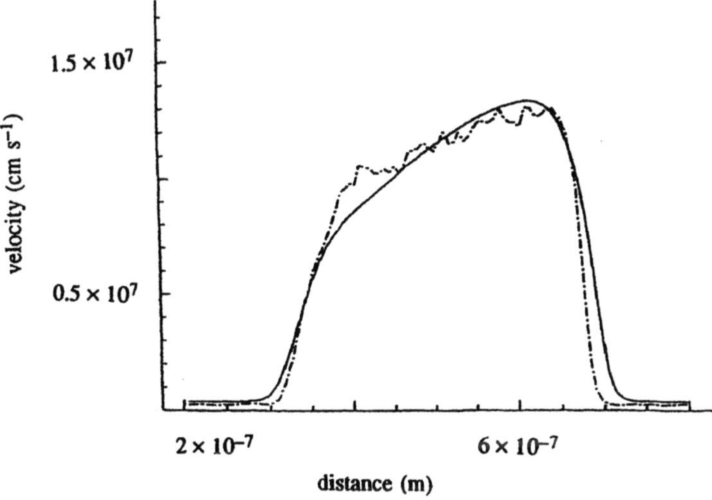

Fig. 14.2. Velocity profiles in the n^+–n–n^+ silicon diode obtained, respectively, by Monte Carlo simulations (dotted line) and the hydrodynamical model of Anile and Pennisi [14.9].

The presence of ballistic electrons gives rise to several peaks in the velocity distribution function, and as a consequence a high number of moments are required instead of only a few ones as in systems close to equilibrium. Therefore, the technique developed in Sect. 5.7 based on an infinite number of moments could be useful to study charge transport in a way similar to that followed in Chap. 10 to describe ballistic heat transport.

14.3.2 Transport in Plasmas

Hydrodynamical models are not exclusive of semiconductors: they were already used by Bloch in 1933 for the description of electron gases and have been often exploited since that time [14.11]. As an illustration of the importance of going beyond local-equilibrium hydrodynamics [14.12], we present here the derivation of the frequency dispersion relation for plasma waves.

In the local-equilibrium version, one takes the electron density n, the velocity v, the pressure p (or the temperature T), and the electrostatic potential ϕ as independent variables. The corresponding set of evolution equations are the continuity equation, and Euler and Poisson equations. They result in a dispersion relation for plasma longitudinal waves of the form

$$\omega^2(k) = \omega_p^2 + v_0^2 k^2, \tag{14.34}$$

with k being the wavevector and ω_p the plasma frequency. For degenerate Fermi gases with Fermi velocity v_F, it turns out that the parameter v_0 appearing in (14.34) is given by $v_0^2 = \frac{1}{3} v_F^2$ instead of the correct microscopic and experimental result $v_0^2 = \frac{3}{5} v_F^2$. The reason of this discrepancy is that electron gases are often found in the collisionless (Vlasov) regime, with very long relaxation times, which classical hydrodynamics does not include in its description. It turns out that introduction of $\mathbf{P}^\nu$ as a further independent variables is sufficient to yield the correct plasma dispersion relation up to the order in k^4. Here, we will restrict our attention to the order k^2 as in (14.34).

We follow here the presentation of reference [14.12]. The evolution equations for n, v, p and $\mathbf{P}^\nu$ are

$$\frac{\mathrm{d}n}{\mathrm{d}t} + n\nabla \cdot v = 0, \tag{14.35}$$

$$mn\frac{\mathrm{d}v}{\mathrm{d}t} + \nabla p + \nabla \cdot \mathbf{P}^\nu - en\nabla\varphi = 0, \tag{14.36}$$

$$\frac{\mathrm{d}p}{\mathrm{d}t} + \frac{5}{3}p\nabla \cdot v + \frac{2}{3}\mathbf{P}^\nu : \nabla v = 0, \tag{14.37}$$

$$\frac{\mathrm{d}\mathbf{P}^\nu}{\mathrm{d}t} + \mathbf{P}^\nu(\nabla \cdot v) + \mathbf{P}^\nu \cdot \nabla v + (\nabla v)^T \cdot \mathbf{P}^\nu - \frac{2}{3}\mathbf{P}^\nu : \nabla v$$

$$+ p(\nabla v + (\nabla v)^T - \frac{2}{3}(\nabla \cdot v)\mathbf{U}) = 0. \tag{14.38}$$

Note that (14.37) is the energy balance equation and that in (14.38) the right-hand side, which we usually take of the form $-(1/\tau_2)\mathbf{P}^\nu$, is assumed to vanish since we are considering the collisionless regime, for which the relaxation time τ_2 is very long.

The linearized equations for the perturbations δn, δv, δp, $\delta \mathbf{P}^\nu$, and $\delta \phi$ are

$$\frac{\partial \delta n}{\partial t} + n_0 \nabla \cdot \delta v = 0, \tag{14.39}$$

$$mn_0 \frac{\partial \delta v}{\partial t} + \nabla \delta p + \nabla \cdot \delta \mathbf{P}^\nu - en_0 \nabla \delta \varphi = 0, \tag{14.40}$$

$$\frac{\partial \delta p}{\partial t} + \frac{5}{3}p_0 \nabla \cdot \delta v = 0, \tag{14.41}$$

$$\frac{\partial \delta \mathbf{P}^\nu}{\partial t} + p_0 \left[(\nabla \delta v) + (\nabla \delta v)^T - \frac{2}{3}\mathbf{U}(\nabla \cdot \delta v) \right] = 0, \tag{14.42}$$

$$\nabla^2 \delta\varphi = 4\pi e \delta n. \tag{14.43}$$

To obtain the dispersion relation, we assume plane-wave solutions where v and $\mathbf{P}^v$ have only one non-vanishing component, v_x and P^v_{xx} respectively. Differentiation of (14.41) with respect to x yields

$$-\frac{\partial^2 \delta n}{\partial t^2} + \frac{1}{m}\frac{\partial^2 \delta p}{\partial x^2} + \frac{1}{m}\frac{\partial^2 \delta P^v_{xx}}{\partial x^2} - \frac{e n_0}{m}\frac{\partial^2 \varphi}{\partial x^2} = 0 \tag{14.44}$$

From (14.33), (14.41), and (14.42) it is inferred that

$$\delta p = \frac{5}{3}\frac{p_0}{n_0}\delta n, \qquad \delta P^v_{xx} = \frac{4}{3}\frac{p_0}{n_0}\delta n. \tag{14.45}$$

Introduction of (14.45) into (14.44) together with (14.43) gives an expression like (14.34) with

$$v_0^2 = \frac{5p_0}{3mn_0} + \frac{4p_0}{3mn_0} = \frac{3p_0}{mn_0}. \tag{14.46}$$

The first contribution arises from δp, which is the unique appearing in the local-equilibrium theory, whereas the second one comes from the addition of $\mathbf{P}^v$ as independent variable. For a Fermi degenerate gas, $p_0 = \frac{1}{3}mn_0 v_F^2$ and the result (14.46) is the correct one, in good agreement with microscopic formalisms and experiments. A more general presentation of the hydrodynamics of Fermi liquids, would include a hierarchy of higher-order fluxes as that presented in Sect. 5.7.

14.4 Dielectric Relaxation of Polar Liquids

Up to now we have dealt with charge transport from one place to another. Dielectric relaxation, instead, is a local effect in the sense that electric charges move inside globally neutral particles (usually macromolecules) which remain themselves at rest. The electric polarization P of such molecules relaxes to its final equilibrium value, which depends on the electric field E applied to the system. The simplest model to describe this relaxation is Debye's model, according to which [1.5]

$$\tau_1 \frac{dP}{dt} + P = \chi_0 E \tag{14.47}$$

where χ_0 is the electric susceptibility and τ_1 the relaxation time. This equation gives a satisfactory description of low-frequency phenomena, but it is insufficient at high frequencies, and a generalisation is therefore required.

A possible way to describe dielectric relaxation in polarizable media is to introduce one or several polarization vectors as internal independent variables (see for instance [14.13]). In the classical description, the state variables are the density of internal energy ρu and the density of polarization charges ρ_p (defined as minus the divergence of the polarization vector P), whose time evolutions are given by the energy and charge balance laws

$$\rho \dot{u} = E \cdot J_p, \qquad \dot{\rho}_p = -\nabla \cdot J_p, \tag{14.48}$$

where $J_p = \mathrm{d}P/\mathrm{d}t$ is the polarization current. In the classical approach, J_p is given by the constitutive law (14.47) instead of being considered as an independent variable.

The first step beyond the classical approach would be to include J_p as an additional independent variable, as was done, for instance, with the heat flux in Sect. 2.2. However, experimental observations on ultrafast dielectric response of dense polar liquids requires a higher level of precision so that, here, we will take not only J_p but also its 'flux' $\Lambda = \mathrm{d}^2 P/\mathrm{d}t^2$ as additional independent variables. It must be stressed that the meaning of the 'fluxes' J_p and Λ is conceptually different from that of the fluxes of heat or mass diffusion introduced earlier. Indeed, the latter refer to displacements of particles (or quasiparticles) in space, while the former are taken at a given point of space and describe the variation of a given quantity, respectively P and its time derivative, in the course of time. Although in EIT the transport fluxes receive special attention, purely relaxational 'fluxes' may also be incorporated in the formalism, and the experience gained in the analysis of the former may be applied to the latter. This assertion will be illustrated by the foregoing considerations.

It could be asked why 'fluxes' of higher orders, like $\mathrm{d}^n P/\mathrm{d}t^n$, are not introduced by analogy with the continued-fraction description of Sect. 5.7: the equivalent of (5.110) would correspond in this case to the Mori formalism [8.6]. However, as seen in Sect. 10.3, one obtains generally very reasonable results by taking only the heat flux and the flux of the heat flux as variables. By analogy, and because of the satisfactory agreement with the experimental results in the ultrafast dielectric response, we limit here the hierarchy of variables to the second order in time derivatives.

The generalised Gibbs equation corresponding to the extended description incorporating the four variables u, ρ_p, J_p, and Λ is [14.14]

$$\rho \dot{s} = T^{-1} \rho \dot{u} + T^{-1} \psi \dot{\rho}_p + (\alpha_1 J_p + \alpha_2 \Lambda) \cdot \dot{J}_p + (\alpha_3 J_p + \alpha_4 \Lambda) \cdot \dot{\Lambda}, \tag{14.49}$$

where $\alpha_2 = \alpha_3$ by the integrability condition, and ψ is the electric potential, related to the electrical field $\boldsymbol{E}$ by $\boldsymbol{E} = -\nabla\psi$. Classical irreversible thermodynamics corresponds to the particular situation where all the coefficients α_i $(i = 1, 2, 3)$ vanish.

We take for the entropy flux $\boldsymbol{J}^s$ the expression

$$\boldsymbol{J}^s = (\psi/T)\boldsymbol{J}_p , \qquad (14.50)$$

which is the flux of internal electric energy divided by the absolute temperature [14.14]. Note that this 'flux' does not represent a transport of entropy from one place to another, but the 'flux' of entropy between different internal states of the molecules of the system. By combining (14.48–50), it is easy to check that the entropy production is given by

$$\sigma^s = \left[\frac{1}{T}(\boldsymbol{E} - \boldsymbol{E}_0) + \alpha_1\frac{\mathrm{d}\boldsymbol{J}_p}{\mathrm{d}t} + \alpha_3\frac{\mathrm{d}\boldsymbol{\Lambda}}{\mathrm{d}t}\right]\cdot\boldsymbol{J}_p + \left[\alpha_2\frac{\mathrm{d}\boldsymbol{J}_p}{\mathrm{d}t} + \alpha_4\frac{\mathrm{d}\boldsymbol{\Lambda}}{\mathrm{d}t}\right]\cdot\boldsymbol{\Lambda}, \quad (14.51)$$

where $\boldsymbol{E}_0$ is the value of $\boldsymbol{E}$ at equilibrium $(\boldsymbol{E}_0 = -\nabla\psi)$. In the linear approximation, one may formulate the evolution equations for the fast variables $\boldsymbol{J}_p$ and $\boldsymbol{\Lambda}$ as follows

$$\frac{1}{T}(\boldsymbol{E} - \boldsymbol{E}_0) + \alpha_1\frac{\mathrm{d}\boldsymbol{J}_p}{\mathrm{d}t} + \alpha_3\frac{\mathrm{d}\boldsymbol{\Lambda}}{\mathrm{d}t} = \mu_1\boldsymbol{J}_p + \mu_3\boldsymbol{\Lambda},$$

$$\alpha_2\frac{\mathrm{d}\boldsymbol{J}_p}{\mathrm{d}t} + \alpha_4\frac{\mathrm{d}\boldsymbol{\Lambda}}{\mathrm{d}t} = \mu_3\boldsymbol{J}_p + \mu_2\boldsymbol{\Lambda}, \qquad (14.52)$$

where Onsager's reciprocal relations, expressed by the presence of the same coefficient μ_3 in both equations, are taken for granted. The above equations may be written in terms of $\boldsymbol{P}$ and $\boldsymbol{\Lambda}$ in the form

$$-\tau_2\frac{\mathrm{d}^2\boldsymbol{P}}{\mathrm{d}t^2} + \tau_1\delta\left(1 + \lambda_2\frac{\mathrm{d}}{\mathrm{d}t}\right)\boldsymbol{\Lambda} - (\boldsymbol{P} - \chi_0\boldsymbol{E}) = 0, \qquad (14.53)$$

$$\left(1 + \lambda_3\frac{\mathrm{d}}{\mathrm{d}t}\right)\boldsymbol{\Lambda} + \frac{1}{\delta}\left(1 + \lambda_4\frac{\mathrm{d}}{\mathrm{d}t}\right)\frac{\mathrm{d}\boldsymbol{P}}{\mathrm{d}t} = 0, \qquad (14.54)$$

with the following identifications: $\tau_2 = -T\chi_0\alpha_1$, $\tau_1 = \frac{1}{4}\mu_3^2 T\chi_0/\mu_2$, $\delta = 2\mu_2/\mu_3$, $\lambda_2 = \lambda_4 = -2\alpha_3/\mu_3$, and $\lambda_3 = -\alpha_4/\mu_2$.

If the relaxation times λ_2, λ_3, and λ_4 are small, (14.54) reduces to

$$\frac{\mathrm{d}\boldsymbol{\Lambda}}{\mathrm{d}t} = -\frac{\mu_3}{2\mu_2}\frac{\mathrm{d}\boldsymbol{P}}{\mathrm{d}t} \qquad (14.55)$$

and (14.53) takes the form

$$\tau_2\frac{d^2P}{dt^2}+\tau_1\frac{dP}{dt}+\left(P-\chi_0E\right)=0,\tag{14.56}$$

which is a first generalisation of the Debye equation (14.47) for P.

Relevant experimental information can be obtained from the autocorrelation function for the polarization vector P, given by

$$\phi_a(t)=\frac{\left\langle\delta P_a(t)\delta P_a(0)\right\rangle}{\left\langle\delta P_a(0)\delta P_a(0)\right\rangle}\tag{14.57}$$

where $a=x$ or z (x is the transverse mode and z the longitudinal mode, when E is assumed to be directed along the z axis). Indeed, the dielectric susceptibility $\chi(\omega)$ and the complex dielectric constant $\varepsilon(\omega)$ are given by the relations

$$\frac{\chi(\omega)-\chi_\infty}{\chi(0)-\chi_\infty}=\frac{\varepsilon(\omega)-\varepsilon_\infty}{\varepsilon(0)-\varepsilon_\infty}\frac{\varepsilon(0)}{\varepsilon(\omega)}=L(-\dot\phi_z)\quad\text{(longitudinal mode)}$$

$$\tag{14.58}$$

$$\frac{\chi(\omega)-\chi_\infty}{\chi(0)-\chi_\infty}=\frac{\varepsilon(\omega)-\varepsilon_\infty}{\varepsilon(0)-\varepsilon_\infty}=L(-\dot\phi_x)\qquad\text{(transverse mode)}$$

where $L(...)$ denotes the Laplace–Fourier transform, while index ∞ corresponds to the limit of infinite frequency. From the Laplace–Fourier transforms of (14.53–54), namely

$$-\left(1+\tau_2\omega^2\right)P+\tau_1\delta\left(1+i\lambda_2\omega\right)\Lambda=-\chi_0E,$$

$$\tag{14.59}$$

$$\delta^{-1}\left(1+i\lambda_4\omega\right)i\omega P+\left(1+i\lambda_3\omega\right)\Lambda=0,$$

it follows that the Laplace–Fourier transform of (14.57) has the form

$$\phi_a(\omega)=\cfrac{1}{-i\omega-\cfrac{M_0}{-i\omega+\cfrac{M_0+\gamma_2}{-i\omega-\cfrac{\gamma_3}{M_0}}}}\tag{14.60}$$

with $M_0=-\gamma_3/\gamma_1$, $\gamma_1=(\lambda_1+2\lambda_2\tau_1)(\lambda_1\lambda_3+\lambda_2^2\tau_1)^{-1}$; $\gamma_2=(\lambda_3+\tau_1)(\lambda_1\lambda_3+\lambda_2^2\tau_1)^{-1}$; and $\gamma_3=(\lambda_1\lambda_3+\lambda_2^2\tau_1)^{-1}$.

If $\lambda_2 = 0$, this expression reduces to the classical model of Kivelson and Keyes [14.15]. The parameter λ_2 (or μ_3) represents the coupling between two dynamical effects described by J_p and Λ, as exhibited by (14.52). Del Castillo et al. [14.14b] have compared (14.47) with the experimental data on the transverse mode of chloroform in the Cole–Cole diagram, representing the imaginary versus the real part of the complex dielectric constant $\varepsilon(\omega)$. They find that for chloroform at 295 K ($\varepsilon(0) = 4.71$, $\varepsilon_\infty = 2.13$), the choice $\tau_2 = 6.33 \times 10^{-12}$ s, $\tau_1 = 0.38 \times 10^{-12}$ s, $\lambda_3 = 0.33 \times 10^{-12}$ s and $\lambda_2 = 0.012 \times 10^{-12}$ s yields a much better agreement for the high-frequency knob of the Cole–Cole plot than the model corresponding to $\lambda_2 = 0$.

Note that the coupling expressed by λ_2 is analogous to that between the flux of the heat flux $\mathbf{Q}$ and the heat flux $\boldsymbol{q}$ in equations (10.34–36). This exhibits the importance of the cross-coupling effects which are easily taken into account in a macroscopic formalism but which are more difficult to implement in a microscopic approach. It can therefore be stated that it is the same general unifying scheme which is applicable to the descriptions of non-classical heat transport and ultrafast dielectric relaxation.

Problems

14.1 *Channels in a membrane.* Assume that there are N channels per unit area in a membrane, and that each of them may be either open or closed. Across the membrane there exists a perpendicular electric field E, and each channel has an electric dipolar moment p and an electric conductance ξ. Consider the following two simple models: (1) the energy of the open state is $H_{open} = -pE$ and the energy of the closed state is $H_{closed} = 0$; (2) $H_{open} = 0$, $H_{closed} = -pE$.

(a) Show that the characteristic curves I-E at constant temperature, I being the electric current per unit area of the membrane, are

$$I = \xi E N \frac{\exp\left[pE/(k_B T)\right]}{1 + \exp\left[pE/(k_B T)\right]} \qquad \text{(case 1)},$$

$$I = \frac{\xi E N}{1 + \exp\left[pE/(k_B T)\right]} \qquad \text{(case 2)}.$$

(b) Assume that the contribution of the fluxes to the free energy of the pores is given by

$$g = g_{eq}(N_{open}, T, E) + a N_{open} J_o^2 \,,$$

with J_o being the electric current through an open channel and $a = \tau v (\sigma A^2)^{-1} = \tau (\xi l)^{-1}$, v the volume of a pore, τ the relaxational time of the current in a pore, A and l the cross-sectional area and the length of the pore, and σ the conductivity of the pore, which is linked to the conductance ξ by $\xi = \sigma(A/l)$. Show that the previous expressions can be modified to

$$I = \xi E N \frac{\exp\left[(k_B T)^{-1}[pE - (a\xi^2/2)E^2]\right]}{1 + \exp\left[(k_B T)^{-1}[pE - (a\xi^2/2)E^2]\right]} \qquad \text{(case 1)},$$

$$I = \frac{\xi E N}{1 + \exp\left[(k_B T)^{-1}[pE - (a\xi^2/2)E^2]\right]} \qquad \text{(case 2)}.$$

(*Hint*: determine the chemical potential of the open and closed pores, use $\mu_{open} = \mu_{closed}$ in equilibrium and remember that in the absence of fluxes $\mu_i(T, E, N_i) = g_i(T, E) + k_B T \ln(N_i/N)$ both for open and closed pores. [See D. Jou et al., J. Chem. Phys. **85** (1986) 5314].)

14.2 Assume that the generalised entropy $s(u,n,i)$ describing electric conduction includes fourth-order terms in the fluxes, instead of being limited to second-order terms as in the text; i.e. assume that

$$s = s_{eq}(u,n) - \alpha i^2 - \beta i^4 ,$$

with u being the internal energy, n the number of electrons per unit volume, and i the electric current density. The coefficients α and β are related to the second-order and fourth-order moments of the fluctuations by

$$\langle (\delta i)^2 \rangle = \frac{k_B}{2\alpha} ,$$

$$\langle (\delta i)^2 \rangle = \frac{3k_B^2}{4\alpha^2} - \frac{3\beta k_B^3}{2\alpha^4} .$$

(a) Show that for $i = \int e v f(v) dv$ and for Maxwell–Boltzmann statistics one has $\alpha = k_B (2e^2 n^2 k_B T)^{-1}$ and $\beta = k_B m/(8 e^4 n^4 k_B^2 T^2)$ with e and m being the charge and mass of the electron. (b) Find out the value of i for which $(\partial^2 s/\partial i^2)_{u,n}$ vanishes.

14.3 *Non-equilibrium Einstein relation.* Around equilibrium, the classical Einstein relation between the diffusion coefficient D and the mobility v reads as $qD/v = k_BT$. When the electric field becomes important, some corrections appear, so that $qD/v = k_BT$ $(1 + a\,E^2 + ...)$ [see, for instance, F. J. Uribe and E. A. Mason, Chem. Phys. **133** (1989) 335]. Starting from the general expression for the Einstein relation $qD/v = n(\partial\mu/\partial n)_{T,V}$, with μ the chemical potential of the carriers and n the carrier density, and the expression for non-equilibrium chemical potential $\mu = \mu_{eq} + [\partial(\tau V/2\sigma)/\partial n\,]J^2$, show that $qD/v \approx k_BT\,[1 + (\tau^2 q^2/mk_BT)E^2]$. (*Hint*: use the Drude formula for the electrical conductivity σ_e).

14.4 Consider a one-dimensional circuit with electrical resistance R and inductance L. The expression relating the intensity I of the electrical current to the electromotive force ξ is a relaxational equation given by

$$\xi = IR + L\frac{dI}{dt},$$

with relaxation time $\tau_e = L/R$. The intensity I is related to the flux of electric current i as $I = iA$, with A the cross-section of the conductor. The magnetic energy stored in the inductor is given by $U_m = \frac{1}{2}LI^2$. Consider the total internal energy $U_{tot} = U + U_m$, with U the internal energy of the material. Show that the Gibbs equation

$$dS = T^{-1}dU + T^{-1}pdV$$

may be rewritten as

$$dS = T^{-1}dU + T^{-1}pdV - \frac{\tau_e V}{\sigma_e T}\,idi, \tag{14.15}$$

which is a relation reminiscent of the Gibbs equation (14.2) proposed in EIT. (Hint: recall that $R = (\sigma_e A)^{-1}l$, with l the length of the circuit).

14.5 *Ballistic devices.* Much interest is paid to ballistic electronic devices, i.e. devices shorter than the mean-free path, where electrons move without collisions [C. W.J. Beenaker, Rev. Mod. Phys. **69** (1997) 731]. The electric current between two reservoirs in which there is a difference $\Delta\mu$ in electrochemical potential of electrons is given by

$$j = -ev\frac{dn}{d\mu}\Delta\mu$$

At low temperature, only the carriers on the Fermi surface will move. (a) Show that in this case, $dn/d\mu = (\pi\hbar v)^{-1}$, with v the Fermi velocity. (b) Show that the conductance, defined as $G = j/\Delta V$, is $G = (e^2/\pi\hbar)$. If the transmission coefficient is T instead of 1 one has $G = (e^2/\pi\hbar)\mathrm{T}$, which is the Landauer formula [see S. Godoy and L.S. García-Colín, Physica A **268** (1999) 65, for a discussion of this equation in the context of persistent random walks introduced in Sect. 13.2.1].

14.6 (a) Obtain (14.52), taking into account that E_0 is defined as $E_0 = -\nabla\psi$. (b) Obtain (14.53–54) and, from them, (14.60).

Chapter 15

Rheological Materials

In ordinary incompressible fluids the flow and transport phenomena are fairly well described by Newton's linear constitutive equation for the viscous pressure tensor $\mathbf{P}^v$

$$\mathbf{P}^v = -2\eta\mathbf{V}, \qquad (15.1)$$

$\mathbf{V}$ is the symmetric (and traceless, because of incompressibility) part of the velocity gradient and η the dynamic viscosity, which may depend on the temperature and the pressure, but not on the velocity gradient. However, it has been observed that there exists a wide class of materials, such as polymers, soap solutions, some honeys, asphalts, and physiological fluids, that fail to obey (15.1): these materials are generally referred to as viscoelastic materials. They behave as fluids with a behaviour reminiscent of solids by exhibiting elastic effects. In ordinary fluids, the relaxation of the pressure tensor is very short, in elastic bodies it is infinite so that no relaxation is observed: viscoelastic materials are characterized by relaxation times between these two limits. Materials with the above property are also called non-Newtonian in the technical literature. The terms 'viscoelastic' and 'non-Newtonian' are used rather loosely; here we shall reserve the term 'non-Newtonian' for any material described by a non-linear constitutive relation between the pressure tensor and the velocity gradient tensor, and the term 'linear viscoelastic' will be used for systems exhibiting both viscous and elastic effects in the linear regime and described by material coefficients independent of the velocity gradient.

Historically, the linear viscoelastic models were the first to be proposed. Although very simple, they have proved to be useful for describing a wide range of phenomena and for providing the first step towards more realistic descriptions. It is our purpose to show that linear viscoelasticity is easily interpreted within the framework of EIT. In particular, the classical models of Maxwell, Kelvin–Voigt, Poynting–Thomson, and Jeffreys are recovered as particular cases. The description of viscoelastic materials by means of thermodynamic theories is not new: many papers have been devoted to the subject but most of them were inspired either by classical irreversible thermodynamics (CIT) [15.1–3] or rational thermodynamics (RT) [15.4–8] by including several kinds of internal variables related to the microstructure of the fluid. Here we propose a derivation based on extended irreversible thermodynamics (EIT) [15.9–14], which uses the viscous pressure tensor as additional variable, rather than internal variables.

It may also be asked what are the consequences of introducing a whole spectrum of relaxational modes for the pressure tensor, instead of working with one single mode. This problem was partially solved by Rouse and Zimm in the 1950s [15.15]. Their molecular models, based on a whole relaxational spectrum, are very useful for describing dilute polymer solutions. We explore to what extent the Rouse and Zimm models can be incorporated into an EIT description. The macroscopic results are also grounded on kinetic foundations.

As stated earlier, EIT is by no means restricted to linear developments. To illustrate the ability of EIT to cope with non-linear situations, two EIT models describing non-linear generalised Newtonian fluids are proposed. The first one reproduces fairly well the dependence of the viscometric functions on the strain rate in simple steady shearing flows. The second is nothing else than Giesekus model [15.16]. For simplicity, thermal effects are ignored throughout this chapter.

15.1 Rheological Models

For the sake of completeness, we briefly recall in this section the main properties of the current models used in rheology. Among the simplest are the linear viscoelastic models.

15.1.1 Linear Viscoelastic Models

The oldest linear viscoelastic model is that of Maxwell, given, in differential form, by

$$\tau_\varepsilon \frac{\partial \mathbf{P}}{\partial t} + \mathbf{P} = -2\eta \frac{\partial \boldsymbol{\varepsilon}}{\partial t}, \tag{15.2}$$

where τ_ε is a constant relaxation time, η the viscosity, and $\boldsymbol{\varepsilon}$ the symmetric strain tensor defined by

$$\varepsilon = \tfrac{1}{2}[\nabla u + (\nabla u)^T], \tag{15.3}$$

u being the displacement vector. The Maxwell model may be regarded as being composed of a linear spring and a dashpot placed in series and may be generalised by coupling N parallel Maxwell elements: denoting by $\mathbf{P}_\alpha$ the pressure tensor in sub-element α, one may write $\mathbf{P} = \sum_\alpha \mathbf{P}_\alpha$, with $\mathbf{P}_\alpha$ obeying the relation

$$\tau_{\varepsilon_\alpha} \frac{\partial \mathbf{P}_\alpha}{\partial t} + \mathbf{P}_\alpha = -2\eta_\alpha \frac{\partial \varepsilon}{\partial t} \qquad (\alpha = 1,2,...,N). \tag{15.4}$$

Such a description has been corroborated by Rouse and Zimm [15.15] on microscopic bases.

Another interesting linear viscoelastic model is that of Kelvin and Voigt:

$$\mathbf{P} = -2G\varepsilon - 2\eta \frac{\partial \varepsilon}{\partial t}, \tag{15.5}$$

where G stands for the elastic constant. Somewhat more complicated models involving a supplementary third constant τ_σ are these of Poynting and Thomson, and Jeffreys, respectively given by

$$\tau_\varepsilon \frac{\partial \mathbf{P}}{\partial t} + \mathbf{P} = -2G\left(\varepsilon + \tau_\sigma \frac{\partial \varepsilon}{\partial t}\right) \quad \text{(Poynting–Thomson)}, \tag{15.6}$$

$$\tau_\varepsilon \frac{\partial \mathbf{P}}{\partial t} + \mathbf{P} = -2\eta\left(\frac{\partial \varepsilon}{\partial t} + \tau_\sigma \frac{\partial^2 \varepsilon}{\partial t^2}\right) \quad \text{(Jeffreys)}. \tag{15.7}$$

15.1.2 Non-Newtonian Models

In this subsection, only incompressible materials without bulk effects are considered. For later use, it is interesting to examine the response of a fluid under a simple shearing flow between two parallel horizontal planes separated by a distance h. If one of the planes, for instance the upper one, is moved in the x direction with a constant velocity v_0, the velocity field of the fluid is $v_x = \dot{\gamma}y$, $v_y = v_z = 0$, $\dot{\gamma} = v_0/h$, with y in the direction normal to the planes, z the neutral direction, and $\dot{\gamma}$ the constant shear rate.

For a Newtonian incompressible fluid, the components of the pressure tensor, also called stress tensor by many authors, are given by $P_{xz} = P_{yz} = 0$, $P_{xx} = P_{yy} = P_{zz} = p$, and $P_{xy} = -\eta\,\dot{\gamma}$. Defining the first normal stress N_1 and the second normal stress N_2 as $N_1 = P_{xx} - P_{yy}$, $N_2 = P_{yy} - P_{zz}$, one sees straightforwardly that for a Newtonian fluid both N_1 and N_2 are zero. The linear constitutive relation (15.1) together with a constant viscosity and the absence of normal stresses are the main properties of a Newtonian fluid.

In contrast, non-Newtonian fluids are characterized by a shear-rate-dependent viscosity and non-vanishing normal stresses. It is common to introduce the 'apparent' viscosity μ and the first and second normal stress coefficients Ψ_1 and Ψ_2 by

$$P_{xy} = -\mu(\dot{\gamma})\dot{\gamma} \quad (\mu > 0),$$

$$N_1 = -\Psi_1(\dot{\gamma})\dot{\gamma}^2 \quad (\Psi_1 > 0), \tag{15.8}$$

$$N_2 = -\Psi_2(\dot{\gamma})\dot{\gamma}^2 \quad (\Psi_2 < 0);$$

where μ, Ψ_1, and Ψ_2 are generally referred to as the viscometric functions. Experimental data indicate that Ψ_1 is positive and Ψ_2 negative. While the dependence of μ with respect to $\dot{\gamma}$ gives rise to velocity profiles totally unusual in Newtonian flows, the normal stress coefficients are responsible for the well-known Weissenberg rod-climbing effect [15.17–19]. Non-zero values of Ψ_2 have been detected for several substances but usually Ψ_2 remains very small; $\Psi_2 = 0$ is known as Weissenberg's hypothesis.

A widely used model for the apparent viscosity μ is the Ostwald power law

$$\mu = m\dot{\gamma}^{n-1}, \tag{15.9}$$

where m and n are parameters characteristic of a given fluid allowed to depend on the temperature. However, the power law (15.9) is not realistic because it gives $\mu = \infty$ for $\dot{\gamma} = 0$ when $n < 1$, which is the most usual situation. To circumvent this difficulty, Carreau [15.20] has proposed the following viscosity equation:

$$\frac{\mu - \mu_\infty}{\mu_0 - \mu_\infty} = \left[1 + (\tau\dot{\gamma})^2\right]^{\frac{n+1}{2}}, \tag{15.10}$$

involving four parameters: a zero-shear-rate viscosity μ_0, an infinite shear-rate viscosity μ_∞, a time constant τ, and a supplementary constant n. This formula describes fairly well most engineering experiments.

The main problem in non-Newtonian fluid mechanics is that of formulating a constitutive equation for **P** containing the smallest number of parameters and applicable to the widest range of fluid responses.

The simplest generalisation of Newton's law is the Reiner–Rivlin non-linear fluid model,

$$\mathbf{P} = p\mathbf{U} - 2\eta\mathbf{V} - 4\alpha_1\left[\mathbf{V}\cdot\mathbf{V} - \tfrac{1}{3}(\mathbf{V}:\mathbf{V})\mathbf{U}\right], \tag{15.11}$$

with $p = \frac{1}{3}\text{Tr }\mathbf{P}$. Expression (15.11) contains two material functions η and α_1 that depend on the second and third principal invariants of $\mathbf{V}$: $\text{II}_v = \mathbf{V}\!:\!\mathbf{V}$, $\text{III}_v = \det \mathbf{V}$. The quantities η and α_1 do not depend on the first invariant $\text{I}_v = \nabla \cdot \mathbf{v}$, since it vanishes for an incompressible fluid. On continuum mechanics grounds, it can be shown [15.19] that η and α_1 cannot be simultaneously taken constant, unless $\alpha_1 = 0$, in which case Newton's law is recovered.

Going back to the Reiner–Rivlin model, it is directly shown that for a steady shear flow, the viscometric functions are related to the material coefficients η and α_1 by $\mu = \eta$, $\Psi_1 = 0$, and $\Psi_2 = \alpha_1$. The result $\Psi_1 = 0$ is at variance with experimental observations and this means that the Reiner–Rivlin model does not provide a realistic description of non-Newtonian fluids.

To correct for these undesirable features, one can use the more sophisticated model of Rivlin–Ericksen [15.4]. To meet the principle of frame indifference, Rivlin and Ericksen introduced the following tensors, which reflect the above requirement:

$$\mathbf{A}^{(0)} = \mathbf{U},$$

$$\mathbf{A}^{(1)} = 2\mathbf{V},$$

$$\mathbf{A}^{(n)} = \frac{\mathrm{d}\mathbf{A}^{(n-1)}}{\mathrm{d}t} + (\nabla v) \cdot \mathbf{A}^{(n-1)} + \mathbf{A}^{(n-1)} \cdot (\nabla v)^T \quad (n = 2,3,\ldots). \tag{15.12}$$

The tensors $\mathbf{A}^{(n)}$ are the Rivlin–Ericksen tensors. The simplest Rivlin–Ericksen fluid is the second-order fluid

$$\mathbf{P} = p\mathbf{U} - \eta\mathbf{A}^{(1)} - \alpha_2\mathbf{A}^{(2)} - \alpha_1\left[\mathbf{A}^{(1)} \cdot \mathbf{A}^{(1)} - \tfrac{1}{3}\left(\mathbf{A}^{(1)} : \mathbf{A}^{(1)}\right)\mathbf{U}\right], \tag{15.13}$$

that depends on three material coefficients, η, α_1, and α_2, which may be functions of the temperature but not of the velocity gradient. It is found experimentally that this assumption is not very satisfactory. By setting α_2 equal to zero in (15.13), one obtains again the Reiner–Rivlin model. It is instructive to consider what predictions follow from (15.13) under a steady shearing flow. The material constants η, α_1, and α_2 are related to the usual viscometric functions by $\eta = \mu$, $\alpha_1 = \Psi_1 + \Psi_2$, and $\alpha_2 = -\tfrac{1}{2}\Psi_1$. Working within the framework of rational thermodynamics, Dunn and Fosdick [15.21] have shown that the second principle of thermodynamics and stability of equilibrium require that

$$\alpha_2 > 0 \quad (\text{or } \Psi_1 < 0), \qquad \alpha_1 + \alpha_2 = 0 \quad (\text{or } \Psi_2 = -\tfrac{1}{2}\Psi_1). \tag{15.14}$$

However, both from experimental data and theoretical considerations it can be concluded that Ψ_1 should be positive and that Ψ_2 is usually not equal to $-\tfrac{1}{2}\Psi_1$.

In the face of these contradictory results, we are confronted with two alternatives. Either the Rivlin–Ericksen model is at variance with thermodynamics and cannot pretend to model real fluids for which $\Psi_1 > 0$, or rational thermodynamics contains some deficiencies that rule out a realistic physical model. We shall come back to this problem at the end of Sect. 15.4.1.

The Rivlin–Ericksen second-order fluid has been extended in several ways. A natural generalisation is that of Criminale, Ericksen, and Filbey [15.22], who used equation (15.13) but with shear-rate dependent viscometric coefficients. Higher-order expansions in terms of the Rivlin–Ericksen tensors $\mathbf{A}^{(n)}$ ($n > 2$) have also been proposed but they lead to a proliferation of the parameters and are not very useful in practical problems.

An alternative approach consists in expressing the viscous pressure tensor by means of an integral dependence of the whole strain history,

$$\mathbf{P}^v = \int_{-\infty}^{t} K(t - t')\mathbf{V}(t')\mathrm{d}t', \tag{15.15}$$

where $K(t - t')$ is the memory kernel. A typical history dependence is provided by the relation [15.23]

$$\mathbf{P}^v = \int_{-\infty}^{t} K_1(t - t')\mathbf{V}(t')\mathrm{d}t'$$

$$+ \int_{-\infty}^{t} \mathrm{d}t'' \int_{-\infty}^{t} K_2(t - t', t - t'')[\mathbf{V}(t') \cdot \mathbf{V}(t'') + \mathbf{V}(t'') \cdot \mathbf{V}(t')]\mathrm{d}t'. \tag{15.16}$$

All the aforementioned models are explicit in the pressure tensor, meaning that $\mathbf{P}^v$ can be substituted in the momentum equation to give a flow problem for the velocity.

Another possibility, still mentioned earlier, is to describe the behaviour of fluids exhibiting elastic effects by means of rate equations. Examples of rate-equation models are provided by the generalised Maxwell equation

$$\tau \mathrm{D}\mathbf{P}^v + \mathbf{P}^v = -2\eta\mathbf{V}, \tag{15.17}$$

where D is an objective time derivative; the substitution of D for $\partial/\partial t$ is dictated by the objectivity criterion. An example of an objective derivative is Jaumann's derivative

$$\mathrm{D}_J\mathbf{P}^v = \dot{\mathbf{P}}^v + \mathbf{W} \cdot \mathbf{P}^v - \mathbf{P}^v \cdot \mathbf{W}. \tag{15.18}$$

For some materials, it may be convenient to use the upper-convected or the lower-convected time derivative whose expressions are to be found in (1.91) and (1.92).

Oldroyd [15.24] suggested an extension of (15.17) by including all possible non-linear terms involving the products of pressure tensors and velocity gradients as well as products of velocity gradients with each other. This yields the well established 8-constant Oldroyd model:

$$\mathbf{P}^v + \lambda_1 D\mathbf{P}^v + \lambda_2\left(\mathrm{Tr}\,\mathbf{P}^v\right)\mathbf{V} + \lambda_0\left(\mathbf{P}^v \cdot \mathbf{V} + \mathbf{V} \cdot \mathbf{P}^v\right) + \lambda_4\left(\mathbf{P}^v : \mathbf{V}\right)\mathbf{U}$$

$$= \lambda_5\left[\mathbf{V} + \lambda_6 D\mathbf{V} + \lambda_7(\mathbf{V} \cdot \mathbf{V}) + \lambda_8(\mathbf{V} : \mathbf{V})\mathbf{U}\right]. \tag{15.19}$$

Flow problems involving non-Newtonian fluids demand in general lengthy numerical solutions: these methods are the object of several papers and books (e.g. [15.25]) and will not be reviewed here. We shall no longer discuss the properties of the several rheological models found in the literature. We wish only to stress that none of these models is applicable to the study of all the motion states of a given particular non-Newtonian fluid. Several constitutive equations may be needed to describe the behaviour of one single fluid, depending on the particular circumstances under which the motion takes place. For instance, blood is rather well represented by Newton's law in large capillaries, while in narrow capillaries, its flow properties require a non-Newtonian description.

15.2 EIT Description of Linear Viscoelasticity

After this brief review of some typical rheological models, it will be shown that they can easily and systematically be derived within the framework of EIT. In the present section, only linear viscoelasticity is considered. The following hypotheses are taken for granted: the deformations are infinitesimally small and the material is isotropic. Generalisation to anisotropic systems undergoing large deformations, although not a trivial matter, should not raise fundamental difficulties.

15.2.1 Constitutive and Evolution Equations

The choice of the variables is inspired by earlier results derived for viscous fluids. Let us recall that in the latter case, the set of classical variables is complemented by including the viscous part of the pressure tensor. Similarly, in the case of viscoelastic fluids, we shall decompose the pressure tensor, assumed to be symmetric, into an elastic part, $\mathbf{P}'$ and an inelastic part, $\mathbf{P}''$ so that $\mathbf{P} = \mathbf{P}' + \mathbf{P}''$, and shall insert $\mathbf{P}''$ among the set of variables; $\mathbf{P}'$ obeys Hooke's law

$$\mathbf{P}' = -2G\varepsilon + \lambda^L(\mathrm{Tr}\,\varepsilon)\mathbf{U}, \tag{15.20}$$

where λ^L and G are the Lamé coefficients. The reasons for selecting $\mathbf{P}''$ as a supplementary variable are found in the general axioms underlying EIT. Indeed, the latter demands that the set of conserved classical variables be completed by variables taking the form of fluxes. Clearly, $\mathbf{P}''$ meets this requirement.

It is also assumed that there exists a non-equilibrium entropy that depends on the inelastic pressure tensor in addition to the classical variables (internal energy u and strain tensor $\boldsymbol{\varepsilon}$): $s = s(u, \varepsilon, \overset{\scriptscriptstyle 0}{\varepsilon}, p'', \overset{\scriptscriptstyle 0}{\mathbf{P}''})$; the quantities ε, p'' and $\overset{\scriptscriptstyle 0}{\varepsilon}$, $\overset{\scriptscriptstyle 0}{\mathbf{P}''}$ are respectively the isotropic and deviatoric parts of $\boldsymbol{\varepsilon}$, $\mathbf{P}''$. Defining by analogy p', $\overset{\scriptscriptstyle 0}{\mathbf{P}'}$ as the isotropic and deviatoric parts of $\mathbf{P}'$, the corresponding Gibbs equation is

$$ds = \frac{1}{T}du + \frac{p'}{3\rho T}d\varepsilon + \frac{\overset{\scriptscriptstyle 0}{\mathbf{P}'}}{\rho T}:d\overset{\scriptscriptstyle 0}{\varepsilon} - \frac{c}{\rho T}dp'' - \frac{\overset{\scriptscriptstyle 0}{\mathbf{C}}}{\rho T}:d\overset{\scriptscriptstyle 0}{\mathbf{P}''}, \tag{15.21}$$

where the coefficients c and $\overset{\scriptscriptstyle 0}{\mathbf{C}}$ have to be specified by means of constitutive relations. By combining the linearized energy balance

$$\rho\frac{\partial u}{\partial t} = -\mathbf{P}:\frac{\partial \varepsilon}{\partial t} \tag{15.22}$$

with the partial derivative of $s = s(u, \varepsilon, \overset{\scriptscriptstyle 0}{\varepsilon}, p'', \overset{\scriptscriptstyle 0}{\mathbf{P}''})$ with respect to the time, one has

$$\rho\frac{\partial s}{\partial t} = -\frac{1}{T}p''\frac{\partial \varepsilon}{\partial t} - \frac{1}{T}\overset{\scriptscriptstyle 0}{\mathbf{P}''}:\frac{\partial \overset{\scriptscriptstyle 0}{\varepsilon}}{\partial t} - \frac{c}{T}\frac{\partial p''}{\partial t} - \frac{1}{T}\overset{\scriptscriptstyle 0}{\mathbf{C}}:\frac{\partial \overset{\scriptscriptstyle 0}{\mathbf{P}''}}{\partial t}. \tag{15.23}$$

Within the linear approximation in the fluxes, the coefficients c and $\overset{\scriptscriptstyle 0}{\mathbf{C}}$ are given by the expressions $c = \beta p''$ and $\overset{\scriptscriptstyle 0}{\mathbf{C}} = \alpha \overset{\scriptscriptstyle 0}{\mathbf{P}''}$, where α and β may eventually depend on u. After substitution of these results into the entropy balance equation (15.23), one is led to

$$\rho T\frac{\partial s}{\partial t} = -\overset{\scriptscriptstyle 0}{\mathbf{P}''}:\left(\frac{\partial \overset{\scriptscriptstyle 0}{\varepsilon}}{\partial t} + \alpha\frac{\partial \overset{\scriptscriptstyle 0}{\mathbf{P}''}}{\partial t}\right) - p''\left(\frac{\partial \varepsilon}{\partial t} + \beta\frac{\partial p''}{\partial t}\right). \tag{15.24}$$

It was shown in Chap. 2 that, in the absence of heat flux, the entropy flux is zero, whereby the right-hand side of (15.24) may be identified with the entropy production. The latter has the structure of a bilinear form in thermodynamic fluxes and forces at the condition to identify the forces conjugated to the fluxes $\overset{\scriptscriptstyle 0}{\mathbf{P}''}$ and p'' respectively as

$$\overset{\scriptscriptstyle 0}{\mathbf{X}} = \frac{\partial \overset{\scriptscriptstyle 0}{\varepsilon}}{\partial t} + \alpha\frac{\partial \overset{\scriptscriptstyle 0}{\mathbf{P}''}}{\partial t}, \tag{15.25a}$$

$$X = \frac{\partial \varepsilon}{\partial t} + \beta \frac{\partial p''}{\partial t}. \tag{15.25b}$$

Assuming linearity between fluxes and forces, one obtains

$$\frac{\partial \overset{0}{\varepsilon}}{\partial t} + \alpha \frac{\partial \overset{0}{\mathbf{P}}''}{\partial t} = -a \overset{0}{\mathbf{P}}'',$$

$$\frac{\partial \varepsilon}{\partial t} + \beta \frac{\partial p''}{\partial t} = -bp'', \tag{15.26}$$

where the coefficients a and b may depend on u. Setting $1/a = 2\eta$, $\frac{1}{2}\alpha/\eta = \tau_\varepsilon$, $1/b = \zeta$, and $\beta/\zeta = \tau'_\varepsilon$, (15.26) take the more familiar form

$$\tau_\varepsilon \frac{\partial \overset{0}{\mathbf{P}}''}{\partial t} + \overset{0}{\mathbf{P}}'' = -2\eta \frac{\partial \overset{0}{\varepsilon}}{\partial t}, \tag{15.27a}$$

$$\tau'_\varepsilon \frac{\partial p''}{\partial t} + p'' = -\zeta \frac{\partial \varepsilon}{\partial t}. \tag{15.27b}$$

These are the required linear evolution equations of the new variables p'' and $\overset{0}{\mathbf{P}}''$. It is possible to obtain evolution equations for the total pressure $\mathbf{P} = \mathbf{P}' + \mathbf{P}''$ by using (15.20) and (15.27). This operation leads to

$$\tau_\varepsilon \frac{\partial \overset{0}{\mathbf{P}}}{\partial t} + \overset{0}{\mathbf{P}} = -2G\left(\overset{0}{\varepsilon} + \tau_\sigma \frac{\partial \overset{0}{\varepsilon}}{\partial t} \right), \tag{15.28a}$$

$$\tau'_\varepsilon \frac{\partial p}{\partial t} + p = -3K\left(\varepsilon + \tau'_\sigma \frac{\partial \varepsilon}{\partial t} \right), \tag{15.28b}$$

where K, τ_σ, and τ'_σ stand respectively for

$$K = \lambda^L + \frac{2}{3}G, \quad \tau_\sigma = \frac{\eta}{G} + \tau_\varepsilon, \quad \tau'_\sigma = \frac{\zeta}{3K} + \tau'_\varepsilon, \tag{15.29}$$

while $\overset{0}{\mathbf{P}}$ and p are respectively the deviatoric and isotropic parts of $\mathbf{P}$. It is worth noticing that relations (15.28) are the constitutive equations for a Poynting–Thomson body and that these arise naturally from the EIT model.

The following particular cases are also of interest. By setting $\tau_\varepsilon = 0$ into (15.28a), one obtains

$$\overset{0}{\overset{\circ}{\mathbf{P}}} = -2G\overset{0}{\boldsymbol{\varepsilon}} - 2\eta\frac{\partial\overset{0}{\boldsymbol{\varepsilon}}}{\partial t}, \tag{15.30}$$

which is representative of a Kelvin–Voigt body.

If in (15.28a), one assumes that $G = 0$, one recovers the basic equation of Maxwell's model, namely

$$\tau_\varepsilon\frac{\partial\overset{0}{\overset{\circ}{\mathbf{P}}}}{\partial t} + \overset{0}{\overset{\circ}{\mathbf{P}}} = -2\eta\frac{\partial\overset{0}{\boldsymbol{\varepsilon}}}{\partial t}. \tag{15.31}$$

It should be emphasized that the material coefficients appearing in the evolution equations for the pressure tensor components are subject to some constraints imposed by the second law of thermodynamics and the condition that entropy is a convex function at equilibrium. These restrictions are now examined.

15.2.2 Restrictions Placed on the Material Coefficients

By writing the rate of dissipated energy derived from (15.24) in the bilinear form

$$T\sigma^s = -\overset{0}{\mathbf{X}} : \overset{0}{\overset{\circ}{\mathbf{P}}}'' - X p'' \tag{15.32}$$

and imposing the condition that it is a non-negative quantity, one derives interesting information about the sign of the material coefficients τ_ε, τ'_ε, η, and ζ introduced into (15.27). Substitution of the expressions (15.25) for $\overset{0}{\mathbf{X}}$ and X into (15.32) results in

$$T\sigma^s = \frac{1}{2\eta}\overset{0}{\mathbf{P}}'' : \overset{0}{\mathbf{P}}'' + \frac{1}{\zeta}(p'')^2, \tag{15.33}$$

after making use of (15.27). Positiveness of $T\sigma^s$ demands that $\eta > 0$, $\zeta > 0$.

Supplementary restrictions are imposed by the requirement of stability of equilibrium. Expanding s around equilibrium up to second-order terms in the fluxes, for fixed values of energy and strain, one obtains

$$s = s_{eq} + \left(\frac{\partial s}{\partial\overset{0}{\mathbf{P}}''}\right)_{eq} : \overset{0}{\mathbf{P}}'' + \frac{1}{2}\left(\frac{\partial^2 s}{\partial\overset{0}{\mathbf{P}}'' : \partial\overset{0}{\mathbf{P}}''}\right)_{eq}\overset{0}{\mathbf{P}}'' : \overset{0}{\mathbf{P}}'' + \left(\frac{\partial s}{\partial p''}\right)_{eq} p'' + \frac{1}{2}\left(\frac{\partial^2 s}{\partial p''^2}\right)_{eq}(p'')^2.$$

After using Gibbs' equation (15.23) and the property that s is extremum at (local) equilibrium, it is found that

$$s = s_{eq} - \frac{1}{2}\frac{\alpha}{\rho T}\overset{0}{\mathbf{P}}'' : \overset{0}{\mathbf{P}}'' - \frac{1}{2}\frac{\beta}{\rho T}(p'')^2. \tag{15.34}$$

Recalling that at equilibrium entropy is a maximum for fixed values of u and $\boldsymbol{\varepsilon}$ one can infer that $\alpha > 0$ and $\beta > 0$. Combining these inequalities with the positive character of η and τ and the definitions $\frac{1}{2}\alpha/\eta = \tau_\varepsilon$ and $\beta/\zeta = \tau'_\varepsilon$ it is found that the relaxation times are positive quantities: $\tau_\varepsilon > 0$, $\tau'_\varepsilon > 0$. The requirement that the entropy production is positive definite has thus led to the important result that the viscosity coefficients η and ζ are positive, while from the convexity property of entropy it is concluded that the relaxation times τ_ε and τ'_ε are positive as well.

15.2.3 The Gibbs Equation for Viscoelastic Bodies

In view of the above results, the Gibbs equation can be written as

$$T\mathrm{d}s = \mathrm{d}u + \frac{1}{\rho}\mathbf{P}' : \mathrm{d}\boldsymbol{\varepsilon} - \frac{\tau_\varepsilon}{2\rho\eta}\overset{0}{\mathbf{P}}'' : \mathrm{d}\overset{0}{\mathbf{P}}'' - \frac{\tau'_\varepsilon}{\rho\zeta}p''\mathrm{d}p''. \tag{15.35}$$

It is interesting to formulate this equation for some particular cases. For simplicity, let us omit bulk effects, so that the stress and strain tensors are traceless; this allows us to drop everywhere the superscript zero.

A Kelvin–Voigt material is defined by $\tau_\varepsilon = 0$; accordingly, (15.35) reduces to

$$T\mathrm{d}s = \mathrm{d}u + \frac{1}{\rho}\mathbf{P}' : \mathrm{d}\boldsymbol{\varepsilon}. \tag{15.36}$$

To obtain the Gibbs equation for a Poynting–Thomson body, let us introduce an inelastic strain tensor $\boldsymbol{\varepsilon}''$ defined as

$$\frac{\partial \boldsymbol{\varepsilon}''}{\partial t} = -\frac{1}{2\eta}\mathbf{P}''. \tag{15.37}$$

Making use of the evolution equation (15.27a), Gibbs equation (15.35), written in terms of time derivatives, reads as

$$T\frac{\partial s}{\partial t} = \frac{\partial u}{\partial t} + \frac{1}{\rho}\mathbf{P}' : \frac{\partial \boldsymbol{\varepsilon}}{\partial t} - \frac{1}{2\rho\eta}\mathbf{P}'' : \left(-\mathbf{P}'' - 2\eta\frac{\partial \boldsymbol{\varepsilon}}{\partial t}\right). \tag{15.38}$$

Combining the terms in $\partial \varepsilon / \partial t$ and using definition, (15.37) leads to

$$T\frac{\partial s}{\partial t} = \frac{\partial u}{\partial t} + \frac{1}{\rho}\mathbf{P}:\frac{\partial \varepsilon}{\partial t} - \frac{1}{\rho}\mathbf{P}'':\frac{\partial \varepsilon''}{\partial t}, \tag{15.39}$$

which is usually referred to as the Gibbs equation for a Poynting–Thomson body. The results (15.36) and (15.39) were derived a few years ago by Kluitenberg [15.2] and Lambermont and Lebon [15.27] on completely different grounds.

15.2.4 Sinusoidal Oscillations

It may be instructive to derive the relationship between the pressure tensor and strain when the material is subjected to external sinusoidal oscillations. In this subsection, the analysis is restricted to one-dimensional deformations.

Assume that the strain and the pressure are of the form

$$\varepsilon = \hat{\varepsilon}\exp(i\omega t), \quad p = \hat{p}\exp(i\omega t), \tag{15.40}$$

where a circumflex refers to the amplitude and ω is the frequency of the oscillations. Defining the complex modulus G^* by $\hat{p} = -2G^*\hat{\varepsilon}$ and substituting (15.40) into (15.28a), one finds that

$$G^*(\omega) = \frac{1+i\omega\tau_\sigma}{1+i\omega\tau_\varepsilon}G. \tag{15.41}$$

At zero frequency, G^* is equal to the Lamé coefficient G, while at infinite frequency one has

$$G^*(\infty) = \frac{\tau_\sigma}{\tau_\varepsilon}G = G\left(1+\frac{\eta}{G\tau_\varepsilon}\right). \tag{15.42}$$

After splitting G^* into real and imaginary parts, $G^* = G' + iG''$, one obtains

$$G' = G + \eta\tau_\varepsilon\frac{\omega^2}{1+\omega^2\tau_\varepsilon^2}, \quad G'' = \eta\frac{\omega}{1+\omega^2\tau_\varepsilon^2}. \tag{15.43}$$

Defining a complex viscosity η^* through $\hat{p} = -2i\omega\eta^*\hat{\varepsilon}$ and setting $\eta^* = \eta' - i\eta''$, one can readily verify that

$$\eta' = \eta\frac{1}{1+\omega^2\tau_\varepsilon^2}, \quad \eta'' = \frac{G}{\omega} + \eta\tau_\varepsilon\frac{\omega}{1+\omega^2\tau_\varepsilon^2}, \tag{15.44}$$

where η', the dynamic viscosity, reduces to the ordinary viscosity for $\omega = 0$. The phase shift δ between the shear and pressure signals is given by

$$\tan\delta = \frac{G''}{G'} = \frac{\omega}{(G/\eta) + \omega^2\tau_\varepsilon[1+(G\tau_\varepsilon/\eta)]}. \tag{15.45}$$

In the particular case $\tau_\varepsilon = \eta = 0$, which characterizes an elastic body, one finds that $\delta = 0$, while for $\tau_\varepsilon = G = 0$, which corresponds to a viscous fluid, one has $\delta = \pi/2$. Finally, by setting $G = 0$, it follows that $\tan\delta = 1/\omega\tau_\varepsilon$, which is typical of a Maxwell body.

15.2.5 Propagation of Plane Harmonic Waves

We examine the propagation of plane harmonic waves in viscoelastic materials described by the general relations (15.28). The set of relations governing the propagation of waves is given by the momentum equation

$$\rho\frac{\partial^2 u}{\partial t^2} = -\nabla \cdot \mathbf{P}, \tag{15.46}$$

completed by the evolution equations (15.28) and the constitutive relation

$$\mathbf{P} = -K\varepsilon\mathbf{U} - 2G\overset{0}{\varepsilon} + \tfrac{1}{3}p''\mathbf{U} + \overset{0}{\mathbf{P}''}; \tag{15.47}$$

u stands as usual for the displacement vector. Select as reference solution $u = \text{constant}$, $p'' = 0$, and $\overset{0}{\mathbf{P}''} = 0$, and superpose upon this basic solution a plane harmonic wave $\exp[i(\omega t - k \cdot r)]$, where ω is the frequency and k the wave vector. Equations (15.28), (15.46), and (15.47) form a system of algebraic equations that can be separated into longitudinal and transverse components. The longitudinal component is obtained after projecting the displacement vector u on the direction of propagation $k/|k|$. The corresponding dispersion relation is

$$\rho\omega^2 = \frac{K(1+i\omega\tau'_\sigma)}{1+i\omega\tau'_\varepsilon}k^2 + \frac{4}{3}\frac{G(1+i\omega\tau'_\sigma)}{1+i\omega\tau_\varepsilon}k^2, \tag{15.48}$$

where k^2 stands for $k \cdot k$. From now on it is assumed that ω is real and k complex, with $k = k' + ik''$. After splitting (15.48) into real and imaginary parts and making use of the classical definitions for the phase velocity v_p $(= \omega/k')$ and attenuation factor Γ_a $(= -k''/k')$, one obtains the following algebraic equations for v_p and Γ_a:

$$\rho^2 \frac{(C+D)^2}{\left[(A+B)^2+(C+D)^2\right]^2} v_p^4 + \rho \frac{A+B}{(A+B)^2+(C+D)^2} v_p^2 - 1 = 0, \quad (15.49a)$$

$$\Gamma_a = \frac{\rho}{2} \frac{C+D}{(A+B)^2+(C+D)^2} v_p^2; \qquad (15.49b)$$

the quantities A, B, C, and D stand for

$$A = K\frac{1+\omega^2\tau'_\varepsilon\tau'_\sigma}{1+\omega^2\tau'^2_\varepsilon}, \qquad B = \frac{4}{3}G\frac{1+\omega^2\tau_\varepsilon\tau_\sigma}{1+\omega^2\tau^2_\varepsilon},$$

$$C = K\omega\frac{\tau'_\sigma-\tau'_\varepsilon}{1+\omega^2\tau'^2_\varepsilon}, \qquad D = \frac{4}{3}G\omega\frac{\tau_\sigma-\tau_\varepsilon}{1+\omega^2\tau^2_\varepsilon}.$$

Within the approximation $\omega\tau_\varepsilon \ll 1$ and $\omega\tau'_\varepsilon \ll 1$, the solutions of (15.49) in the lowest order in ω are given by

$$v_p^2 = \frac{1}{\rho}\left(K + \tfrac{4}{3}G\right), \qquad (15.50a)$$

$$\Gamma_a = \frac{1}{6}\frac{\zeta+4\eta}{K+\tfrac{4}{3}G}\omega, \qquad (15.50b)$$

whereas in the limit $\omega\tau_\varepsilon \gg 1$ and $\omega\tau'_\varepsilon \gg 1$, up to the first order in $1/\omega$,

$$v_p^2 = \frac{1}{\rho}\left(K\frac{\tau'_\sigma}{\tau'_\varepsilon} + \frac{4}{3}G\frac{\tau_\sigma}{\tau_\varepsilon}\right), \qquad (15.51a)$$

$$\Gamma_a = \frac{1}{6\omega}\frac{\zeta/\tau'^2_\varepsilon + 4\eta/\tau^2_\varepsilon}{K\tau'_\sigma/\tau'_\varepsilon + (4/3)G\tau_\sigma/\tau_\varepsilon}. \qquad (15.51b)$$

The phase velocity and attenuation factor corresponding to the Maxwell (respectively Kelvin–Voigt) body are derived by letting G and K (respectively τ_ε and τ'_ε) tend to zero in (15.50) and (15.51).

It is an easy matter to repeat the above procedure for shear waves. Now, the general dispersion relation is written as

$$\rho\omega^2 = G\frac{1+i\omega\tau_\sigma}{1+i\omega\tau_\varepsilon}k^2;$$

the corresponding phase velocities and attenuation factors are given by

$$\lim_{\omega\tau_\varepsilon\to 0} v_p = \left(\frac{G}{\rho}\right)^{1/2}, \qquad \lim_{\omega\tau_\varepsilon\to\infty} v_p = \left(\frac{G\tau_\sigma}{\rho\tau_\varepsilon}\right)^{1/2},$$

$$\lim_{\omega\tau_\varepsilon\to 0} \Gamma_a = \frac{\zeta}{3G}\omega, \qquad \lim_{\omega\tau_\varepsilon\to\infty} \Gamma_a = \frac{\eta}{3G\tau_\sigma\tau_\varepsilon}\frac{1}{\omega}.$$

$$(15.52)$$

The above results are useful because they indicate that the parameters ζ, η, τ_ε, and τ_ε' are directly accessible to experiments by measurements of the velocity propagation and the attenuation factor of the longitudinal and shear waves.

The aim of this section was to show that linear viscoelasticity is easily and naturally incorporated into EIT. It is worth stressing that EIT encompasses a wider class of materials than Maxwell's bodies, in particular the Kelvin–Voigt and Poynting–Thomson bodies are recovered. The various material coefficients can be determined experimentally by measurements of wave velocity and attenuation. Generalisation to non-linear models involving non-linear objective time derivatives and non-linear constitutive equations is quite straightforward, as shown in Sect. 15.4, and in references [15.11, 28, 29].

15.3 The Rouse–Zimm Relaxational Model

Kinetic models are very helpful in gaining insight into the type of rheological equation that is needed. It is expected that a kinetic interpretation enlightens and complements a purely continuum approach. Recently, a vast amount of work devoted to microscopic descriptions has appeared; the results are collected in the classical books by Bird et al. [15.30] and Doi and Edwards [15.31]. It is not our purpose to review the most significant kinetic models but we wish to stress that most of them exhibit a whole relaxational spectrum for the pressure tensor. This important property is fairly well described by the beads-springs chain models associated with the names of Rouse and Zimm [15.15]. For many years these models have played a key role in the interpretation of rheological phenomena. There have been many extensions of Rouse and Zimm's work, but although most of these generalisations give a better fitting of the experimental data by introducing more parameters, they do not alter radically the foundations underlying Zimm and Rouse's theory. Our purpose here is to show that EIT is capable of coping with the relaxational spectrum of the pressure tensor exhibited in the Rouse–Zimm model.

15.3.1 The Rouse–Zimm Model

The current molecular models regard the polymer macromolecules as being formed by assemblies of beads, springs, and rods. In the Rouse model, the molecules are modelled by a chain of N point-mass beads connected by $N - 1$ Hookean springs. The forces acting on each bead can be classified as follows:

(a) The hydrodynamic drag force which is the force experienced by the bead as it moves through the polymeric solution. According to Stokes, it is proportional to the difference between the bead velocity and the centre-of-mass velocity of the solution. By hydrodynamic interaction is meant the perturbation of the velocity field of one particular bead by the motion of the other beads. In the Rouse model, in which it is admitted that the motions of the N beads are uncorrelated, this kind of interaction is ignored; it is accounted for in the Zimm model.

(b) The Brownian force caused by the thermal fluctuations in the solution.

(c) The intramolecular force, which is exerted on one bead through the spring force.

(d) The external forces, such as the gravitational or electrical forces.

The Rouse and Zimm models are appropriate for describing dilute polymer solutions and were primarily used in linear viscoelasticity. It is supposed that the roles of the solvent and the polymer can be separated. Accordingly, the viscous pressure tensor is decomposed into two parts, $\mathbf{P}^{v} = \mathbf{P}_0 + \mathbf{P}_p$, where subscripts 0 and p denote the contribution of the solvent and the polymer respectively; next, $\mathbf{P}_p$ is decomposed into N normal modes,

$$\mathbf{P}_p = \sum_{\alpha=1}^{N} \mathbf{P}_\alpha, \tag{15.53}$$

and each individual mode α is assumed to satisfy, in the linear regime, a relaxational dynamics of the form

$$\tau_\alpha \frac{\partial \mathbf{P}_\alpha}{\partial t} + \mathbf{P}_\alpha = -2\eta_\alpha \mathbf{V}. \tag{15.54}$$

In the Rouse model, the coefficients τ_α and η_α are given by $\tau_\alpha = \frac{1}{2}\xi/Ha_\alpha$ and $\eta_\alpha = nk_BT\tau_\alpha$; k_B is the Boltzmann constant, T the absolute temperature, n the number of molecules per unit volume, ξ the friction coefficient related to the drag force experienced by the bead, H the Hookean spring constant, and a_α the eigenvalues of the Rouse matrices given by

$$a_\alpha = 4\sin^2\left(\frac{\pi\alpha}{2N}\right). \tag{15.55}$$

In Zimm's model, the a_α are replaced by eigenvalues $\bar{a}_\alpha$ of the modified Rouse matrices, which are found in tabular form or given by approximate expressions [15.30].

15.3.2 The EIT description of the Rouse–Zimm model

The procedure is similar to that developed in Sect. 15.2, but the set of independent variables is now extended in order to include the contributions arising from the N normal modes. In this way, the entropy will depend on p_α and $\overset{0}{\mathbf{P}}_\alpha$ ($\alpha = 0,\ 1,...,\ N$) besides the usual variables u (internal energy) and v (specific volume): $s = s(u,\ v,\ p_\alpha,\ \overset{0}{\mathbf{P}}_\alpha)$. As usual, the pressure tensor has been decomposed into a bulk and a deviatoric part. The polymer concentration is supposed to remain constant and therefore has not been introduced into the set of variables. Assuming that the entropy s is analytic in the fluxes, one may write

$$ds = \frac{1}{T}du + \frac{p}{T}dv - \frac{1}{\rho T}\sum_{\alpha=0}^{N}\frac{\tau_\alpha'}{\zeta_\alpha}p_\alpha dp_\alpha - \frac{1}{\rho T}\sum_{\alpha=0}^{N}\frac{\tau_\alpha}{2\eta_\alpha}\overset{0}{\mathbf{P}}_\alpha : d\overset{0}{\mathbf{P}}_\alpha, \qquad (15.56)$$

on condition that we neglect the coupling between the various modes. A more complete treatment in which this coupling is included can be found in [15.32].

Up to second-order terms in the velocity gradients and pressure tensors, the entropy production can be straightforwardly calculated:

$$T\sigma^s = -\sum_{\alpha=0}^{N}p_\alpha\left(\nabla\cdot v + \frac{\tau_\alpha'}{\zeta_\alpha}\frac{\partial}{\partial t}p_\alpha\right) - \sum_{\alpha=0}^{N}\overset{0}{\mathbf{P}}_\alpha : \left(\overset{0}{\mathbf{V}} + \frac{\tau_\alpha}{2\eta_\alpha}\frac{\partial}{\partial t}\overset{0}{\mathbf{P}}_\alpha\right). \qquad (15.57)$$

This takes the form of a bilinear relation between fluxes and forces, suggesting the following linear evolution equations:

$$\tau_\alpha'\frac{\partial p_\alpha}{\partial t} + p_\alpha = -\zeta_\alpha\nabla\cdot v, \qquad (\alpha = 0,1,...,N), \qquad (15.58a)$$

$$\tau_\alpha\frac{\partial \overset{0}{\mathbf{P}}_\alpha}{\partial t} + \overset{0}{\mathbf{P}}_\alpha = -2\eta_\alpha\overset{0}{\mathbf{V}}, \qquad (\alpha = 0,1,...,N). \qquad (15.58b)$$

Assuming with Rouse that the solvent is a Newtonian fluid, which means that $\tau_0' \ll \tau_\alpha'$ and $\tau_0 \ll \tau_\alpha$ ($\alpha = 1, 2,..., N$), one obtains for the solvent the classical Newton–Stokes equations

$$p_0 = -\zeta_0 \nabla \cdot v, \qquad\qquad (15.59a)$$

$$\overset{\circ}{\mathbf{P}}_0 = -2\eta_0 \overset{\circ}{\mathbf{V}}, \qquad\qquad (15.59b)$$

while expressions (15.58) remain unchanged, provided α runs from 1 to N. Relations (15.58b) and (15.59b) are identical with the Rouse–Zimm equations, which were derived from kinetic arguments under the condition that the identifications $\tau_\alpha = \frac{1}{2}\xi/Ha_\alpha$ and $\eta_\alpha = nk_BT\tau_\alpha$ are valid.

It must be stressed that (15.58a) and (15.59a) have no counterpart in the Rouse–Zimm formalism. A clarifying treatment of bulk effects in polymer solutions was due to Metiu and Freed [15.33]. Bulk effects are important in interpreting experiments on absorption of longitudinal sound waves. According to Metiu and Freed, the polymer contribution to the bulk viscosity is equal to two thirds of its contribution to the shear viscosity. In our macroscopic model, this result in recovered by taking $\zeta_\alpha = \frac{1}{3}\eta_\alpha$ and $\tau'_\alpha = \tau_\alpha$.

15.3.3 Comparison with the Kinetic Theory

Our objective is to ground the macroscopic results obtained in the previous subsection on kinetic foundations. The present discussion complements the general results of Chap. 5 on the kinetic theory of ideal monatomic gases. It is shown that the evolution equations and the entropy expression predicted by EIT are confirmed; incidentally, a new insight into the physical meaning of partial pressure tensors is gained.

To calculate the polymer contribution to the entropy, one starts from the Boltzmann formula

$$\rho s = -nk_B \int f \ln f \, d\Gamma, \qquad\qquad (15.60)$$

where f is the polymer distribution function and $d\Gamma$ the volume in the phase space consisting of the positions $r_1, ..., r_N$ and the velocities $\dot{r}_1, ..., \dot{r}_N$ of the N beads forming the chain. It is common to take as variables the centre-of-mass coordinate vector r_c and the relative positions vectors $Q_1, ..., Q_{N-1}$, with Q_i starting from particle i and ending at particle $i+1$. It is also usual to split the distribution function f into two factors, one depending only on the velocities and the other only on the configuration:

$$f = \Xi(\dot{r}_c, \dot{Q}_1, ..., \dot{Q}_{N-1}) \Psi(r_c, Q_1, ..., Q_{N-1}). \qquad\qquad (15.61)$$

Under the three assumptions that the distribution of polymers in the solution is uniform, the velocity gradient is homogeneous and external forces are position independent, the configuration distribution function Ψ may be expressed as $n\psi(Q_1,...,Q_{N-1})$, with ψ normalized to unity [15.30]. Moreover, it is supposed that the velocity-dependent part of the distribution function is given by the Maxwell–Boltzmann local-equilibrium distribution function, from which follows that all the non-equilibrium effects are due to changes in the configurational distribution. This amounts to assuming that the relaxation time of the velocity distribution is much shorter than those associated with the configurational changes.

It is convenient to write the expression (15.60) for the entropy in terms of the equilibrium distribution function f_{eq}:

$$\rho s = -nk_B \int f \ln f_{eq} \, \mathrm{d}\Gamma - nk_B \int f \ln(f/f_{eq}) \mathrm{d}\Gamma. \qquad (15.62)$$

The equilibrium distribution function in the canonical ensemble is given by

$$f_{eq} = [1/Z(T,V)]\exp[-E(\Gamma)/k_B T], \qquad (15.63)$$

where $E(\Gamma)$ is the energy of the configuration Γ and $Z(T, V)$ the equilibrium partition function with V the total volume. Substituting (15.63) into the first term on the right-hand side of (15.62) results in

$$\rho s = \rho u / T + nk_B \ln Z - nk_B \int f \ln(f/f_{eq}) \mathrm{d}\Gamma. \qquad (15.64)$$

In equilibrium, one has

$$\rho s_{eq}(u_{eq}) = \rho u_{eq} / T + nk_B \ln Z, \qquad (15.65)$$

whereby (15.64) becomes

$$\rho s - \rho s_{eq}(u_{eq}) = \rho \frac{u - u_{eq}}{T} - nk_B \int f \ln(f/f_{eq}) \mathrm{d}\Gamma. \qquad (15.66)$$

It should be noticed that in the kinetic theory of monatomic ideal gases, it is assumed that $u = u_{eq}$ (see (5.32)) and the first term on the right-hand side of (15.66) is lacking.

Since the non-equilibrium effects arise from changes in the configurational distribution function, the non-equilibrium contribution to s may be written as

$$\rho s - \rho s_{eq}(u_{eq}) = \rho \frac{u - u_{eq}}{T} - nk_B \int \psi \ln(\psi/\psi_{eq}) \mathrm{d}Q_1...\mathrm{d}Q_{N-1}. \qquad (15.67)$$

Instead of using the relative position vectors $Q_1, ..., Q_{N-1}$, we shall introduce the normal coordinates $Q'_1, ..., Q'_{N-1}$ of the chain variables. In the Rouse model, the configurational distribution function simply reads [15.30]

$$\psi(Q'_1,...,Q'_{N-1},t) = \prod_{j=1}^{N-1} \psi_j(Q'_j,t), \tag{15.68}$$

where ψ_j is the distribution function corresponding to the normal mode j. According to Boltzmann's statistics, ψ_j is given in equilibrium by

$$\psi_{jeq}(Q'_j) = \left(\frac{H}{2\pi k_B T}\right)^{3/2} \exp\left(-\frac{H}{2k_B T} Q'_j \cdot Q'_j\right), \tag{15.69}$$

in which H is an elastic constant characterizing the intramolecular interactions.

The non-equilibrium correction to ψ_j may be derived from its evolution equation. The latter is viewed as a continuity equation in the configuration space, taking into account the various forces acting on the beads. In the absence of external forces, this equation [15.30] is

$$\frac{\partial \psi_j}{\partial t} = -\frac{\partial}{\partial Q'_j} \cdot \left[(\nabla v) \cdot Q'_j \psi_j - \frac{k_B T a_j}{\xi} \frac{\partial \psi_j}{\partial Q'_j} - \frac{a_j}{\xi} F_j^c \psi_j\right]; \tag{15.70}$$

F_j^c is the Hookean force acting between the beads (in our case $F_j^c = -H Q'_j$), ξ is the friction coefficient describing the hydrodynamic drag force on a bead, and a_j is given by (15.56). Under a given velocity gradient, the steady solution of (15.70) is

$$\psi_j = \psi_{jeq}\left(1 + \frac{\xi}{4a_j k_B T} \mathbf{V} : Q'_j Q'_j\right), \tag{15.71}$$

as it is easily verified [15.30].

When this result is introduced into (15.67), the entropy for a plane Couette flow with shear rate $\dot{\gamma}$ is

$$\rho s(u,\dot{\gamma}) - \rho s_{eq}(u_{eq}) = \frac{1}{2} n k_B \sum_{j=1}^{N-1} (\lambda_j \dot{\gamma})^2, \tag{15.72}$$

where $\lambda_j = (\xi/2Ha_j)$, which shows explicitly the dependence of s on $\dot{\gamma}$. Expression (15.72) could seem paradoxical in the sense that $s(u,\dot{\gamma})$ is higher in non-equilibrium than in equilibrium. The solution to this apparent paradox is that the internal energy at equilibrium and at a given temperature differs from the internal energy under shear at the same temperature, because under shear the polymer stores an elastic energy as a result of the stretching of the chains. The elastic energy u_{el} is easily evaluated to be

$$\rho u_{el} = \int \sum_{j=1}^{N-1} \tfrac{1}{2} H Q'_j \cdot Q'_j \, \psi(Q'_1,...,Q'_{N-1},t) \, dQ'_1...dQ'_{N-1}. \tag{15.73}$$

By taking (15.71) and (15.68) into account, one finds that

$$\rho(u - u_{eq}) = nk_B T \sum_{j=1}^{N-1} (\lambda_j \dot{\gamma})^2. \tag{15.74}$$

Combining (15.72) and (15.74) results in

$$\rho s(u,\dot{\gamma}) - \rho s_{eq}(u_{eq}) = \frac{\rho}{T}(u - u_{eq}) - \frac{1}{2} nk_B \sum_{j=1}^{N-1} (\lambda_j \dot{\gamma})^2, \tag{15.75}$$

or, what is the same,

$$\rho s(u,\dot{\gamma}) - \rho s_{eq}(u_{eq}) = \rho s_{eq}(u) - \frac{1}{2} nk_B \sum_{j=1}^{N-1} (\lambda_j \dot{\gamma})^2, \tag{15.76}$$

after use of the expansion

$$s_{eq}(u) - s_{eq}(u_{eq}) = \left(\frac{\partial s}{\partial u}\right)_{eq} (u - u_{eq}) + O(2). \tag{15.77}$$

According to the macroscopic theory (see (15.56)), the contribution of the fluxes to the entropy is

$$\rho s(u,\dot{\gamma}) = s_{eq}(u) - \frac{1}{4} \sum_{j=1}^{N} \frac{\tau_j}{\rho \eta_j T} \overset{0}{\mathbf{P}}{}^v_j : \overset{0}{\mathbf{P}}{}^v_j. \tag{15.78}$$

To check whether the microscopic expression (15.76) is compatible with the macroscopic result (15.78) predicted by EIT, a microscopic evaluation of the relaxation times and the viscosities is needed. To proceed in the simplest way, we recall two equivalent expressions of the polymeric contribution to the viscous pressure tensor, namely the Kramers and Giesekus tensors. The former reads [15.30]

$$\overset{0}{\mathbf{P}}{}^{v} = -nH \sum_{j=1}^{N-1} \langle Q_j' Q_j' \rangle + (N-1)nk_B T \mathbf{U}, \tag{15.79}$$

with $\langle ... \rangle$ standing for an average over the configuration space, i.e.

$$\langle Q_j' Q_j' \rangle = \int \psi(Q_1', ..., Q_{N-1}', t)\, Q_j' Q_j'\, \mathrm{d}Q_1' ... \mathrm{d}Q_{N-1}'. \tag{15.80}$$

Expression (15.79) exhibits the microscopic meaning of the splitting of $\overset{0}{\mathbf{P}}{}^{v}$ as a sum of several terms: each of them is related to the contribution of one normal mode. To the $N-1$ modes appearing in (15.79) should be added the contribution of the motion of the centre of mass. The sum over the modes should therefore really extend to N instead of $N-1$.

We now turn to the Giesekus expression for $\overset{0}{\mathbf{P}}{}^{v}$ [15.30]:

$$\overset{0}{\mathbf{P}}{}^{v} = \frac{1}{2} n\xi \sum_{j=1}^{N-1} \frac{1}{a_j} D^{\uparrow} \langle Q_j' Q_j' \rangle, \tag{15.81}$$

where $D^{\uparrow}$ denotes the contravariant convected time derivative. After applying $D^{\uparrow}$ to (15.79) and using definition (15.81), one obtains

$$\overset{0}{\mathbf{P}}{}^{v}_{j} + \lambda_j D^{\uparrow} \overset{0}{\mathbf{P}}{}^{v}_{j} = -2nk_B T \lambda_j \mathbf{V}, \tag{15.82}$$

where the relation $D^{\uparrow} \mathbf{U} = -2\mathbf{V}$ has been taken into account. Comparison of (15.82) with (15.58b) allows us to identify the relaxation time and the viscosity related to each partial viscous pressure tensor as $\tau_j = \lambda_j$, $\eta_j = nk_B T \lambda_j$. With these identifications in mind and in the particular case of a plane Couette flow for which $\overset{0}{\mathbf{P}}{}^{v}_{j} : \overset{0}{\mathbf{P}}{}^{v}_{j} = 2\eta_j^2 \dot{\gamma}^2$, it is seen that the microscopic (15.76) and macroscopic (15.78) expressions are identical.

Whereas in the Rouse model the hydrodynamic interactions are neglected, in the Zimm model they are introduced in an average form. But this changes only the values of the eigenvalues a_j, in such a way that the basic formalism remains the same, and the agreement with extended irreversible thermodynamics is still achieved [15.34].

15.4 EIT Description of Second-Order Non-Newtonian Fluids

In the previous sections, only linear situations were investigated. However, as repeatedly mentioned, EIT is by no means restricted to the linear range. Our aim in this section is to show that a description of second-order non-Newtonian fluids stems naturally from EIT. Two particular models are derived: a three-parameter model and Giesekus four-parameter model. In particular, the simple three-parameter description is seen to yield the correct signs of the material coefficients of the Rivlin–Ericksen model, in agreement with observations. We also examine to what extent the three-parameter model is able to cope with experimental dependence of the material functions on shear rate in steady shearing flows.

Heat effects are assumed to be negligible, and only isotropic and incompressible fluids are considered. For incompressible fluids, bulk viscous effects are negligible in the linear theory, but they must be taken into account when non-linear contributions (of the order of $\dot{\gamma}^2$) are included.

15.4.1 A Simple Non–linear Model

The basic variables are selected to be the same as for an ordinary Newtonian fluid, namely u and $\mathbf{P}^v$, and the corresponding Gibbs equation is

$$T\mathrm{d}s = \mathrm{d}u - \frac{1}{\rho}A\mathbf{P}^v : \mathrm{d}\mathbf{P}^v, \tag{15.83}$$

where A is a scalar coefficient which may depend on u and the invariants of $\mathbf{P}^v$. The convexity of s implies that $\partial^2 s/(\partial\mathbf{P}^v : \partial\mathbf{P}^v) < 0$, from which it follows that $A > 0$. Combining Gibbs' equation with the balance law of energy yields the following expression of the entropy production:

$$T\sigma^s = -\mathbf{P}^v : (\mathbf{V} + A\dot{\mathbf{P}}^v).$$

We identify the generalised thermodynamic force $\mathbf{X}$ as the quantity between the brackets:

$$\mathbf{X} = \mathbf{V} + A\dot{\mathbf{P}}^v. \tag{15.84}$$

It is noted that $T\sigma^s$ takes the form of a bilinear expression in the flux $\mathbf{P}^v$ and the force $\mathbf{X}$.

On the other hand, we can formulate another expression for $\mathbf{X}$ if it is imposed that $\mathbf{X}$ must be constructed as the most general second-order traceless tensor depending on the quantities u and $\mathbf{P}^v$; accordingly, we are allowed to write $\mathbf{X}$ as

$$\mathbf{X} = -a_1\mathbf{P}^v - a_2(\mathbf{P}^v \cdot \mathbf{P}^v - \pi^v\mathbf{U}), \tag{15.85}$$

where $\pi^v = \frac{1}{3}(\mathbf{P}^v : \mathbf{P}^v)$. The minus signs in front of a_1 and a_2 are introduced for convenience; a_1 and a_2 are functions of u and the invariants of $\mathbf{P}^v$. The positiveness of $T\sigma^s$ requires that $a_1 > 0$. Equating the right-hand sides of (15.84) and (15.85), one gets the evolution equation for the viscous pressure tensor

$$\frac{A}{a_1}\dot{\mathbf{P}}^\nu = -\frac{1}{a_1}\mathbf{V} - \mathbf{P}^\nu - \frac{a_2}{a_1}\left(\mathbf{P}^\nu \cdot \mathbf{P}^\nu - \pi^\nu \mathbf{U}\right). \tag{15.86}$$

For latter purpose, we set $A/a_1 = \tau$, $1/a_1 = 2\eta$, and $a_2/a_1 = a$, where τ and η have the dimensions of time and viscosity respectively and a is a coefficient that takes account of the non-linear correction to Newton's law. To satisfy objectivity, the time derivative in (15.86) must be replaced by an objective time derivative. Since a macroscopic theory cannot discriminate amongst the several objective time derivatives, we choose the Jaumann derivative D_J, which guarantees the form invariance of the entropy production. It follows that the final expression for the evolution equation reads

$$\tau D_J \mathbf{P}^\nu = -2\eta \mathbf{V} - \mathbf{P}^\nu - a(\mathbf{P}^\nu \cdot \mathbf{P}^\nu - \pi^\nu \mathbf{U}). \tag{15.87}$$

This relation is the cornerstone of the model; a similar result was obtained by Müller and Wilmanski [15.35], who based their analysis on Müller and Liu's version of EIT [15.36] (see also Sect. 2.5). For negligible values of τ (in ordinary fluids, τ is of the order of the collision time, i.e. 10^{-12} s) and $a = 0$, one recovers Newton's law.

Combining the constraints $A > 0$ and $a_1 > 0$, one obtains, in virtue of the above definitions of τ and η, that $\tau > 0$, $\eta > 0$ .This result indicates that the evolution equation (15.87) is characterized by a positive viscosity and a positive relaxation time. No information about the sign of a is available. The positiveness of τ and η together with (15.87) provide the essential features of the model. The latter involves three parameters, η, τ, and a, to be determined from experimental observations or molecular theories.

It is interesting to investigate the transition from the rate-type equation (15.87) to constitutive equations like those of Rivlin–Ericksen and Reiner–Rivlin. To this end, let us rewrite (15.87) in terms of the Rivlin–Ericksen time derivative $\mathbf{P}^{(2)}$ defined by (15.13): one finds that

$$\mathbf{P}^\nu = -2\eta \mathbf{V} - \tau \mathbf{P}^{(2)} - a(\mathbf{P}^\nu \cdot \mathbf{P}^\nu - \pi^\nu \mathbf{U}) + \tau(\mathbf{P}^\nu \cdot \mathbf{V} + \mathbf{V} \cdot \mathbf{P}^\nu), \tag{15.88}$$

where use has been made of the result

$$D_J \mathbf{P}^\nu = \mathbf{P}^{(2)} - (\mathbf{P}^\nu \cdot \mathbf{V} + \mathbf{V} \cdot \mathbf{P}^\nu).$$

In the first-order approximation and in the limit $\tau = 0$, (15.88) reduces to

$$_{(1)}\mathbf{P}^\nu = -2\eta \mathbf{V}, \tag{15.89}$$

which is the Navier-Stokes relation; subscript (1) refers to the first-order approximation.

For a non-vanishing relaxation time, still in the first-order approximation, (15.88) becomes

$$_{(1)}\mathbf{P}^{\nu} = -2\eta\mathbf{V} - \tau\frac{\partial_{(1)}\mathbf{P}^{\nu}}{\partial t},$$

and this is the basic equation of Maxwell's model.

The second-order approximation with $\tau = 0$ is obtained by substituting $\mathbf{P}^{\nu}$ by its value (15.89) into the right-hand side of (15.88). This operation leads to the Reiner–Rivlin equation,

$$_{(2)}\mathbf{P}^{\nu} = -2\eta\mathbf{V} - 4\eta^2 a\left[\mathbf{V}\cdot\mathbf{V} - \tfrac{1}{3}(\mathbf{V}:\mathbf{V})\mathbf{U}\right].$$

If the relaxation time does not vanish, one obtains from (15.88)

$$_{(2)}\mathbf{P}^{\nu} = -2\eta\mathbf{V} + 2\tau\eta\mathbf{V}^{(2)} - 4\eta(\eta a + \tau)\left[\mathbf{V}\cdot\mathbf{V} - \tfrac{1}{3}(\mathbf{V}:\mathbf{V})\mathbf{U}\right]. \tag{15.90}$$

Comparison between (15.90) and the Rivlin–Ericksen equation (15.13) allows one to express the coefficients Ψ_1 and Ψ_2 [see (15.8)] in terms of the parameters η, τ, and a:

$$\Psi_1 + \Psi_2 = \eta\tau + \eta^2 a,$$

$$\Psi_1 = 2\eta\tau. \tag{15.91}$$

Recalling that τ and η are positive, it is clear from the second equation in (15.91) that Ψ_1 is a positive quantity, in accordance with experiments. Moreover, combining the two equations in (15.91) leads to $\Psi_2 = -\tfrac{1}{2}\Psi_1 + \eta^2 a$, which shows that Ψ_2 is generally different from $-\tfrac{1}{2}\Psi_1$. As experiments suggest that $-\Psi_2/\Psi_1$ takes values between 0.1 and 0.4, it is expected that the coefficient a is positive.

By working in the framework of rational thermodynamics, Dunn and Fosdick [15.21] have found that $\Psi_1 < 0$ and $\Psi_2 = -\tfrac{1}{2}\Psi_1$, in contradiction with experiments. Therefore, it can be concluded that by considering the Rivlin–Ericksen equation to be an approximation of the more general rate-type model (15.88), one avoids the problems raised by Dunn and Fosdick's work. The essential conclusions are that $\Psi_1 > 0$ is not in contradiction with the second law of thermodynamics and that EIT is a good candidate for describing second-order non-Newtonian fluids.

A further check of the quality of the model can be obtained by calculating the viscometric functions μ, Ψ_1, and Ψ_2 and comparing them with experimental results. This is done in the next section, where steady shearing flows of polymer solutions are studied.

15.4.2 Steady shearing flows

Let us determine the dependence of the viscometric functions μ, Ψ_1, and Ψ_2 with respect to the shear rate for a steady shearing flow in the x direction. Inserting the velocity field $v_x = \dot{\gamma} y$ into the evolution equation (15.87), and making use of the definitions (15.8) of the viscometric functions, one obtains the following set of non-linear algebraic equations:

$$4\tau\mu^2 - \Psi_1\left(2\eta - \tau\Psi_1\dot{\gamma}^2\right) = 0,$$

$$\tau\mu(\Psi_1 + 2\Psi_2) + a\Psi_1\Psi_2(\Psi_1 + \Psi_2)\dot{\gamma}^2 - a\mu^2\Psi_1 = 0, \qquad (15.92)$$

$$3\Psi_1 - 6\mu\tau - a\Psi_1(\Psi_1 + 2\Psi_2)\dot{\gamma}^2 = 0.$$

The unknowns in (15.92) are μ, Ψ_1, and Ψ_2, while $\dot{\gamma}$ is the independent variable. To calculate μ, Ψ_1, and Ψ_2 it is necessary to know the behaviour of η, τ, and a. Preliminary information is obtained by assuming that τ and η are constant and that a is negligibly small. In view of (15.92) it follows that

$$\mu = \frac{\eta}{1 + \tau^2\dot{\gamma}^2}, \qquad \Psi_1 = \frac{2\tau\eta}{1 + \tau^2\dot{\gamma}^2}, \qquad \Psi_2 = -\frac{1}{2}\Psi_1.$$

These results indicate that the three viscometric functions are even functions of $\dot{\gamma}$, and that Ψ_1 is positive and Ψ_2 is negative: although these conclusions are corroborated experimentally, the model is too crude. At high shear rates, it is found that η behaves like $\dot{\gamma}^{-2}$, which is not very realistic. On the other hand the asymptotic behaviour of Ψ_1 is proportional to $\dot{\gamma}^{-2}$, which is closer to experimental observations. It is also seen that the ratio $-\Psi_2/\Psi_1$ is equal to 0.5, higher than the experimental values, which lie between 0.1 and 0.4.

A more realistic description is obtained by supposing that the material coefficients η, τ and a are allowed to depend on the invariants of $\mathbf{P}^v$ [15.40]. Moreover, we do not alter the generality by considering that η, τ, and a are functions of $\dot{\gamma}$ rather than of $\mathbf{P}^v$, because in steady motions both quantities are related by (15.8a).

The problem is of course to find the most adequate dependence of the parameters with respect to $\dot{\gamma}$. To achieve this task, we rely on the fact that, for steady shearing, it is generally accepted that the apparent viscosity μ and the first normal stress coefficient Ψ_1 are well accounted for by Ostwald's power laws (15.9) [15.30,37]. It is also widely admitted that Ψ_2 is proportional to Ψ_1 and can therefore be described by a power law as well. We are thus allowed to write

$$\mu = \mu_0\left(\frac{\dot{\gamma}}{\dot{\gamma}_0}\right)^{n_\mu}, \qquad \Psi_1 = \Psi_0^1\left(\frac{\dot{\gamma}}{\dot{\gamma}_0}\right)^{n_1}, \qquad \Psi_2 = \Psi_0^2\left(\frac{\dot{\gamma}}{\dot{\gamma}_0}\right)^{n_2}; \qquad (15.93)$$

$\dot{\gamma}_0$ represents a reference value for $\dot{\gamma}$ and is introduced for dimensional reasons; μ_0, n_μ, Ψ_0^1, n_1, Ψ_0^2, and n_2 are undetermined constants.

Substitution of (15.93) into (15.92) shows that the parameters η, τ, and a have also to satisfy the following Ostwald power laws:

$$\eta = \eta_0\left(\frac{\dot{\gamma}}{\dot{\gamma}_0}\right)^{n_\eta}, \qquad \tau = \tau_0\left(\frac{\dot{\gamma}}{\dot{\gamma}_0}\right)^{n_\tau}, \qquad a = a_0\left(\frac{\dot{\gamma}}{\dot{\gamma}_0}\right)^{n_a}. \qquad (15.94)$$

The assumption that τ is shear dependent is a direct consequence of the present formalism but is also confirmed by other rheological models, such as the White–Metzner model [15.20,30,35].

After introducing (15.93) and (15.94) into (15.92), we observe that the 12 unknown constant parameters appearing in (15.93) and (15.94) are not independent but are related by $n_1 = n_2 = n_\mu - 1$, $n_a = -1 - n_\eta$, $n_\tau = -1$, $n_\mu = n_\eta$, plus three relations identical to (15.92), with every quantity affected by a subscript 0. Finally, bearing all these results in mind, we are left with four independent adjustable constants that can be selected as μ_0, Ψ_0^1, Ψ_0^2, and n_η. This choice is dictated by the fact that these quantities are easily obtained from experimental measurements.

Comparison with experimental data is performed with three polymeric solutions: a 2.5% solution of polyacrylamide in a 50% water and 50% glycerine solution (PAA), a 1.1% polyisobutylen solution in decahydronaphtalene (Oppanol B200), and an 8.7% polyisobutylen-decalin solution 'D2b' (Oppanol B50). The reason for choosing the PAA and Oppanol B200 solutions is that in these cases all three viscometric functions η, Ψ_0^1, and Ψ_0^2 have been determined experimentally [15.38]. We have also checked our model on Oppanol B50 because it is, to our knowledge, the only polymeric solution for which η and Ψ_0^1 have been measured at very large values of $\dot{\gamma}$ up to $10^6\,\mathrm{s}^{-1}$ [15.39]; for this particular solution we have imposed $\Psi_2/\Psi_1 = -0.1$ because no experimental data are available for the viscometric function Ψ_2. The values of the adjustable constants are reported in Table 15.1 and given in SI units [15.40].

Table 15.1. Some values of adjustable constants in (15.93) for viscometric coefficients

Solution	μ_0	Ψ_0^1	Ψ_0^2	n_μ
PAA	6.01×10^1	4.12×10^2	-2.41×10^1	-0.774
Oppanol B200	1.71×10^{-1}	5.53×10^{-2}	-8.02×10^{-4}	-0.302
Oppanol B50	1.14×10^{-1}	2.66×10^{-5}	-2.66×10^{-6}	-0.323

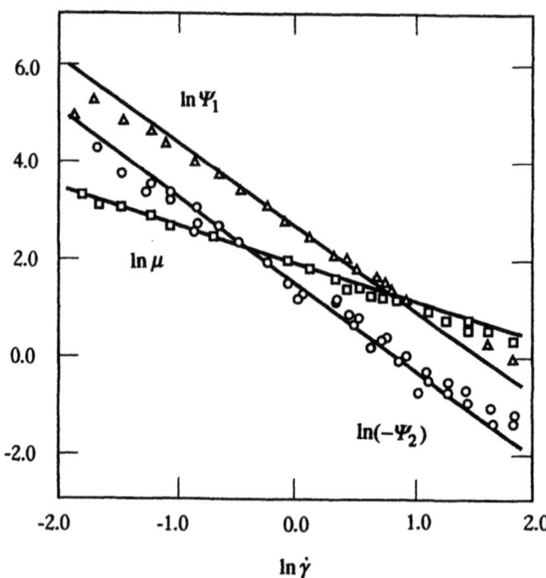

Fig. 15.1. Dependence of the viscometric functions on the shear rate for PAA: comparison between theoretical and experimental results. Squares, triangles, and open circles represent, respectively, experimental data for the viscosity, and the first and second normal stress coefficients. The solid lines are the theoretical results.

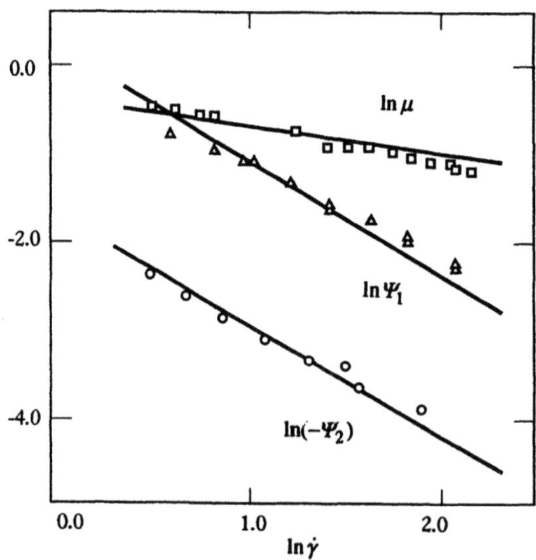

Fig. 15.2. Dependence of the viscometric functions on the shear rate for Oppanol B200: comparison between theoretical and experimental results. Squares, triangles, and circles represent experimental data for μ, Ψ_1, and Ψ_2, respectively; the solid lines are the theoretical results.

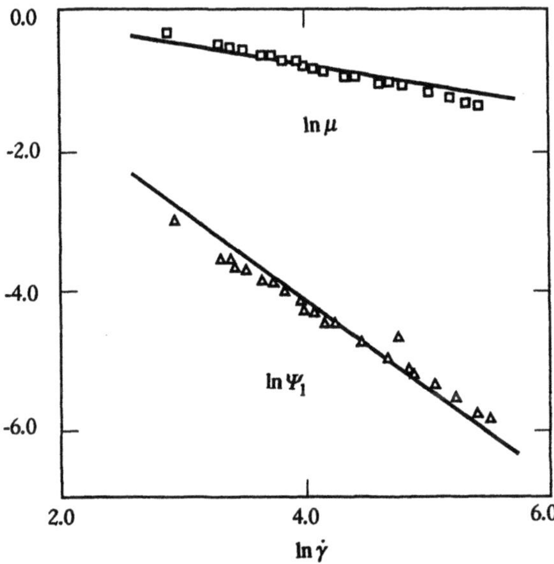

Fig. 15.3. The viscosity and the first normal stress coefficient as a function of the shear rate for Oppanol B50: comparison between theoretical and experimental results. Squares and triangles correspond, respectively, to experimental data for the viscosity and the first normal stress. Solid lines are the theoretical results.

Comparison between experimental and theoretical results is shown in Figs. 15.1–3. A very satisfactory agreement is achieved in all the cases: it is worth noticing that even at very high values of the shear rate, as in Oppanol B50 (Fig. 15.3), the model accounts for the experimental data to a very high degree of accuracy.

15.4.3 The Giesekus Four-parameter Model

In the previous subsections, we analysed a non-Newtonian model that was described by means of the three parameters η, τ, and a. We show here that more complicated descriptions, such as the Giesekus four-parameter model, can easily be generated if it is admitted that the total viscous pressure tensor $\mathbf{P}^v$ is the sum of a contribution $\mathbf{P}_0$ from the solvent and a contribution $\mathbf{P}_p$ from the polymer chains: $\mathbf{P}^v = \mathbf{P}_0 + \mathbf{P}_p$. As in the Rouse–Zimm model, it is supposed that the solvent is an incompressible Newtonian fluid, so that $\mathbf{P}_0$ is related to $\mathbf{V}$ by Newton's law, $\mathbf{P}_0 = -2\eta_0\mathbf{V}$, and η_0 is the shear viscosity of the solvent. As a consequence, $\mathbf{P}_0$ will not be considered as independent variable and the Gibbs equation will take the same form as (15.83), with $\mathbf{P}^v$ replaced by $\mathbf{P}_p$:

$$Tds = du - \frac{1}{\rho} A_p \mathbf{P}_p : d\mathbf{P}_p. \tag{15.95}$$

From the convexity property of s, it is inferred that $A_p > 0$. The corresponding entropy production is

$$T\sigma^s = -\mathbf{P}_0 : \mathbf{V} - \mathbf{P}_p : \mathbf{X}, \tag{15.96}$$

where $\mathbf{X}$ is given by (15.84), with $\mathbf{P}^v$ replaced by $\mathbf{P}_p$. Repeating the procedure followed in Subsect. 15.4.1, it is found that $\mathbf{P}_p$ satisfies the evolution equation

$$\tau_p D_J \mathbf{P}_p = -2\eta_p \mathbf{V} - \mathbf{P}_p - a_p\left(\mathbf{P}_p \cdot \mathbf{P}_p - \pi_p \mathbf{U}\right), \tag{15.97}$$

which contains the three parameters τ_p, η_p, and a_p, and the quantity π_p given by $\pi_p = \frac{1}{3}(\mathbf{P}_p : \mathbf{P}_p)$. Eliminating $\mathbf{P}_p$ in terms of $\mathbf{P}^v$ and $\mathbf{P}_0$ and using $\mathbf{P}_0 = -2\eta_0 \mathbf{V}$, one obtains the following evolution equation for the total viscous pressure:

$$\tau_p D_J \mathbf{P}^v + \mathbf{P}^v - \frac{a\tau_p}{\eta}\left(\mathbf{P}^v \cdot \mathbf{P}^v - \pi^v \mathbf{U}\right) - 2a\lambda\left(\mathbf{V} \cdot \mathbf{P}^v + \mathbf{P}^v \cdot \mathbf{V} - \bar{\pi} \mathbf{U}\right)$$

$$= -2\eta\left[\mathbf{V} + \lambda D_J \mathbf{V} - 2a\frac{\lambda^2}{\tau_p}\left(\mathbf{V} \cdot \mathbf{V} - \bar{\bar{\pi}} \mathbf{U}\right)\right], \tag{15.98}$$

where the unidentified coefficients η, a, λ, $\bar{\pi}$, and $\bar{\bar{\pi}}$ stand respectively for $\eta = \eta_0 + \eta_p$, $a = -(\eta a_p/\tau_p)$, $\lambda = (\eta_0 \tau_p/\eta)$, $\bar{\pi} = \frac{2}{3}(\mathbf{P}^v : \mathbf{V})$, and $\bar{\bar{\pi}} = \frac{1}{3}(\mathbf{V} : \mathbf{V})$.

Expression (15.98) is the same as the Giesekus constitutive equation [15.16,30] except for the terms in π^v, $\bar{\pi}$, and $\bar{\bar{\pi}}$, which appear here as a consequence of the hypothesis of the absence of bulk effects. The adjustable parameters in Giesekus' model are η_0, a_p, η_p, and τ_p, or equivalently η, a, λ, and τ_p. The result (15.98) is particularly promising, since it allows one to derive the Giesekus equations from very simple macroscopic considerations.

It is interesting to mention that by setting into (15.98) the coefficient $a = 0$ one recovers the Jeffreys model

$$\tau_p D_J \mathbf{P}^v + \mathbf{P}^v = -2\eta(\mathbf{V} + \lambda D_J \mathbf{V}). \tag{15.99}$$

It is worth stressing that the above results were obtained by simply requiring that the pressure tensor is selected as an independent variable and that it obeys a non-linear evolution equation of the relaxation type. It is possible to complicate the model by assuming that a non-Newtonian solvent or by introducing viscous bulk effects. This would result in more realistic descriptions and generate other rheological models, such as the Oldroyd eight-constants model [15.41]. This is of course a further step in the formulation of the theory but it does not raise any fundamental difficulty. Indeed, the thermodynamical formalism presented here exhibits such a flexibility and power of generalisation that it can successfully deal with more sophisticated systems.

Problems

15.1 Consider a dilute polymeric solution modelled by rigid dumbbells, each of them characterized by a director vector u. The orientational equilibrium distribution function is the isotropic distribution $\psi_{eq} = 1/4\pi$. In a plane Couette flow with shear rate $\dot{\gamma} = (\partial v_x/\partial y)$, the steady state distribution function is, up to the first order in $\dot{\gamma}$,

$$\psi(u) = \psi_{eq}[1 + 3u_x u_y \,\tau\dot{\gamma}]\ ,$$

with the relaxation time τ given by $\tau = \zeta L^2 (12k_B T)^{-1}$, L being the length of the dumbbell and ζ the friction coefficient between the beads of the dumbbell and the solvent. If the entropy of the dumbbells is given by

$$s = -k_B \int \psi \ln \psi \, du,$$

show that the non-equilibrium entropy at steady shear flow is

$$s(u,\dot{\gamma}) = s_{eq}(u) - \tfrac{3}{10} n k_B (\tau\dot{\gamma})^2\ .$$

[J. Camacho and D. Jou, J. Chem. Phys. **92** (1990) 1339.]

15.2 The steady-state viscosity of a dilute solution of rigid dumbbells is $\eta = n k_B T \tau$, with n the number of dumbbells per unit volume and τ the relaxation time as given in Problem 15.1. (a) Compare the expression for the non-equilibrium entropy obtained in the above problem with the following expression

$$s = s_{eq} - \frac{\tau}{2T}(\eta\dot{\gamma})^2\ .$$

Is this result satisfactory? (b) The viscous pressure tensor for a rigid dumbbell solution has the form

$$\mathbf{P}^v = -2\eta_s \mathbf{V} + \mathbf{P}_1^v + \mathbf{P}_2^v,$$

with

$$\mathbf{P}_1^v = -3nk_BT\left[\langle \mathbf{u}\mathbf{u}\rangle - \tfrac{1}{3}\mathbf{U}\right]$$

and

$$\mathbf{P}_2^v = -6nk_BT\tau\langle \mathbf{u}\mathbf{u}\mathbf{u}\mathbf{u}\rangle : \mathbf{V}.$$

The quantity η_s is the viscosity of the pure solvent, $\mathbf{V}$ the symmetric part of the velocity gradient, and $\mathbf{u}$ the director vector of the rigid dumbbells. It may be shown that the part $\mathbf{P}_1^v$ has a relaxation time $\tau_1 = \tau$ and that the corresponding shear viscosity is $\eta_1 = \tfrac{3}{5}nk_BT\tau$, while the part $\mathbf{P}_2^v$ has a relaxation time $\tau_2 = 0$ and a viscosity $\eta_2 = \tfrac{2}{5}nk_BT\tau$ [R. B. Bird et al., *Dynamics of Polymeric Liquids* (Wiley, New York, 1977) vol. 2]. Show that the above decomposition of the viscous pressure tensor is consistent with the following expression for the entropy:

$$s = s_{eq} - \frac{1}{2T}(\tau_1\eta_1 + \tau_2\eta_2)\dot{\gamma}^2.$$

15.3 In ideal gases and in polymeric solutions described by rigid dumbbells, a shear flow does not change the internal energy of the system at a given temperature T. However, for elastic dumbbells, the shear flow produces a stretching of the dumbbells and consequently induces a change of the internal energy at constant T. The elastic potential energy of the dumbbells may be obtained from

$$u_d = \int \tfrac{1}{2}HQ^2\psi(\mathbf{Q})\,d\mathbf{Q},$$

where H is the elastic constant of the dumbbells, $\mathbf{Q}$ the bead-to-bead vector, and $\psi(\mathbf{Q})$ the distribution function of the dumbbells. Up to the first order in the shear rate $\dot{\gamma}$, the distribution function is

$$\psi(\mathbf{Q},\dot{\gamma}) = \psi_{eq}(\mathbf{Q})\left(1 + \frac{\zeta}{4k_BT}\mathbf{V}:\mathbf{Q}\mathbf{Q}\right),$$

where $\psi_{eq}(\mathbf{Q})$ is the equilibrium distribution function (15.69), $\mathbf{V}$ the traceless symmetric part of the velocity gradient, and ζ the friction coefficient between the beads and the solvent. Assuming constant n (number of dumbbells per unit volume) and T (temperature), show that the difference in internal energy per unit volume between a steady state and an equilibrium state is

$$u(n,T,\dot{\gamma}) - u_{eq}(n,T) = nk_B T(\tau\dot{\gamma})^2,$$

with the relaxation time τ given by $\tau = \zeta/4H$.

15.4 The entropy of a dilute solution of Hookean dumbbells under shear $\dot{\gamma}$ is

$$s(u,\dot{\gamma}) = s_{eq}(u_{eq}) + nk_B \ln\left\{\left[1 + (\tau\dot{\gamma})^2\right]^{1/2}\right\}.$$

(a) Find the expression of $s(u,\dot{\gamma}) - s_{eq}(u_{eq})$ at low $\tau\dot{\gamma}$. Should it be surprising that the non-equilibrium entropy $s(u,\dot{\gamma})$ is higher than the equilibrium entropy $s_{eq}(u_{eq})$? (b) Taking into account the expression for $u - u_{eq}$ found in Problem 15.3, calculate the difference $s(u,\dot{\gamma}) - s_{eq}(u)$. (c) Compare the expressions for $s(u,\dot{\gamma})$, $s_{eq}(u)$ and $s_{eq}(u_{eq})$.

15.5 The evolution equation for $\mathbf{P}^v$ in the upper convected Maxwell model is

$$\dot{\mathbf{P}}^v - (\nabla v)^T \cdot \mathbf{P}^v - \mathbf{P}^v \cdot (\nabla v) = -\frac{1}{\tau}\mathbf{P}^v - \frac{2\eta}{\tau}\mathbf{V}$$

with $\mathbf{V}$ being the symmetric part of the velocity gradient. (a) Show that, in a steady pure shear flow, $\mathbf{P}^v$ is given by

$$\mathbf{P}^v = \begin{pmatrix} -2\tau\eta\dot{\gamma}^2 & -\eta\dot{\gamma} & 0 \\ -\eta\dot{\gamma} & 0 & 0 \\ 0 & 0 & 0 \end{pmatrix}.$$

(b) Show that the normal stress coefficients are $\Psi_1(\dot{\gamma}) = 2\tau\eta$ and $\Psi_2(\dot{\gamma}) = 0$.

15.6 Determine the expression for $\mathbf{P}^v$ in a steady shear flow if the lower-convected derivative (1.91) or the Jaumann derivative (15.19) are used in Problem 15.5 instead of the upper-convected derivative.

15.7 A planar extensional flow is defined by a velocity gradient of the form

$$\nabla v = \begin{pmatrix} \dot{\varepsilon} & 0 & 0 \\ 0 & -\dot{\varepsilon} & 0 \\ 0 & 0 & 0 \end{pmatrix},$$

with $\dot{\varepsilon}$ being the extensional rate. Show that for the upper convected Maxwell model in the steady state the viscous pressure tensor reads

$$\mathbf{P}^{v} = \begin{pmatrix} -2\eta\dot{\varepsilon}(1-2\tau\dot{\varepsilon})^{-1} & 0 & 0 \\ 0 & 2\eta\dot{\varepsilon}(1+2\tau\dot{\varepsilon})^{-1} & 0 \\ 0 & 0 & 0 \end{pmatrix}.$$

15.8 Show that the constitutive equations of the Maxwell, Kelvin–Voigt, and Poynting–Thomson bodies are easily interpretable in terms of the following mechanical models involving coupling of springs and dashpots:

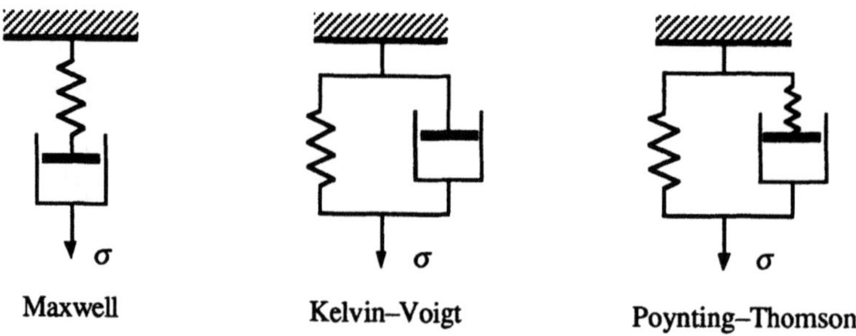

Maxwell   Kelvin–Voigt   Poynting–Thomson

15.9 Repeat the demonstration leading to (15.91) when, in the evolution equation (15.87) for $\mathbf{P}^{v}$, the Jaumann derivative is replaced by the lower-convected and the upper-convected time derivatives, respectively.

15.10 Determine the dependence of the viscometric functions μ, Ψ_1 and Ψ_2 in terms of the supposed constant parameters η, τ, and a for a material described by an evolution equation of the form (15.87) for: (a) a steady shearing flow $v_x = \dot{\gamma}y$; (b) a time-dependent shear flow $v_x = \dot{\gamma}(t)y$; (c) a small amplitude oscillatory shearing flow.

Chapter 16

Thermodynamics of Polymer Solutions Under Shear Flow

In Chaps. 3 and 10 we studied effects of a heat flux on the thermodynamics of non-equilibrium steady states. Here, we deal with another interesting situation, that of polymer solutions under shear flow. The presence of a shear results not only in drastic changes in the phase diagram of the solution but is also able to modify the kinetics of chemical reactions taking place within it [16.1]. Various thermodynamic approaches have been concerned with the study of the interaction between the flow and the thermodynamic properties of a fluid with internal structure; some of them are based on internal variables [16.2] and others on Hamiltonian formalisms [16.3]. The difference between extended irreversible thermodynamics (EIT) and other theories is that the former uses as variables the shear viscous pressure or the shear rate, whereas the latter prefer variables related with the internal structure of the fluid, as for instance, the configuration tensor. The relation between the two classes of variables has already been discussed in Sect. 15.3. The EIT variables are especially useful in macroscopic analyses while the internal variables are more suitable for a microscopic understanding. Let us also mention that EIT predicts modifications of the thermodynamic equations of state not only for fluids with internal structure but also for ideal monatomic gases or for liquids of spherical particles. These systems have been the subject of many computer simulations, some of which have been discussed in Sect. 9.5, and of several kinds of microscopic analyses, ranging from the kinetic theory of gases to renormalization group techniques [16.4].

The study of polymer solutions under shear is particularly interesting because the effects of shear are more perceptible than in ordinary liquids or gases. Besides its theoretical interest, this study is also important in engineering, since most of polymer processings take place under motion. The phase diagrams established for equilibrium situations cannot be trusted in the presence of flows since the latter may enhance or reduce the solubility of the polymer and the conditions under which the phase separation occurs.

Many works in laboratories and in industry have been devoted to flow-induced changes in phase diagrams of polymer solutions [16.5]. It is obvious that the classical local-equilibrium thermodynamics must be modified in such a way that the equations of state incorporate explicitly the influence of the flow. Moreover, the equilibrium thermodynamic stability conditions cannot be blindly extrapolated to non-equilibrium steady states, but a justification based on dynamic arguments is needed and shall be found in Sect. 16.4. Another problem is the effect of a shear on chemical equilibrium, which is treated in Sect. 16.3, and applied to thermodynamically-induced polymer degradation under flow. Polymer degradation under elongational flow has been extensively studied from kinetic arguments [16.6] .

Several authors bypass thermodynamics and analyse phase separation or phase homogenisation under shear by writing directly the relevant dynamical equations and investigating afterwards the stability of their solutions [16.7]. At first sight, it may be thought that such a procedure has a wider range of potentialities than a pure thermodynamic analysis, as the latter cannot provide details about the process of segregation of the different phases or the changes in viscosity observed during such a process. However, it is our opinion that the coexistence of both approaches should be fostered because the dynamical method is helpful to specify the range of validity of thermodynamics and, reciprocally, the dynamical analysis asks for equations of state which are supplied by thermodynamics.

As the chemical potential plays a prominent role in all these problems, we examine in Sect. 16.1 the effects of a shear on the expression of this potential and the consequences on phase diagrams of polymeric solutions.

16.1 The Chemical Potential Under Shear: Flow-Induced Changes in the Phase Diagram of Solutions

A basic quantity in multicomponent mixtures is the chemical potential. Recall that in Chap. 3 thermal and caloric equations of state for a fluid in non-equilibrium were proposed. In this section, we shall determine the expressions of the Gibbs function and the chemical potential of a fluid under shear. Assume that temperature and pressure remain constant and that heat flux, diffusion flux and bulk viscous pressure are negligible. Define the Gibbs function g per unit mass as

$$g = u - \theta s + \pi v. \tag{16.1}$$

The corresponding Gibbs equation at constant temperature and pressure reads as

$$dg = \sum_k \mu_k dc_k + \frac{v\tau}{2\eta} \mathbf{P}^v : d\mathbf{P}^v \tag{16.2}$$

and, after integration,

$$g(c_i, \mathbf{P}^v) = g_{eq}(c_i) + (\tau v / 4\eta)\mathbf{P}^v : \mathbf{P}^v, \tag{16.3}$$

where τ designates the relaxation time of the viscous pressure tensor. Note that for the sake of commodity, the constant pressure and temperature have not been written explicitly.

In a steady shear, the components of the viscous pressure tensor for the Maxwell upper-convected model (see Problem 15.5) are

$$\mathbf{P}^v = \begin{pmatrix} -2\tau\eta\dot{\gamma}^2 & -\eta\dot{\gamma} & 0 \\ -\eta\dot{\gamma} & 0 & 0 \\ 0 & 0 & 0 \end{pmatrix}, \tag{16.4}$$

from which it follows that (16.3) takes the form

$$g(c_i, \dot{\gamma}) = g_{eq}(c_i) + \tfrac{1}{2}\tau\eta v\dot{\gamma}^2 + O(\dot{\gamma}^4), \tag{16.5a}$$

or equivalently

$$g(c_i, P_{12}^v) = g_{eq}(c_i) + \tfrac{1}{2}Jv(P_{12}^v)^2 + O((P_{12}^v)^4), \tag{16.5b}$$

where the quantity $J = \tau/\eta$ is the so-called steady-state compliance.

More information about the non-equilibrium free energy can be obtained from microscopic considerations: the Rouse–Zimm model predicts that, for arbitrary laminar motions, the flow contribution to the Helmholtz free energy f is [16.8a]

$$\Delta f = -\tfrac{1}{2}nk_B T \left\{ \mathrm{Tr}\,(\mathbf{P}^v)' + \ln |\det[(\mathbf{U} - (\mathbf{P}^v)')]| \right\}, \tag{16.6}$$

where Δf stands for $\Delta f = f(T, v, c, (\mathbf{P}^v)') - f(T, v, c, 0)$, n is the number of macromolecules per unit volume and $(\mathbf{P}^v)' = (nk_B T)^{-1}\mathbf{P}^v$. Note that $\Delta g = \Delta f + p\Delta v$ at constant pressure, with, according to (16.3),

$$\Delta v = \frac{\partial}{\partial p}\left(\frac{1}{4}Jv\right)\mathbf{P}^\nu : \mathbf{P}^\nu. \tag{16.7}$$

For an incompressible fluid ($\partial v/\partial p = 0$), J does not depend on pressure, and therefore $\Delta v = 0$, and $\Delta g = \Delta f$.

In a Maxwell model, the viscosity η and the relaxation time τ are related by $\eta = nk_BT\tau$. Substitution of (16.4) into (16.6) gives

$$\Delta g = \frac{1}{2}\left(2\tau\eta\dot{\gamma}^2 - nk_BT\ln\left|1 + \frac{2\tau\eta}{nk_BT}\dot{\gamma}^2 - \frac{\eta^2}{(nk_BT)^2}\dot{\gamma}^2\right|\right). \tag{16.8}$$

For small values of $\dot{\gamma}$ one has, up to the second order in $\dot{\gamma}$, $\Delta g = \frac{1}{2}\tau\eta\dot{\gamma}^2 = \frac{1}{2}J(P_{12}^\nu)^2$ which is equivalent to (16.5). At high values of the shear rate, the first term in (16.8) is dominant and therefore $\Delta g = \tau\eta\dot{\gamma}^2 = J(P_{12}^\nu)^2$.

The presence of non-equilibrium contributions may drastically change the phase diagram and the value of the critical point of polymer solutions. At equilibrium, the Gibbs free energy of mixing is given by the well-known Flory–Huggins formula [16.1, 5],

$$(RT)^{-1}(\Delta G)_{FH} = n_1\ln(1-\phi) + n_2\ln\phi + \chi N\phi(1-\phi), \tag{16.9}$$

where G is the free energy per unit volume, ϕ the volume fraction of the polymer, and n_1 and n_2 the number of moles per unit volume of the polymer and the solvent, respectively; N is given by $N = n_1 + mn_2$, with m the ratio of the molar volume of the polymer and the molar volume of the solvent, so that $\phi = n_2m[n_1 + mn_2]^{-1}$. The interaction parameter χ in (16.9) is assumed to depend on the temperature according to

$$\chi = \frac{1}{2} + \psi\left(\frac{\Theta}{T} - 1\right), \tag{16.10}$$

with ψ being a constant and Θ the so-called theta temperature of the quiescent solution.

In non-equilibrium situations, expression (16.5b) suggests to add to (16.9) a corrective term of the form

$$\Delta G_f = v_1NJ(P_{12}^\nu)^2, \tag{16.11}$$

with v_1 the molar volume of the solvent. Note that, in general, J may depend on ϕ and will be determined from experiments.

The chemical potential of the ith component is defined as usually by

$$\mu_i = \left(\frac{\partial G}{\partial n_i}\right)_{T,p,Z} = \left(\frac{\partial G}{\partial \phi}\right)\left(\frac{\partial \phi}{\partial n_i}\right)_{T,p,Z}, \tag{16.12}$$

with Z a non-equilibrium parameter, either the shear rate or the viscous pressure, depending on the external constraints acting on the system. Note that different choices of Z lead to different expressions for the chemical potential, so that the selection of Z is important in formulating a coherent theory (see Problem 16.2).

From (16.9) and (16.11) it is inferred that the changes in the chemical potentials of the polymer and the solvent due to mixing are

$$\frac{\Delta\mu_1}{RT} = \ln(1-\phi) + \left(1 - \frac{1}{m}\right)\phi + \chi\phi^2 + \frac{\mu_{1f}}{RT},$$

$$\frac{\Delta\mu_2}{RT} = \ln\phi + (1-m)(1-\phi) + \chi m(1-\phi)^2 + \frac{\mu_{2f}}{RT}. \tag{16.13}$$

In (16.13) the subscripts 1 and 2 refer as above to the polymer and the solvent, respectively, and use was made of $\partial\phi/\partial n_1 = -(\phi/N)$ and $\partial\phi/\partial n_2 = (m/N)(1-\phi)$, with μ_{1f} and μ_{2f} the non-equilibrium contributions to the change of the chemical potential due to mixing, i.e. $\mu_{if} = (\partial\Delta G_f/\partial n_i)_{T,p,Z}$.

The spinodal line in the T, ϕ plane is given by the relation

$$\left(\frac{\partial\Delta\mu_1}{\partial\phi}\right)_{T,p,Z} = 0, \tag{16.14}$$

and the critical point, corresponding to the maximum of the spinodal line, is specified by the supplementary relation

$$\left(\frac{\partial^2\Delta\mu_1}{\partial\phi^2}\right)_{T,p,Z} = 0. \tag{16.15}$$

Since these expressions are directly borrowed from equilibrium thermodynamics, their applicability to non-equilibrium situations must be conforted by dynamical arguments. We leave such discussion to Sect. 16.4.

From (16.13) one obtains for the spinodal line and the critical point

$$\frac{1}{RT}\left(\frac{\partial\Delta\mu_1}{\partial\phi}\right) = -\frac{1}{1-\phi} + \left(1 - m^{-1}\right) + 2\chi\phi + \frac{1}{RT}\left(\frac{\partial\mu_{1f}}{\partial\phi}\right) = 0, \tag{16.16a}$$

$$\frac{1}{RT}\left(\frac{\partial^2 \Delta \mu_1}{\partial \phi^2}\right) = -\frac{1}{(1-\phi)^2} + 2\chi + \frac{1}{RT}\left(\frac{\partial^2 \mu_{1f}}{\partial \phi^2}\right) = 0. \qquad (16.16b)$$

The coordinates of the critical point of the quiescent solution are obtained from (16.15) by setting $\mu_{1f} = 0$; this leads to

$$\phi_c = (1 + m^{1/2})^{-1}, \qquad \chi_c = \tfrac{1}{2}(1 + m^{-1/2})^2. \qquad (16.17)$$

When m is very high (high molecular mass), ϕ_c tends to zero and χ_c to $\tfrac{1}{2}$, i.e. the critical temperature T_c tends to the theta temperature. The parameters θ and ψ in the original Flory–Huggins model for a solution of PS in TD are [16.9] $\psi = 0.50$ and $\theta = 294.4$ K. In addition, starting from the values for the molar volumes given in [16.3], namely, $V_1 = 1.586 \times 10^3$ m^3 mol^{-1} and $V_2 = 0.486$ m^3 mol^{-1} when the PS mass is 520 kg mol^{-1}, one finds $m = 3\,064$. With these values of ψ, θ, and m, one obtains from the Flory–Huggins theory, $T_c = 284.1$ K for the equilibrium critical temperature.

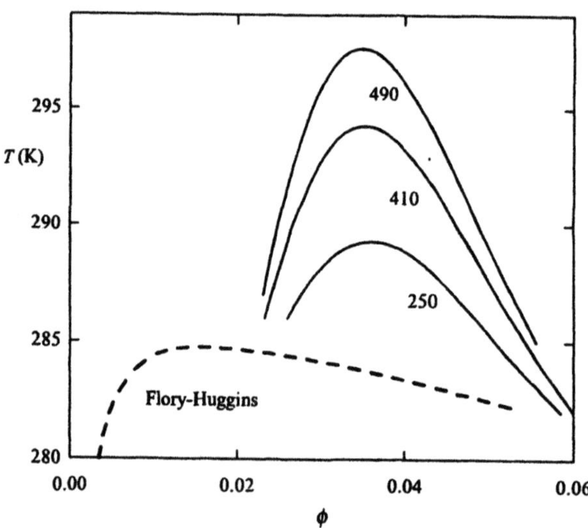

Fig. 16.1. Phase diagram of a binary solution of polystyrene in dioctylphtalate for several values of P_{12}^v (expressed in N/m^2). The dashed curve is the equilibrium spinodal line ($P_{12}^v = 0$) [16.5a].

The shear contribution μ_{1f} modifies not only the positions of the critical point, but also the spinodal and the coexistence lines (Fig. 16.1). To obtain them, explicit expressions for J as a function of ϕ are needed; these have been established by Rangel-Nafaile et al. [16.5a] for polystyrene in dioctylphtalate and by Wolf for polystyrene in transdecalin [16.5b] on experimental bases. For polystyrene in dioctylphtalate, the experimental shifts of the critical temperature with respect to the quiescent solution are 4 K at $P_{12}^{v} = 100$ N/m^2, 14 K at $P_{12}^{v} = 200$ N/m^2, and 24 K at $P_{12}^{v} = 400$ N/m^2. In the aforementioned examples, the critical temperature is enhanced by the shear flow, but this is not a general rule. As shown in the next section and confirmed by Wolf observations [16.5], for low-molecular-weight polymers there is a decrease in the critical temperature, whereas for high molecular weights there is an increase.

16.2 Explicit Solution for the Rouse–Zimm Model

It is well known that the equations of state for real polymer solutions are obtained experimentally; however, in some occasions the experimental data are scarce and, frequently, their analytical fit is not an easy task. For these reasons, we shall illustrate the general considerations of Sect. 16.1 by appealing to the well-known Rouse–Zimm microscopic model [15.30–31], which was already introduced in Sect. 15.3. We also analyse the influence of the nature of hydrodynamic interactions on the shift of the critical point, and compare the phase diagram at constant shear viscous pressure and at constant shear rate to emphasize the role of different constraints [16.9]. The price to be paid for making use of this simple model is a restriction to relatively low values of the shear rate.

Our purpose is to modify the Flory–Huggins equilibrium model (16.9) taking into account the non-equilibrium correction. To do that, it is necessary to know how the steady-state compliance J depends on the composition of the mixture. This dependence is easily obtained by assuming that the polymer behaves as a linear generalised Maxwell model (see Sect. 15.1), for which

$$J = \frac{C M_1}{c RT}\left(1 - \frac{\eta_s}{\eta}\right)^2, \tag{16.18}$$

where c is the polymer concentration expressed in terms of mass per unit volume, η the viscosity of the solution, η_s the solvent viscosity, M_1 the polymer molar mass and C a parameter whose value shall be discussed below.

It is important to have in mind that η depends on the concentration. The latter dependence is crucial in the analysis of the shift of the critical point: predictions based on the hypothesis that $J \sim c^{-1}$ yield unsatisfactory results (see Problem 16.4). It is usually assumed that $\eta(c)$ takes the form [16.5,8]

$$\frac{\eta}{\eta_s} = 1 + [\eta]c + k[\eta]^2 c^2, \tag{16.19}$$

where k is the so-called Huggins constant and $[\eta]$ stands for the intrinsic viscosity, defined, as usually, by

$$[\eta] = \lim_{c \to 0} \frac{\eta - \eta_s}{\eta_s}. \tag{16.20}$$

For the system TD/PS with a polymer molecular mass of 520 kg mol^{-1}, $k = 1.40$ [16.5b], $[\eta] = 0.043$ m^3 kg^{-1}, and the solvent viscosity $\eta_s = 0.0023$ Pa s. With these values, (16.19) describes fairly well the viscosity in a concentration range for which the product $c[\eta]$ is less than 1. Since the critical density of the system is found in this range, this limitation is not too restrictive.

When hydrodynamic interactions are ignored, as in the Rouse model, the parameter C in (16.18) takes the value 0.4, whereas accounting for hydrodynamic interactions, as in the Zimm model, the value of C is 0.206. A comparison with experimental data shows that Zimm's theory is useful at low concentrations, whereas at high concentrations, the behaviour of the system is well described by the Rouse model. There is thus a shift from the Zimm to Rouse model with increasing polymer concentration, which may be described, by assuming that C is a function of the composition, via a reduced concentration $\tilde{c}$ defined by $\tilde{c} = [\eta]c$. Since the functional dependence $C = C(\tilde{c})$ is unknown we propose to describe the gradual transition from the Zimm to the Rouse behaviour by means of

$$C(\tilde{c}) = 0.206 + \frac{0.194\,\alpha\tilde{c}}{1 + \alpha\tilde{c}}, \tag{16.21}$$

where α is a parameter which measures how steep is the transition.

We may summarize our hypotheses by the following expression for (16.11)

$$\frac{(\Delta G)_f}{RT} = B\,C(\tilde{c})\frac{(P_{12}^v)^2}{T^2} N F(\tilde{c}), \tag{16.22}$$

where B is a constant given by $B = V_1 M_1 [\eta]/R^2$ and $F(\tilde{c})$ a function of the reduced concentration defined as

$$F(\tilde{c}) = \tilde{c}\left(\frac{1 + k\tilde{c}}{1 + \tilde{c} + k\tilde{c}^2}\right)^2. \tag{16.23}$$

At this point, let us open a parentheses to discuss the consequences of keeping constant either the shear rate $\dot{\gamma}$ or the shear viscous pressure P_{12}^v. These different constraints will lead to different non-equilibrium phase diagrams. Let $(\mu_1)_{P_{12}^v}$ be the chemical potential at constant shear viscous pressure and $(\mu_1)_{\dot{\gamma}}$ the chemical potential at constant shear rate, defined respectively as

$$(\mu_1)_{P_{12}^v} = \left(\frac{\partial \Delta G^{(s)}}{\partial n_1} \right)_{P_{12}^v}, \qquad (\mu_1)_{\dot{\gamma}} = \left(\frac{\partial \Delta G^{(s)}}{\partial n_1} \right)_{\dot{\gamma}}. \qquad (16.24)$$

At constant P_{12}^v the explicit expression for the chemical potential is

$$\frac{1}{RT}(\mu_1)_{P_{12}^v} = -\frac{2B(P_{12}^v)^2}{T^2} C \frac{P_5(\tilde{c})}{P_6(\tilde{c})} - \frac{B(P_{12}^v)^2}{T^2} C' \tilde{c} F(\tilde{c}), \qquad (16.25)$$

where C' stands for the derivative of $C(\tilde{c})$ with respect to $\tilde{c}$, with the auxiliary functions $P_5(\tilde{c})$ and $P_6(\tilde{c})$ given by

$$P_5(\tilde{c}) = (k-1)\tilde{c}^2 + (k^2 - 3k)\tilde{c}^3 - 3k^2\,\tilde{c}^4 - k^3\,\tilde{c}^5, \quad P_6(\tilde{c}) = (1 + \tilde{c} + k\tilde{c}^2)^3. \qquad (16.26)$$

At constant shear rate $\dot{\gamma}$, the explicit expression for the chemical potential is

$$\frac{(\mu_1)_{\dot{\gamma}}}{RT} = \frac{(\mu_1)_{P_{12}^v}}{RT} - \frac{2B\eta_s^2\dot{\gamma}^2}{T^2} C\,\Omega(\tilde{c}), \qquad (16.27)$$

where Ω is defined as $\Omega(\tilde{c}) = \tilde{c}(1 + \tilde{c} + k\tilde{c}^2)(1 + 2k\tilde{c})F(\tilde{c})$. It is easy to show that the expression for the chemical potential at constant $\dot{\gamma}$ is smaller than the chemical potential at constant P_{12}^v, because the corrective term in (16.27) is positive.

We are now in position to examine the consequences issued from the thermodynamic criterion for the limit of stability. From $(\partial\mu/\partial\tilde{c}) = 0$, we obtain the following expression for the spinodal line

$$\frac{\partial}{\partial\tilde{c}}\left(\frac{\mu_1}{RT} \right) = \frac{\partial}{\partial\tilde{c}}\left(\frac{\mu_1^{(0)}}{RT} \right) + \frac{\partial}{\partial\tilde{c}}\left(\frac{(\mu_1)_s}{RT} \right) = 0 \qquad (s = P_{12}^v, \dot{\gamma}), \qquad (16.28)$$

where the first derivative on the right-hand side is given by

$$\frac{\partial}{\partial\tilde{c}}\left(\frac{\mu_1^{(0)}}{RT} \right) = -\frac{v}{1 - v\tilde{c}} + \left(1 - \frac{1}{m} \right)v + 2\chi v^2\tilde{c}, \qquad (16.29)$$

with v a constant defined as $v = V_1/M_1[\eta]$. For $(\mu_1)_{P_{12}^v}$ and $(\mu_1)_{\dot\gamma}$ the corrective terms in (16.22) are respectively

$$\frac{\partial}{\partial \tilde c}\left(\frac{1}{RT}(\mu_1)_{P_{12}^v}\right) = -\frac{2B(P_{12}^v)^2}{T^2}\left[C\frac{P_6 P_5' - P_5 P_6'}{P_6^2} + C'\left(\frac{P_5}{P_6}+F\right)+C''\tilde c F\right], \quad (16.30)$$

$$\frac{\partial}{\partial \tilde c}\left(\frac{1}{RT}(\mu_1)_{\dot\gamma}\right) = \frac{\partial}{\partial \tilde c}\left(\frac{1}{RT}(\mu_1)_{P_{12}^v}\right)$$

$$-\frac{2B\eta_s^2\dot\gamma^2}{T^2}\left[C\frac{\Omega_2(1+\tilde c+k\tilde c^2)-\Omega_1(1+2k\tilde c)}{(1+\tilde c+k\tilde c^2)^2}+C'\Omega\right]. \quad (16.31)$$

In the latter expressions, P_n' stands for the derivative of P_n with respect to $\tilde c$, and C' and C'' are the two first derivatives of (16.21) with respect to $\tilde c$. Furthermore, we have introduced the auxiliary functions

$$\Omega_1(\tilde c) = \tilde c^2+4k\tilde c^3+5k^2\tilde c^4+2k^3\tilde c^5,$$

$$(16.32)$$

$$\Omega_2(\tilde c) = 2\tilde c+12k\tilde c^2+20k^2\tilde c^3+10k^3\tilde c^4.$$

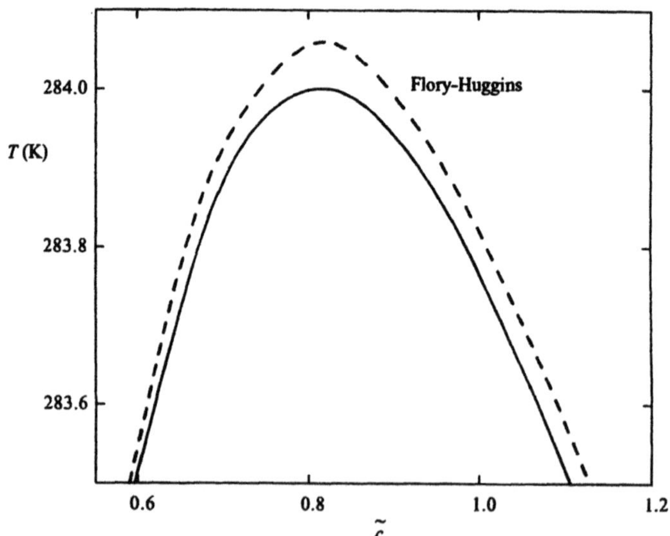

Fig. 16.2. Spinodal curve of a solution of polystyrene in transdecalin at equilibrium (*dashed line*) and in a non-equilibrium steady state at constant shear rate $\dot\gamma$ =1500 s^{-1} (*continuous line*).

To obtain numerical values, consider a solution of polystyrene of molecular mass 520 kg mol^{-1} in transdecalin. The corresponding equilibrium and non-equilibrium spinodal lines are shown in Fig. 16.2, where it is seen that at constant $\dot{\gamma}$ the presence of a shear rate lowers the critical temperature, in agreement with experiments [16.5].

However, when the same system is submitted to a constant P_{12}^{v}, the conclusion drawn from (16.28–31) is that the critical temperature is increased, as seen in Fig. 16.3 for different values of the shear viscous pressure. Since the temperature shifts are important, it is clear that equilibrium thermodynamics is inadequate for describing polymer solutions in the presence of a shear pressure. However, the above theoretical predictions about the temperature shifts are still lower than the actual values; the agreement is nevertheless improved when the effects of shear-enhanced concentration fluctuations are taken into account in a dynamical approach [16.11].

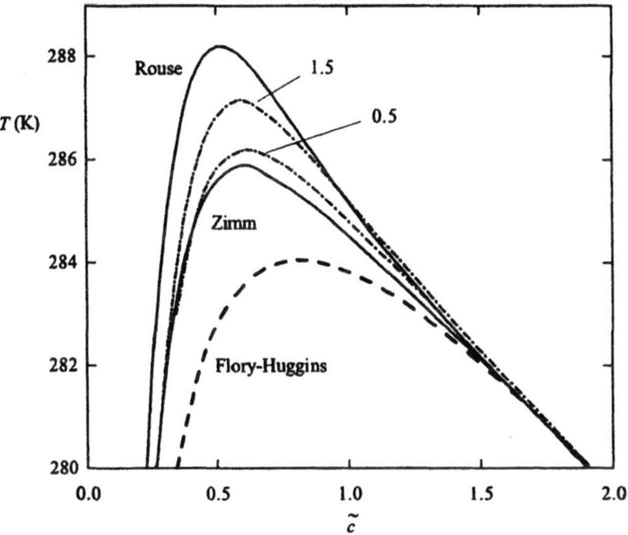

Fig. 16.3. Spinodal curve of a Rouse–Zimm binary solution at equilibrium (Flory–Huggins) and in a non-equilibrium steady state at constant shear viscous pressure $P_{12}^{v} = 150\,\mathrm{Nm^{-2}}$. The figure collects the results corresponding to the Rouse (no hydrodynamic interactions), the Zimm (hydrodynamic interactions) and two intermediate models corresponding to the values $\alpha = 0.5$ and $\alpha = 1.5$ of the parameter α describing the transition from the Rouse model to the Zimm model for increasing values of the concentration according to Eq. (16.21). It is seen that the higher the hydrodynamic interactions, the lower the shift of the critical temperature.

Another interesting result derived from Fig. 16.3 is the effect of hydrodynamic interactions on the shift of the critical point: when they are neglected, as in the Rouse model, the shift is maximum, while it is minimum in the Zimm model, where they are relevant. Our model shows that hydrodynamic interactions are more important in the shift (increase) of the critical temperature T_c than in the shift (decrease) of the critical concentration.

16.3 Chemical Reactions Under Flow

It is expected that the application of a shear pressure will modify the properties of a chemical reaction taking place in the system, and in particular the chemical equilibrium constant. After general considerations, we shall examine as an illustration the interesting problem of the thermodynamic degradation of polymers under flow.

16.3.1 Shear-Induced Effect on the Affinity

The change of chemical potential due to the presence of a shear pressure will be reflected in the expression of the affinity, and consequently in that of chemical equilibrium constant. To see that, consider a chemical reaction in a system submitted to a steady shear flow; the generalised Gibbs equation will take the form

$$dG = -SdT + Vdp + \sum_k \mu_k dN_k + (V\tau/2\eta)\mathbf{P}^v : d\mathbf{P}^v, \qquad (16.33)$$

where G is the global Gibbs function. By equating the cross derivatives

$$\left(\frac{\partial \mu_k}{\partial \mathbf{P}^v}\right)_{T,p,N_j} = \frac{\partial}{\partial N_k}\left(\frac{V\tau}{2\eta}\right)_{T,p,N_j} \mathbf{P}^v, \qquad (16.34)$$

and after integration, one obtains for the generalised chemical potential

$$\mu_k(T,p,N_j,\mathbf{P}^v) = \mu_{k,eq}(T,p,N_j) + \frac{1}{4}\frac{\partial(V\tau/\eta)}{\partial N_k}\mathbf{P}^v : \mathbf{P}^v, \qquad (16.35)$$

where $\mu_{k,eq}$ is the local-equilibrium chemical potential, independent of $\mathbf{P}^v$; it is usual to write $\mu_{k,eq}$ as

$$\mu_{k,eq}(T,p,N_j) = \mu_{k,eq}^0(T,p) + RT\ln a_k, \qquad (16.36)$$

with a_k being the activity of component k. If we use $dN_k = \nu_k d\xi$, where ν_k is the stoichiometric coefficient of component k and ξ the rate of advancement of the reaction, (16.33) reads as

$$dG = -SdT + Vdp - \mathcal{A}d\xi + (V\tau/2\eta)\mathbf{P}^v : d\mathbf{P}^v, \qquad (16.37)$$

where

$$\mathcal{A} = -(\partial G/\partial\xi)_{T,p,\mathbf{P}^v} = -\sum_k v_k\mu_k \qquad (16.38)$$

denotes the affinity of the reaction. At chemical equilibrium in the absence of a shear pressure, it is well known that G is minimum from which it follows that $\mathcal{A} = 0$, at constant T and p. According to expression (16.37), the same properties hold in the presence of a shear pressure but now the minimum of G corresponds to fixed values of T, p and $\mathbf{P}^v$, and is generally shifted compared with the classical case where $\mathbf{P}^v = 0$. As a consequence, we are still allowed to put $\mathcal{A} = 0$, so that

$$\sum_k v_k\mu_k = 0. \qquad (15.39)$$

Substituting in this relation the expression for the chemical potential given by (16.35) and taking (16.36) into account, one obtains

$$\sum_k v_k\left[\frac{\mu_{k,eq}^0}{RT} + \ln a_k + \frac{1}{4RT}\frac{\partial}{\partial N_k}(V\tau/\eta)\mathbf{P}^v : \mathbf{P}^v\right] = 0. \qquad (16.40)$$

Defining the equilibrium constant K by

$$\ln K(T,p) = -\frac{1}{RT}\sum_k v_k\mu_{k,eq}^0 \qquad (16.41)$$

and a function λ describing the effects of the viscous shear by

$$\ln\lambda = -\frac{1}{4RT}\sum_k v_k\left[\frac{\partial}{\partial N_k}(V\tau/\eta)\right]\mathbf{P}^v : \mathbf{P}^v, \qquad (16.42)$$

it is found that at chemical equilibrium under a shear flow

$$K(T,p) = \frac{1}{\lambda}\prod_k a_k^{v_k}. \qquad (16.43)$$

Note that when $\mathbf{P}^v = 0$, $\lambda = 1$ and one recovers the classical expression relating the values of the activities a_k in chemical equilibrium to the equilibrium constant. The above result is important, as it shows that even chemical equilibrium may be influenced by non-equilibrium constraints.

16.3.2 Illustration: Polymer Degradation Under Flow

As an illustration of the above general considerations, we shall examine the problem of polymer degradation. The latter is usually solved by chemical kinetics [16.6]. The first hypothesis underlying any kinetic mechanism for polymer degradation is to assume that each chain is broken only at one point (i.e. only two fragments are produced per each broken macromolecule). The second hypothesis states that each elementary reaction

$$P_j \rightarrow P_i + P_{j-i} \tag{16.44}$$

follows a first-order kinetics, i.e.

$$\frac{dn_i}{dt} = k_{ij} n_j, \tag{16.45}$$

where n_i and n_j stand for the number of macromolecules with i and j degrees of polymerization, respectively, and k_{ij} is the kinetic constant describing the breaking of a macromolecule with i monomers into a macromolecule with j monomers and another with $i - j$ monomers.

Besides numerical difficulties, the only (and non trivial) problem is to know the values of the kinetic constants k_{ij}. Three situations are often analysed [16.10]: (1) the simplest one is to assume that all bonds have the same breaking probability, so that the value of k_{ij} is independent of i and j and the macromolecular fragmentation is a random process; (2) a slightly more complicated situation occurs when the kinetic constant depends on the length of the chain which is broken, but not on the length of the ensuing fragments, i.e. the value of k_{ij} depends on i but not on j; (3) the most complicated case arises when k_{ij} depends on the position of the breaking point: this situation is usually dealt with by assuming that the value of the kinetic constant for the breaking of any chain is given by a Gaussian function whose parameters are obtained by fitting experimental data.

The kinetic mechanism proposed in (16.44) is not the most general, and it may be considered as a particular case of a degradation-combination scheme of the form

$$P_j \rightleftarrows P_i + P_{j-i} \tag{16.46}$$

In this event, the mass action law (16.45) will contain a new kinetic constant κ_{ij} which corresponds to the recombination of the chains P_i and P_{j-i} and which is related to the chemical equilibrium constant K_{ij} by $K_{ij} = k_{ij}/\kappa_{ij}$.

In the present approach, the system is viewed as a multicomponent mixture, constituted by the solvent (component 1) and a set of polymeric species P_i (with i the degree of polymerization), whose chemical potentials for macromolecules with different lengths will be written as

$$\mu_j = \mu_j^{eq} + RT\left(\frac{\partial \Delta G_s}{\partial n_j}\right)_{n_k,\dot{\gamma}},$$

(16.47)

where ΔG_s is the non-equilibrium contribution to the Gibbs function and j refers to the number of monomeric units in the chain.

Furthermore, the average character of the polymeric molecular mass in the solution amounts to admit that M_1 appearing in (16.18) is a function of n_i. It is also assumed that in equilibrium the n_i are distributed according to some well-known equilibrium distribution functions of the number of particles, as for instance the so-called most probable distribution [16.9, 10].

Applying the result (16.35) to reaction (16.44) and considering for simplicity an ideal mixture, one may write

$$\frac{N_j}{N_i N_{j-1}} = \frac{N_j^{(0)}}{N_i^{(0)} N_{j-1}^{(0)}} \lambda(i,j)^{-1},$$

(16.48)

where N_k is the number of chains with k units while the superscript (0) refers to values with $\dot{\gamma} = 0$. The function λ that takes into account the non-equilibrium effects is obtained from (16.42) and it is given explicitly in [16.11] for a solution of polystyrene in transdecalin.

The results are summarized in Fig. 16.4, where the differential mass fraction $W(i)$ (the polymeric mass with a degree of polymerization between i and $i + di$, considering i as a continuous variable) is plotted as a function of i for a solution of polystyrene of mean molecular mass 1170 kg/mol in transdecalin. To describe the form of $W(i)$ at equilibrium we have used the so-called most-probable distribution $W(i) = \alpha^2 i(1-\alpha)^{i-1}$, where α is a parameter describing the width of the distribution. The parameters of the equilibrium distributions without shear rate and with shear rate are related by the expression $\alpha(1-\alpha)^{-1} = \alpha^{(0)}(1-\alpha^{(0)})^{-1}\lambda$. Fig. 16.4 exhibits a tendency towards an enhancement of the number of smaller particles, in such a way that the mean molecular mass diminishes by 5% for a shear rate $\dot{\gamma} = 10^4$ s^{-1} A detailed analysis shows a slight dependence of the results with the form used to describe the distribution function and with the ratio between the weight average and the number average [16.10].

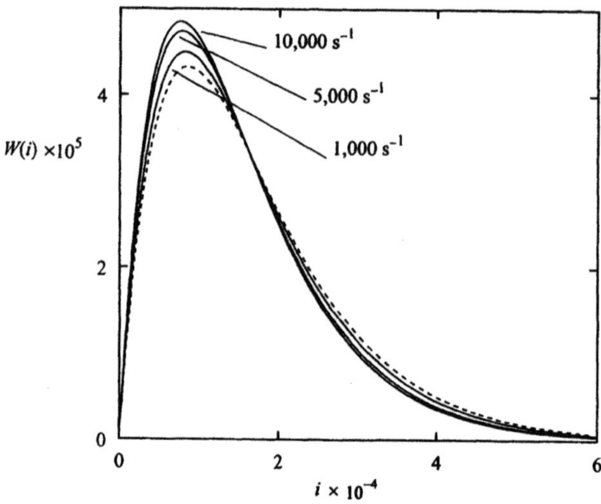

Fig. 16.4. Differential mass fraction of polystyrene in a polystyrene–transdecalin solution for an equilibrium mean molecular weight of 1 170 Kg/mol under several shear rates. The dashed line corresponds to a quiescent situation. It has been assumed that the form of the curve is given by the so-called most-probable distribution [16.11a].

16.4 Dynamical Approach

Our purpose in this section is to justify from dynamical bases the use of the thermodynamic stability criterion (16.14) in non-equilibrium situations [16.12]. Let us start from the general results of Sect. 13.4 about the coupling between viscous pressure and diffusion. According to (13.38–40), the diffusion flux and the viscous pressure in a steady state are related by

$$J_1 = -D'\nabla\tilde{\mu} + D'T\nabla \cdot (\beta \mathbf{P}^v),$$
(16.49)

$$\mathbf{P}^v = -2\eta(\nabla v)^s + 2\eta T\beta(\nabla J_1)^s,$$
(16.50)

with $\tilde{\mu} = \mu_1 - \mu_2$ being the difference between the specific chemical potentials of the solute and the solvent. The coefficient D' is related to the diffusivity D through $D = D'(\partial\tilde{\mu}_{eq}/\partial n_1)_{T,p}$, where n_1 is the concentration of the solute.

When diffusion effects are negligible compared with those of $\mathbf{P}^v$ (for instance, when the system is subjected to a relatively high shear pressure and the diffusion flux is very weak), one has (see (16.35))

$$\tilde{\mu}(n_1, \mathbf{P}^v) = \tilde{\mu}_{eq}(n_1) + \frac{1}{4}\left[\frac{\partial}{\partial n_1}\left(\frac{v\tau}{\eta}\right)\right]_{T,p} \mathbf{P}^v : \mathbf{P}^v. \tag{16.51}$$

Our purpose is to determine under which conditions the classical thermodynamic stability criterion $(\partial\tilde{\mu}/\partial n_1) \geq 0$ used in (16.14) to obtain the spinodal curve in the presence of a shear, is supported by a dynamical theory.

First of all, in the absence of coupling effects, i.e. when $\beta = 0$, (16.49) reduces to

$$J_1 = -D'(n_1, \dot{\gamma})\nabla\tilde{\mu}(n_1, \dot{\gamma}). \tag{16.52}$$

Note that both $\tilde{\mu}$ and D' may depend not only on n_1 but also on $\dot{\gamma}$. For a homogeneous shear rate, i.e. an uniform value of $\dot{\gamma}$, (16.52) can be written as

$$J_1 = -D'(n_1, \dot{\gamma})(\partial\tilde{\mu}/\partial n_1)_{T,\dot{\gamma}}\nabla n_1. \tag{16.53}$$

Expression (16.53) allows one to identify an effective diffusion coefficient D_{eff} by

$$D_{eff} = D'(n_1, \dot{\gamma})(\partial\tilde{\mu}/\partial n_1)_{T,\dot{\gamma}}. \tag{16.54}$$

According to the positiveness of entropy production, $D'(n_1, \dot{\gamma})$ is always positive. When D_{eff} becomes negative the inhomogeneities are amplified so that the homogeneous state becomes unstable. Thus, for $(\partial\tilde{\mu}/\partial n_1)_{T,\dot{\gamma}} > 0$, one has $D_{eff}>0$, and the states are stable whereas when $(\partial\tilde{\mu}/\partial n_1)_{T,\dot{\gamma}}$ is negative, D_{eff} is also negative and the homogeneous states are unstable. As a consequence, it can be said that the criterion $(\partial\tilde{\mu}/\partial n_1)_{T,\dot{\gamma}} = 0$ yields the border between stable and unstable states, both from the thermodynamical and the dynamical points of view. Of course, the dynamical approach may predict the details of phase separation (droplets, percolating structures, etc.) and the rate of separation, in contrast with the purely thermodynamic analysis. An illustration of the above considerations for a sheared suspension flow was treated by Nozières and Quemada [16.13].

When the coefficient β is different from zero, the diffusion flux is coupled with the viscous pressure tensor. Take a steady state and a plane shear flow with velocity distribution of the form $v_x(y)$; then (16.49) for the component J_y of the diffusion flux normal to the velocity, which is responsible for the appearance of inhomogeneities in the fluid, can be written as

$$J_y = -D'(n_1, \dot{\gamma})\frac{\partial}{\partial y}\left[\tilde{\mu}(n_1, \dot{\gamma}) - T\beta P_{yy}^v\right], \tag{16.55}$$

provided that the quantities appearing in (16.49) do not change along the direction of the flow.

As in (13.47), one may identify a generalised chemical potential $\tilde{\mu}''$ as

$$\tilde{\mu}''(n_1,\dot{\gamma}) = \tilde{\mu}(n_1,\dot{\gamma}) - T\beta P_{yy}^v = \tilde{\mu}(n_1,\dot{\gamma}) + \overline{v}P_{yy}^v. \tag{16.56}$$

Then, (16.55) can simply be expressed as

$$J_y = -D'(n_1,\dot{\gamma})\frac{\partial}{\partial y}\tilde{\mu}''(n_1,\dot{\gamma}). \tag{16.57}$$

In this case, the dynamical stability condition of a positive effective diffusion coefficient is

$$\frac{\partial \tilde{\mu}''}{\partial n_1} > 0, \tag{16.58}$$

instead of $(\partial\tilde{\mu}/\partial n_1) > 0$. Condition (16.58) is equivalent to that obtained by Onuki [16.7] from a full hydrodynamic analysis. This means that when the normal viscous pressure P_{yy}^v is zero (as, for instance, in Maxwell upper-convected models, where the normal pressure along the x direction is different from zero but the normal pressure along the y and z directions are zero), the chemical potential $\tilde{\mu}''(n_1,\dot{\gamma})$ coincides with $\tilde{\mu}(n_1,\dot{\gamma})$. Since this condition is satisfied in the analysis presented in Sect. 16.2, the results presented there are valid.

In general terms, it is clear that more experimental and theoretical work is needed. The spinodal curves should be examined for a larger variety of flows and materials. The composition of the individual phases should also be measured carefully. Spectrographic techniques rather than simple visual observation of the turbidity should be used in order to improve the accurateness of the data. An open problem is to formulate some general criteria to predict, for a particular material, whether shear-induced solubility or shear-induced phase separation will occur.

16.5 Shear-Induced Migration of Polymers

The coupling between shear effects and diffusion is nowadays one of the most active topics in rheology. In particular, shear-induced migration of polymers deserves a great attention, both for its practical aspects (chromatography, separation techniques, flow through porous media) and its theoretical implications in non-equilibrium thermodynamics and transport theory [16.14]. The subject is especially attractive, as it concerns the coupling of vectorial fluxes and tensorial forces, which lies outside the range of applicability of classical irreversible thermodynamics, and because it puts forward a testing ground for non-equilibrium equations of state. The basic frame of this section is provided by equations (16.49–50).

Recently, MacDonald and Muller [16.15], to whom we refer for a wide bibliography on this topic, have studied the evolution of the concentration profile of a polymer under the effect of a shear pressure in a cone-and-plate configuration. Their conclusions are extremely challenging, as they observe a discrepancy of two to three orders of magnitude between theoretical predictions and experimental observations.

The constitutive equation for the diffusion flux J proposed by MacDonald and Muller is

$$J = -D\nabla n - \frac{D}{RT}\nabla \cdot \mathbf{P}^v, \tag{16.59}$$

where n is the polymer concentration (in moles per unit volume), D the diffusion coefficient and R the ideal gas constant. This constitutive equation has been examined from different macroscopic and microscopic points of view [16.16].

MacDonald and Muller have applied equation (16.59) in the particular case of a cone-and-plate configuration, where the only non-zero components of the viscous pressure tensor are given by

$$P^v_{\phi\phi} = -2RTn(\tau\dot{\gamma})^2, \qquad P^v_{\phi\theta} = -RTn\tau\dot{\gamma}, \tag{16.60}$$

where $\eta = nRT\tau$, r, ϕ and θ refer to the radial, axial and azymuthal directions respectively, and τ is the polymer relaxation time.

By combining (16.59) and (16.60) with the mass balance equation

$$\frac{dn}{dt} = \frac{\partial n}{\partial t} + v \cdot \nabla n = -\nabla \cdot J, \tag{16.61}$$

and since the convection term vanishes identically in this geometry, one has

$$\frac{\partial n}{\partial t} = \frac{D}{r^2}\frac{\partial}{\partial r}\left(r^2\frac{\partial n}{\partial r} + \beta_{MM} r n\right), \tag{16.62}$$

with the parameter $\beta_{MM} = 2(\tau\dot{\gamma})^2$. To obtain (16.60) it has been assumed that the steady viscometric flow has only one non-zero component of the velocity (the ϕ component) which depends on the r and θ coordinates, that the shear-induced flux in the θ direction is negligible when the angle between the cone and the plate is small (as occurred in the experiments), and that the contribution of convection of molecules by migration to the viscous pressure is negligible. The term in β_{MM}, arising from the second term in the right-hand side of (16.62), induces a flux of solute towards the apex of the cone (designated by arrows 1 in Fig. 16.5) and it is usually believed to be responsible for induced migration.

But, according to MacDonald and Muller, this contribution cannot explain by itself the actual rate of migration. Indeed, they have calculated a short-time solution of (16.62) for $n(r, t)$ and have compared it with their experimental results for polystyrene macromolecules of different molecular weights 2×10^3 and 4×10^3 kg/mol (denoted by 2M and 4M respectively) in a solvent of oligomeric polystyrene molecules of 0.5 kg/mol, in a cone-and-plate configuration sheared at $\dot{\gamma} = 2$ s^{-1}. The initial homogeneous concentration of the molecules of each solution was 0.20 and 0.12% in weight for the 2M and 4M solutions, respectively. Taking an average value of τ obtained from steady-state data for the shear viscosity, they found for the 2M and 4M solutions the theoretical values $\beta_{2MM} = 240$ and $\beta_{4MM} = 1500$ respectively.

However, when MacDonald and Müller compare the result obtained from (16.62) with the experimental concentration profiles, they notice a dramatic difference with a migration motion much faster than that predicted by equation (16.62). They tried to fit the data by allowing either β_{MM} or D to be adjustable parameters: to comply with the data it is necessary to take $\beta_{2MM} = 200\ 000$ and $\beta_{4MM} = 1\ 100\ 000$, instead of the values mentioned previously.

It is interesting to stress that (16.49), instead, alleviates this severe discrepancy. We write it as

$$J = -D'\nabla\mu - \frac{D}{RT}\nabla \cdot \mathbf{P}^v, \tag{16.63}$$

where $D' = (\partial\mu_{eq}/\partial n)D$. The essential point is that in the presence of a non-vanishing $\mathbf{P}^v$, μ is not only a function of n but may contain contributions of $\mathbf{P}^v$. This gives rise to a new coupling neglected up to now, between viscous effects and diffusion, besides the term in $\nabla \cdot \mathbf{P}^v$.

To be explicit, μ may be expressed as

$$\mu = \mu_{eq} + \frac{1}{V}(1 - V'n)\left(\frac{\partial G}{\partial n}\right)_{T,p,\mathbf{P}^v}, \tag{16.64}$$

where $V' = \partial V / \partial N$ is the partial molar volume of the solute and $N = nV$. The terms within parentheses in (16.63) take into account that a variation of N at constant p produces a change in the total volume V.

According to (16.11) and (16.64), the chemical potential of the solute is

$$\mu = \mu_{eq} + \frac{1}{4V}(1 - V'n)\frac{\partial}{\partial n}(JV)\mathbf{P}^v : \mathbf{P}^v. \tag{16.65}$$

The use of the generalised chemical potential suggests to introduce an effective diffusion coefficient as $D_{eff} = D'(\partial\mu/\partial n)$ or, by writing D' in terms of the classical diffusion coefficient D,

$$D_{eff} = \frac{D}{\left(\partial\mu_{eq}/\partial n\right)}\left(\frac{\partial\mu}{\partial n}\right). \tag{16.66}$$

Introduction of (16.65) into (16.66), leads to

$$D_{eff} = D\left\{1 + \frac{1}{\left(\partial\mu_{eq}/\partial n\right)}\frac{\partial}{\partial n}\left[\frac{(1 - V'n)}{4V}\frac{\partial}{\partial n}(JV)\mathbf{P}^v : \mathbf{P}^v\right]\right\}. \tag{16.67}$$

If the contribution of the term in $\mathbf{P}^v : \mathbf{P}^v$ is negative and in absolute value larger than one, so that $D_{eff} < 0$, it induces a flow of solute towards higher solute concentrations, i.e. in opposite direction of the usual Fickian diffusion. Such an effect coupled to the contribution of the term containing $\nabla \cdot \mathbf{P}^v$ in (16.63) will drastically fasten the migration process of the solute towards the center, as shown in Fig. 16.5.

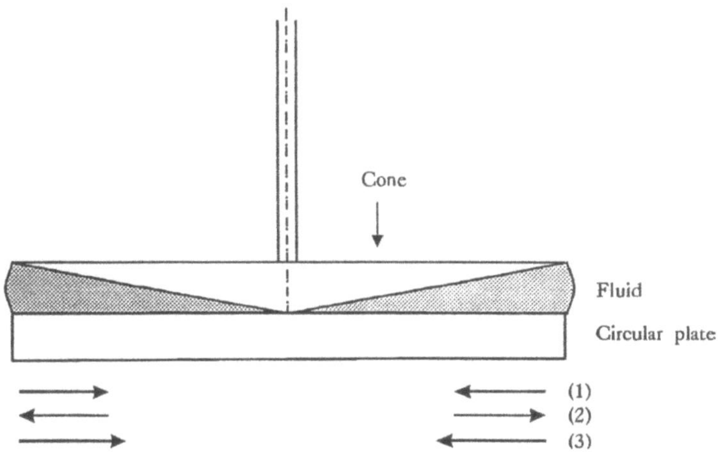

Fig. 16.5. The flows of matter are indicated in the cone-and-plate configuration. Arrows (1) shows the shear-induced flow described by the second term in the right-hand side of equations (16.59) and (16.63); arrows (2) and (3) indicate the diffusion flux corresponding to the first term in the right-hand side of equation (16.63). When the effective diffusion coefficient (16.67) is positive, the diffusion flux (2) is opposite to the shear-induced flow (1), whereas when it is negative, the diffusion flux (3) enhances the shear-induced effects.

To see whether the non-equilibrium contribution to D_{eff} is negative requires a detailed knowledge of μ_{eq}, V' and J with respect to the concentration. We consider polystyrene in transdecalin, whose non-equilibrium chemical potential under flow has been studied in detail [16. 1, 9]. In Fig. 16.6 is represented $(\partial\mu/\partial n)$ as a function of n for a given $\mathbf{P}^v$. To obtain this figure we have used for the equilibrium chemical potential the Flory–Huggins model and to evaluate the non-equilibrium contribution we have taken for J the formula (16.18). The explicit expression for the non-equilibrium contribution to the chemical potential of the solute is given by (16.25).

It is seen in Fig. 16.6 that for sufficiently low values of the concentration, $(\partial\mu/\partial\tilde{c})$, and therefore D_{eff}, is positive, whereas for higher concentrations $(\partial\mu/\partial\tilde{c})$ and consequently D_{eff} is negative; this transition is observed for $\tilde{c} \approx 0.5$. This value corresponds to a mass percentage of 1.3%.

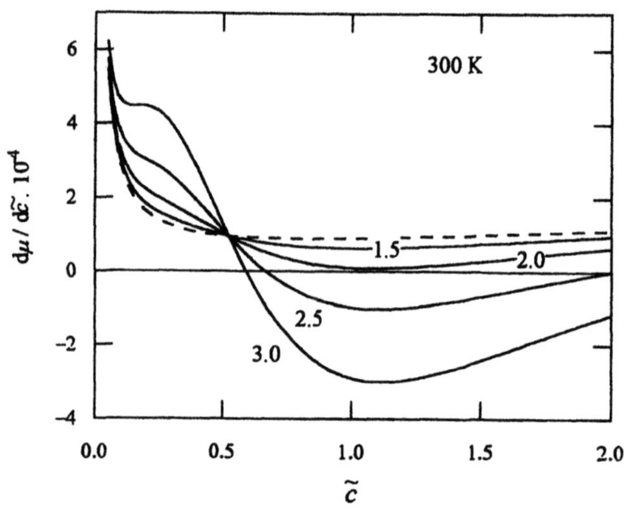

Fig. 16.6. The variation of $(\partial\mu/\partial\tilde{c})$ –which is proportional to the polymer effective diffusivity– in terms of the reduced concentration $\tilde{c}$ is presented for different values of $\tau\dot{\gamma}$, for polystyrene in transdecalin solution. The dashed line corresponds to the equilibrium Flory–Huggins contribution. For $\tilde{c}$ higher than a critical value approximately equal to 0.5, and for high values of $\tau\dot{\gamma}$ the effective diffusivity becomes negative.

An estimation of the order of magnitude of $D_{eff}/D = (\partial\mu/\partial n)/(\partial\mu_{eq}/\partial n)$ for the situations described in Fig. 16.6, i.e. for $\tau\dot{\gamma} \approx 12$ and $\tau\dot{\gamma} \approx 25$ (corresponding to the experiments by MacDonald and Muller), yields values of order 10^3 and 10^4 respectively, which are consistent with the experimental observations [16.17]. These results confirm that the corrections introduced in our analysis are all but negligible.

To conclude, let us emphasize that for a given concentration (higher than a critical value), the effective diffusion coefficient is reduced when the shear viscous pressure increases, and may become negative. In this regime, the non-equilibrium contribution to the chemical potential considerably enhances the polymer migration, and makes it much faster. In contrast, at low shear rates, the only thermodynamic force leading to migration is the second term on the right-hand side of (16.63), and migration is very slow. While many authors generalise the transport equations far from equilibrium, most of them do not modify the equations of state. The example in this section is highly representative of the shortcomings of this procedure, as it indicates clearly and explicitly that including the fluxes in the equations of state is not merely an academic exercise for systems out of equilibrium, but their contribution may be decisive to account for the experimental observations.

Problems

16.1 (a) Show that, according to Problem 15.5, one may write

$$\mathrm{Tr}\,\mathbf{P}^{\nu} = -2J(P_{12}^{\nu})^2,$$

with $J = \tau/\eta$ the steady-state shear compliance. (b) Starting from the microscopic expression for the non-equilibrium contribution to the Gibbs free energy

$$\Delta G = -\tfrac{1}{2} N k_B T \left[\mathrm{Tr}(\mathbf{P}'^{\nu}) + \ln|\det(\mathbf{U} - \mathbf{P}'^{\nu})| \right],$$

where N is the total number of macromolecules in the volume V and $P_i'^{\nu} = (nk_BT)^{-1}P_i^{\nu}$, show that for high values of the shear rate, when the second term may be neglected in this expression for ΔG, one has $G = G_{eq} + \tfrac{1}{2} V \mathrm{Tr}\,\mathbf{P}^{\nu}$.

16.2 Starting from the definition of chemical potential,

$$\mu_i = \left(\frac{\partial g}{\partial n_i}\right)_{T,p,n_j},$$

with n_i the number of moles of the species i and the non-equilibrium Gibbs free energy

$$g(T,p,c_i,P_{12}^{\nu}) = g_{eq}(T,p,c_i) + \tfrac{1}{2}J(P_{12}^{\nu})^2$$

show that the expressions for the chemical potentials at constant viscous pressure $\mathbf{P}^{\nu}$ and at constant shear rate $\dot{\gamma}$ for a polymeric solution satisfying the Zimm scaling relations $\tau \propto M^{3/2}$, $\eta \propto cM^{1/2}$, and $J \propto c^{-1}M$ are $\mu_{P^v} = \mu_{eq} - \frac{1}{2}(J/c)(P_{12}^v)^2$ and $\mu_{\dot\gamma} = \mu_{eq} + \frac{1}{2}(\tau\eta/c)\dot\gamma^2 = \mu_{eq} + \frac{1}{2}(J/c)(P_{12}^v)^2$ respectively.

16.3 A polymer solution has an equilibrium critical temperature of $T_c = 285$ K. Under a shear pressure, the critical temperature is shifted according to the law $T_c = 5 \times 10^{-2}$ (K m^2/N)$(P_{xy}^v)^2$, while the critical composition remains practically unchanged. A solution of critical composition is kept at 300 K and has a viscosity of 10^2 N s m^{-2}. What is the maximum rate flow of the solution through a cylindrical tube of radius 10^{-2} m without splitting into two phases of different composition?

16.4 Show that if instead of the expression (16.18) for the steady-state compliance J one uses the simple scaling-law relation $J = CM_1/(cRT)$, equations (16.14–5) yield a negative value for the shear-induced shift of the critical temperature. Thus, it is seen how the detailed form of the equation of state for J is important in the predictions of shear-induced thermodynamic effects.

16.5 A chemical reaction $A + B \rightleftarrows C$ takes place in a viscous solvent subjected to a shear flow, at constant shear rate $\dot\gamma$. Evaluate the influence of the shear rate on the equilibrium constant. (*Hint*: Express the influence of the shear flow on the chemical potential of each species from the result

$$G(T,p,n_i,\dot\gamma) = G_{eq}(T,p,n_i) + \sum_i V\tau_i n_i \dot\gamma^2 .$$

Take $\eta_i = \eta[1 + \frac{5}{2}(\frac{4}{3}\pi r_i^3 n_i)]$ and $\tau_i = m_i(6\pi\eta r)^{-1}$, where η is the viscosity of the pure solvent and r_i the radius of molecules of species i, and n_i the number of molecules per unit volume. Make numerical estimations for globular macromolecules A, B, and C with a radius of the order of 10^{-6} m in water at $\dot\gamma = 10$ s^{-1}.)

16.6 A dynamical interpretation of the effects of a shear flow on the equilibrium chemical composition in ideal gases may be obtained from the kinetic theory. Assume a reaction $A + A \rightarrow B + C$ for a dilute gas. The rate of the chemical reaction may be expressed in microscopic terms as

$$J = -\iiint f(r,c,t)f(r,\tilde c,t)|c - \tilde c|\sigma_r(c - \tilde c)\,dc\,d\tilde c\,d\Omega ,$$

where all the symbols have the same meaning as in Sect. 5.1, and where $\sigma_r(c - \tilde c)$ is the cross section for reactive collisions. Assume that $\sigma_r(c - \tilde c) = 0$ for $\varepsilon < \varepsilon^*$ and

$\sigma_r(c-\tilde{c})=\sigma[1-(\varepsilon/\varepsilon^*)]$ for $\varepsilon > \varepsilon^*$, where $\varepsilon = \frac{1}{4}m(c-\tilde{c})^2$ is the kinetic energy of the collision in the centre of mass reference frame and ε^* is an activation energy. Show that in the presence of a viscous pressure $\mathbf{P}^v$, and if one uses expression (5.36) (with $q = 0$) for the distribution function f, then the reaction rate is given by

$$J = -4n^2\sigma^2\sqrt{\frac{\pi k_B T}{m}}\exp\left[-\frac{\varepsilon^*}{k_B T}\right]\left\{1-\frac{2}{15(k_B T n)^2}\left[\left(\frac{\varepsilon^*}{k_B T}\right)^2-\frac{\varepsilon^*}{k_B T}-\frac{1}{4}\right]\mathbf{P}^v:\mathbf{P}^v\right\}.$$

This shows how $\mathbf{P}^v$ changes the reaction rate. If the change on the rate of the direct and the inverse reactions is different, the equilibrium constant of the reaction will be also changed (see A. S. Cukrowski and J. Popielawski, Chem. Phys. **109** (1986) 215 and B. C. Eu and K. W. Li, Physica **88 A** (1977) 135).

16.7 By using an approach analogous to that followed in (5.84–86), show that in a plane Couette flow with vanishing heat flux and $P_{12}^v = -\eta\dot{\gamma}$, the second moments of the fluctuations of the viscous pressure tensor and the correlation of the energy fluctuations with the fluctuations of the viscous pressure are given by

$$\langle\delta P_{12}^v\delta P_{12}^v\rangle = \frac{k_B\eta T}{\tau_2 V}\left(1+\frac{20}{3}\tau^2\dot{\gamma}^2\right),$$

$$\langle\delta u\,\delta P_{12}^v\rangle = -2\frac{k_B T}{M}\tau\dot{\gamma}.$$

For a more general analysis of non-equilibrium fluctuations of the shear stress in visco-elastic liquids see J. C. Dyre, Phys. Rev. E **48** (1993) 400; and for a molecular dynamics analysis of the effects of a shear flow on fluctuations see A. Baranyiai and P. T. Cummings, Phys. Rev. E **52** (1995) 2198.

16.8 *Shear-induced chromatography.* A practical application of shear-induced diffusion is shear-induced chromatography. i.e. the separation of macromolecules of different molecular mass under the action of a flow on a polymer solution. Taking into account the explicit dependence of J on the molecular mass M given in (16.18), discuss which macromolecules will exhibit a higher accumulation near the apex of the cone in the cone-and-plate experiment discussed in Sect. 16.5. [For a more detailed analysis including the dependence of the viscosity on the macromolecular mass, see D. Jou, M. Criado-Sancho, L. F. del Castillo and M. Criado-Sancho, Polymer **42** (2001)].

16. 9 In the entangled regime, the steady-state compliance of a polymer solution is given by

$$J = \frac{2M_e}{ck_BT}\left(1 - \frac{\eta_s}{\eta}\right)^2$$

The main differences with the dilute regime discussed in (16.18) are the following ones: M_e is the mass of the average length of polymer between successive entanglements and, consequently, it decreases when the concentration increases, as $M_e \cong c^{-2}$; the viscosity η_s of the solute is negligible as compared to the total viscosity η of the solution, in such a way that the parenthesis may be equated to 1. (a) Estimate the form of the non-equilibrium chemical potential. (b) Show that a shear flow under a constant pressure increases the stability of the solution, in contrast with the behaviour observed in the dilute regime.

Chapter 17

Relativistic Formulation

The standard first-order relativistic thermodynamic theories developed by Eckart [17.1] and Landau and Lifshitz [17.2] suffer from two main drawbacks: firstly, they predict an infinite speed for the propagation of thermal and viscous signals. This fact may be uncomfortable in classical thermodynamics, because of its inconsistency with the principle of causality and some experimental data, but it is certainly intolerable in any relativistic theory. Secondly, the transport equations of the first-order theory lead to some undesirable generic instabilities: as a matter of fact, small-amplitude disturbances from equilibrium diverge exponentially with time on very short time scales [17.3].

These shortcomings are avoided when use is made of the relativistic version of extended irreversible thermodynamics (EIT). It can be shown that within a given range of values of the parameters, the equilibrium is stable and the characteristic speeds are lower than the speed of light *in vacuo*. This has prompted an intensive activity on relativistic EIT [17.3–13]. In Chap. 2 we recalled that the finite value of propagation of thermal and viscous signals was one of the first motivations for EIT. In fact, the present chapter deals with a basic general aspect of the theory, and could have been included in the introductory chapters. However, we have deferred the relativistic presentation to the last chapters for pedagogical purposes and in order to provide a joint exposition of macroscopic aspects and kinetic foundations.

The motivations for a relativistic formulation are not purely conceptual: applications are found in several outstanding fields as nuclear physics, astrophysics and cosmology. In Chap. 18 we pay a special attention to viscous cosmological models, where bulk viscosity provides a source of dissipation which may have relevant effects on the dynamics of the Universe. In Sect. 17.4 we briefly comment some aspects of the extended Gibbs relation for nuclear matter, which may be of interest in the hydrodynamic analysis of high-energy collisions of atomic nuclei. In astrophysics, the role of thermal conductivity and viscosity of a gas of neutrinos is essential in phenomena like stellar collapse.

17.1 The Macroscopic Theory

In this section, we briefly summarize the main characteristics of the relativistic formulation of EIT. In contrast with classical relativistic irreversible thermodynamics [17.1–2], one assumes that the dissipative fluxes are independent variables. The general formulation parallels that of the classical non-relativistic theory. In the theory of relativity, the basic variables are the four-current vector J^μ and the second-order symmetric energy–momentum tensor $T^{\mu\nu}$. For a local observer at rest with respect to the average motion of the particles, J^μ and $T^{\mu\nu}$ are related to the four-velocity vector u^μ, the heat flux vector q^μ, the specific internal energy ε, and the pressure tensor $P^{\mu\nu}$ by

$$J^\mu = \rho u^\mu, \tag{17.1a}$$

$$T^{\mu\nu} = \rho\varepsilon u^\mu u^\nu + 2c^{-1}q^{(\mu}u^{\nu)} + P^{\mu\nu}. \tag{17.1b}$$

Greek superscripts or subscripts run from 1 to 4, c is the velocity of light *in vacuo*, ρ the mass density, and u^μ a dimensionless hydrodynamic velocity satisfying $u^\mu u_\mu = -1$; in the reference frame attached to a comoving observer, u^μ reduces to the usual velocity. Parentheses around a set of indices denote symmetrization. Both q^μ and $P^{\mu\nu}$ are of spatial type,

$$q^\mu = \Delta^\mu_\nu q^\nu \quad \text{and} \quad P^{\mu\nu} = \Delta^\mu_\rho \Delta^\nu_\sigma P^{\rho\sigma},$$

i.e. they obey $q^\mu u_\mu = u_\mu P^{\mu\nu} = P^{\mu\nu} u_\nu = 0$, where $\Delta^{\mu\nu}$ is the symmetric spatial projector $\Delta^{\mu\nu} = g^{\mu\nu} + u^\mu u^\nu$, and $g^{\mu\nu}$ the metric tensor of the spacetime. The pressure tensor splits into

$$P^{\mu\nu} = p\Delta^{\mu\nu} + \Pi\Delta^{\mu\nu} + \Pi^{\mu\nu},$$

with p and Π being the equilibrium pressure and the bulk viscous pressure respectively, and $\Pi^{\mu\nu}$ the traceless viscous pressure tensor. We prefer this notation, instead of p^ν

and $\overset{0}{P}{}^\nu$ used throughout the book, because it is much more usual in the relativistic literature; therefore, the researchers interested by these chapters will feel more at ease with this notation without being a difficulty for the reader who has read the previous chapters.

The decomposition (17.1b) is not unique. A different splitting can be devised by an observer drifting slowly in the direction of the heat flux, with velocity adjusted in such a way that the mass–energy counterflow of particles relative to his frame cancels exactly the heat flux [17.6]. It turns out that in the frame attached to such an observer, the heat flux vanishes (see Landau and Lifshitz [17.2]). Here, we choose to follow the scheme consistent with (17.1), first proposed by Eckart. The two descriptions are nearly equivalent, although their decomposition of the invariant fields J^μ and $T^{\mu\nu}$ is different.

The conservation equations read

$$J^\mu{}_{;\mu} = 0 \tag{17.2}$$

(or equivalently $\rho\dot{v} = u^\mu{}_{;\mu}$) and

$$T^{\mu\nu}{}_{;\nu} = 0, \tag{17.3}$$

where the semicolon stands for the covariant derivative. The temporal and the spatial parts of (17.3) correspond to the energy and the momentum conservation laws respectively. They can be explicitly written as

$$\rho(\dot{\varepsilon} + p\dot{v}) = -\frac{1}{c}\left[q^\mu{}_{;\mu} + q^\mu \dot{u}_\mu\right] - \left[\Pi\Delta^{\mu\nu} + \Pi^{\mu\nu}\right]u_{(\mu;\nu)} \tag{17.4}$$

and

$$\rho\varepsilon\dot{u}^\lambda + \frac{1}{c}\left[\dot{q}^\lambda - u^\lambda q_\mu \dot{u}^\mu + q^\lambda u^\nu{}_{;\nu} + u^\lambda{}_{;\nu}q^\nu\right] + \Delta^\lambda_\mu P^{\mu\nu}{}_{;\nu} = 0, \tag{17.5}$$

where the upper dot denotes differentiation along the four-velocity, i.e. $\dot{A}^\nu = u^\mu A^\nu{}_{;\mu}$.

To determine the evolution equations of the fluxes, we postulate, in the spirit of EIT, the existence of a non-equilibrium entropy flux vector S^μ depending on the whole set of variables. Up to the second order in the fluxes, one has

$$S^\mu = \left(s_{eq} - v\alpha_1 T^{-1}q^\nu q_\nu - v\alpha_0 T^{-1}\Pi^2 - v\alpha_2 T^{-1}\Pi^{\lambda\nu}\Pi_{\lambda\nu}\right)u^\mu$$

$$+ \frac{1}{cT}q^\mu + \beta_0\Pi q^\mu + \beta_2\Pi^{\lambda\mu}q_\lambda, \tag{17.6}$$

The standard form of the entropy balance equation reads

$$S^{\mu}{}_{;\mu} = \sigma^s \geq 0. \tag{17.7}$$

The four-divergence of S^{μ} may be calculated from (17.6) by taking into account that

$$\dot{s}_{eq} = T^{-1}\dot{\varepsilon} + T^{-1}p\dot{v} \tag{17.8}$$

and by making use of (17.4). Then, by neglecting second-order contributions to the equations of state, one can write the entropy production in the bilinear form

$$\sigma^s = T^{-1}\left(q^{\mu}X^1_{\mu} + \Pi X^0 + \Pi^{\mu\nu}X^2_{\mu\nu}\right), \tag{17.9}$$

with

$$X^1_{\mu} = \Delta^{\nu}_{\mu}\left[-(cT)^{-1}(T_{,\nu} + T\dot{u}_{,\nu}) - 2\alpha_1\dot{q}_{\nu} + T\beta_0\Pi_{;\nu} + T\beta_2\Pi^{\lambda}_{\nu;\lambda}\right],$$

$$X^0 = -u^{\mu}{}_{;\mu} - 2\alpha_0\dot{\Pi} + T\beta_0 q^{\mu}{}_{;\mu},$$

and

$$X^2_{\mu\nu} = \Delta^{\lambda}_{(\mu}\Delta^{\rho}_{\nu)}\left[u_{(\lambda;\rho)} - 2\alpha_2\dot{\Pi}_{\lambda\rho} + T\beta_2 q_{(\lambda;\rho)}\right].$$

Here, the semicolon means partial derivative, and the quantities β_0 and β_2 are assumed to be constant; this is consistent with the approximation of neglecting non-linear terms, made in the non-relativistic description.

In analogy with the treatment of Chap. 2, we may take X^1_{μ}, X^0, and $X^2_{\mu\nu}$ proportional to q_{μ}, Π, and $\Pi_{\mu\nu}$, respectively, with positive proportionality coefficients to ensure the positive semi-definite character of the entropy production. As a consequence, the evolution equations of the fluxes read

$$\tau_1\dot{q}_{\mu} = -q_{\mu} - \lambda T\Delta^{\nu}_{\mu}\left[(cT)^{-1}T_{,\nu} + T\dot{u}_{\nu} + T\beta_0\Pi_{,\nu} + T\beta_2\Pi^{\lambda}_{\nu;\lambda}\right], \tag{17.10}$$

$$\tau_0\dot{\Pi} = -\Pi - \zeta u^{\mu}{}_{;\mu} - T\beta_0\zeta q^{\mu}{}_{;\mu}, \tag{17.11}$$

$$\tau_2\dot{\Pi}_{\mu\nu} = \langle -\Pi_{\mu\nu} - 2\eta(u_{\lambda;\rho} + T\beta_2 q_{\lambda;\rho})\rangle, \tag{17.12}$$

where the angular brackets $\langle...\rangle$ mean the symmetric, trace-free spatial part of the corresponding tensor. Also the left-hand side of (17.10) and (17.12) have to be understood as the projected part of the corresponding quantities. These equations generalise the non-relativistic ones (2.64–66). Note that in the classical limit the relativistic equation for the heat flux does not reduce exactly to the classical Fourier law, but contains a supplementary term which depends on the acceleration $\dot{u}^\mu$, as first noted by Eckart [17.1]. Its physical meaning is connected with the inertia of heat; effectively, a heat flux will set in through accelerated matter even in the absence of temperature gradients [17.14].

Table 17.1. Coefficients of (17.10–12) for an ultrarelativistic ideal gas and for a mixture of photons and electrons (β stands for mc^2/k_BT and a is the radiation constant)

Fluid	λ	ζ	η	β_0	β_2
Relativistic ideal gas	$\dfrac{8}{5}\dfrac{p}{\rho T}\tau_1$	$\dfrac{1}{108}\dfrac{p\beta^4}{\rho}\tau_0$	$\dfrac{4}{3}\dfrac{p}{\rho}\tau_2$	$\dfrac{6}{pT\beta^2}$	$-\dfrac{1}{4pT}$
Photons and electrons	$\dfrac{4}{3}aT^3c^2\tau_1$	$4aT^4\Omega^2\tau_0$	$\dfrac{4}{15}aT^4\tau_2$	$4aT^5\left(\dfrac{\Omega}{\Omega_m}\right)^2$	$-\dfrac{3}{4aT^5}$

In Table 17.1 are reported the values of the coefficients appearing in (17.10–12) as a function of the relaxation times τ_1, τ_0, and τ_2. Two systems are considered: an ideal monatomic gas in the ultrarelativistic limit ($\beta = mc^2/k_BT \ll 1$) and a radiative fluid consisting of a mixture of photons and electrons, the above results are drawn from the kinetic theory. The quantities Ω and Ω_m are given by $\Omega = \frac{1}{3} - (\partial p/\partial T)_n (\partial \rho/\partial T)_n^{-1}$ and $\Omega_m = \frac{1}{3} - (\partial p_m/\partial T)_n (\partial \rho_m/\partial T)_n^{-1}$, where p and ρ are the pressure and density of the total fluid and p_m and ρ_m the pressure and density of the matter. The values of the relaxation times depend of the kind of theory used to describe the interaction between matter and radiation. For pure Thomson scattering (i.e. scattering of electromagnetic radiation by free charged particles in the classical non-relativistic theory), the values of the relaxation times are $\tau_1 = \frac{9}{10}\tau_2$ and $\tau_0 = \infty$.

17.2 Characteristic Speeds

A relevant calculation is that of the characteristic speeds associated with (17.10–12), to ascertain whether their values remain finite [17.15]. To this end, we use the linear equations (17.10–12). These nine equations, together with the five conservation laws (17.2–3), provide a set of 14 equations for the 14 unknowns ρ, ε, u_μ, Π, q_μ, and $\Pi_{\mu\nu}$.

The complete set of first-order and quasi-linear equations can be expressed in the simplified form

$$a^{AB\mu}(\alpha,\beta,u_\nu)\partial_\mu y_B = D^A(y) \qquad (A,B = 1,...,14).$$

(17.13)

The right-hand side contains all the collision terms, and the coefficients $a^{AB\mu}$ are purely thermodynamical functions. The wavefront speeds are determined by the condition

$$\det\left[a^{AB\mu}(\partial_\mu\phi)\right] = 0,$$

(17.14)

where $\phi(x_\mu) = $ constant is a characteristic surface of (17.13), that is, a 3-dimensional space across which the variables y_A are continuous but their first derivatives are allowed to exhibit discontinuities $[\partial_\mu y_A]$ normal to the surface, i.e. $[\partial_\mu y_A] = Y_A(\partial_\mu\phi)$. The characteristic velocities are independent of the details of the collisions.

To solve the characteristic equation (17.14) consider a coordinate system (x^1, x^2, x^3, x^4) chosen in such a way that at any point in the fluid the system of reference is orthogonal and comoving, with the property

$$g^{\mu\nu}\partial_\mu\partial_\nu = -(\partial_1)^2 + (\partial_2)^2 + (\partial_3)^2 + (\partial_4)^2, \qquad u^\mu\partial_\mu = \partial_1,$$

(17.15)

and assume that ϕ is a function of only (x^1, x^2). In this coordinate system, the characteristic equation simplifies to

$$\det\left[v'a^{AB1} - a^{AB2}\right] = 0,$$

(17.16)

where v' is the characteristic velocity defined by $v' = -(\partial_1\phi)/(\partial_2\phi)$. The form of the characteristic matrix simplifies further if use is made of the following set of 14 perturbation variables: $y_B = (T\delta\theta, T^{-1}\delta T, \delta\Pi, \delta u^2, \delta q^2, \delta\Pi^{22}, \delta u^3, \delta q^3, \delta\Pi^{32}, \delta u^4, \delta q^4, \delta\Pi^{42}, \delta\Pi^{33}, -\delta\Pi^{44}, \delta\Pi^{34})$, where $\theta = (\rho T)^{-1}(\varepsilon + p) - s$ [17.15]. With this choice of variables, and after some lengthy calculations, the characteristic matrix may be written as [17.15]

$$v'a^1 - a^2 = \begin{pmatrix} \mathbf{Q} & 0 & 0 & 0 \\ 0 & \mathbf{R} & 0 & 0 \\ 0 & 0 & \mathbf{R} & 0 \\ 0 & 0 & 0 & \mathbf{S} \end{pmatrix}.$$

(17.17)

The blocks $\mathbf{Q}$, $\mathbf{R}$ and $\mathbf{S}$ are respectively given by

$$\mathbf{Q} = \begin{pmatrix} \dfrac{v'}{T}\left(\dfrac{\partial \rho}{\partial \theta}\right)_T & \dfrac{v'}{T}\left(\dfrac{\partial \rho}{\partial T}\right)_\theta & 0 & -\rho & 0 & 0 \\[2mm] \dfrac{v'}{T}\left(\dfrac{\partial \varepsilon}{\partial \theta}\right)_T & \dfrac{v'}{T}\left(\dfrac{\partial \varepsilon}{\partial T}\right)_\theta & 0 & -(\varepsilon + p) & -1 & 0 \\[2mm] 0 & 0 & -\alpha_0 v' & -1 & -\beta_0 & 0 \\[2mm] -\rho & -(\varepsilon + p) & -1 & -(\varepsilon + p)v' & -v' & -1 \\[2mm] 0 & -1 & -\beta_0 & -v' & -\alpha_1 v' & -\beta_2 \\[2mm] 0 & 0 & 0 & -1 & -\beta_2 & \tfrac{3}{2}\alpha_2 v' \end{pmatrix},$$

$$\mathbf{R} = \begin{pmatrix} v'(\varepsilon + p) & v' & -1 \\[2mm] v' & \alpha_1 v' & \beta_2 \\[2mm] -1 & \beta_2 & 2\alpha_2 v' \end{pmatrix}, \qquad\qquad (17.18)$$

$$\mathbf{S} = \begin{pmatrix} 2\alpha_2 v' & 0 \\[2mm] 0 & \alpha_2 v' \end{pmatrix}.$$

The total determinant now reduces to

$$\det\left[v'a^1 - a^2\right] = (\det \mathbf{Q})(\det \mathbf{R})^2 (\det \mathbf{S}). \qquad\qquad (17.19)$$

The roots of the characteristic equation (17.14) are the ensemble of roots obtained after setting to zero the determinant of each of the matrix blocks. The main results may be summarized as follows.

1. The determinant of $\mathbf{R}$ is given by

$$\det \mathbf{R} = v'\left\{2\alpha_2\big[(\varepsilon + p)\alpha_1 - 1\big]v'^2 - \big[(\varepsilon + p)\beta_2^2 + 2\beta_2 + \alpha_1\big]\right\}. \qquad (17.20)$$

This yields a vanishing and two non-vanishing characteristic velocities

$$v_t'^2 = \frac{(\varepsilon + p)\beta_2^2 + 2\beta_2 + \alpha_1}{2\alpha_2\big[\alpha_1(\varepsilon + p) - 1\big]}. \qquad\qquad (17.21)$$

These are transverse velocities because they refer to $(\delta q^3, \delta \Pi^{32}, \delta u^3)$ and $(\delta q^4, \delta \Pi^{42}, \delta u^4)$, the components of the perturbation variables which are tangent to the characteristic surface, and therefore transverse to the direction of propagation.

2. The determinant of matrix $\mathbf{Q}$, related to the variables $(T\delta\theta, T^{-1}\delta T, \delta\Pi, \delta u^2, \delta q^2, \delta\Pi^{22})$, is

$$\det \mathbf{Q} = \frac{3}{2}v'^2\left(Av'^4 + Bv'^2 + C\right)\left[\left(\frac{\partial p}{\partial \theta}\right)_T\left(\frac{\partial \varepsilon}{\partial T}\right)_\theta - \left(\frac{\partial p}{\partial T}\right)_\theta\left(\frac{\partial \varepsilon}{\partial \theta}\right)_T\right], \qquad (17.22)$$

with

$$A = \alpha_0\alpha_2\left[\alpha_1(\varepsilon + p) - 1\right], \qquad B = -(\varepsilon + p)D - \alpha_1 E - 2F, \qquad C = (DE - F^2)(\alpha_0\alpha_2)^{-1},$$

$$D = \alpha_0\alpha_2\left[\frac{\beta_0^2}{\alpha_0} + \frac{2\beta_2^2}{3\alpha_2} + \frac{1}{\rho T^2}\left(\frac{\partial T}{\partial s}\right)_n\right], \qquad E = \alpha_0\alpha_2\left[\frac{1}{\alpha_0} + \frac{2}{3\alpha_2} + (\varepsilon + p)\left(\frac{\partial p}{\partial \varepsilon}\right)_s\right],$$

$$F = \alpha_0\alpha_2\left[\frac{\beta_0}{\alpha_0} + \frac{2\beta_2}{3\alpha_2} - \frac{p}{T}\left(\frac{\partial T}{\partial p}\right)_s\right].$$

It is checked that $\mathbf{Q}$ has two vanishing and four non-vanishing characteristic velocities. The latter are the roots of

$$Av_l'^4 + Bv_l'^2 + C = 0, \qquad (17.23)$$

and correspond to longitudinal velocities, the first pair to sound propagation and the second pair to temperature waves or second sound.

3. The determinant of the matrix S corresponding to the variables $(\delta\Pi^{33}, -\delta\Pi^{44}, \delta\Pi^{34})$ is simply

$$\det \mathbf{S} = 2\left(\alpha_2 v'\right)^2, \qquad (17.24)$$

and the two associated characteristic velocities vanish.

We are thus able to express the characteristic velocities in terms of thermodynamic parameters. Hiscock and Lindblom [17.15] have shown that when thermodynamic stability conditions are satisfied, the characteristic speeds are lower than the speed of light. Explicit values for the velocities may be obtained by using the equations of state derived from kinetic theory. Stewart [17.7] has found that for a Boltzmann gas in the ultrarelativistic limit, the velocities of the transverse modes increase monotonically between $\frac{2}{5}k_BT/m \le v_t^2 \le \frac{1}{5}c^2$, whereas the two longitudinal speeds satisfy $1.35(k_BT/m) \le v_{l1}^2 \le \frac{1}{3}c^2$ and $5.18(k_BT/m) \le v_{l2}^2 \le \frac{2}{3}c^2$. Other analyses of the properties of waves in relativistic hyperbolic heat-conducting and viscous fluids can be found in [17.17–18].

17.3 Relativistic Kinetic Theory

Supplementary information about the coefficients appearing in the entropy and the entropy flux can be drawn from a kinetic approach. Here we follow the work of Israel and Stewart [17.7]; other significant references on the connection between relativistic kinetic theory and EIT may be found in [17.4,10,12,13]. We study here the general case of quantum relativistic gases, following the main lines of Grad's classical analysis, and for the sake of simplicity, we will skip all the details of calculations which are rather cumbersome.

The microscopic expression for the four-dimensional current vector J^μ, the energy–momentum tensor $T^{\mu\nu}$, and the entropy four-vector S^μ are

$$J^\mu = c \int dP\, p^\mu f(x^\nu, p^\lambda), \qquad (17.25a)$$

$$T^{\mu\nu} = c \int dP\, p^\mu p^\nu f(x^\rho, p^\lambda), \qquad (17.25b)$$

$$S^\mu = -k_B c \int dP\, p^\mu \left[f \ln\frac{f}{y} \mp y\left(1 \pm \frac{f}{y}\right)\ln\left(1 \pm \frac{f}{y}\right)\right], \qquad (17.25c)$$

where p^μ is the four-momentum of a particle, with $p^\mu p_\mu = m^2 c^2$, $dP = (-g)^{1/2}dp/p_0$, the invariant volume element (g is the determinant of $g_{\mu\nu}$), and $y = (2s + 1)/h^3$, with s being the spin of the particle. In (17.25c) upper sign refers to bosons and lower sign to fermions. Remember that $T^{\mu\nu}$ may be decomposed in terms of the heat flux q^μ, viscous pressure Π, and viscous pressure tensor $\Pi^{\mu\nu}$:

$$T^{\mu\nu} = \varepsilon u^\mu u^\nu + (p + \Pi)\Delta^{\mu\nu} + 2c^{-1}q^{(\mu}u^{\nu)} + \Pi^{\mu\nu}. \qquad (17.26)$$

The distribution function $f(x^\mu, p^\nu)$ gives the number density of particles in the element $dxdp$ at x^μ, p^ν. In equilibrium, $f(x^\mu, p^\nu)$ is given by the Jüttner expression

$$f_{eq} = y\left[\exp\left(\frac{\alpha}{k_B} + \frac{u_\mu p^\mu}{k_B T}\right) \mp 1\right]^{-1}. \qquad (17.27)$$

Introduction of (17.27) into (17.25c) leads to the equilibrium entropy expression

$$S^\mu_{eq} = p(\alpha, \beta)\beta^\mu - \alpha J^\mu - \beta_\lambda T^{\lambda\mu}, \qquad (17.28)$$

with $\beta^\mu = 1/(k_B T)u^\mu$, $\beta = mc^2/(k_B T)$ and $\alpha = -\mu/T$. Out of equilibrium, f obeys the relativistic Boltzmann equation

$$p^\mu \frac{\partial f}{\partial x^\mu} - \Gamma^\mu_{\nu\lambda} p^\nu p^\lambda \frac{\partial f}{\partial p^\mu} = \mathcal{J}(f), \tag{17.29}$$

with $\Gamma^\mu_{\nu\lambda}$ being the Christoffel symbols and $\mathcal{J}(f)$ the collision operator. It is not necessary to solve (17.29) to obtain an expression for the stationary distribution function. Rather, we shall follow an approach similar to the non-relativistic Grad formalism. To this end we set

$$\ln\left[\frac{f}{y}\left(1 \pm \frac{f}{y}\right)^{-1}\right] = \Delta \tag{17.30}$$

and expand Δ in its first moments with respect to p^μ:

$$\Delta = \Delta_{eq} + \varepsilon(x) + m^{-1}\varepsilon_\lambda(x)p^\lambda + m^{-2}\varepsilon_{\lambda\mu}(x)p^\lambda p^\mu . \tag{17.31}$$

The fourteen parameters $\varepsilon(x)$, $\varepsilon_\lambda(x)$ and $\varepsilon_{\lambda\mu}(x)$ will be related to the dissipative fluxes. Note that the non-equilibrium distribution function may be written in terms of Δ as

$$f - f_{eq} = \left[f_{eq}\left(1 \pm \frac{f_{eq}}{y}\right)\right]^{-1}\delta\Delta, \tag{17.32}$$

with $\delta\Delta = \Delta - \Delta_{eq}$ given by (17.31). Accordingly,

$$f - f_{eq} = \frac{\varepsilon(x) + m^{-1}\varepsilon_\lambda(x)p^\lambda + m^{-2}\varepsilon_{\lambda\mu}(x)p^\lambda p^\mu}{f_{eq}\left[1 \pm (f_{eq}/y)\right]}. \tag{17.33}$$

We assume that the tensor $\varepsilon_{\lambda\mu}$ is trace-free, i.e. $\varepsilon_{\lambda}^{\lambda} = 0$. The actual non-equilibrium distribution function is thus specified by (17.33). Introduction of (17.33) into (17.29) leads to the evolution equations for ε, ε_λ, and $\varepsilon_{\lambda\mu}$.

We now calculate the change of J^μ, $T^{\mu\nu}$, and S^μ with respect to their equilibrium values in terms of $f - f_{eq}$. First of all, we define $\phi(f)$ as

$$\phi(f) = \frac{f}{y}\ln\frac{f}{y} \mp \left(1 \pm \frac{f}{y}\right)\ln\left(1 \pm \frac{f}{y}\right), \tag{17.34}$$

and expand it around $\phi(f_{eq})$ up to the second order, it is found that

$$\phi(f) = \phi(f_{eq}) + \phi'(f_{eq})(f - f_{eq}) + \tfrac{1}{2}\phi''(f_{eq})(f - f_{eq})^2, \qquad (17.35)$$

where

$$\phi'(f) = \ln\frac{f/y}{1 \pm (f/y)} \qquad (17.36a)$$

and

$$\phi''(f) = \frac{1}{[1 \pm (f/y)](f/y)} \qquad (17.36b)$$

are the first and second functional derivatives of ϕ with respect to f. Introducing (17.35) into expression (17.25c) for the entropy yields

$$S^\mu = -k_B c \int dP\,\phi(f)p^\mu = S_{eq}^\mu - \frac{1}{T}q^\mu - \frac{1}{2}k_B c \int dP\,\phi''(f_{eq})(f - f_{eq})^2 p^\mu. \quad (17.37)$$

Taking into account (17.33), one finds for the non-equilibrium expressions of J^μ and $T^{\mu\nu}$

$$J^\mu = \varepsilon c \int dP\,\phi''(f_{eq})p^\mu + \varepsilon_\lambda \frac{c}{m} \int dP\,\phi''(f_{eq})p^\mu p^\lambda$$

$$+ \varepsilon_{\lambda\nu} \frac{c}{m^2} \int dP\,\phi''(f_{eq})p^\lambda p^\nu p^\mu, \qquad (17.38)$$

$$T^{\mu\nu} = T_{eq}^{\mu\nu} + \varepsilon c \int dP\,\phi''(f_{eq})p^\mu p^\nu + \varepsilon_\lambda \frac{c}{m} \int dP\,\phi''(f_{eq})p^\mu p^\lambda p^\nu$$

$$+ \varepsilon_{\lambda\rho} \frac{c}{m^2} \int dP\,\phi''(f_{eq})p^\mu p^\nu p^\lambda p^\rho, \qquad (17.39)$$

respectively. We require that the local-equilibrium particle and energy densities $\rho(\alpha, \beta)$ and $\varepsilon(\alpha, \beta)$ equal their actual densities, given by $u_\mu J^\mu$ and $u_\mu u_\nu T^{\mu\nu}$ respectively (this is equivalent to the requirements (5.31) in classical kinetic theory) and we fix the nine remaining variables in terms of q^μ, Π, and $\Pi^{\mu\nu}$. From (17.38) and (17.39) it follows that [17.7]

$$\varepsilon_{\lambda\mu} = B_0(\Delta_{\lambda\mu} + 3u_\lambda u_\mu)\Pi + \tfrac{1}{2}B_1(u_\lambda q_\mu + q_\lambda u_\mu) + B_2\Pi_{\lambda\mu}, \qquad (17.40a)$$

$$\varepsilon_\mu = D_0 u_\mu \Pi + D_1 q_\mu, \qquad \varepsilon = E_0 \Pi. \tag{17.40b}$$

The coefficients in (17.40) are given by

$$B_0 = -\frac{m}{4k_B T \zeta' \Omega}, \quad B_1 = -\frac{1}{k_B T \Lambda n}, \quad B_2 = -\frac{m}{2k_B T \zeta'}, \quad D_0 = -3B_0 C_2,$$

$$D_1 = B_1 \frac{J_{41}}{J_{31}}, \quad E_0 = 3B_0 C_1, \quad \zeta' = (\varepsilon + p)\frac{k_B T}{m}, \quad \Omega = 3\left(\frac{\partial \ln \zeta'}{\partial \ln \rho}\right)_{\zeta/\rho} - 5,$$

$$\Lambda = \frac{D_{31}}{J_{21}^2}, \quad C_1 = -1 - \frac{4J_{31}J_{30} - J_{41}J_{20}}{D_{20}}, \quad C_2 = \frac{J_{31}J_{20} - J_{41}J_{10}}{D_{20}}$$

with

$$J_{qn} = \frac{4\pi y m^2}{(2q+1)!!}\int_0^\infty d\chi \left(\frac{f_{eq}}{y}\right)\left(1 + \frac{f_{eq}}{y}\right) \sinh^{2(q+1)}\chi \cosh^{n-2q}\chi, \tag{17.41}$$

where χ is obtained from $m\cosh\chi = -p_\mu u^\mu$. In terms of (17.27), f_{eq}/y takes the simple form

$$\frac{f_{eq}}{y} = \frac{1}{\exp(\beta\cosh\chi - \alpha)\pm 1}, \tag{17.42}$$

with α and β as defined in (17.28). This is a usual variable transform in relativistic kinetic theory, in terms of which $dP = 4\pi m^2 \sinh^2\chi\, d\chi$. Though the procedure is simple from a conceptual point of view, to obtain the above coefficients is very intricate. The interested reader is advised to go to the first of references [17.7].

Once ε, ε_μ, and $\varepsilon_{\lambda\mu}$ have been expressed in terms of the fluxes, one may substitute (17.33) in (17.37) to obtain the non-equilibrium entropy

$$S^\mu = S_{eq}^\mu + T^{-1}q^\mu - u^\mu v T^{-1}\left[\alpha_0 \Pi^2 + \alpha_1 q^\lambda q_\lambda + \alpha_2 \Pi^{\rho\lambda}\Pi_{\lambda\rho}\right]$$

$$+ \beta_0 \Pi q^\mu + \beta_2 \Pi^{\mu\nu} q_\nu, \tag{17.43}$$

where the coefficients α_i, β_i stand for

$$\alpha_1 = \frac{1}{2vT}D_{41}nm\Lambda^2 J_{21}J_{31}, \qquad \alpha_2 = \frac{1}{4vT}\beta\zeta'^2 J_{52}, \tag{17.44a}$$

$$\alpha_0 = \frac{3}{2vT\Omega^2}\,\beta\zeta'\left\{5J_{52} - 3\frac{J_{31}(J_{30}J_{31} - J_{41}J_{20}) + J_{41}(J_{41}J_{10} - J_{31}J_{20})}{D_{20}}\right\}, \quad (17.44b)$$

$$\beta_0 = (D_{41}D_{20} - D_{31}D_{30})T\Lambda\zeta'J_{21}J_{31}, \quad \beta_2 = -(J_{41}J_{42} - J_{31}J_{52})T\Lambda\zeta'J_{21}J_{31}, \quad (17.44c)$$

with D_{qn} defined by $D_{qn} = J_{q+1,n}J_{q-1,n} - J_{qn}^2$. In the limit of non-degenerate gases, i.e. $\alpha \gg 1$ in (17.42), the asymptotic expressions for these coefficients in the classical gas $[\beta \gg 1$ in (17.42)$]$ are $v\alpha_1 = \frac{1}{3}m(pk_BT)^{-1}$, $v\alpha_2 = \frac{1}{4}p^{-1}$, and $\beta_2 = -\frac{2}{5}(pT)^{-1}$, which are nothing but Grad's results. From the other hand, $v\alpha_0 = \frac{3}{5}m^2(k_B^2Tp)^{-1}$ and $\beta_0 = \frac{4}{5}m(k_BT^2p)^{-1}$, but these coefficients will play no role as they are multiplied by the bulk viscous pressure, which is zero in this limit. In the ultrarelativistic case $(\beta \ll 1)$, one obtains $v\alpha_1 = \frac{5}{8}p^{-1}$, $v\alpha_2 = \frac{3}{8}p^{-1}$, $\beta_2 = -\frac{1}{4}(pT)^{-1}$, $v\alpha_0 = 108(p\beta^4)^{-1}$, and $\beta_0 = 6(pT\beta^2)^{-1}$.

The knowledge of the coefficients (17.44) allows one to write λ, ζ, and η in terms of the relaxation times τ_0, τ_1, and τ_2. Such relaxation times depend themselves on the collision term of the Boltzmann equation, and their derivation is an extremely laborious task. For the present purpose, it suffices to say that the kinetic theory corroborates the form of the evolution equations for the fluxes (17.10–12) and provides explicit values for the coefficients (17.44).

17.4 Nuclear Matter

Relativistic hydrodynamics is the root of the theory of collisions of heavy ions, and the analysis of these collisions is helpful to determine the nuclear equations of state and, to search for the phase transition from the hadronic matter to the quark-gluon plasma [17.21]. It must be noted, however, that generally the duration of the collisions between heavy nuclei is only one order of magnitude higher than the mean free time of nucleons collisions inside the nuclei. As a consequence, relaxational effects are important in describing the dynamics of such phenomena. Furthermore, during collisions the nuclei are far from equilibrium, so that the non-equilibrium corrections of the equations of state may considerably affect the energy required for the phase transition in nuclear matter.

To formulate the state equations, we need an explicit expression of the Gibbs equation which means that we must know the expressions for the transport coefficients and the relaxation times. They have been derived by Danielewicz [17.22] from the kinetic theory by expanding the Uhlenbeck–Uehling equation, the quantum version of the Boltzmann equation: the following expressions for the shear viscosity η and the thermal conductivity λ have been obtained

$$\eta = \frac{1700}{(k_B T)^2}\left(\frac{n}{n_0}\right)^2 + \frac{22}{1+10^{-3}(k_B T)^2}\left(\frac{n}{n_0}\right)^{0.7} + \frac{5.8(k_B T)^{1/2}}{1+160(k_B T)^{-2}}, \qquad (17.45)$$

$$\lambda = \frac{0.15}{k_B T}\left(\frac{n}{n_0}\right)^{1.4} + \frac{0.02}{1+10^{-6}(k_B T)^4/7}\left(\frac{n}{n_0}\right)^{0.4} + \frac{0.225(k_B T)^{1/2}}{1+160(k_B T)^{-2}}, \qquad (17.46)$$

where $k_B T$ is expressed in MeV, η in MeV/fm^{2}c, λ in c/fm^2, the nucleon number n in fm^{-3} and where $n_0 = 0.145$ fm^{-3} (here, fm stands for fermi, with 1 fm = 10^{-15} m and c for the speed of light). The corresponding relaxation times for the viscous pressure tensor $\mathbf{P}^\nu$ and the heat flux q are [17.22]

$$\tau_2 = \left(\frac{850}{(k_B T)^2}\right)\left(\frac{n}{n_0}\right)^{1/3}\left[1+0.04 k_B T\frac{n_0}{n}\right] + \frac{38(k_B T)^{-1/2}}{1+160(k_B T)^{-2}}\left(\frac{n_0}{n}\right), \qquad (17.47)$$

$$\tau_1 = \left(\frac{310}{(k_B T)^2}\right)\left(\frac{n}{n_0}\right)^{0.4}\left[1+0.1 k_B T\frac{n_0}{n}\right] + \frac{57(k_B T)^{-1/2}}{1+160(k_B T)^{-2}}\left(\frac{n_0}{n}\right), \qquad (17.48)$$

where τ_i ($i = 1, 2$) is expressed in fm/c. The last terms in the right-hand side correspond to binary colllisions amongst nucleons, whereas the first terms describe collisions of nucleons with the surface of the nuclei. For $k_B T > 6$ MeV, binary collisions are more frequent whereas at lower temperatures, a ballistic regime with collisions against the boundaries of the nucleus is observed.

The generalised non-equilibrium entropy up to the second-order contributions in the fluxes is given by

$$s(u,v,q,\mathbf{P}^\nu) = s_{eq}(u,v) - \frac{\tau_1}{2\lambda\rho T^2}q\cdot q - \frac{\tau_2}{4\eta\rho T}\mathbf{P}^\nu : \mathbf{P}^\nu. \qquad (17.49)$$

Combining (17.47–48) with (17.49) results in an explicit expression for the non-equilibrium entropy. Since the general expression is very cumbersome, we particularize the results for high temperature, which is the regime of interest in the analysis of high-energy collisions. In this case, the expression of entropy which will serve as a basis to determine the nonequilibrium nuclear equations of state is

$$s = s_{eq} - \frac{1.26}{mT^3} 10^3 \frac{n_0}{n^2} \boldsymbol{q} \cdot \boldsymbol{q} - \frac{1.64}{mT^2} \frac{n_0}{n^2} \mathbf{P}^\nu : \mathbf{P}^\nu. \qquad (17.50)$$

Another subject of interest is the problem of the energy transfer, which is also affected by the relaxational effects. Supplementary applications are concerned with thermodynamic stability requirements on the nuclear equations of state [17.23]. It can thus be concluded that a promising field of application of relativistic EIT in the near future will be nuclear physics, where rather exciting topics are presently developed.

Problems

17.1 In the steady state and in the absence of viscous effects, the evolution equation (17.10) for the heat flux is written as

$$q^\mu = -\lambda \Delta^{\mu\nu}(\nabla T + T a_\nu),$$

with a_n being the acceleration four-vector. In the local instantaneous rest frame, $\Delta^{\mu\nu}$ is diag $(0,1,1,1)$, and the spatial components of the equation for the heat flux take the form

$$q = -\lambda(\nabla T + Tc^{-2}a).$$

Here λ is the thermal conductivity, q the three-dimensional heat flux, c the velocity of light *in vacuo*, and a the acceleration. In the presence of a gravitational field ϕ, one has $a = -\nabla\phi$. (a) Show that in the presence of a gravitational field the equilibrium condition is not T uniform but T^* uniform, with $T^* = T(1 + c^{-2}\phi)$ up to the first order in ϕ. (b) Calculate the difference in the temperature T between a point located on the earth surface, at 300 K, and another one 100 m above. Determine the value of a so that the difference is 1 K. [See Pavón et al., Phys. Lett. A **78** (1980) 317.]

17.2 In general relativity, the conditions of thermal and material equilibrium are

$$T\sqrt{g_{00}} = \text{const}, \qquad \mu\sqrt{g_{00}} = \text{const}$$

with g_{00} the temporal component of the metric tensor (L. D. Landau and E. M. Lifshitz, *Statistical Physics*, 3rd ed. Part 1, Pergamon, Oxford, 1980). In a weak gravitational field, $g_{00} = 1 + (2\phi/c^2)$, with ϕ being the gravitational potential. (a) Show the result of the Problem 17.1 for thermal equilibrium in terms of the temperature T^* introduced there. (b) Taking into account that in the classical limit the relativistic chemical potential μ tends to $\mu_0 + mc^2$, where μ_0 is the usual equilibrium chemical potential and m the mass of the particles, show that the condition of material equilibrium reduces in the classical limit to $\mu_0 + m\phi = \text{const}$.

17.3 In relativistic transformations, it is generally assumed that entropy remains constant, $S = S_0$, while volume V changes as $V = \gamma^{-1}V_0$. Here, the subscript 0 denotes the values measured by an observer at rest, the quantities without the subscript correspond to those measured by an observer moving at speed v with respect to the system, and $\gamma = \left[1 - (v/c)^2\right]^{-1/2}$, with c the speed of light *in vacuo*. Show that the non-equilibrium entropy density (per unit volume) for radiation with energy density e and energy flux J obtained in (7.64) in the context of information theory, which may be rewritten as

$$s(e,J) = \tfrac{1}{2} s_{eq}(e)(2 + x)^{1/2}(x - 1)^{1/4},$$

with $x = [4 - 3(J/ce)^2]^{1/2}$, may be alternatively obtained from the equilibrium entropy density of radiation at rest, with energy density e_0 and $J = 0$, i.e. $s_{eq}(e_0) = \tfrac{4}{3}a^{1/4}e_0^{3/4}$ through a Lorentz transformation corresponding to the average speed of the photons in the direction of the energy flux, which is given by

$$v = c\left(3\frac{2 - x}{2 + x}\right)^{1/2}.$$

(*Hints*: The internal energy density transforms as $e = \gamma^2(e_0 + p_0) - p_0$, and $p_0 = \tfrac{1}{3}e_0$.)
(R. Domínguez-Cascante, J. Phys. A: Gen. Math. **30** (1997) 7707).

17.4 There has been a debate about the relativistic transformation of temperature (see, for instance, P. T. Landsberg, Nature **212** (1966) 571; **214** (1967) 903; H. B. Callen and G. Horwitz, Am. J. Phys. **39** (1971) 938). The initial proposals by Einstein and Planck were that $T = T_0\gamma^{-1}$, with γ given in Problem 17.3, while in the 1960s Ott claimed that $T = T_0\gamma$. Show that the expression for the non-equilibrium temperature of radiation with energy density e and energy flux J obtained in (7.62), namely

$$T(e) = T_{eq}(e)2\frac{(x - 1)^{3/4}}{(2 + x)^{1/2}}$$

may be obtained by the mentioned Einstein–Planck criterion from the Lorentz transformation discussed in Problem 17.3.

Chapter 18

Viscous Cosmological Models

Thermodynamics plays an important role in cosmological models because the energy–momentum tensor appearing in the equations of general relativity must be specified according to the contents of the Universe. The earlier classical cosmological models do not include dissipation, as they postulate a reversible adiabatic expansion of the Universe. However, the attention paid to dissipative phenomena in cosmology and astrophysics is growing up rapidly. The interaction between matter and radiation, and other phenomena, discussed later in this chapter, depend on dissipative effects mainly due to bulk viscosity.

Bulk viscous effects play a determinant role as entropy-producing processes in the early universe; as an example, let us mention galaxy formation, wherein the typical galactic mass is the mass of the smallest density fluctuation which could grow against dissipative effects during the period just prior to hydrogen recombination; another illustration is its possible driving role in inflationary periods (i.e. periods with exponential expansion) independently of the details of the phase transitions of the grand unified theories. Furthermore, relaxational effects have to be included when the relaxation time of some components of the contents of the Universe become very long, as a consequence of cooling and expansion of the Universe.

At a first stage, these bulk viscous effects were described by the classical Newton–Stokes equation relating the bulk viscous pressure and the rate of change of volume. However, this equation is inconsistent with the postulates of relativity theory, as it yields infinite speed of propagation for viscous signals. Therefore, relaxational effects granting upper bounds to such speed must be taken into account, as postulated in extended irreversible thermodynamics (EIT). In this chapter we give an overview of this active topic of research.

18.1 Cosmological Applications: Viscous Models

Thermodynamics enters in cosmological problems through the energy–momentum tensor $T_{\mu\nu}$ which appears in the right-hand side of the Einstein equation of general relativity governing the spacetime metric tensor $g_{\mu\nu}$ [18.1]:

$$R_{\mu\nu} - \tfrac{1}{2}g_{\mu\nu}\mathcal{R} = 8\pi G T_{\mu\nu}, \tag{18.1}$$

where $R_{\mu\nu}$ is the Ricci tensor associated to $g_{\mu\nu}$, $\mathcal{R}$ the scalar curvature, and G the Newtonian gravitational constant. In some occasions, a further term proportional to $g_{\mu\nu}$ is added to the right-hand side: it is related to the so-called gravitational constant (see Problem 18.9), but we will ignore it for the sake of simplicity.

These equations are extremely intricate, but they may be simplified by assuming isotropy and homogeneity at large scales. These properties follow from cosmic observations, which have revealed the isotropy properties of our group of galaxies, and the cosmological principle, which states that, on the average, any two sufficiently large and otherwise arbitrary portions of the Universe look very similar. The hypothesis of isotropy is strongly supported by the 2.7 K cosmic microwave background radiation the spectrum of which deviates from the Planckian one by less than one in ten thousand. These hypotheses are also confirmed by the observed abundances of light elements and redshifts of distant galaxies [18.2]. Accordingly, one may adopt the Friedmann–Robertson–Walker (FRW) metric

$$ds^2 = -c^2 dt^2 + R^2(t)\left[(1-\kappa r^2)^{-1}dr^2 + r^2(d\theta^2 + \sin^2\theta d\phi^2)\right], \tag{18.2}$$

which is the most general representation of the spacetime separation between two arbitrary events in a homogeneous and isotropic space. Here r, θ, and ϕ are comoving coordinates and $R(t)$ is the cosmic scale factor, whose determination is one of the main aims of the model. The parameter κ is a constant related to the spatial curvature of the hypersurfaces of homogeneity. An appropriate scaling limits κ to only three possible values: $\kappa = 0$ (flat space sections), $\kappa = -1$ (negative spatial curvature, spatial sections are 3-dimensional hyperboloids) and $\kappa = +1$ (positive spatial curvature, spatial sections are 3-dimensional spheres).

To calculate $R(t)$ from the Einstein equations, an expression for the energy–momentum tensor is needed. Its most general form compatible with homogeneity and isotropy is

$$T_{\mu\nu} = \varepsilon u_\mu u_\nu + (p + \Pi)\Delta_{\mu\nu},$$

where p and Π are the equilibrium pressure and the bulk viscous pressure respectively, the four-velocity vector u^μ and the symmetric spatial projector $\Delta_{\mu\nu}$ have been defined in Sect. 17.1. In the reference frame attached to a comoving observer, the components of the energy–momentum tensor $T_{\mu\nu}$ may be written explicitly as

$$
T^{\mu\nu} =
\begin{pmatrix}
-\varepsilon & 0 & 0 & 0 \\
0 & p+\Pi & 0 & 0 \\
0 & 0 & p+\Pi & 0 \\
0 & 0 & 0 & p+\Pi
\end{pmatrix}.
$$

Unlike the perfect fluid hypothesis, which states that viscous pressure Π vanishes, here it is assumed that Π does not necessarily vanish. In this chapter, as in the preceding one, we use Π instead of p^v to denote the viscous pressure, because this is a more usual notation in cosmology.

Under the above assumptions, the non-trivial components of the Einstein equation (18.1) take the simple form

$$
\left(\frac{\dot R}{R}\right)^2 + \frac{\kappa}{R^2} = \frac{8\pi G}{3}\varepsilon,
\tag{18.3a}
$$

$$
\frac{\ddot R}{R} = -\frac{4\pi G}{3}\left[\varepsilon + 3(p+\Pi)\right].
\tag{18.3b}
$$

The first of these is known as Friedmann's equation. To complement the set of equations (18.3) one needs an equation of state for p and a constitutive equation for Π, and for this we select

$$
p = (\gamma_0 - 1)\varepsilon, \qquad \tau_0 \dot\Pi + \Pi = -3\zeta \frac{\dot R}{R}.
\tag{18.4}
$$

The first relation (18.4) is standard, the second one is nothing but a simplified version of the EIT equation (17.11), in which use is made of $u^\mu{}_{;\mu} = 3(\dot R/R)$ and $\nabla \cdot v = 3H$, that follows from the observational result $v = dr/dt = Hr$, with r being the position vector of the particles and H (Hubble factor) independent of position. The coefficient γ_0 i s supposed to satisfy $1 \le \gamma_0 \le 2$. The value $\gamma_0 = \frac{4}{3}$ corresponds to radiation ($p = \frac{1}{3}\varepsilon$) and $\gamma_0 = 1$ to an ideal non-relativistic gas ($k_B T \ll mc^2$); recalling that the pressure is proportional to T, one may put $p = 0$. The bulk viscosity ζ and the relaxation time τ are assumed to depend on ε through the following relations [18.4]:

$$\zeta = \alpha \varepsilon^v, \qquad \tau = \zeta / \varepsilon = \alpha \varepsilon^{v-1}. \tag{18.5}$$

With this choice, the viscous signals propagate with speed equal to c, since the speed of propagation of viscous signals is $v = [\zeta/(\varepsilon\tau)]^{1/2}$. If they propagated more slowly, τ should be higher than the value given by the second of equations in (18.5). By setting $v = 1$, (18.5) describes a radiative fluid, whereas the choice $v = \frac{3}{2}$ corresponds to a string-dominated universe [18.5].

Elimination of ε in (18.3) yields, for a flat spacetime ($\kappa = 0$),

$$\frac{\ddot{R}}{R} + \frac{3\gamma_0 - 2}{2}\left(\frac{\dot{R}}{R}\right)^2 = -4\pi G\Pi. \tag{18.6}$$

In the standard cosmological model, viscous effects are ignored ($\Pi = 0$), and one is led to $R \sim t^{1/2}$ and $R \sim t^{2/3}$ for radiation and matter-dominated universes, respectively [18.1]. The presence of bulk viscosity may have several consequences, depending on the relation between ζ and ε. If ζ is proportional to ε, the big-bang singularity with infinite spacetime curvature would not occur [18.7a]. However, more realistic models are based on $\zeta \sim \varepsilon^v$, with $0 \le v \le \frac{1}{2}$, in which case the big-bang singularity still persists [18.5].

The initial motivations behind the introduction of bulk viscous effects were, for instance [18.5–9 and references therein]: (1) to study the action of the entropy-producing processes in the early universe; (2) to analyse the initial density fluctuations, which are of pivotal importance in galaxy formation; (3) shear viscosity could provide a mechanism to eliminate initial anisotropies in the universe, and transform them in heat [18.1]; (4) bulk viscous effects could drive an inflationary period independently of the details of the phase transitions of grand unified theories [18.5].

Viscous dissipation finds its origin in several phenomena. Following a chronological order, let us first mention the appearance of four-dimensional superstrings right after the compactification of the extra spatial dimensions (around the Planck time $t_P = hG/(2\pi c^5) \approx 10^{-44}$ s) [18.6a]. Another possibility is the quantum creation of particles by the intense gravitational field [18.6b]. Likewise, mini black holes could spontaneously condense from the radiation at very high temperature [18.6]. These black holes behave as non-relativistic particles [$k_B T/(mc^2) \ll 1$], whereby the mixture might result in a radiative fluid of non-vanishing viscosity. However, a careful study [18.7] has revealed that this is not the case, for at that stage of the Universe evolution the interaction time between these mini black holes and the surrounding radiation is greater than the expansion time of the Universe. In the context of grand unified theories (GUT), bulk viscosity may arise from the production of superheavy gauge bosons and magnetic monopoles in the broken phase of the GUT phase transition, or from the existence of a primordial bosonic charge (the difference between boson and antiboson numbers) initially confined to the Bose–Einstein condensed ground state [18.5]. Neutrinos (whether massless or massive) interacting with matter represent an efficient mechanism of dissipation owing to their long mean free path [18.6]. Likewise, photons in contact with matter, e.g. electrons, constitute a radiative fluid able to generate dissipation [18.3].

In most of these models, bulk viscosity is introduced by means of classical transport equations, i.e. $\Pi = -3\zeta\dot{R}/R$, but this is inconsistent, since it leads to infinite speeds and unstable solutions. Belinskii et al. [18.4] studied several consequences of the introduction of the relaxation term in (18.4), which are important when the product of the relaxation time and the Hubble parameter is higher or comparable to unity. This may happen at several stages of the expansion of the Universe, because different components of the cosmic fluid may have different collision times. Indeed, there are several decoupling stages in which some components decouple from the others, because their collision time becomes comparable to the inverse of the expansion rate of the Universe.

We analyse here the simple model based on (18.3b) [18.5]. This equation may be written in terms of the Hubble parameter $H = \dot{R}/R$ and for a flat spacetime $\kappa = 0$ adopts the form

$$\dot{H} + \tfrac{3}{2}\gamma_0 H^2 = -4\pi G \Pi. \tag{18.7}$$

After combining this relation with (18.4) one has

$$\tau \ddot{H} + \dot{H}\left(3\tau\gamma_0 H + 1\right) + H\left(\tfrac{3}{2}\gamma_0 H - 12\pi G\zeta\right) = 0. \tag{18.8}$$

Resolution of (18.8) implies the knowledge of two initial conditions for H and $\dot{H}$. In the standard theory, only one initial condition for H is required and is usually taken to be $H(0) = \infty$. In the present case, the initial value of H can be finite, even negative. Therefore the analysis of the first moments of the Universe needs not only a unified theory of the basic interactions, but also a thermodynamic description of the collective effects, which are usually ignored in standard approaches.

Inserting into (18.8) the relationships (18.5) for ζ and τ, and using (18.3a) (with $\kappa = 0$), one obtains

$$\alpha\beta^n H^{2n}\ddot{H} + \dot{H}\left(3\alpha\beta^n\gamma_0 H^{2n+1} + 1\right) + H^2\left(\frac{3}{2}\gamma_0 - \frac{12\pi G\alpha}{\beta^v}H^{2n+1}\right) = 0, \tag{18.9}$$

with $n = v - 1$ and β a shorthand for $\tfrac{3}{8}(\pi G)^{-1}$. Upon introducing the quantities

$$h = \frac{H}{H_0}, \qquad H_0 = \left[\frac{\gamma_0}{3\alpha}\left(\frac{8\pi G}{3}\right)^n\right]^{1/(2n+1)}, \qquad \nu \neq \frac{1}{2}, \tag{18.10}$$

and rescaling the time as $t^* = H_0 t$, (18.9) may be cast into the simpler form

$$\gamma_0 h^{2n}\ddot{h} + 3\dot{h}(\gamma_0 h^{2n+1} + 1) + \tfrac{9}{2}\gamma_0 h^2(1 - h^{2n+1}) = 0. \tag{18.11}$$

From now on an upper dot will denote derivation with respect to t^*; the asterisks will be dropped whenever no confusion arises. The second-order differential equation (18.11) which governs the evolution of the reduced Hubble parameter h has two different stationary trivial solutions $h = 1$ and $h = 0$. The first implies inflationary expansion with a constant rate given by H_0.

One may study the behaviour of h near the steady solutions analytically. Setting $h = 1 + \chi$, with $|\chi| \ll 1$, one obtains after linearization

$$\gamma_0 \ddot{\chi} + 3\dot{\chi}(\gamma_0^2 + 1) - \tfrac{3}{2}\gamma_0(2n+1)\chi = 0. \tag{18.12}$$

Its solution reads as

$$\chi(t) = \chi_+ \exp(\lambda_+ t) + \chi_- \exp(\lambda_- t), \tag{18.13}$$

where χ_+ and χ_- are constants which depend on $h(0)$ and $\dot{h}(0)$, while λ_+ and λ_- are the roots

$$\lambda_\pm = \frac{3(1+\gamma_0^2)}{2\gamma_0}\left[-1 \pm \sqrt{1 + 2\gamma_0^2\frac{2n+1}{(1+\gamma_0^2)^2}}\,\right]. \tag{18.14}$$

Note that for $n + 1 < -(1 + \gamma_0^4)/4\gamma_0^2$, the quantity under the square root of the right-hand side of (18.14) is negative, so that h shows an oscillatory damped behaviour of frequency

$$\omega = \tfrac{3}{2}\gamma_0^{-1}\sqrt{(1+\gamma_0^2)^2 + 2\gamma_0^2(2n+1)} \tag{18.15}$$

around $h = 1$. This kind of behaviour has not been explored in the literature. Note also that if $\nu > \frac{1}{2}$, one of the roots of (18.14) will be positive, and therefore the state with $h = 1$ will be unstable.

Non-stationary solutions of (18.11) are obtained via numerical integration. The most significant results are shown in Fig. 18.1 and compared with the behaviour corresponding to a zero relaxation time. Some comments are in order.

1. At short times, the solutions depend strongly on the initial conditions $h(0)$ and $\dot{h}(0)$.

2. At short times, there are quantitative and qualitative differences between the solutions, leading to changes in the chronology of the successive eras of the primordial Universe. At long times, both sets of solutions attain the same stationary value.

3. For $0 < v < \frac{1}{2}$, and an initial value of $|\dot{h}|$ not very high, the solutions tend to the constant value $h = 1$, whether or not the initial value of h is greater or lower than unity.

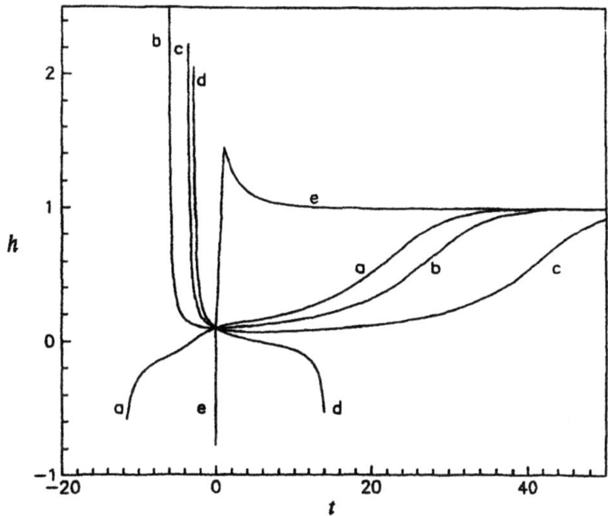

Fig. 18.1. Behaviour of the reduced Hubble factor h as a function of the reduced time t^* for a fluid with $\gamma_0 = 1.5$ and $v = 0.3$ according to (18.11). The initial value $h(0)$ is 0.1. The different curves correspond to initial conditions for $\dot{h}$: (a) $\dot{h}(0) = 0.02$; (b) $\dot{h}(0) = 0$; (c) $\dot{h}(0) = -0.02$; (d) $\dot{h}(0) = -0.04$; (e) $\dot{h}(0) = 9.0$. Curves (a) and (e) start with a contracting phase ($h < 0$) and evolve to a de Sitter expansion. Curves (b) and (c) begin with a Friedmann-like behaviour and tend to a de Sitter one for long times. Curve (d) starts with a Friedmann-like expansion and evolves towards a contracting phase. [From [18.5d].]

4. For $v > \frac{1}{2}$ and $h(0) > 1$, the solutions diverge if $|\dot{h}|$ is large enough; otherwise they tend to $h = 0$. For $h(0) < 1$, they also tend to $h = 0$.

5. If $v > 0$ and the initial value $|\dot{h}(0)|$ is large enough, the solutions of (18.11) diverge, irrespective of the value of $h(0)$.

It was also shown by Belinskii et al. [18.4] that, in the relaxed model, the effect of matter creation near the initial singularity is preserved, just as in the absence of relaxation, but that the tendency towards isotropy during the contraction disappears. The cosmological singularity still persists, but it may belong to a new type connected to the accumulation of elastic energy. Furthermore, the Friedmann solutions are shown to be unstable in the vicinity of the singularity. Fustero and Pavón [18.8] have made use of the relaxational equations to explore some properties of the entropy production during the leptonic era of the Universe. They have shown that, in that period, the relaxation term $\tau_0 \dot{\Pi}$ and the term $3\xi H$ in the second of equations (18.4) are of the same order and that the entropy production as calculated from EIT is larger, but of the same order of magnitude, than its classical counterpart.

Note that (18.4b) for Π is not the most general evolution equation obtained from EIT. Indeed, if in expression (17.6) for the entropy four-vector one assumes $q^\mu = P^{\mu\nu} = 0$, it is found that

$$S^\mu = \left(s_{eq} - \frac{\tau\Pi^2}{2\zeta T} \right) u^\mu.$$

(18.16)

Taking the full derivative of this expression, one obtains for the entropy production

$$S^\mu_{;\mu} = -\frac{\Pi}{T} \left(u^\mu_{;\mu} + \frac{\tau}{\zeta} \dot{\Pi} + \frac{1}{2} \Pi T \frac{\tau}{\zeta T} u^\mu_{;\mu} \right),$$

(18.17)

which implies, instead of (18.4b), the non-truncated equation

$$\Pi + \tau\dot{\Pi} = -3\zeta H - \frac{1}{2} \Pi\tau \left(3H + \frac{\dot{\tau}}{\tau} - \frac{\dot{\zeta}}{\zeta} - \frac{\dot{T}}{T} \right),$$

(18.18)

where we have used the result $u^\mu_{;\mu} = 3H$ from the FRW model. The influence of the non-linear terms has been studied by several authors [18.9–10]. The behaviour of the cosmological models depends both on the equations of state for p, T, τ, and ζ and on the transport equation being used. For instance, Hiscock and Salmonson, starting from (18.18) together with the equations of state for the Boltzmann gas, found no inflationary phase. With the simpler equations of state $\Pi = \lambda\rho$, $\tau = \alpha\rho^m$, $\zeta = \tau p$ other authors have found that an inflationary phase remains possible, but with a different expansion rate than in the truncated version (18.4). Both the values of H_0 and the behaviour of the temperature during this inflationary period have been the subject of recent analyses [18.9–10].

To summarize, when viscous effects are introduced in cosmological models, it is imperative to include relaxational equations compatible with causality. This is a new topic of research in cosmology, whose importance has only been recognized recently.

18.2 Particle Production and Effective Bulk Viscosity

Another topic of interest related to bulk viscosity concerns particle production, which in some situations may be described by means of an apparent or effective bulk viscosity [18.11–12]. Such a decay or production of particles finds its origin in several effects, as for instance the decay of scalar particles, production of relativistic particles in the reheating phase of inflationary eras, electron–positron annihilation after neutrino decoupling, decay of heavy bosons to quarks and leptons, or creation of particles by the gravitational field. Prigogine et al. [18.12] were able to frame these effects in a thermodynamic description by formulating a balance equation for the number density of created particles in addition to the Einstein field equations. Another way out is to model the loss and source terms by an effective viscous pressure [18.13].

To be explicit, assume that the number of particles in the Universe is not constant, but that there is a particle production with a rate per unit volume $\dot{\xi}$. The evolution equation for the particle number density n is then

$$\dot{n} = -3n\frac{\dot{R}}{R} + \dot{\xi}. \tag{18.19}$$

This is a particular case of the general mass balance equation

$$\dot{n} = -n\nabla \cdot v + \dot{\xi}, \tag{18.20}$$

where $n\nabla \cdot v$ describes the change of density due to the variation in volume and v expresses the particle production rate per unit volume. In the situation of expanding universe, one has $v = HR$ and therefore $\nabla \cdot v = 3H = 3(\dot{R}/R)$. Introduction of this result into (18.20) yields (18.19).

Let us now assume that each created particle carries an entropy s. To obtain the entropy production, we start from the classical Gibbs equation

$$Tds = de + pdv = de + pd(1/n) \tag{18.21}$$

where s is the entropy per particle and e the energy per particle (i.e. $e = \varepsilon/n$, since ε is the energy density per unit volume). Taking the time derivative of (18.21) yields

$$nT\dot{s} = (\varepsilon/n)^{\cdot} + p(1/n)^{\cdot} = (1/n)\dot{\varepsilon} - (\varepsilon + p)(\dot{n}/n^2). \tag{18.22}$$

Since the non-trivial components of Einstein's equations (18.3a–b) are

$$3H^2 = 8\pi G\varepsilon, \tag{18.23a}$$

$$2\dot{H} = -8\pi G(\varepsilon + p - 3\zeta H), \tag{18.23b}$$

the energy balance equation will be written as

$$\dot{\varepsilon} = -3H(\varepsilon + p)\left(1 - \frac{3\zeta}{\varepsilon + p}H\right). \tag{18.24}$$

Introducing into (18.22) the balance equations (18.19) and (18.24) leads to

$$\dot{s} = -\frac{\varepsilon + p}{n^2 T}\left(-\frac{9\zeta}{\varepsilon + p}H^2 + \frac{\dot{\xi}}{n}\right). \tag{18.25}$$

Two particular cases are worth to be discussed: (a) non-vanishing bulk viscosity ($\zeta \neq 0$) but vanishing particle production ($\dot{\xi} = 0$); in this example one has simply

$$\dot{s}_1 = \frac{9\zeta}{n^2 T}H^2; \tag{18.26}$$

(b) vanishing bulk viscosity ($\zeta = 0$) but non-vanishing particle production ($\dot{\xi} \neq 0$), then

$$\dot{s}_2 = -\frac{\varepsilon + p}{n^2 T}\frac{\dot{\xi}}{n}. \tag{18.27}$$

Comparison of (18.26) and (18.27) shows that one can modelize the entropy production due to the particle production in terms of an effective bulk viscosity ζ_{eff} by writing

$$\dot{s}_2 = -\frac{\varepsilon + p}{n^2 T}\frac{\dot{\xi}}{n} \equiv \frac{9\zeta_{eff}}{n^2 T}H^2. \tag{18.28}$$

This yields for the effective bulk viscosity the result

$$\zeta_{eff} = \frac{\varepsilon + p}{9}\frac{\dot{\xi}}{n}\left(\frac{\dot{R}}{R}\right)^{-2}. \tag{18.29}$$

Spacetime geometries more complicated than the FRW model have been studied, for instance, Bianchi I and II universes include bulk and shear viscous stresses, both in the truncated and the non-truncated models, and other descriptions include peculiar velocities of the galaxies with respect to the global cosmic motion. Several of these macroscopic results have been confirmed by the kinetic theory of gases in an expanding universe [18.14].

18.3 Extended Thermodynamics and Cosmological Horizons

The qualitative differences between Eckart's version of relativistic thermodynamics and extended relativistic thermodynamics also appear in the context of the de Sitter model of the Universe. The latter possesses an event horizon endowed with an entropy and a temperature of quantum origin [18.15] given by

$$S_h = \frac{8\pi^2}{H^2}, \qquad T_h = \frac{H}{2\pi}, \tag{18.30}$$

where the subscript h refers to horizon. (In this section we use units in which $8\pi G = h/(2\pi) = c = k_B = 1$.) The horizon is a closed spherical surface purely geometrical in nature such that events at one side of it cannot enter into causal contact with events at the other side. This follows from the spacetime metric of this model, which can be found in the specialized literature [18.2].

It is currently thought that, at some very early epoch, our Universe undergoes a de Sitter phase of exponential inflation (i.e. $R \sim \exp{(Ht)}$, with H being a positive constant) driven by the energy density ε_v and pressure p_v of the quantum vacuum, which dominates at that time all other forms of energy (subscript v stands for vacuum). Roughly speaking, a de Sitter universe should contain no other forms of energy than that of the vacuum. However, in the case of our Universe one can take for certain that other forms of energy may contaminate the mentioned vacuum. As a consequence, the Hubble parameter deviates slightly from its otherwise steady value.

In virtue of (18.30), the horizon entropy change is

$$\dot{S}_h = -16\pi^2 \frac{\dot{H}}{H^3}, \tag{18.31}$$

and will increase or decrease depending on the sign of $\dot{H}$. By using Eckart's theory, Davies [18.16] has shown that, if the fluid (let us say matter plus radiation) perturbing the vacuum obeys a reasonable equation of state of the form of the first of equations (18.4), then the time derivative of the total entropy of the fluid plus the horizon $\dot{S}_h$, will be positive or zero so long as $\gamma_0 > 0$, a relationship not violated by any known kind of fluid. Note that the first of equations (18.4) implies that the fluid fulfils the dominant energy condition [18.1]

$$\varepsilon_f + p_f \geq 0, \tag{18.32}$$

with subscript f referring to the fluid. Moreover, if the fluid has a bulk viscosity of the form $\zeta = \alpha \varepsilon_f$ with α being a positive constant, it was shown [18.16] that the entropy of the fluid plus the horizon satisfies

$$\dot{S} = \dot{S}_f + \dot{S}_h = \frac{8\pi^2}{H^3}\gamma_0 \varepsilon_f, \tag{18.33}$$

which is positive, provided that $\gamma_0 \geq 0$. It is worth noting that (18.33) is met whether the dominant energy condition, which for a viscous fluid in standard thermodynamics reads

$$(\gamma_0 - 3\alpha H)\varepsilon_f \geq 0, \tag{18.34}$$

is satisfied or not.

The rate of variation of total entropy S by using relativistic EIT instead of the traditional one was calculated by Pavón [18.17]. The relevant relations are the Friedmann equation, the first of equations (18.3) with $\kappa = 0$ and the energy conservation equation

$$\dot{\varepsilon} + 3(\rho + p) = 0, \tag{18.35}$$

with $\varepsilon = \varepsilon_v + \varepsilon_f$ and $p = p_v + p_f$ where units have been chosen so that $8\pi G = 1$. These have to be supplemented by appropriate equations of state for the vacuum $p_v = -\varepsilon_v$ and for the material fluid $p_f = (\gamma_0 - 1)\varepsilon_v + \Pi_f$, where the bulk viscous pressure obeys a relaxational equation, the second of equations (18.4), with $\zeta = \alpha \varepsilon_f$. Combining these equations results in

$$\dot{\varepsilon} = -3\left[(\gamma_0 - 3\alpha H)\varepsilon_f - \tau \dot{\Pi}_f\right]H. \tag{18.36}$$

Substituting in the rate of entropy variation associated with the evolution of the horizon,

$$\dot{S}_h = -\tfrac{8}{3}\pi^2 \dot{\varepsilon} H^4, \tag{18.37}$$

leads to

$$\dot{S}_h = \frac{8\pi^2}{H^3}\left[(\gamma_0 - 3\alpha H)\varepsilon_f - \tau \dot{\Pi}_f\right]. \tag{18.38}$$

On the other hand, the rate of entropy variation associated to the material fluid

$$\dot{S}_f = \frac{R^3}{T_f \zeta}\Pi_f^2,$$ (18.39)

becomes

$$\dot{S}_f = \frac{R^3}{\alpha \varepsilon_f T_f}\left(3\alpha \rho_f H + \tau \dot{\Pi}_f\right)^2.$$ (18.40)

The term $\tau \dot{\Pi}_f$ in (18.40) may bear either sign. For decreasing bulk viscous pressure rate, $\dot{\Pi}_f < 0$, since causality demands $\tau > 0$, hence $\dot{S}_h > 0$, and the rate of variation of total entropy is positive. However, if $\dot{\Pi}_f$ is increasing, $\dot{S}_f$ could be negative, which implies that $\dot{S}_f + \dot{S}_h$ could become negative, in clear contradiction with the second law of thermodynamics.

The total rate of entropy production, with $T_f = T_h$, is given by

$$\dot{S} = \frac{8\pi^2}{H^3}\left(\gamma_0 \varepsilon_f + \frac{\tau^2 \dot{H}_f^2}{3\alpha \varepsilon_f H} + \tau \dot{\Pi}_f\right).$$ (18.41)

This is a positive quantity unless the last term on the right-hand side becomes dominantly negative. Although ordinary fluids are not expected to present very large values for $\tau \dot{\Pi}_f$, this cannot be ruled out in principle, especially at the very early stages of cosmic expansion where exotic quantum fields may play a crucial role.

Nevertheless, if the dominant energy condition holds, which in the formalism of EIT reads as

$$\gamma_0 \varepsilon_f \geq \tau \dot{\Pi}_f + 3\alpha \varepsilon_f H,$$ (18.42)

then combination of (18.41) and (18.42) implies that

$$\dot{S} \geq \frac{8\pi^2}{3\alpha \varepsilon_f H^4}\left(\tau \dot{\Pi}_f + 3\alpha \varepsilon_f H\right)^2 \geq 0.$$ (18.43)

In other terms, it can be stated that the second law ($\dot{S} \geq 0$) is guaranteed if and only if the dominant energy condition (18.42) is satisfied. This contrasts with the result of Davies, according to which the second law holds irrespectively of the dominant energy condition (18.34).

18.4 Astrophysical Problems: Gravitational Collapse

The relaxational terms in the transport laws have been exploited to interpret fast explosions or implosions in astrophysics [18.18]. In rapid processes, as in the fast collapse phase preceding neutron star formation, the relaxational effects are important and the fluid is far from hydrostatic equilibrium. Such effects cannot be accounted for in classical models wherein the evolution of the star is regarded as a sequence of hydrostatic states. In contrast with cosmological problems, two different metrics must be introduced, one for the interior and the other for the exterior of the star. Furthermore, the inhomogeneity generates a heat flux, which plays a decisive role. Among other results, it turns out that the behaviour of the system depends strongly on the observers: for an observer at rest at infinity, the total mass loss of the star is the same but with greater speed (i.e. higher luminosity) than in the non-viscous case, whereas the comoving observer perceives a larger mass loss [18.18]. Another result is that the energy of the neutrinos at the surface of the star is not correlated with that at the interior. In [18.18] it was shown that when the heat flux relaxes, the evolution of the star depends critically on the thermal relaxation time, thermal conductivity, proper energy density and pressure.

Starting from a Maxwell–Cattaneo equation, Herrera and Falcón [18.18c] have studied the secular stability of nuclear burning in accreting neutron star envelopes, which is at the basis of X-ray bursters models, for times shorter than the effective relaxation time of the heat flux. The characteristic time scales for the growth of the nuclear burning instability spans a wide period, from milliseconds to minutes. The relaxational effects of the heat flux (its relaxation time is of the order of milliseconds for a highly degenerate matter) produce a fast oscillation in the luminosity as a precursor of the X-ray burst.

Problems

18.1 Compare the entropy in the 2.7 K microwave background radiation with the entropy in the baryons to conclude that the ratio between the number of photons and baryons in the Universe is about 10^8. Assume that the expansion of the Universe is adiabatic and that the present number density of baryons is roughly 1 m^{-3} and that the entropy per baryon is the Boltzmann constant.

18.2 Using the data of Problems 6.4 and 6.5, find the bulk viscosity of a gas composed of photons, electrons, and protons at 3000 K, with $n = 4000$ electrons/m^3. Neglect the collisions between photons and protons, and between the particles themselves.

18.3 (a) Starting from (18.7) and assuming that $\Pi = -3\zeta H$, show that a constant bulk viscosity leads to inflationary behaviour of the universe (i.e. behaviour with $H = $ constant and therefore an exponential increase of R with time). (b) Study the stability of the solutions in terms of the dimensionless quantity $c^2 H/(G\zeta)$. (c) Compare the results with those obtained under the assumption that the bulk viscosity is of the form $\zeta = \alpha \varepsilon$ with α a constant. [See G. L. Murphy, Phys. Rev. D **8** (1973) 4231.]

18.4 Show that, for a viscous fluid with $p = \lambda \varepsilon$ and with a bulk viscosity $\zeta = a\varepsilon^{1/2}$, the Einstein equations yield a behaviour for the scale factor of the Universe of the form $R \sim t^a$, with $a^{-1} = \frac{3}{2}[1 + \lambda - 2a(6\pi G)^{1/2}]$.

18.5 The mixture of neutrino and electrons is especially relevant at the so-called leptonic era, which lasted between 10^{-3} s and 10 s, with temperatures ranging from 10^{12} K to 10^{10} K. The cross-section of the neutrino–electron collisions due to weak interactions is

$$\sigma_{wk} = (2\pi G_F \, k_B T/h^2 c^2)^2,$$

with the Fermi constant $G_F = 1.4 \times 10^{-62}$ kg m^{5}s^{-2} (a) Evaluate the neutrino mean free path at the initial and final stages of the leptonic era, for an electron density given by $n = (2\pi k_B T/hc)^3$. (b) Determine the neutrino collision time τ and τH, i.e. the ratio of τ to the characteristic expansion time of the universe H^{-1}. (c) The neutrinos are said to decouple from the electrons when $\tau H = 1$; find the decoupling temperature of neutrinos (remember that in a radiation-dominated era, $R \propto t^{1/2}$ and $T \propto t^{-1/2}$).

18.6 The bulk viscosity of a mixture of neutrinos and nuclei in a collapsing stellar core has been estimated to be of the order of 10^{23} kg m^{-1} s^{-1}. The compression rate $\nabla \cdot v$ is of the order of 10^3 s^{-1}. Evaluate the entropy production per unit time per baryon for a temperature of 1 MeV and a density of 10^{15} kg m^{-3}. If the time interval during which the bulk viscosity is effective is 1 s, what is the total entropy production per particle?

18.7 The classical entropy production has the form

$$\sigma_{C\Pi} = -T^{-1}\Pi \nabla \cdot v,$$

whereas the entropy production found in EIT is given by

$$\sigma_{EIT} = -T^{-1}\Pi [\nabla \cdot v - (\tau/\zeta)\dot\Pi].$$

This expression suggests writing $\tau \dot\Pi + \Pi = -3\zeta H$. (a) Show that for constant ζ and up to the first order in τ one may write $\Pi = -3\zeta(H - \tau\dot H)$ for $\nabla . v = 3H$. (b) Evaluate the ratio $\sigma_{E\Pi}/\sigma_{C\Pi}$ at the beginning of the leptonic era and at the neutrino decoupling time, by using the data of Problem 18.5 (recall that in a radiation-dominated era $R \propto t^{1/2}$).

18.8 Show that if one uses the non-truncated transport equation (18.18) instead of the truncated equation (18.4b), one obtains (for $0 < v < \frac{1}{2}$) a de Sitter expansion with constant reduced Hubble factor $h = H/H_0 = [2/(2-\gamma_0)]^{1/(2v-1)}$ instead of $h = 1$, with H_0 and v defined in (18.10) and (18.5) respectively (see V. Romano and D. Pavón, Phys. Rev. D **50** (1994) 2572).

18.9 Einstein introduced a 'cosmological constant' Λ by writing instead of (18.1)

$$R_{\mu v} - \tfrac{1}{2} g_{\mu v} \mathcal{R} = 8\pi G T_{\mu v} - \Lambda g_{\mu v}.$$

A possible solution to the puzzle of the cosmological constant (i.e. to understand its low present value in contrast with its possibly high values in the past) has been to introduce a variable cosmological 'constant' in the Einstein equations. (a) Show that in the presence of a time-dependent cosmological constant, the entropy production reads as

$$S^\alpha_{;\alpha} = \frac{9\zeta H^2}{T} - \frac{c^5}{8\pi G} \frac{\dot{\Lambda}}{T}.$$

(V. Méndez and D. Pavón, Gen. Rel. Grav. **28** (1996) 679). (b) Show that if Λ depends on t as $\Lambda \sim t^{-2}$, its effects are similar to those of a bulk viscosity in a fluid with null cosmological constant in the radiation era (A. Beesham, Phys. Rev. D **48** (1993) 3539).

18.10 Assume that instead of the classical equation $\Pi = -3\zeta H$ one has a non-linear relation yielding a saturation behaviour as the flux limiters used in Sect. 10.6, i.e.

$$\Pi = -\frac{3\zeta H}{\sqrt{1+(aH)^2}}.$$

Find which requirements should obey the parameter a in order that the fluid fulfils the dominant energy condition (18.21) at high values of H. (See R. Maartens and V. Méndez, Phys. Rev. D **55** (1997) 1937, for the use of a slightly different model of flux limiter).

References

Complete bibliographical details of most of these references, as well as an updated bibliography on EIT and related topics may be found at the website www.uab.es/dep-fisica/eit.

Chapter 1

1.1   L. Onsager, Phys. Rev. **37** (1931) 405 and **38** (1931) 2265.

1.2   C. Eckart, Phys. Rev. **58** (1940) 267, 269, 919.

1.3   J. Meixner and H. Reik, Thermodynamik der Irreversiblen Prozesse (*Handbuch der Physik III/2*), (S. Flugge, ed.), Springer, Berlin, 1959.

1.4   I. Prigogine, *Introduction to Thermodynamics of Irreversible Processes*, Interscience, New York, 1961.

1.5   S.R. de Groot and P. Mazur, *Non-equilibrium Thermodynamics*, North-Holland, Amsterdam, 1962.

1.6   R. G. Denbigh, *The Thermodynamics of the Steady State*, Wiley, New York, 1950.

1.7   D.D. Fitts, *Non-equilibrium Thermodynamics*, McGraw-Hill, New York, 1962.

1.8   R. Haase, *Thermodynamics of Irreversible Processes*, Addison-Wesley, Reading, 1969.

1.9   I. Gyarmati, *Non-equilibrium Thermodynamics*, Springer, Berlin, 1970.

1.10   L.C. Woods, *The Thermodynamics of Fluid Systems*, Clarendon Press, Oxford, 1975.

1.11   B.H. Lavenda, *Thermodynamics of Irreversible Processes*, Macmillan, London, 1978.

1.12   H.J. Kreuzer, *Nonequilibrium Thermodynamics and Its Statistical Foundations*, Clarendon, Oxford, 1981.

1.13   B.D. Coleman, Arch. Rat. Mech. Anal. **17** (1964) 1.

1.14   C. Truesdell, *Rational Thermodynamics*, McGraw-Hill, New York, 1969; 2nd en larged edition, Springer, New York, 1984.

1.15   W. Noll, *The Foundations of Mechanics and Thermodynamics*, Springer, Berlin, 1974.

1.16   C. Truesdell and R.A. Toupin, The Classical Field Theories (*Handbuch der Physik, III*) (S. Flugge, ed.), Springer, Berlin, 1960.

1.17   A. Bedford, Acta Mechanica **30** (1978) 275.

1.18   J. Lambermont and G. Lebon, Int. J. Non-Linear Mech. **9** (1973) 55.

1.19   I. Prigogine, Physica **15** (1949) 272.

1.20   J. Meixner, Ann. Phys. (Leipzig) **43** (1943) 244.

1.21   C. Truesdell and W. Noll, The Non-Linear Field Theories (*Handbuch der Physik, III/3*) (S. Flugge, ed.), Springer, Berlin, 1965.

1.22   H.B.G. Casimir, Rev. Mod. Phys. **17** (1945) 843.

1.23   D.G. Miller in *Foundations of Continuum Thermodynamics* (J.J. Delgado-Domingos, M.N.R. Nina and J.H. Whitelaw, eds.), Macmillan, London, 1974.

1.24   O. Reynolds, Phil. Mag. **20** (1885) 469.

1.25   L.C. Woods, Inst. of Math. and Appl. **17** (1981) 98.

1.26   I.S. Liu, Arch. Rat. Mech. Anal. **46** (1972) 131.

1.27   B.D. Coleman and D. Owen, Arch. Rat. Mech. Anal. **54** (1974) 1; M. Silhavy, *The Mechanics and Thermodynamics of Continuous Media*, Springer, Berlin, 1997

1.28   W. Day, Acta Mechanica **27** (1977) 251.

1.29   A.E. Green and P. Naghdi, Proc. Roy. Soc. London A **357** (1977) 253.

1.30   I. Müller, Arch. Rat. Mech. Anal. **45** (1972) 241.

1.31   D.G.B. Edelen and J. McLennan, Int. J. Engn. Sci. **11** (1973) 813.

1.32   G. Lebon and M.S. Boukary, Int. J. Engn. Sci. **26** (1988) 471.

1.33   J. Lumley, J. Appl. Mech. **50** (1983) 1097.

1.34   R.B. Bird and P.G. de Gennes, Physica A **118** (1983) 43.

1.35   W.G. Hoover, B. Moran, R.H. More and A.J. Ladd, Phys. Rev. A **24** (1981) 2109.

1.36   G. Ryskin, Phys. Rev. A **32** (1985) 1239.

1.37   G. Ryskin, Phys. Rev. Lett. **61** (1988) 1442.

1.38   L.S. Soderholm, Int. J. Engn. Sci. **14** (1976) 523.

1.39   A.I. Murdoch, Arch. Rat. Mech. Anal. **83** (1983) 183.

Chapter 2

2.1   J.C. Maxwell, Philos. Trans. Roy. Soc. London **157** (1867) 49.

2.2   C. Cattaneo, Atti Seminario Univ. Modena **3** (1948) 33.

2.3   S. Machlup and L. Onsager, Phys. Rev. **91** (1953) 1512.

2.4   R.E. Nettleton, Phys. Fluids **2** (1959) 256; Phys. Fluids **3** (1960) 216.

2.5   I. Müller, Z. Phys. **198** (1967) 329; I.S. Liu and I. Müller, Arch. Rat. Mech. Anal. **83**

(1983) 285; I. Müller, *Thermodynamics*, Pitman, London, 1985; I. Müller and T. Ruggeri, *Rational Extended Thermodynamics* (2nd ed.), Springer, New York, 1998.

2.6   J. Lambermont and G. Lebon, Phys. Lett. A **42** (1973) 499.

2.7   M. Kranys, Nuovo Cimento B **50** (1967) 48; I. Müller, Arch. Rat. Mech. Anal. **34** (1969) 259; W. Israel, Ann. Phys. (NY) **100** (1976) 310.

2.8   I. Gyarmati, J. Non-Equilib. Thermodyn. **2** (1977) 233.

2.9   D. Jou, J. Casas-Vázquez and G. Lebon, J. Non-Equilib. Thermodyn. **4** (1979) 349; G. Lebon, D. Jou and J. Casas-Vázquez, J. Phys. A **13** (1980) 275; J. Casas-Vázquez, D. Jou and G. Lebon (eds.), *Recent Developments in Nonequilibrium Thermodynamics* (Lecture Notes in Physics, 199), Springer, Berlin, 1984.

2.10   B.C. Eu, J. Chem. Phys. **71** (1979) 4832; J. Chem. Phys. **73** (1980) 2958; *Kinetic Theory and Irreversible Thermodynamics*, Wiley, New York, 1992; *Nonequilibrium Statistical Mechanics: Ensemble Method*, Kluwer, Dordrecht, 1998; S. Sieniutycz, Int. J. Heat Mass Transfer **24** (1981) 1759, **29** (1986) 651; S. Sieniutycz and R.S. Berry, Phys. Rev. A **43** (1991) 2807.

2.11   L.S. García-Colín, M. López de Haro, R.F. Rodríguez, J. Casas-Vázquez and D. Jou, J. Stat. Phys. **37** (1984) 465; L.S. García-Colín, Rev. Mex. Física **34** (1988) 344; L.S. García-Colín and F. Uribe, J. Non-Equilib.Thermodyn. **16** (1991) 89.

2.12   D. Jou, J. Casas-Vázquez and G. Lebon, Rep. Prog. Phys. **51** (1988) 1105; **62** (1999) 1035; J. Non-Equilib. Thermodyn. **17** (1992) 383; **21** (1995) 103; **23** (1998) 277; S. Sieniutycz and P. Salamon (eds.), *Extended Thermodynamic Systems*, Taylor and Francis, New York, 1992.

2.13   G. Lebon, D. Jou, J. Casas-Vázquez and W. Muschik, J. Non-Equilib. Thermodyn. **23** (1998) 176.

2.14   M. Grmela, Physica D **21** (1986) 179; J. Phys. A **20** (1989) 4375; M. Grmela and G. Lebon, J. Phys. A **23** (1990) 3341; M. Grmela and D. Jou, J. Phys. A **24** (1991) 741; S. Sieniutycz and R. S. Berry, Phys. Rev. A **40** (1989) 348.

2.15   D. Jou, J. Casas-Vázquez and G. Lebon, Int. J. Thermophys. **14** (1993) 671.

2.16   J. Kestin, Int. J. Solids Structures **29** (1992) 1826; G.A. Maugin, *The Thermo-mechanics of Nonlinear Irreversible Behaviours*, World Scientific, Singapore, 1999; G.A. Maugin and W. Muschik, J. Non-Equilib. Thermodyn. **19** (1994) 217, 250.

2.17   M. Criado-Sancho and J.E. Llebot, Phys. Rev. E **47** (1993) 4104.

Chapter 3

3.1   B.C. Eu, *Kinetic Theory and Irreversible Thermodynamics*, Wiley, New York; 1992: Phys. Rev. E **51** (1995) 768; J. Chem. Phys. **102** (1995) 7169.

3.2   M. Grmela, Phys. Rev. E **48** (1993) 919; M. Chen, J. Math. Phys. **38** (1997) 5153.

3.3   D. Jou and J. Casas-Vázquez, Phys. Rev. A **45** (1992) 8371; J. Casas-Vázquez and D. Jou, Phys. Rev. E **49** (1994) 1040.

3.4   J. Fort, Physica A **243** (1997) 275; J. Fort, D. Jou and J.E. Llebot, Physica A **248** (1998) 97.

3.5   K. Henjes, Phys. Rev E **48** (1992) 3199; W.G. Hoover, B.L. Holian and H.A. Posch, Phys. Rev E **48** (1992) 3196; R.E. Nettleton, Can. J. Phys. **72** (1994); 106; B.C. Eu and L.S. García-Colín, Phys. Rev. E **54** (1996) 2510.

3.6   Z. Banach, Arch. Mech. **48** (1996) 791; L.S. García-Colín and V. Micenmacher, Molec. Phys. **86** (1995) 697.

3.7   J.J. Brey and A. Santos, Phys. Rev. A **45** (1992) 8566.

3.8   J. Meixner in *Foundations of Continuum Thermodynamics* (J.J. Delgado-Domingos, M.N.R. Nina, and J.H. Whitelaw, eds.), Wiley, New York, 1973.

3.9   I. Müller, Arch. Rat. Mech. Anal. **41** (1971) 319.

3.10   W. Muschik, Arch. Rat. Mech. Anal. **66** (1977) 379.

3.11   J. Keizer, J. Chem. Phys. **82** (1985) 2751.

3.12   R. Luzzi, A.R. Vasconcellos, J. Casas-Vázquez and D. Jou, Physica A **234** (1997) 699.

3.13   R. Domínguez and D. Jou, Phys. Rev. E **51** (1995) 158; J. Non-Equilib. Thermodyn. **20** (1995) 263; R.E. Nettleton, Phys. Rev. E **53** (1996) 1241; R. Domínguez-Cascante and J. Faraudo, Phys. Rev. E **54** (1996) 1933.

3.14   J. Camacho and D. Jou, Phys. Rev. E **52** (1995) 3490.

3.15   K.O. Friedrichs and P.O. Lax, Proc. Natl. Acad. Sci. (USA) **68** (1971) 1686.

3.16   T. Ruggeri, Acta Mechanica **47** (1983) 163; G. Boillat and A. Strumia, J. Math. Phys. **22** (1981) 1824; K. Wilmanski, *Thermomechanics of Continua*, Springer, Berlin, 1998.

3.17   M. Criado-Sancho and J.E. Llebot, Phys. Lett A **177** (1993) 323.

Chapter 4

4.1   M. Grmela, Physica D **21** (1986) 179.

4.2   M. Grmela, J. Phys. A: Math. Gen. **20** (1989) 4375; Phys. Rev. E **47** (1993) 351; **48** (1993) 919.

4.3   B.J. Edwards, J. Non-Equilib. Thermodyn. **23** (1998) 301; B.J. Edwards, A.N. Beris and H.C. Öttinger, J. Non-Equilib. Thermodyn. **23** (1998) 334.

4.4   M. Grmela and G. Lebon, J. Phys. A: Math. Gen. **23** (1990) 3341; M. Grmela and D. Jou, J. Phys. A: Math. Gen. **24** (1991) 741; M. Grmela, A. Elafif and G. Lebon, J. Non-Equilib. Thermodyn. **23** (1998) 376.

4.5   A.N. Beris and B.J. Edwards, *Thermodynamics of Flowing Fluids with Internal Microstructure*, Oxford University Press, New York, 1994.

4.6   S. Sieniutycz, *Conservation Laws in Variational Thermohydrodynamics*, Kluwer, Dordrecht, 1994; F. Markus and K. Gambar, Phys. Rev. E **50** (1994) 1227; J. Non-Equilib. Thermodyn. **16** (1991) 27; F. Vázquez and J.A. del Río, Phys. Rev. E **47** (1993) 178.

4.7   M. Grmela and H.C. Öttinger, Phys. Rev. E **56** (1997) 6620; H.C. Öttinger and M. Grmela, Phys. Rev. E **56** (1997) 6633.

4.8   V.I. Arnold, *Mathematical Methods of Classical Mechanics*, 2nd ed., Springer, New York, 1989; J.E. Marsden, *Lectures on Mechanics*, Cambridge University Press, Cambridge, 1992.

4.9   D. ter Haar, (ed.), *Collected papers of L. D. Landau*, Pergamon, Oxford, 1965.

4.10   M. Grmela, D. Jou and J. Casas-Vázquez, J. Chem. Phys. **108** (1998) 7937; J. Non-Equilib. Thermodyn. **23** (1998) 203.

4.11   M. Grmela and J. Teichmann, Int. J. Eng. Sci. **21** (1983) 197; M. Grmela and G. Lebon, Physica A **248** (1998) 428.

4.12   M. Grmela, Phys. Rev. E **47** (1993) 351.

Chapter 5

5.1   S. Chapman and T.G. Cowling, *The Mathematical Theory of Non-uniform Gases*, Cambridge University Press, Cambridge, 1970.

5.2   H. Grad, Principles of the Kinetic Theory of Gases, (*Handbuch der Physik XII*), (S. Flugge, ed.), Springer, Berlin, 1958.

5.3   S. Harris, *An Introduction to the Theory of the Boltzmann Equation*, Holt, Rinehart and Winston, New York, 1971; L.C. Woods, *Kinetic Theory of Gases and Magneto-plasmas*, Oxford University Press, 1994.

5.4   E. Ikenberry and C. Truesdell, J. Rat. Mech. Anal. **5** (1956) 1.

5.5   B.C. Eu, J. Chem. Phys. **73** (1980) 2958; J. Chem. Phys. **74** (1981) 6362; J. Chem. Phys. **102** (1995) 7169; *Kinetic Theory and Irreversible Thermodynamics*, Wiley, New York, 1992; Phys. Rev. E **51** (1995) 768.

5.6   G. Lebon, Bull. Soc. Roy. Belgique, Clas. Sci. **64** (1978) 456.

5.7   L.S. García-Colín and M. López de Haro, J. Non-Equilib. Thermodyn. **7** (1982) 95; R.F. Rodríguez, M. López de Haro and L.F. del Castillo, J. Non-Equilib. Thermodyn. **10** (1985) 1.

5.8   Z. Banach, J. Stat. Phys. **48** (1987) 813; Physica A **159** (1989) 343.

5.9   T.F. Nonnenmacher, J. Non-Equilib. Thermodyn. **5** (1980) 361.

5.10   J.W. Dufty and J.A. McLennan, Phys. Rev. A **9** (1974) 1256; M.H. Ernst, E.H. Hauge, and J.M.J. van Leeuwen, J. Stat. Phys. **15** (1976) 23; L. Reichl, *A Modern Course in Statistical Mechanics*, Texas University Press, Austin, 1982.

5.11   H.S. Green, *The Molecular Theory of Fluids*, Dover, New York, 1969.

5.12   S.A. Rice and P. Gray, *The Statistical Mechanics of Simple Liquids*, Wiley-Interscience, New York, 1965.

5.13   S. Hess, Phys. Rev. A **22** (1980) 2844.

5.14   D. Jou and V. Micenmacher, J. Phys. A. **20** (1987) 6519.

5.15   R.F. Rodríguez, L.S. García-Colín, and L.F. del Castillo, J. Chem. Phys. **86** (1987) 4208.

5.16   J. Camacho and D. Jou, J. Chem. Phys. **92** (1990) 1339.

5.17   B.C. Eu, J. Chem. Phys. **74** (1981) 3006, 6362.

5.18   P.J. Clause and R. Balescu, Plasma Physics **24** (1982) 1429; B.C. Eu, Phys. Fluids **28** (1985) 222.

5.19   D.K. Bhattacharya and B.C. Eu, Phys. Rev. A **35** (1987) 821; R.E. Khayat and B.C. Eu, Phys. Rev. A **38** (1988) 2492.

5.20   A Santos and J.J. Brey, Physica A **174** (1991) 355.

5.21   M. Grmela, J. Phys. A **22** (1989) 4375; R. Salmon, Ann. Rev. Fluid Mech. **20** (1988) 225; M. Grmela and D. Jou, J. Phys. A **24** (1991) 741.

5.22   R.M. Velasco and L.S. García-Colín, Phys. Rev. A **44** (1991) 4961, J. Non-Equilib. Thermodyn. **18** (1993) 157.

5.23   C. Pérez-García and D. Jou, J. Phys. A **19** (1986) 2881; D. Jou and D. Pavón, Phys. Rev. A **44** (1991) 6496.

5.24   S. Hess, Z. Naturforsch. **32a** (1977) 678.

5.25   H. Mori, Prog. Theor. Phys. **33** (1965) 432; **34** (1965) 399; K. Nagano, T. Karasudani, and H. Mori, Prog. Theor. Phys. **63** (1980) 1904.

5.26   P. Giannozzi, G. Grosso, S. Moroni and G. Pastori-Parravicini, Appl. Num. Math. **4** (1988) 273.

5.27   V.I. Roldughin, J. Non-Equilib. Thermodyn. **16** (1991) 13, 285; L. Waldmann and H. Vestner, Physica A **80** (1975) 528; H. Vestner, M. Kubel, and L. Waldmann, Nuovo Cimento B **25** (1975) 405.

5.28   H. Struchtrup and W. Weiss, Phys. Rev. Lett. **80** (1998) 5048; H. Struchtrup, Ann. Phys. **266** (1998) 1.

Chapter 6

6.1   H. B. Callen, *Thermodynamics*, Wiley, New York, 1960.

6.2   A. Einstein, Ann. Phys. (Leipzig) **33** (1910) 1277.

6.3   D. Jou, J.M. Rubí, and J. Casas-Vázquez, Physica A **101** (1980) 588.

6.4   D. Jou and J. Casas-Vázquez, J. Non-Equilib. Thermodyn. **5** (1980) 91.

6.5   M. Lax, Rev. Mod. Phys. **32** (1960) 25.

6.6   L.D. Landau and E.M. Lifshitz, *Statistical Physics,* 3rd ed. Part 1, Pergamon, Oxford, 1980.

6.7   D. Jou, Physica A **155** (1989) 221.

6.8   L.D. Landau and E.M. Lifshitz, *Fluid Mechanics*, Pergamon, Oxford, 1985.

6.9   M. Bixon and R. Zwanzig, Phys. Rev. **187** (1969) 267; R.F. Fox and G. Uhlenbeck, Phys. Fluids **13** (1970) 1893, 2881; J. Logan and M. Kac, Phys. Rev. A **13** (1976) 458.

6.10   L. Onsager and S. Machlup, Phys. Rev. **91** (1953) 1505; L. Tisza and I. Manning, Phys. Rev. **105** (1957) 1695.

6.11   J. Keizer, J. Stat. Phys. **31** (1983) 485.

6.12   D. Jou, J.E. Llebot, and J. Casas-Vázquez, Physica A **109** (1981) 208.

6.13   H. Sato, J. Phys. Soc. Japan **44** (1978) 49; R. Voss and J. Clarke, Phys. Rev. B **13** (1976) 287; T.G.M. Kleipenning, Physica B **48** (1976) 353.

6.14   D. Pavón, D. Jou, and J. Casas-Vázquez, J. Phys. A **16** (1983) 775; D. Jou and D. Pavón, Astrophys. J. **291** (1985) 447.

6.15   M.H. Ernst, Am. J. Phys. **38** (1970) 908.

6.16   C.K. Wong, J.A. McLennan, M. Lindenfeld, and J.W. Dufty, J. Chem. Phys. **68** (1978) 1563; J.J. Brey, J. Chem. Phys. **79** (1983) 4585.

Chapter 7

7.1   E.T. Jaynes in *Statistical Physics* (W.K. Ford, ed.), Benjamin, New York, 1963.

7.2   R.D. Levine and M. Tribus (eds.), *The Maximum Entropy Formalism*, MIT Press, Cambridge, Mass., 1979.

7.3   D.N. Zubarev, V. Morozov and G. Röpke, *Statistical Mechanics of Nonequilibrium Processes*, Vol 1: *Basic Concepts, Kinetic Theory*, Vol. 2: *Relaxation and Hydrodynamic Processes* Akademie Verlag, Berlin, 1997.

7.4   W.T. Grandy, Jr., Phys. Rep. **62** (1980) 175; F. Schlögl, Phys. Rep. **62** (1980) 267; B.C. Eu, *Nonequilibrium Statistical Mechanics: Ensemble Method*, Kluwer, Drodrecht, 1998.

7.5   J.G. Ramos, A.R. Vasconcellos, and R. Luzzi, Fortschr. Phys. **43** (1995) 265;
R. Luzzi, A.R. Vasconcellos and J.G. Ramos, *Statistical Foundations of Irreversible Thermodynamics*, Teubner Verlag, Berlin, 2000; *Foundations of a Nonequilibrium Ensemble Formalism*, Kluwer, Dordrecht, in press.

7.6   B. Robertson, Phys. Rev. **160** (1967) 175.

7.7   A.R. Vasconcellos, R. Luzzi, and L.S. García-Colín, Phys. Rev. A **43** (1991) 6622, 6633; R. Luzzi, A.R. Vasconcellos, J. Casas-Vázquez and D. Jou, Physica A **248** (1998) 111.

7.8   M. Ichiyanagi, J. Phys. Soc. Jpn. **59** (1990) 1970; Prog. Theor. Phys. **84** (1991) 810.

7.9   Z. Banach, Physica A **159** (1987) 343; R.E. Nettleton, Phys. Rev. A **42** (1990) 4622; J. Chem. Phys **93** (1990) 81; B.C. Eu, Physica A **171** (1991) 285.

7.10   A.B. Corbet, Phys. Rev. A **9** (1974) 1371; R.M. Nisbet and W.S.C. Gurney, Phys. Rev. A **10** (1974) 720.

7.11   D. Jou, C. Pérez-García, and J. Casas-Vázquez, J. Phys. A **17** (1984) 2799; R.E. Nettleton, J. Phys. A **21** (1988) 3939.

7.12   R. Domínguez-Cascante and D. Jou, J. Non-Equilib. Thermodyn. **20** (1995) 263.

7.13   H. Bidar, D. Jou and M. Criado-Sancho, Physica A **233** (1997) 163; D. Jou, J. Casas-Vázquez and M. Criado-Sancho, *Thermodynamics of Fluids under Flow*, Springer, Berlin, 2000.

7.14   R. Domínguez and D. Jou, Phys. Rev. E **51** (1995) 158.

7.15   M. Ferrer and D. Jou, Am. J. Phys. **63** (1995) 237.

7.16   D. Mihalas and B.W. Mihalas, *Foundations of Radiation Hydrodynamics*, Oxford University Press, Oxford, 1985; A.M. Anile, S. Pennisi, and M. Sanmartino, J. Math. Phys. **32** (1991) 544; C.D. Levermore, J. Stat. Phys. **83** (1996) 331; R. Domínguez-Cascante and J. Faraudo, Phys. Rev. E **54** (1996) 6933.

7.17   W. Larecki, Nuovo Cimento D **14** (1993) 141.

7.18   H. Spohn and J.L. Lebowitz, Comm. Math. Phys. **54** (1977) 97; B.N. Miller and P.M. Larson, Phys. Rev A **20** (1979) 1717.

7.19   H. Goldstein, *Classical Mechanics*, 2nd edn, Addison-Wesley, Reading, Mass., 1975.

7.20   J. Camacho, Phys. Rev. E **51** (1995) 220.

7.21   H. Fröhlich, Riv. Nuovo Cimento **7** (1977) 399, H. Fröhlich and F. Kremer (eds.), *Coherent Excitations in Biological Systems*, Springer, Berlin, 1983.

7.22   A.M.S. Tremblay, E.D. Siggia, and M.R. Arai, Phys. Rev. A **23** (1981) 1451; R. Schmitz, Phys. Rep. **171** (1988) 1; A.M.S. Tremblay in *Recent Developments in Nonequilibrium Thermodynamics* (J. Casas-Vázquez, D. Jou, and G. Lebon, eds.), Springer, Berlin, 1984.

7.23   D. Jou, J.E. Llebot, and J. Casas-Vázquez, Phys. Rev. A **25** (1982) 508;

D. Jou and T. Careta, J. Phys. A **15** (1982) 3195; R.E. Nettleton, J. Chem. Phys. **81** (1984) 2458.

7.24   E. Jahnig and P.H. Richter, J. Chem. Phys. **64** (1976) 4645; J. Keizer, *Statistical Thermodynamics of Nonequilibrium Processes*, Springer, Berlin, 1987.

Chapter 8

8.1   P. Résibois and M. de Leener, *Classical Kinetic Theory of Fluids*, Wiley, New York, 1977.

8.2   D.A. McQuarrie, *Statistical Mechanics*, Harper and Row, New York, 1976.

8.3   B.J. Berne in *Statistical Mechanics (Part B: Time-Dependent Processes)* (B.J. Berne, ed.), Plenum, New York, 1977.

8.4   H. Grabert, *Projection Operator Techniques in Non-equilibrium Statistical Mechanics*, Springer, Berlin, 1981.

8.5   A.Z. Akcasu and E. Daniels, Phys. Rev. A **2** (1970) 962; M. Grant and R. Desai, Phys. Rev. A **25** (1982) 2727.

8.6   H. Mori, Prog. Theor. Phys. **33** (1965) 432; **34** (1965) 399.

8.7   J. Madureira, A.R. Vasconcellos, R. Luzzi, J. Casas-Vázquez and D. Jou, J. Chem. Phys. **108** (1998) 7568, 7580.

Chapter 9

9.1   D.J. Evans and G.P. Morriss, *Statistical Mechanics of Non-equilibrium Liquids*, Academic Press, London, 1990.

9.2   B. Hafskjold and S. Kjelstrup Ratkje, J. Stat. Phys. **78** (1995) 463.

9.3   S. Nosé, J. Chem. Phys. **81** (1984) 511; W.G. Hoover, Phys. Rev. A **31** (1985) 1695.

9.4   W. Loose and S. Hess, Rheol. Acta **28** (1989) 91.

9.5   A. Baranyiai and D.J. Evans, Phys. Rev. A **40** (1989) 3817.

9.6   H.J.M. Hanley and D.J. Evans, J. Chem. Phys. **76** (1982) 3225.

9.7   D.J. Evans, J. Stat. Phys. **57** (1989) 745.

9.8   C. Pérez-García and D. Jou, Phys. Lett. A **95** (1983) 23; M. Grmela, Phys. Lett. A **120** (1987) 276.

9.9   R.E. Nettleton, J. Non-Equilib. Thermodyn. **12** (1987) 273.

9.10   B.J. Ackerson and N.A. Clark, Phys. Rev. Lett. **46** (1981) 123; H.J.M. Hanley, J.C. Rainwater, N.A. Clark, and B.J. Anderson, J. Chem. Phys. **79** (1983) 4448.

9.11   H. Bidar, J. Non-Equilib. Thermodyn. **22** (1997) 156.

9.12   B.D. Todd, P.J. Daivis and D.J. Evans, Phys. Rev. E **51** (1995) 4362.

9.13   B.D. Todd, D.J. Evans and P.J. Daivis, Phys. Rev. E **52** (1995) 1627.

9.14   B.D. Todd and D.J. Evans, J. Chem. Phys. **103** (1995) 9804.

9.15   A. Baranyai, D.J. Evans and P.J. Daivis, Phys. Rev. A **46** (1992) 7593.

9.16   A. Baranyai, Phys. Rev. E **61** (2000) R 3306.

9.17   W.G. Hoover, D.J. Evans, R.B. Hickman, A.J.C. Ladd, W.T. Ashurst, and B. Moran, Phys. Rev. A **22** (1980) 1690.

9.18   D.J. Evans and G.P. Morriss, Phys. Rev. A **30** (1984) 1528.

Chapter 10

10.1   C. Cattaneo, C. R. Acad. Sci. Paris **247** (1958) 431.

10.2   P. Vernotte, C. R. Acad. Sci. Paris **246** (1958) 3154.

10.3   C.C. Ackerman and R.A. Guyer, Ann. Phys. (NY) **50** (1986) 128.

10.4   S.J. Roger, Phys. Rev. B **3** (1971) 1440; R.J. von Gutfeld in *Physical Acoustics* (W. P. Mason, ed.) vol. V, Academic Press, London, 1968.

10.5   D.Y. Tzou, *Macro-to-Microscale Heat Transfer. The Lagging Behaviour*, Taylor and Francis, New York, 1997; D.D. Joseph and L. Preziosi, Rev. Mod. Phys. **61** (1989) 41; **62** (1990) 375; R.A. Mac Donald and D.H. Tsai, Phys. Rep. **46** (1978) 1; D.E. Glass, M.N. Ozisik, and B.Vick, Int. J. Heat Mass Transfer **30** (1987) 1623.

10.6   A.V. Luikov, V.A. Bubnov, and A. Soloviev, Int. J. Heat Mass Transfer **19** (1976) 245; V.A. Bubnov, Int. J. Heat Mass Transfer **19** (1976) 175.

10.7   M. Chester, Phys. Rev. **131** (1963) 2013; Phys. Rev. **145** (1966) 76.

10.8   R.A. Guyer and J.A. Krumhansl, Phys. Rev. **133** (1964) 1411; **148** (1966) 778.

10.9   I.F. Mikhail and S. Simons, J. Phys. C **8** (1975) 3068, 3087.

10.10   H.E. Jackson and C.T. Walker, Phys. Rev. B **3** (1971) 1428; V. Narayanamurti and R.C. Dynes, Phys. Rev. Lett. **28** (1972) 1461.

10.11   B.D. Coleman, M. Fabrizio, and D.R. Owen, Arch. Rat. Mech. Anal. **80** (1982) 135; B.D. Coleman and D. Newmann, Phys. Rev. B **37** (1988) 1492.

10.12   W. Dreyer and H. Struchtrup, Continuum Mech. Thermodyn. 5 (1993) 3.

10.13   P.M. Morse and H. Feshbach, *Mathematical Methods of Theoretical Physics*, McGraw-Hill, New York, 1953, pp 865-869; B. Vick and M.N. Ozisick, J. Heat Transfer **105** (1983) 902.

10.14   W.S. Kim, L.G. Hector, Jr., and M.N. Ozisik, J. Appl. Phys. **68** (1990) 5478.

10.15   D. Jou and C. Pérez-García, Physica A **104** (1980) 320; F. Bampi, A. Morro, and D. Jou, Physica A **107** (1981) 393; D. Jou, J.E. Llebot, and J. Casas-Vázquez, Physica A **109** (1981) 208; G. Lebon and P.C. Dauby, Phys. Rev. A **42** (1990) 4710.

10.16   J. Casas-Vázquez and D. Jou, J. Phys. A **14** (1981) 1225; D. Jou and
J. Casas-Vázquez, J. Phys. A **20** (1987) 5371.

10.17   J. Casas-Vázquez and D. Jou, Acta Phys. Hung. **66** (1989) 99; D. Jou and
J. Casas-Vázquez, Phys. Rev. A **45** (1992) 8371.

10.18   J. Meixner in *Foundations of Continuum Thermodynamics* (J.J. Delgado-Domingos,
M.N.R. Nina, and J.H. Whitelaw, eds.), Wiley, New York, 1973.

10.19   I. Müller, Arch. Rat. Mech. Anal. **41** (1971) 319.

10.20   W. Muschik, Arch. Rat. Mech. Anal. **66** (1977) 379.

10.21   J. Keizer, J. Chem. Phys. **82** (1985) 2751.

10.22   D. Jou and J. M. Rubí, J. Non-Equilib. Thermodyn. **5** (1980) 125.

10.23   D. Jou, J. Casas-Vázquez, and G. Lebon, Int. J. Thermophys. **14** (1993) 671;
M. Criado-Sancho and J.E. Llebot, Phys. Rev. E **47** (1993) 4104.

10.24   D. Mihalas and B.W. Mihalas, *Radiation Hydrodynamics*, Oxford University Press,
1984.

10.25   C.D. Levermore, J. Quant. Spectr. Rad. Transfer **31** (1984) 149; C.D. Levermore and
G.C. Pomraning, Astrophys. J. **348** (1981) 321.

10.26   D. Sharts, J. Delehrez, R. McCrory, and C.P. Verdon, Phys. Rev. Lett **47** (1981) 247.

10.27   A.M. Anile, S. Pennisi, and M. Sammartino, J. Math. Phys. **32** (1991) 544.

10.28   D. Jou and M. Zakari, J. Phys. A **28** (1995).

10.29   T. Q. Qiu and C. L. Tien, Int. J. Heat Mass Transfer **35** (1992) 719; ASME J. Heat
Transfer **115** (1993) 835, 842.

10.30   D.Y. Tzou, ASME J. Heat Transfer **111** (1989) 232, Int. J. Heat Mass Transfer
33 (1992) 877.

10.31   S.L. Sobolev, Sov. Phys. Usp **34** (1991) 217, J. Low-Temp. Phys. **83** (1991) 307.

10.32   L.D. Landau and E.M. Lifshitz, *Fluid Mechanics*, Addison Wesley, Reading, 1958;
J. Wilks, *An Introduction to Liquid Helium*, Clarendon, Oxford, 1970.

10.33   G. Lebon and D. Jou, J. Non-Equilib. Thermodyn. **4** (1979) 259.

10.34   A. Greco and I. Müller, Arch. Rat. Mech. Anal. **85** (1984) 279.

10.35   L. Lindblom and W.A. Hiscock, Phys. Lett. A **131** (1988) 280.

10.36   M.S. Mongiovi, J. Non-Equilib. Thermodyn. **16** (1991) 225; **18** (1993) 19;
Phys. Rev. B **48** (1993) 6276.

Chapter 11

11.1   D.D. Joseph, O. Riccius, and M. Arney, J. Fluid Mech. **171** (1986) 309; D.D. Joseph
and K. Chen, J. Non-Newtonian Fluid Mech. **28** (1988) 47.

11.2   M. Greenspan, J. Acoust. Soc. Amer. **28** (1956) 644; *Physical Acoustics* (vol 2A) (W.P. Mason, ed.), Academic Press, New York, 1965; J.D. Foch and G.W. Ford, in *Studies in Statistical Mechanics* (vol. 5), North-Holland, Amsterdam, 1970.

11.3   E. Meyer and G. Sessler, Z. für Phys. **149** (1957) 15.

11.4   M. Carrassi and A. Morro, Nuovo Cimento B **9** (1972) 321; **13** (1973) 281.

11.5   A.M. Anile and S. Pluchino, Meccanica **19** (1984) 104; J. Mécanique **3** (1984) 167.

11.6   G. Lebon and A. Cloot, Wave Motion **11** (1989) 23.

11.7   I. Müller in *Recent Developments in Nonequilibrium Thermodynamics* (J. Casas-Vázquez, D. Jou, and G. Lebon, eds.), Springer, Berlin, 1984, p. 32.

11.8   L.C. Woods and H. Troughton, J. Fluid Mech. **100** (1980) 321.

11.9   H.J. Bauer in *Physical Acoustics* (vol. 2A) (W.P. Mason, ed.), Academic Press, New York, 1965.

11.10   H.O. Knesser in *Physical Acoustics* (vol. 2A) (W.P. Mason, ed.), Academic Press, New York, 1965.

11.11   M. Kranys, J. Phys. A **10** (1977) 189.

11.12   D. Jou, F. Bampi, and A. Morro, J. Non-Equilib. Thermodyn. **8** (1982) 127.

11.13   D. Gilbarg and P. Paolucci, J. Rat. Mech. Anal. **2** (1953) 617.

11.14   A.M. Anile and A. Majorana, Meccanica **16** (1981) 149; T.S. Olson and W.A. Hiscock, Ann. Phys. NY **204** (1990) 331.

11.15   G. Boillat and T. Ruggeri, Acta Mechanica **35** (1980) 271; G. Boillat, C. R. Acad. Sci. Paris A **283** (1976) 409.

11.16   D. Jou and D. Pavón, Phys. Rev. A **44** (1991) 6496.

11.17   T. Ruggeri, Phys. Rev. E **47** (1993) 4135.

11.18   W. Weiss, Phys. Rev. E **52** (1995) R5760; Phys. Fluids A **8** (1996) 1689.

Chapter 12

12.1   B.J. Berne and R. Pecora, *Dynamic Light Scattering with Applications to Chemistry, Biology and Physics*, Wiley, New York, 1976.

12.2   J.P. Boon and S. Yip, *Molecular Hydrodynamics*, McGraw-Hill, New York, 1980.

12.3   D. Jou, C. Pérez-García, L.S. García-Colín, M. López de Haro, and R.F. Rodríguez, Phys. Rev. A **31** (1985) 2502; C. Pérez-García, D. Jou, L.S. García-Colín, M. López de Haro, and R.F. Rodríguez, Physica A **135** (1986) 2502.

12.4   J. Frenkel, *Kinetic Theory of Liquids*, Dover, New York, 1955.

12.5   W.E. Alley and B.J. Alder, Phys. Rev. A **27** (1983) 3158.

12.6   C.H. Chung and S. Yip, Phys. Rev. **182** (1969) 323.

12.7   R.M. Velasco and L.S García-Colín, J. Stat. Phys. **69** (1992) 217; J. Non-Equilib. Thermodyn. **20** (1995) 1.

12.8   W. Weiss and I. Müller, Continuum Mech. Thermodyn. **7** (1995) 123.

Chapter 13

13.1   M.S. Boukary and G. Lebon, Physica A **137** (1986) 546; R.F. Rodríguez,
L.S. García-Colín, and M. López de Haro, J. Chem. Phys. **83** (1985) 4099;
R. E. Nettleton, J. Phys. A **21** (1988) 1079.

13.2   D. Lhuillier, Physica A **165** (1990) 303; N. Depireux and G. Lebon, J. Non-
Newtonian Fluid Mech. **96** (2000) 105.

13.3   S. Goldstein, Q.J. Mech. Appl. Math. **4** (1951) 129; H.D. Weyman, Am. J. Phys.
35 (1965) 488; M.O. Hongler and L. Streit, Physica A **165** (1990) 196.

13.4   J. Masoliver, J.M. Porrà, and G.H. Weiss, Phys. Rev. A **45** (1992) 2222;
J. Masoliver and G.H. Weiss, Physica A **183** (1992) 537; J.Masoliver,
K. Lindenberg, and G.H. Weiss, Physica A **157** (1989) 593; W. Horthemske and
R. Lefever, *Noise-induced Phase Transitions*, Springer, Berlin, 1983.

13.5   P. Rosenau, Phys. Rev. E **48** (1993) R 655.

13.6   J. Camacho and D. Jou, Phys. Lett. A **171** (1992) 26; M. Grmela and D. Jou,
J. Math. Phys. **34** (1993) 2290; H. Vlad and J. Ross, Phys. Lett. A **184** (1993) 403.

13.7   M.A. Schweizer, Can. J. Phys. **63** (1985) 956.

13.8   G.I. Taylor, Proc. Roy. Soc. London, Ser A **219** (1953) 186, **223** (1954) 446;
R. Aris, Proc. R. Soc. London, Ser A **235** (1956) 67.

13.9   P.C. Chatwin, J. Fluid Mech **43** (1970) 321; W.N. Gill and R. Sankasubramanian,
Proc. Roy. Soc. London, Ser. A **316** (1970) 341, **322** (1971) 101; V.I. Maron,
Int. J. Multiphase Flow **4** (1977) 339; R. Smith, J. Fluid Mech. **105** (1981) 469; **175**
(1987) 201; **182** (1987) 447; W.R. Young and S. Jones, Phys. Fluids A **3** (1991) 1087.

13.10   J. Camacho, Phys. Rev. E **47**, 1049 (1993); **48** (1993) 310, 1844.

13.11   P. Rigord, Ph. D Thesis, Université de Paris 6, 1990.

13.12   H.L. Frisch, Polym. Engn. Sci. **20** (1980) 1; H.B. Hopfenberg and V. Stannett in
The Physics of Glassy Polymers (R. N. Howard, ed.), Appl. Sci. Publish., London,
1973.

13.13   J. Crank, *The Mathematics of Diffusion*, Clarendon, Oxford, 1975.

13.14   H.B. Hopfenberg, R.M. Holley, and V. Stannett, Polym. Engn. Sci. **9** (1969) 242;
T.K. Kwei and H.M. Zupko, J. Polym. Sci. A 2, **7** (1969) 867; P. Neogi, AIChE J. **29**
(1983) 829.

13.15   N.L. Thomas and A.H. Windle, Polymer **23** (1982) 529; C.J. Durning and M. Tabor,
Macromolecules **19** (1986) 2220.

13.16   V. Méndez, J. Fort and J. Farjas, Phys. Rev. E **60** (1999) 5231; V. Méndez and
J.E. Llebot, Phys. Rev. E **56** (1997) 6557.

13.17   J. Fort and V. Méndez, Phys. Rev. Lett. **82** (1999) 867; L.L. Cavalli-Sforza,
P. Menozzi and A. Piazza, Science **259** (1993) 693.

13.18   S. Fedotov, Phys. Rev. E **58** (1998) 5143.

13.19   V. Méndez and J. Fort, Phys. Rev. E **60** (1999) 6168; H.G. Ohtmer, S.R. Dunbar and
W. Alt, J. Math. Biol. **26** (1988) 263.

13. 20   V. Méndez and A. Compte, Physica A **260** (1998) 90.

Chapter 14

14.1   D. Jou ,J. E. Llebot, and J. Casas-Vázquez, Phys. Rev. A **25** (1982) 3277;
B. Maruszewski and G. Lebon, J. Tech. Phys. **27** (1986) 63.

14.2   A.N. Krall and A.W. Trivelpiece, *Principles of Plasma Physics*, McGraw-Hill, New
York, 1973.

14.3   M.F. Schlesinger, Ann. Rev. Phys. Chem. **39** (1988) 269; D. Jou and J. Camacho,,
J. Phys. A **23** (1990) 4603.

14.4   A.M.S. Tremblay and F. Vidal, Phys. Rev. B **25** (1982) 7562.

14.5   L.J. de Felice, *Introduction to Membrane Noise*, Plenum, New York, 1981; D. Jou,
F. Ferrer-Suquet, and C. Pérez-Vicente, J. Chem. Phys. **85** (1986) 5314.

14.6   B.C. Eu and A.S. Wagh, Phys. Rev. B **27** (1983) 1037; A.M. Anile and S. Pennisi,
Continuum Mech. Thermodyn. **4** (1992) 187; A.R. Vasconcellos, A.C. Algarte,
and R. Luzzi, Physica A **166** (1990) 517.

14.7   J.E. Llebot, D. Jou, and J. Casas-Vázquez, Physica A **121** (1983) 552.

14.8   W. Hänsch, *The Drift-diffusion Equation and its Application in MOSFET Modeling*,
Springer, Berlin, 1991; K. Blotekjaer, IEEE Trans. on Electron Devices **17** (1970)
38; G. Baccarani and M.R. Wordeman, Solid-State Electronics **29** (1982) 970.

14.9   A.M. Anile and S. Pennisi, Phys. Rev. B **46** (1992) 13186; Continuum Mech.
Thermodyn. **4** (1992) 187; A.M. Anile and O. Muscato, Phys. Rev. B **51** (1995)
7628; A.M. Anile, O. Muscato and V. Romano, VLSI Design **10** (2000);
P. Falsaperla and M. Trovato, VLSI Design **8** (1998) 527.

14.10   M. Nekovee, B.J. Geurts, H.M. J. Boots and M.F.H. Shuurmans, Phys. Rev. B **45**
(1992) 6643; M. Heiblum, M.I. Nathan, D.C. Thomas, and C.M. Knoedler,
Phys. Rev. Lett. **55** (1985) 2200; F. Müller, B. Lengeler, Th. Schäpers, J. Appenzeller,
A. Förster, Th. Kloch, and H. Luth, Phys. Rev. B **51** (1995) 5099.

14.11   F. Bloch, Z. Phys. **81** (1933) 363; G. Barton, Rep. Prog. Phys. **42** (1979) 963;
J. Dempsey and B.I. Halperin, Phys. Rev. B **45** (1992) 1719.

14.12   I. V. Tokatly and O. Pankratov, Phys. Rev. B **60** (1999) 15500, **62** (2000) 2759.

14.13   V. Ciancio, L. Restuccia and G.A. Kluitenberg, J. Non-Equilib. Thermodyn. **15** (1990) 157; V. Ciancio and J. Verhas, J. Non-Equilib. Thermodyn. **16** (1991) 57; F. Conforto and S. Giambò, J. Non-Equilib. Thermodyn. **21** (1996) 260.

14.14   L.F. del Castillo and L.A. Dávalos-Orozco, J. Chem. Phys. **93** (1990) 5147; L.F. del Castillo, L.A. Dávalos-Orozco and L.S. García-Colín, J. Chem. Phys. **100** (1997) 2348, L.F. del Castillo and R.F. Rodríguez, J. Non-Equilib. Thermodyn. **14** (1989) 127.

14.15   D. Kivelson and T. Keyes, J. Chem. Phys. **57** (1972) 4599.

Chapter 15

15.1   J. Meixner, Z. Naturforsch. **4a** (1943) 594; **9a** (1954) 654.

15.2   G. Kluitenberg, *Plasticity and Non-equilibrium Thermodynamics* (CISM Course 281), Springer, Wien, 1984; G. Kluitenberg in *Non-equilibrium Thermodynamics, Variational Techniques and Stability* (R. Donnelly, R. Hermann, and I. Prigogine, eds.), University of Chicago Press, Chicago, 1966.

15.3   J. Bataille and J. Kestin, J. Mécanique **14** (1975) 365.

15.4   R.S. Rivlin and J.L. Ericksen, J. Rat. Mech. Anal. **4** (1955) 323.

15.5   W. Noll, J. Rat. Mech. Anal. **4** (1955) 3.

15.6   S. Koh and C. Eringen, Int. J. Engn. Sci. **1** (1963) 199.

15.7   B.D. Coleman, H. Markowitz, and W. Noll, *Viscometric Flows of Non-Newtonian Fluids*, Springer, New York, 1966.

15.8   R.R. Huilgol and N. Phan-Thien, Int. J. Engn. Sci. **24** (1986) 161.

15.9   A. Palumbo and G. Valenti, J. Non-Equilib. Thermodyn. **10** (1985) 209; G. Valenti, Physica A **144** (1987) 211.

15.10   M. López de Haro, L.F. del Castillo, and R.F. Rodríguez, Rheol. Acta **25** (1986) 207.

15.11   B.C. Eu, J. Chem. Phys. **82** (1985) 4683.

15.12   G. Lebon, C. Pérez-García, and J. Casas-Vázquez, Physica **137 A** (1986) 531.

15.13   G. Lebon, C. Pérez-García, and J. Casas-Vázquez, J. Chem. Phys. **88** (1988) 5068; G. Lebon and J. Casas-Vázquez, Int. J. Thermophys. **9** (1988) 1003.

15.14   G. Lebon and A. Cloot, J. Non-Newtonian Fluid Mech. **28** (1988) 61.

15.15   P.E. Rouse, J. Chem. Phys. **21** (1953) 1272; B.H. Zimm, J. Chem. Phys. **24** (1956) 269.

15.16   H. Giesekus, J. Non-Newtonian Fluid Mech. **11** (1982) 69.

15.17   A.S. Lodge, *Elastic Liquids*, Academic Press, New York, 1964.

15.18   J.D. Ferry, *Viscoelastic Properties of Polymers* (3rd edn), Wiley, New York, 1980.

15.19   G. Astarita and G. Marrucci, *Principles of Non-Newtonian Fluid Mechanics*, McGraw-Hill, New York, 1974.

15.20   P.J. Carreau, Trans. Soc. Rheol. **16** (1972) 99.

15.21   J.E. Dunn and R.L. Fosdick, Arch. Rat. Mech. Anal. **56** (1974) 191.

15.22   W.O. Criminale, J.L. Ericksen, and G.K. Filbey, Arch. Rat. Mech. Anal. **2** (1958) 410.

15.23   A.E. Green and R.S. Rivlin, Arch. Rat. Mech. Anal. **1** (1957) 1.

15.24   J.G. Oldroyd, Proc. Roy. Soc. London A **245** (1958) 278.

15.25   M.J. Crochet, A. Davies, and K. Walters, *Numerical Simulation of Non-Newtonian Flow*, Elsevier, Amsterdam, 1984.

15.26   J.C. Maxwell, Phil. Trans. Roy. Soc. London A **157** (1867) 49.

15.27   J. Lambermont and G. Lebon, Int. J. Non-linear Mech. **9** (1974) 55.

15.28   R.F. Rodríguez, M. López de Haro, and O. Manero, Rheol. Acta **27** (1988) 217.

15.29   G. Lebon in *Extended Thermodynamic Systems* (P. Salamon and S. Sieniutycz, eds.), Taylor and Francis, New York, 1990.

15.30   R.B. Bird, R.C. Armstrong, and O. Hassager, *Dynamics of Polymeric Liquids*, 2nd edn. Vol. 1: *Fluid Mechanics*; R.B. Bird, C.F. Curtiss, R.C. Armstrong, and O. Hassager, Vol. 2: *Kinetic Theory*, Wiley, New York, 1987.

15.31   M. Doi and S.F. Edwards, *The Theory of Polymer Dynamics*, Clarendon, Oxford, 1986.

15.32   C. Pérez-García, J. Casas-Vázquez, and G. Lebon, J. Polym. Sci (B. Polym. Phys.) **27** (1989) 1807.

15.33   H. Metiu and K. Freed, J. Chem. Phys. **67** (1977) 3303.

15.34   J. Camacho and D. Jou, J. Chem. Phys. **92** (1990) 1339.

15.35   I. Müller and K. Wilmanski, Rheol. Acta **25** (1986) 335.

15.36   I.S. Liu and I. Müller, Arch. Rat. Mech. Anal. **83** (1983) 285.

15.37   H. Tanner, *Engineering Rheology*, Clarendon, Oxford, 1985.

15.38   R.F. Christiansen and W.R. Leppard, Trans. Soc. Rheol. **18** (1974) 65.

15.39   A.S. Lodge, J. Rheol. **33** (1989) 821.

15.40   G. Lebon, P. Dauby, A. Palumbo, and G. Valenti, Rheol. Acta **29** (1990) 127.

15.41   P.C. Dauby and G. Lebon, Appl. Math. Lett. **33** (1990) 45.

Chapter 16

16.1   D. Jou, J. Casas-Vázquez, and M. Criado-Sancho, Adv. Polym. Sci. **120** (1995) 207; *Thermodynamics of Fluids under Flow*, Springer, Berlin, 2000; A. Onuki, J. Phys.: Condens. Matter **9** (1997) 6119.

16.2   D. Lhuillier and A. Ouibrahim, J. Mécanique **19** (1980) 1; G.A. Maugin and W. Muschik, J. Non-Equilib. Thermodyn. **19** (1994) 2572; J. Verhas, J. Non-Equilib. Thermodyn. **18** (1993) 311.

16.3   M. Grmela and H.C. Öttinger, Phys. Rev. E, **56** (1997) 6620; H.C. Öttinger and
M. Grmela, Phys. Rev E **56** (1997) 6633; A.N. Beris and B.J. Edwards,
Thermodynamics of Flowing Systems with Internal Microstructure, Oxford Science
Publications, Oxford, 1994.

16.4   A. Onuki and K. Kawasaki, Ann. Phys. **121** (1979) 456; S. Hess and W. Loose,
Physica A **162** (1989) 138; W. Loose and S. Hess, Rheol. Acta **28** (1989) 91.

16.5   C. Rangel-Nafaile, A. Metzner, and K. Wissbrun, Macromolecules **17** (1984) 1187;
B.A. Wolf, Macromolecules **17** (1984) 615; H. Kramer and B.A. Wolf,
Makromol. Chem. Rapid Commun. **6** (1985) 21; H. Kramer, H. Schenck and B.A
Wolf, Makromol. Chem. **189** (1988) 1613; **189** (1988) 1627; L.A. Utracki, *Polymer
Alloys and Blends. Thermodynamics and Rheology*, Hanser Publishers, Munich, 1990.

16.6   T.Q. Nguyen and H-H. Kausch, Adv. Polym. Sci. **100** (1992) 73.

16.7   A. Onuki, Phys. Rev. Lett. **62** (1989) 2472; J. Phys. Soc. Jpn. **59** (1990) 3427;
E. Helfand and H. Fredrickson, Phys. Rev. Lett. **62** (1989) 2468.

16.8   R.B Bird, F.C. Curtiss, R.C. Armstrong, and O. Hassager, *Dynamics of Polymeric
Liquids*, 2nd ed., Vol. 2, Wiley, New York, 1987; M. Doi and S.F. Edwards, *The
Theory of Polymer Dynamics*, Clarendon Press, Oxford, 1986.

16.9   M. Criado-Sancho, J. Casas-Vázquez, and D. Jou, Polymer **36** (1995) 4107;
J. Non-Equilib. Thermodyn. **18** (1993) 103; M. Criado-Sancho, D. Jou, and
J. Casas-Vázquez, Macromolecules **24** (1991) 2834; G. Lebon, J. Casas-Vázquez,
D. Jou, and M. Criado-Sancho, J. Chem. Phys. **98** (1993) 7434.

16.10   A.M. Basedow, K.H. Ebert, and H.J. Ederer, Macromolecules **11** (1978) 774.

16.11   M. Criado-Sancho , D. Jou, and J. Casas-Vázquez, J. Non-Equilib. Thermodyn. **19**
(1994) 1; Phys. Rev. E **56** (1997) 1887.

16.12   J. Casas-Vázquez, M. Criado-Sancho, and D. Jou, Europhys. Lett. **23** (1993) 469.

16.13   P. Nozières and D. Quemada, Europhys. Lett. **2** (1986) 129.

16.14   P.G. Larson, Rheol. Acta **31** (1992) 497; J.R. Prakash and R.A. Mashelkar, J. Chem.
Phys. **95** (1991) 3743; J. Rheology 1992, **36**, 789.

16.15   M.J. MacDonald and S.J. Muller, J. Rheology **40** (1996) 259.

16.16   A.N. Beris and J. Mavrantzas, J. Rheol. **38** (1994) 1235; H.C. Öttinger, Rheol. Acta
31 (1992) 14; V.G. Mavrantzas and A.N. Beris, Phys. Rev. Lett. **69** (1992) 273;
A.V. Bhave, R.C. Angstrom and R.A. Brown, J. Chem. Phys. **95** (1991) 2988.

16.17   L.F. del Castillo, M. Criado-Sancho and D. Jou, Polymer **41** (2000) 2633.

Chapter 17

17.1   C. Eckart, Phys. Rev. **58** (1940) 919.

17.2   L.D. Landau and E.M. Lifshitz, *Fluid Mechanics*, Pergamon, Oxford, 1985.

17.3   W.A. Hiscock and L. Lindblom, Phys. Rev. D **31** (1985) 725.

17.4   M. Kranys, Nuovo Cimento B **50** (1967) 48; Nuovo Cimento **88** (1972) 417.

17.5   I. Müller, Arch. Rat. Mech. Anal. **34** (1969) 259.

17.6   W. Israel, Ann. Phys. (NY) **100** (1976) 310; Physica A **106** (1981) 204.

17.7   W. Israel and J.M. Stewart, Ann. Phys. (NY) **118** (1979) 341; J.M. Stewart, Proc. Roy. Soc. London A **357** (1977) 59; N. Udey and W. Israel, Mon. Not. R. Astron. Soc. **199** (1982) 1137.

17.8   D. Pavón, D. Jou, and J. Casas-Vázquez, Ann. Inst. H. Poincaré A **36** (1982) 79; J. Non-Equilib. Thermodyn. **6** (1981) 173; Ann. Inst. H. Poincaré **42** (1985) 31.

17.9   F. Bampi and A. Morro, Phys. Lett. A **79** (1980) 156; J. Math. Phys. **21** (1980) 1201.

17.10   M.A. Schweizer, Astron. Astrophys. **151** (1985) 79.

17.11   W.G. Dixon, *Special Relativity*, Cambridge University Press, Cambridge, 1978.

17.12   I.S. Liu, I. Müller, and T. Ruggeri, Ann. Phys. (NY) **169** (1986) 191; W. Dreyer and W. Weiss, Ann. Inst. H. Poincaré **45** (1986) 401.

17.13   G. Boillat and T. Ruggeri, Continuum Mech. Thermodyn. **10** (1998) 285; H. Struchtrup, Physica A **253** (1998) 555; Z. Banach, Physica A **275** (2000) 405.

17.14   D. Pavón, D. Jou, and J. Casas-Vázquez, Phys. Lett. A **78** (1980) 317.

17.15   W.A. Hiscock and L. Lindblom, Ann. Phys. (NY) **151** (1983) 466.

17.16   G.A. Kluitenberg, S.R. de Groot, and P. Mazur, Physica **19** (1953) 689.

17.17   A. Bressan, *Relativistic Theories of Materials*, Springer, Berlin, 1978.

17.18   C. van Weert, Ann. Phys. (NY) **140** (1982) 133.

17.19   A. Palumbo and P. Pantano, Lett. Nuovo Cimento **41** (1984) 247; M.P. Galipo, Nuovo Cimento **38** (1983) 233, 427, 544.

17.20   S.R. de Groot, W.A. van Leeuwen, and C. van Weert, *Relativistic Kinetic Theory: Principles and Applications,* North-Holland, Amsterdam, 1980.

17.21   J.J. Griffin and K.K. Kan, Rev. Mod. Phys. **48** (1976) 467; H. Stöcker and W. Greiner, Phys. Rep. **137** (1986) 277; L.W. Neise, H. Stöcker and W. Greiner, J. Phys. G **13** (1987) L 181; H.W. Barz, B. Kämpfer, B. Lukács, K. Martinás, and Gy. Wolf, Phys. Lett. B **194** (1987) 15.

17.22   P. Danielewicz, Phys. Lett. B **146** (1984) 168; G. Fai and P. Danielewicz, Phys. Lett. B **373** (1996) 5.

17.23   M. Kozlowski, Nucl. Phys. A **492** (1989) 285; T.S. Olson and W.A. Hiscock, Phys. Rev. C **39** (1989) 1818.

Chapter 18

18.1   B.E. Schultz, *A First Course in General Relativity*, Cambridge University Press, Cambridge, 1985; R. M. Wald, *General Relativity*, The University of Chicago Press, Chicago, 1984; S. Weinberg, *Gravitation and Cosmology*, Wiley, New York, 1972.

18.2   P.J.E. Peebles, *Principles of Physical Cosmology*, Princeton University Press, Princeton, 1993.

18.3   S. Weinberg, Astrophys. J. **168** (1971) 175; M.A. Matzner and C.W. Misner, Astrophys. J. **171** (1972) 415, 433; M. Novello and J.B.S. d'Olival, Acta Phys. Polon. B **11** (1980) 3.

18.4   V.A. Belinskii, S. Nikomarov and I.S. Khalatnikov, Sov. Phys. JETP **50** (1979) 213.

18.5   J.A.S. Lima, R. Portugal and I. Waga, Phys. Rev. D **37** (1988) 2755; H.P. Oliveira and J.M. Salim, Acta Phys. Polon. B **19** (1988) 649; J.D. Barrow, Nucl. Phys. B **310** (1988) 743; D. Pavón, J. Bafaluy, and D. Jou, Class. Quantum Grav. **8** (1991) 347.

18.6   N. Turok, Phys. Rev. Lett. **60** (1988) 549; B.L. Hu, Phys. Lett. A **90** (1982) 375; P.I. Gross, M. Perry and L.G. Yaffe, Phys. Rev. D **25** (1982) 330; C.W. Misner, Phys. Rev. Lett. **19** (1967) 533; M.O. Calvão, H.P. Oliveira, D. Pavón, and J.M. Salim, Phys. Rev. D **45** (1992) 3869.

18.7   G.L. Murphy, Phys. Rev. D **8** (1973) 4231; M.O. Santos, R.S. Dies, and A. Banerjee, J. Math. Phys. **26** (1985) 878; M.O. Calvão and J.M. Salim, Class. Quantum Grav. **9** (1992) 127; G. Hayward and D. Pavón, Phys. Rev. D **40** (1989) 1748.

18.8   X. Fustero and D. Pavón, Lett. Nuovo Cimento **35** (1982) 427.

18.9   W.A. Hiscock and T. Salmonson, Phys. Rev. D **43** (1991) 3249; M. Zakari and D. Jou, Phys. Rev. D **48** (1993) 1597.

18.10   V. Romano and D. Pavón, Phys. Rev. D **50** (1994) 2572; J. Gariel and G. Le Denmat, Phys. Rev D **50** (1994) 2566; R. Maartens, Class. Quantum Grav. **12** (1995) 1455; R.J. van den Hoogen and A. Coley, Class. Quantum Grav. **12** (1995) 2335.

18.11   W. Zimdahl and D. Pavón, Phys. Lett. A **176** (1993) 57; Mont. Not. Roy. Astron. Soc. **266** (1994) 872; Gen. Rel. Grav. **26** (1994) 872; Class. Quantum Grav. **10** (1993) 1775.

18.12   M.O. Calvão, J.A.S. Lima, and I. Waga, Phys. Lett. A 162 (1992) 223; I. Prigogine, J. Geheniau, E. Gunzig and P. Nardone, Gen. Rel. Grav. **21** (1989) 767.

18.13   P. Sudharan and V.B. Johri, Gen. Rel. Grav. **26** (1994) 41; R. A. Sussman, Class. Quantum Grav. **11** (1994) 1445; J. Triginer and D. Pavón, Gen. Rel. Grav. **26** (1994) 513; J. Gariel and G. Le Denmat, Phys. Lett. A **200** (1995) 11; L.R.W. Abramo and J.A.S. Lima, Class. Quantum Grav. **13** (1996) 2953;

D. Pavón, J. Gariel and G. Le Denmat, Gen. Rel. Grav. **28** (1996) 573.

18.14   V. Romano and D. Pavón, Phys. Rev. D **47** (1993) 1396; **50** (1994) 2572;
J. Triginer and D. Pavón, Class. Quantum Grav. **12** (1995) 199, 689;
J. Triginer, W. Zimdalh and D. Pavón, Class. Quantum Grav. **13** (1996) 403,
W. Zimdahl, J. Triginer and D. Pavón, Phys. Rev. D **54** (1996) 1601.

18.15   G.W. Gibbons and S.W. Hawking, Phys. Rev. D **15** (1977) 2738.

18.16   P.C.W. Davies, Class. Quantum Grav. **4** (1987) L 225.

18.17   D. Pavón, Class. Quantum Grav. **7** (1990) 487.

18.18   J. Martínez and D. Pavón, Mon. Not. Roy. Astron. Soc. **268** (1994) 654; J. Martínez,
Phys. Rev. D **53** (1996) 6921; L. Herrera and N. Falcon, Astrophys. Space Sci. **229**
(1995) 105; **234** (1995) 139; L. Herrera, A. di Prisco, J.L. Hernández-Pastora,
J. Martin, and J. Martínez, Class. Quantum Grav. **14** (1997) 2239.

Appendix A

Summary of Vector and Tensor Notation

In general, we have used tensorial notation throughout the book. Tensors of rank 0 (scalars) are denoted by means of italic type letters a; tensors of order 1 (vectors) by means of boldface italic letters $\boldsymbol{a}$ and tensors of rank two and higher orders by capital boldface letters $\mathbf{A}$. In some special circumstances, three-dimensional Cartesian coordinates are used:

$$
\begin{aligned}
&\boldsymbol{a}(a_i) \quad \text{vector,} \\
&\mathbf{A}\ (A_{ij}) \quad \text{tensor of rank 2,} \\
&\mathbf{U}\ (\delta_{ij}) \quad \text{unit tensor } (\delta_{ij} \text{ is Kronecker's symbol}), \\
&\mathbf{J}\ (J_{ijk}) \quad \text{tensor of rank 3.}
\end{aligned}
$$

Symmetric and antisymmetric tensors

Denoting by superscript T the transpose, the symmetric and antisymmetric tensors are respectively defined as

$$
\text{symmetric } \mathbf{A} = \mathbf{A}^T\ (A_{ij} = A_{ji}), \quad \text{antisymmetric } \mathbf{A} = -\mathbf{A}^T\ (A_{ij} = -A_{ji}). \quad (A.1)
$$

The trace of a tensor is defined as the sum of its diagonal components, namely

$$
\text{trace of a tensor } \mathrm{Tr}\,\mathbf{A} = \sum_i A_{ii}. \qquad (A.2)
$$

Decomposition of a tensor

It is customary to decompose second-order tensors into a scalar (invariant) part A, a symmetric traceless part $\overset{0}{\mathbf{A}}$, and an antisymmetric part $\mathbf{A}^a$ as follows

$$\mathbf{A} = \tfrac{1}{3}(\mathrm{Tr}\,\mathbf{A})\mathbf{U} + \overset{0}{\mathbf{A}} + \mathbf{A}^a = \tfrac{1}{3}A\delta_{ij} + \overset{0}{A}_{ij} + A_{ij}^a\,. \tag{A.3}$$

Note that this decomposition implies $\mathrm{Tr}\,\overset{0}{\mathbf{A}} = 0$ $(\sum_i \overset{0}{A}_{ii} = 0)$.

The antisymmetric part of the tensor is often written in terms of an axial vector $\boldsymbol{a}^a$ whose components are defined as

$$a_i^a = \sum_{j,k} \varepsilon_{ijk} A_{jk}^a\,, \tag{A.4}$$

where the permutation symbol ε_{ijk} has the values

$$\varepsilon_{ijk} = \begin{cases} +1 & \text{for even permutations of indices (i.e. 123, 231, 312)} \\ -1 & \text{for odd permutations of indices (i.e. 321, 132, 213)} \\ 0 & \text{for repeated indices.} \end{cases} \tag{A.5}$$

Scalar (or dot) and tensorial (inner) products

We have used for the more common products the following notation:

Dot product between two vectors $\qquad \boldsymbol{a}.\boldsymbol{b} = \sum_i a_i b_i$ (scalar),

$$\begin{aligned} \text{a vector and a tensor} \quad & \mathbf{A}.\boldsymbol{b} = \sum_j A_{ij} b_j \quad \text{(vector)}, \\ \text{a tensor and a vector} \quad & \boldsymbol{b}.\mathbf{A} = \sum_j b_j A_{jk} \quad \text{(vector)}, \\ \text{two tensors} \quad & \mathbf{A}.\mathbf{B} = \sum_k A_{ik} B_{kj} \quad \text{(tensor)}. \end{aligned} \tag{A.6}$$

Double scalar product between tensors $\mathbf{A}:\mathbf{B} = \sum_{i,k} A_{ik} B_{ki}$ (scalar). $\tag{A.7}$

The trace of a tensor may also be written in terms of its double scalar product with the unit matrix as $\mathrm{Tr}\,\mathbf{A} = \mathbf{A}:\mathbf{U}$.

(Inner) tensorial product (also named dyadic product)

between two vectors $\qquad (\boldsymbol{ab})_{ij} = a_i b_j \qquad$ (tensor of rank 2),

$$\tag{A.8a}$$

a vector and a tensor $\qquad (\boldsymbol{a}\mathbf{B})_{ijk} = a_i B_{jk} \qquad$ (tensor of rank 3),

$$a \text{ tensor and a vector} \quad (\mathbf{B}a)_{ijk} = B_{ij}a_k \quad \text{(tensor of rank 3)},$$

$$\text{(A.8b)}$$

$$\text{two tensors} \quad (\mathbf{AB})_{ijkl} = A_{ij}B_{kl} \quad \text{(tensor of rank 4)}.$$

Cross multiplication between two vectors and between a tensor and a vector

$$(a \times b)_k = \sum_{i,j} \varepsilon_{ijk} a_i b_j \text{ (vector)}, \qquad (\mathbf{B} \times a)_{ik} = \sum_{j,l} \varepsilon_{jkl} B_{ij} a_l \text{ (tensor).} \qquad \text{(A.9)}$$

Differentiation

The most usual differential operators acting on tensorial fields may be expressed in terms of the so-called nabla operator, defined in Cartesian coordinates as

$$\nabla = \left(\frac{\partial}{\partial x_1}, \frac{\partial}{\partial x_2}, \frac{\partial}{\partial x_3} \right). \qquad \text{(A.10)}$$

Gradient (defined as dyadic product)

$$(\nabla a)_i = \frac{\partial a}{\partial x_i} \text{ (vector)}, \qquad (\nabla a)_{ij} = \frac{\partial a_j}{\partial x_i} \text{ (tensor of rank 2)},$$

$$(\nabla \mathbf{A})_{jki} = \frac{\partial A_{jk}}{\partial x_i} \text{ (tensor of rank 3).}$$

Divergence (defined as the scalar product)

$$\nabla \cdot a = \sum_i \frac{\partial a_i}{\partial x_i} \text{ (scalar)}, \qquad (\nabla \cdot \mathbf{A})_i = \sum_j \frac{\partial A_{ji}}{\partial x_j} \text{ (vector).} \qquad \text{(A.11)}$$

Rotational or curl (defined as the cross product)

$$(\nabla \times a)_i = \sum_{j,k} \varepsilon_{ijk} \frac{\partial a_k}{\partial x_j} \text{ (vector)}, \quad (\nabla \times \mathbf{A})_{ik} = \sum_{j,l} \varepsilon_{jkl} \frac{\partial A_{ij}}{\partial x_l} \text{ (tensor of rank 2).}$$

The most usual second-order differential operator in tensorial analysis is the Laplacian, defined as

$$\nabla \cdot \nabla = \sum_i \frac{\partial^2}{\partial x_i \partial x_i}. \qquad \text{(A.12)}$$

Tensor invariants

Some combinations of the elements of a tensor remain invariant under changes of co-ordinates. Such invariant combinations are

$$I_1 = \mathrm{Tr}\,\mathbf{A} = \mathbf{A}{:}\mathbf{U} = \sum_i A_{ii},$$

$$I_2 = \mathrm{Tr}\,\mathbf{A}\cdot\mathbf{A} = \mathbf{A}:\mathbf{A} = \sum_{i,j} A_{ij}A_{ji} \qquad\qquad (A.13)$$

$$I_3 = \mathrm{Tr}\,\mathbf{A}\cdot\mathbf{A}\cdot\mathbf{A} = \sum_{i,j,k} A_{ij}A_{jk}A_{ki}$$

Other invariant combinations may also be formed, but they are combinations of I_1, I_2 and I_3; for instance, one often finds the invariants I, II and III defined as

$$I = I_1, \quad II = \tfrac{1}{2}(I_1^2 - I_2), \quad III = \tfrac{1}{6}(I_1^3 - 3I_1I_2 + 2I_3) = \det \mathbf{A}. \qquad (A.14)$$

The invariants I, II and III appear as coefficients in the "characteristic equation"

$$\det(\lambda\mathbf{U} - \mathbf{A}) = 0.$$

It is also possible to form joint invariants of two tensors $\mathbf{A}$ and $\mathbf{B}$ as

$$I_{11} = \mathrm{Tr}\,\mathbf{A}\cdot\mathbf{B}, \quad I_{21} = \mathrm{Tr}\,\mathbf{A}\cdot\mathbf{A}\cdot\mathbf{B}, \quad I_{12} = \mathrm{Tr}\,\mathbf{A}\cdot\mathbf{B}\cdot\mathbf{B}, \quad I_{22} = \mathrm{Tr}\,\mathbf{A}\cdot\mathbf{A}\cdot\mathbf{B}\cdot\mathbf{B}. \quad (A.15)$$

Appendix B

Useful Integrals in the Kinetic Theory of Gases

We present here some useful integrals appearing in several calculations based on the kinetic theory of gases. Let $F(C)$ be any scalar function of the peculiar velocity C such that the integrals appearing below converge, and let C_x and C_y be two components of C. Then

$$\int F(C)C_x^2\,\mathrm{d}C = \frac{1}{3}\int F(C)C^2\,\mathrm{d}C, \tag{B.1}$$

$$\int F(C)C_x^4\,\mathrm{d}C = \frac{1}{5}\int F(C)C^4\,\mathrm{d}C, \tag{B.2}$$

$$\int F(C)C_x^2 C_y^2\,\mathrm{d}C = \frac{1}{15}\int F(C)C^4\,\mathrm{d}C. \tag{B.3}$$

The following definite integrals are also useful

$$\int_0^\infty \exp(-\alpha C^2)C^r\,\mathrm{d}C = \frac{\sqrt{\pi}}{2}\frac{1}{2}\frac{3}{2}\frac{5}{2}\cdots\frac{r-1}{2}\alpha^{-(r+1)/2} \qquad (r\,\text{even}), \tag{B.4}$$

$$\int_0^\infty \exp(-\alpha C^2)C^r\,\mathrm{d}C = \frac{1}{2}[(r-1)/2]!\,\alpha^{-(r+1)/2} \qquad (r\,\text{odd}). \tag{B.5}$$

Appendix C

Some Physical Constants

Boltzmann's constant	k_B	1.38×10^{-23} J K^{-1} = 8.62×10^{-5} eV K^{-1}
Stefan–Boltzmann's constant	σ_0	5.67×10^{-8} W m^{-2} K^{-4}
Radiation constant	$a = (4\sigma_0/c)$	7.56×10^{-16} J m^{-3} K^{-4}
Atomic mass unit	amu	1.66×10^{-27} kg
Electron charge	e	1.60×10^{-19} C
Electron mass	m_e	9.11×10^{-31} kg
Proton mass	m_p	1.673×10^{-27} kg
Planck's constant	h	6.63×10^{-34} J s = 4.14×10^{-15} eV s